PHYSICS RESEARCH AND TECHNOLOGY

POLARONS

RECENT PROGRESS AND PERSPECTIVES

PHYSICS RESEARCH AND TECHNOLOGY

Additional books and e-books in this series can be found on Nova's website
under the Series tab.

PHYSICS RESEARCH AND TECHNOLOGY

POLARONS

RECENT PROGRESS AND PERSPECTIVES

AMEL LAREF
EDITOR

Copyright © 2018 by Nova Science Publishers, Inc.

All rights reserved. No part of this book may be reproduced, stored in a retrieval system or transmitted in any form or by any means: electronic, electrostatic, magnetic, tape, mechanical photocopying, recording or otherwise without the written permission of the Publisher.

We have partnered with Copyright Clearance Center to make it easy for you to obtain permissions to reuse content from this publication. Simply navigate to this publication's page on Nova's website and locate the "Get Permission" button below the title description. This button is linked directly to the title's permission page on copyright.com. Alternatively, you can visit copyright.com and search by title, ISBN, or ISSN.

For further questions about using the service on copyright.com, please contact:
Copyright Clearance Center
Phone: +1-(978) 750-8400   Fax: +1-(978) 750-4470   E-mail: info@copyright.com.

NOTICE TO THE READER

The Publisher has taken reasonable care in the preparation of this book, but makes no expressed or implied warranty of any kind and assumes no responsibility for any errors or omissions. No liability is assumed for incidental or consequential damages in connection with or arising out of information contained in this book. The Publisher shall not be liable for any special, consequential, or exemplary damages resulting, in whole or in part, from the readers' use of, or reliance upon, this material. Any parts of this book based on government reports are so indicated and copyright is claimed for those parts to the extent applicable to compilations of such works.

Independent verification should be sought for any data, advice or recommendations contained in this book. In addition, no responsibility is assumed by the publisher for any injury and/or damage to persons or property arising from any methods, products, instructions, ideas or otherwise contained in this publication.

This publication is designed to provide accurate and authoritative information with regard to the subject matter covered herein. It is sold with the clear understanding that the Publisher is not engaged in rendering legal or any other professional services. If legal or any other expert assistance is required, the services of a competent person should be sought. FROM A DECLARATION OF PARTICIPANTS JOINTLY ADOPTED BY A COMMITTEE OF THE AMERICAN BAR ASSOCIATION AND A COMMITTEE OF PUBLISHERS.

Additional color graphics may be available in the e-book version of this book.

Library of Congress Cataloging-in-Publication Data

ISBN: 978-1-53613-935-8

Published by Nova Science Publishers, Inc. † *New York*

CONTENTS

Preface		vii
Chapter 1	EPR Spectroscopy of Polarons in Conjugated Polymers and Their Nanocomposites *Victor I. Krinichnyi*	1
Chapter 2	Tight-Binding Models for the Charge Transport in Organic Semiconductors *Luiz Antônio Ribeiro Junior, Antonio Luciano de Almeida Fonseca, Jonathan Fernando Teixeira, Wiliam Ferreira da Cunha and Geraldo Magela e Silva*	107
Chapter 3	The Polaron Effect on Charge Transport Property for Organic Semiconductors *Nianduan Lu, Ling Li and Ming Liu*	143
Chapter 4	Polarons in Electrochemically Doped Non-Degenerate Π- Conjugated Polymers *S. S. Kalagi and P. S. Patil*	171
Chapter 5	Anharmonicity, Soliton-Assisted Transport and Electron Surfing as a Generalization of Polaron Transport with Practical Consequences *Manuel G. Velarde and E. Guy Wilson*	197
Chapter 6	Polarons in the Functionalized Nanowires *Victor A. Lykah and Eugene S. Syrkin*	223
Chapter 7	Bound Polarons and Exciton-Phonons Coupling in Semiconductor Nanostructures *Abdelaziz El Moussaouy*	265
Chapter 8	The Electrical and Structural Study of Compounds with a Modulated Scheelite-Type Structure at High Temperature *C. González-Silgo, M. E. Torres, N. P. Sabalisck, I. T. Martín-Mateos, E. Zanardi, A. Mujica, F. Lahoz, J. López-Solano and C. Guzmán-Afonso*	305

Chapter 9	Unravelling the Effects of Polaron Conduction on Mixed Conductivity Glasses *Marisa A. Frechero, Evangelina C. Cardillo, Pablo di Prátula, Soledad Terny, Luis A. Hernandez García, Mariela E. Sola and Magalí C. Molina*	**365**
Chapter 10	Small Polaron Hopping Conduction Mechanism in V_2O_5-Based Glass-Ceramic Nanocomposites *M. M. El-Desoky, M. S. Ayoub, A. E. Harby and A. M. Al-Syadi*	**379**
Chapter 11	Polaron in Perovskite Manganites *Abd El-Moez A. Mohamed and B. Hernando*	**397**
Chapter 12	Correlated Polarons in Mixed Valence Oxides *C. M. Srivastava and N. B. Srivastava*	**409**
Chapter 13	Polarons and Bipolarons in Colossal Magnetoresistive Manganites *Guo-Meng Zhao, J. Labry and Bo Truong*	**429**
Chapter 14	Polaronic High-Temperature Superconductivity in Bismuthates and Cuprates *Guo-Meng Zhao, N. Derimow, J. Labry and A. Khodagulya*	**457**
Chapter 15	Polarons in Ferrites *Madhuri Wuppulluri*	**483**
Chapter 16	Magnetic Polarons in EuTe: Early and Modern Discoveries in Magnetic Order Control in Semiconductors *Flavio C. D. de Moraes*	**495**
About the Editor		**505**
Index		**507**

PREFACE

In this book, the modern developments of polaron and its applications in condensed matter physics and materials sciences, will be exposed. Polarons are an excessively prosperous and profound arenas which are still very active research. The exploration of polarons and bipolarons in magnetic semiconductors and transition metal oxides has renewed interest in the current research related to the contemporary materials. A luminous guidance on various kinds of polaron forms will be discerned theoretically and experimentally in various classes of materials. This book supplies the recent scrutiny and advancement of polaron physics in multifunctional materials, conducting the reader to the comprehension from single-polaron problems to multi-polaron systems. The primary target of the book is to offer the reader a thorough overview about the contemporary advances in the polarons properties of various systems. The book covers broad spectrum of disciplines of polarons performed by well-acknowledged international researchers who are active in this area and will assist scientists with a background in solid state physics and materials sciences. Special emphasis is provided to understand and to describe many interesting phenomena related to polarons problems in advanced materials. Moreover, current importance and ultimate prospects for the supplementary evolution of polarons are addressed. The scope involves a description of many compelling phenomena in manganites, colossal magnetoresistance oxides, high-temperature superconductors, ferromagnetic oxides, conducting polymers, inorganic and organic semiconductors, ferrites, glasses-ceramics, and semiconducting of low-dimensionality, such as, heterostructures, quantum wells, quantum well wires, and quantum dots. The overview of the underlying physical concepts, experimental methods, and applications of polarons are represented. A numerous of new physical phenomena in various materials will be described on the basis of single and multi-polaron theories. Appealing electronic transport and optical mechanisms will be emphasized from the emergence of polarons. The principles prevailing polaron and multi-polaron formations and their applicability into wide spectrum of physical aspects of advanced materials with reduced dimension, like transport through nanowires are also discussed. Novel theoretical models based on the polaron phenomena are needed to promote better comprehension of the fundamental physics of advanced materials. This could aid more practical exploitation in prospective application devices. This book reviews the recent advancement in the field of polarons, going beyond their fundamental description to the numerous of vigorous routes of research. Overall, this book will serve as valuable reference for students, researchers, material and chemical engineers to comprehend the polaron properties in advanced materials and its ongoing breakthrough. The book composes normally into sixteen chapters.

In the chapter 1 of the book, a special emphasis is devoted to Electron Paramagnetic Resonance (EPR) spectroscopy of polarons in conjugated polymers and their nanocomposites. The focus of this chapter is on the utilization of EPR technique combined with the spin label and probe, steady-state microwave saturation, saturation transfer and conductometric methods for the investigation of conjugated polymers. The author reviews the key experimental methodological approaches basically developed for the scrutiny of various organic condensed systems.

Chapter 2 reviews tight-binding models for the charge transport in organic semiconductors. This chapter is dedicated to an exposition of the most basic description regarding the Tight-Binding models of charge transport in organic semiconductors. Moreover, the physical aspects of some valuable applications are succinctly discussed about the recombination process between quasi-particles in organic-based optoelectronic devices.

Chapter 3 is addressed to the manifestations of polaron effect on charge transport property for organic semiconductors. In this chapter, the researchers are primarily concerened to explain the polaron effect on the charge transport of organic semiconductors. Several kinds of theoretical models of polaron effect on the charge transport property in organic semiconductors has been discussed in details. The authors aimed that their contexts can be beneficial for ameliorating polaron effect in organic materials and for serving to the progress of novel organic devices with high performance.

Chapter 4 scrutinizes the recent results about the polarons in electrochemically doped non-degenerate π-–conjugated polymers. The authors reported that the produced quasiparticles, such as polarons and bipolarons can be directly inspected by employing the experimental techniques. The polaron and bipolaron effects on these systems would provide a vision to alter light-to-current conversion processes in organic materials and many other next generation devices of organic electronics.

Chapter 5 focuses on the anharmonicity, soliton-assisted transport and electron surfing, as a generalization of polaron transport with practical consequences. The authors discussed a generality of the polaron theory to involve soliton assisted transport. A novel filed effect transistor based upon the solectron concept (SFET) is shortly exposed. The researchers suggested that SFET provides a thoroughly new idea to create computer elements permitting the switching with three orders a magnitude reduction in energy consumption which would be useful as a valuable energy reductions in the digital computers, server farms, and smart phones.

Chapter 6 delineates the polarons in the functionalized nanowires. The authors discussed their self-consistent theoretical approach that is evolved to portray the electronic spectra in functionalized semiconducting nanowires. The quantization, localization and polaron formation have been explored for an uncompensated charge carrier in a functionalized nanowire.

Chapter 7 describes the bound polarons and exciton-phonons coupling in semiconductor nanostructures. This chapter offers an overview of the contemporary advancement in the description of polaronic effects on confined impurities and excitons in semiconductor nanostructures and their pertinence to ameliorating electronic and excitonic characteristics in these systems. The researchers expose important theoretical frameworks, presently promoted in scrutiny, taking into account the interactions between charge carriers and optical phonons in semiconductor of low-dimensional structures.

Chapter 8 overviews the electrical and structural study of compounds with modulated scheelite-type structure at high temperature. The authors revised and discussed the electrical properties of scheelites and related compounds. They explained the polaronic mechanisms which are correlated with the thermal dependence of the crystal structure.

Chapter 9 reports the unravelling effects of polaron conduction on mixed conductivity glasses. In the current chapter, the authors are principally concerned to discuss the electrical properties of glasses. They characterized and examined the electrical response of uncommon oxide glasses. According to their finding, it was viable to demonstrate the presence of ion-polaron entity in modified tellurite glassy matrices.

Chapter 10 characterizes the small polaron hopping conduction mechanism in based V_2O_5-glass-ceramic nanocomposites. In this chapter, the compositional dependence of the nanostructural and transport properties of V_2O_5-based glasses and corresponding glass-ceramic nanocomposites have been outlined by the researchers in view of Mott's small polaron hopping (SPH) model. They also elucidated the mechanism of electrical conduction of these glasses and the corresponding glass-ceramic nanocomposites.

Chapter 11 explores the polaron in perovskite manganites. This chapter deals with the polaron concept, types and mechanisms in perovskite manganites and its effect on their transport properties. The polaron effect has a main contribution in the magneto-transport correlation in manganites, such as phenomena as colossal magnetoresistance. The authors exhibited the polaron contribution in manganite phenomena and its impact on the magneto-transport correlation.

Chapter 12 discusses the correlated polarons in mixed valence oxides. In this chapter, the authors elucidated the transport properties of some mixed valence oxides compounds by means of correlated polaron theory accounting for the colossal magnetoresistance in manganites. They suggested that the correlated polaron model was extended to transport in the normal state of high temperature copper oxide superconductors.

Chapter 13 covers the polarons and bipolarons in colossal magnetoresistive manganites. In this chapter, the researchers reviewed some unconventional oxygen-isotope effects in doped manganites. According to their results, the detected large unconventional isotope effects notably evince the formation of polarons/ bipolarons to be related to strong electron-phonon coupling, which is pertinent to the fundamental physics of manganites and significant for the emergence of colossal magnetoresistance.

Chapter 14 discusses polaronic high-temperature superconductivity in bismuthates and cuprates. The authors illustrated the experimental evidences for the polaronic Cooper pairs in both bismuthate and cuprate superconductors. They stated that the substantial enhancement of the effective density of states to be connected to the lattice polaronic effect and this will conduct to the increase of the effective electron-phonon coupling constant to a value inevitably for the inspected superconducting transition temperature in bismuthates.

Chapter 15 outlines the polarons in ferrites. These magnetic materials illustrate striking features that are useful in the electronic devices. The researchers discussed their latest results on the ferrite systems of interest with emphasis on polaron models. Ferrites which are ferromagnetic semiconductors opened a novel perspective in material physics and the demand for high resistivity ferrites conducted to the synthesis of various ferrites.

Chapter 16 provides an overview on the magnetic polarons in EuTe: early and modern discoveries in magnetic order control in semiconductors. In this chapter, the authors discuss the evolution of the theoretical modeling for magnetic polarons and the background experimental measurements, with emphasis to the EuTe magnetic polaron. The advancement of spintronic devices, magnetic polarons are also very appealing in a more basic perspective, conveying information about the physics of the exchange interactions and its dynamics, which is a broad area in material science and solid state physics.

In: Polarons: Recent Progress and Perspectives
Editor: Amel Laref

ISBN: 978-1-53613-935-8
© 2018 Nova Science Publishers, Inc.

Chapter 1

EPR SPECTROSCOPY OF POLARONS IN CONJUGATED POLYMERS AND THEIR NANOCOMPOSITES[*]

Victor I. Krinichnyi[†]
Institute of Problems of Chemical Physics,
Chernogolovka, Russian Federation

ABSTRACT

The past two decades have seen extraordinary progress in synthesis and study of organic conjugated polymers and their nanocomposites. This caused by large prospects of utilization of such systems in molecular electronics and spintronics. One of the main scientific goals is to reinforce human brain with computer ability. However, a convenient modern computer technology is based on three-dimensional inorganic crystals, whereas human organism consists of biological systems of lower dimensionality. So, the combination of a future computer based on biopolymers with organic conjugated polymer semiconductors of close dimensionality is expected to increase considerably a power of human apprehension. This is why understanding the major factors determining specific spin charge transfer processes in conjugated polymers is now a hot topic in organic molecular science.

The charge in such systems is transferred by topological excitations, spin polarons and spinless bipolarons, characterized by high mobility along polymer chains. This stipulated the utilization of Electron Paramagnetic Resonance (EPR) spectroscopy as a unique direct tool for more efficient study and monitoring of spin reorganizing, relaxation and dynamics processes carrying out in polymer systems with such charge carriers. It was demonstrated that the method allows to obtain qualitative new information on spin-modified polymer objects and to solve various scientific problems.

The focus of the present chapter is on the use of EPR technique in combination with the spin label and probe, steady-state microwave saturation, saturation transfer and

---

[*] This chapter is dedicated to the veteran of the World War II, designer Grigori T. Rudenko.
[†] Corresponding Author Email: Email: kivirus@gmail.com; Web: http://hf-epr.awardspace.us.

conductometric methods in the study of initial and treated conjugated polymers. It covers a wide range of specific approaches suitable for analyzing of processes carrying out in polymer systems with paramagnetic adducts providing readers with background knowledge and results of the latest research in the field. It reviews the main experimental methodological approaches originally developed for the study of various organic condensed systems.

The chapter is organized as following. The first part includes the fundamental properties of conjugated polymers with topological quasi-particles, polarons and bipolarons, as charge carriers. The second part is devoted to an original data obtained at X-band to D-band (30 − 2-mm, 9.7 − 140 GHz) EPR study of the nature, relaxation and dynamics of polarons stabilized and initiated in widely used conjugated polymers and their nanocomposites. The third part reveals the possibility to handle of charge transport in some polymer composites with spin-spin exchange which can be used in the further creation of novel elements of molecular electronics and spintronics. Finally, theoretical and experimental background necessary for EPR study of the main magnetic resonance, relaxation and dynamics parameters of polaron quasi-particles in organic compounds are described shortly in the Appendix.

1. INTRODUCTION

Basic features of polarons were well recognized a long time ago and have been described in a number of review publications (Emin 2013, Chatterjee and Mukhopadhyay 2018). Nevertheless, interest in the study of polarons has recently increased because they are important to the understanding of properties of the microcosm and even the space.

Fundamental electron–phonon interactions have been shown to be relevant in many inorganic and organic semiconductors, giant magnetoresistance oxides, and transport of Cooper pairs in high-temperature superconductors and free charges through nanowires and quantum dots are often governed by libration displacements of environmental ions. Charge transport in organic semiconductors is also sensitive to polaronic effects, which is particularly relevant in the design of organic components of molecular electronics. When charge carriers, polaron or electron, are formed in a polymer system, the surrounding ions can interact with them. The ions can adjust their positions slightly, balancing their interactions with the charge carriers and the forces that hold the ions in their stable positions. This adjustment of positions leads to a polarization locally centered on the charge carrier. Besides, such a quasi-particle can be formed as topological distortion separating two conformational forms of conjugated polymers (Lu 1988). There can be formed "large" and "small" polarons, defined by whether or not the polarization cloud is much larger than the atomic spacing in the material. Polarons are a useful way to understand charge transport in conjugated polymer semiconductors which are very squishy, deformable systems held together by van-der-Waals rather than covalent bonding.

1.1. Properties of Conjugated Polymers

A wide class of electronic materials, organic conjugated polymers, attracts during the past decades great attention due to their unique capabilities, namely, flexible, solution processable, lightweight, and tunable electronic properties (Wan 2008, Heeger, Sariciftci, and Namdas

2010, Launay and Verdaguer 2013, Kobayashi and Müllen 2015, Kondawar and Sharma 2017). The particular interest to such systems was initiated at 1964 by the Little hypothesis (Little 1964) on principal possibility of synthesis of high-temperature superconductors based on conjugated polymers. This caused by large prospects of utilization of such systems in molecular electronics. Organic semiconductors have been widely investigated in recent years also in the context of spintronics. One of the main scientific goals is to reinforce human brain with computer ability. However, a convenient modern computer technology is based on three-dimensional inorganic crystals, whereas human organism consists of biological systems of lower dimensionality. So, the combination of a future computer based on biopolymers with organic conducting polymers of close dimensionality is expected to increase considerably a power of human apprehension. Understanding the major factors which govern specific spin charge transfer processes in conjugated polymers is now a hot topic in organic molecular science. This is why the investigation of conjugated polymers and their nanocomposites has generated entirely new scientific conceptions and a potential for their perspective application as active material for creation of novel components of organic molecular electronics.

Figure 1 shows room temperature (RT) direct current (*dc*) conductivity σ_{dc} of some conjugated polymers in comparison with that of convenient insulators, semiconductors and conductors. The structures of some conjugated polymers used as active matrix in molecular electronics are schematically presented in the Figure as well. The principal *x*-axis of polymers' chain is shown to be chosen parallel to their longest molecular *c*-axis, the *y*-axis lies in the C-C-C plane, and the *z*-axis is perpendicular to *x*- and *y*-axes. These polymers initially synthesized as films or powders are insulators with *dc* conductivity near $10^{-15} - 10^{-10}$ S/m. This parameter can be varied controllable by more than 12 orders of magnitude by chemical or electrochemical oxidation or reduction of their chains. Electron transfer upon such polymer doping leads to the increase of *dc* conductivity up to ~ $10^{-3} - 10^{2}$ S/m (semiconductor regime) or even up to ~ $10^{4} - 10^{8}$ S/m (metal regime) (Salamone 1996, Nalwa 1997, Scotheim, Elsenbaumer, and Reynolds 1997). The analogous effect can also be achieved by somewhat physical effect on the polymer, e.g., by laser radiation. This leads to the formation of quasi-particles, polarons, on the polymer chains due to their topological distortion (Lu 1988). Conjugated polymers have a highly anisotropic quasi-one-dimensional (Q1D) π-conjugated structure with charge carriers on delocalized polarons which makes such systems fundamentally different from traditional inorganic semiconductors, for example, silicon and selenium, and from well-known convenient insulating polymers, e.g., polyethylene, polyvinyl chloride and polystyrene. Electronic properties of such systems strongly depend on their structure, morphology and quality, whereas the type of their conductivity is governed by the nature of the introduced counter-ion (Menon et al. 1997, Wessling 1997). The introduction of anions, e.g., HSO_4^- BF_4^-, ClO_4^-, I_3^-, $FeCl_4^-$ into a polymer induces a positive charge on a polymer chain and thus leads to *p*-type conductivity of the polymer. Conductivity of *n*-type is realized under polymer doping by Li^+, K^+, Na^+ and ions of other alkali metals.

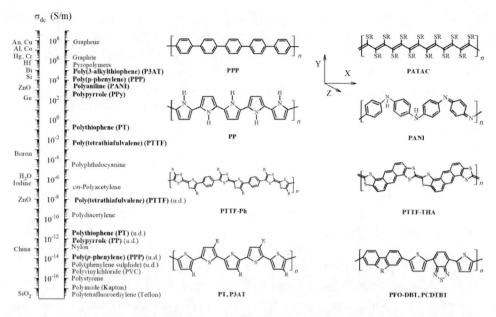

Figure 1. Room temperature (RT) direct current (*dc*) conductivity σ_{dc} of undoped (u.d.) and highly doped conducting polymers (Nalwa 1997) in comparison with that of some convenient polymers and metals. The structures of poly(*p*-phenylene) (PPP). poly(*bis*-alkylthioacetylene) (PATAC, R ≡ CH$_3$, C$_2$H$_5$, C$_3$H$_7$), polypyrrole (PPy), polyaniline (PANI), polytetrathiafulvalene linked *via* phenyl (PTTF-Ph, R ≡ H, CH$_3$, C$_2$H$_5$) and tetrahydroanthracene (PTTF-THA) bridges, polythiophene (PT), poly(3-alkylthophene) (P3AT, R ≡ C$_m$H$_{2m+1}$).poly[2,7-(9,9-dioctylfluorene)-alt-4,7-bis(thiophen-2-yl)benzo-2,1,3-thiadiazole] [PFO-DBT, R ≡ >C=(C$_8$H$_{17}$)$_2$], and poly[N-9'-heptadecanyl-2,7-carbazole-alt-5,5-(4',7'-di-2-thienyl-2',1',3'-benzothiadiazole)] [PCDTBT, R ≡ >N-CH-(C$_8$H$_{17}$)$_2$] are shown schematically.

In traditional 3D inorganic semiconductors, fourfold (or sixfold, etc.) coordination of each atom to its neighbor through covalent bonds leads to a crystalline structure. Electron excitations may be usually considered in the context of this rigid structure, leading to the conventional conception of generalized electrons or holes as dominant charge carriers. The situation with conjugated polymer semiconductors is quite different: once they form bulk nanocomposites with the dopants embedded, their Q1D structure becomes more susceptible to structural distortion. Therefore, electronic properties of conjugated polymers may be conventionally considered in the frames of bands theory (Singleton 2001) as well as of soliton and polaron one (Lu 1988), based on Peierls instability (Peierls 1996) in Q1D systems. Elementary-charge- and energy-transfer phenomena occurring in such materials are crucial to the structural and conformational diversity of their polymer active matrix. So, in order to construct organic electronic elements based on conjugated polymers, the correlations of their morphology, electronics properties, selectivity, sensitivity, etc. with magnetic, relaxation and dynamics properties of spin charge carriers should be obtained and analyzed.

1.2. Polarons and Bipolarons in Conjugated Polymers

The chains of poly(*p*-phenylene) (PPP) and analogous polymers consist of benzene rings linked via para-position (see Figures 1 and 2). In the solid state two successive benzene rings are tilted with respect to one another by torsion (dihedral) angle θ≈ 23° (Brédas 1986). Such angle appears as a compromise between the effect of conjugation and crystal-packing energy, which would lead to a planar morphology, and the steric repulsing between ortho-hydrogen atoms, which would lead to a non-planar morphology (Brédas et al. 1982). The globular structures of PPP consist of packed fibrils with a typical diameter of 100 nm.

The band structure of PPP is obtained as a result of the overlapping of π-orbits of the benzene rings (Figure 2). For this polymer a resonance form can also be derived, which corresponds to a quinoid structure. Calculations have indicated that polaron formation is energetically favorable in all the organic conjugated polymers have so far studied (Brédas and Street 1985). The polaron binding energy is of the order of 0.03 eV in PPP and 0.12 eV in PPy. Benzenoid and quinoid forms are characterized by higher and lower energy, respectively. This provokes the formation on polymer chains of topological excitations, namely polaron with spin $S = \frac{1}{2}$ whose energy levels ΔE_1 and ΔE_2 of ca. 0.5 eV lie in the gap above the valence band (VB} and below the conduction band (CB) edges as it is shown on Figure 2. At intermediate doping level y a second electron is taken out of the chain, the energetically favorable species, bipolarons, are formed. The energy levels associated with the bipolarons are empty and are located closer to the bandgap than those associated with the polarons (Heeger et al. 1988). The binding energy of bipolarons formed, e.g., in PPy is 0.69 eV (Brédas and Street 1985). The width of the polaron and bipolaron depends on the polymer structure and normally is 3 - 5 and 4 – 5.5 monomer units, respectively (Brédas and Street 1985, Elsenbaumer and Shacklette 1986, Devreux et al. 1987, Lu 1988, Westerling, Osterbacka, and Stubb 2002, Niklas et al. 2013). As a doping level y increases, bipolaron states overlap forming bipolaron bands within the gap. At high doping level, these bands tend

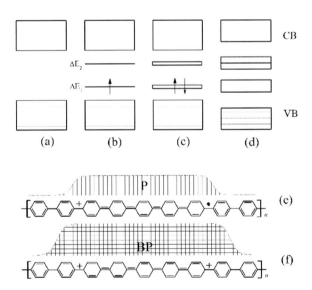

Figure 2. Evolution of an initial band structure of poly(p-phenylene) (a) during the polymer doping: (b) slight doping level with the appearance of polaron states in the mid gap above valence band (VB) and below conducting band (CB) edges, (c) intermediate doping level, with the appearance of states of non-interacting diamagnetic bipolarons, (d) high doping level, where bipolaron states overlap and form filled and semifilled bands with quasi-metallic behavior; the formation of a spin charged polaron P (e) and spinless bipolaron BP (f) on the chain of poly(p-phenylene) is shown. The distribution of the spin on the polaron and the double elemental charge on the bipolaron is shown by appropriate filled extended figures.

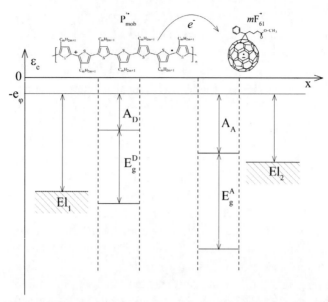

Figure 3. Schematic band diagram of two semiconductors with different electron affinity before making between them the BHJ. The electron donor (A_D) and electron acceptor (A_A) affinities are defined vs. the electron energy in vacuum at the same electrical potential. E_g^D and E_g^A are the band gap energies of the electron donor and electron acceptor, respectively. In the top, the P3AT and PC$_{61}$BM are schematically shown as electron donor and electron acceptor, respectively. The appearance of the polaronic quasi-particle $P_{mob}^{\bullet+}$ with a spin $S = ½$ and an elemental positive charge in a P3AT chain and ion radical $mF_{61}^{\bullet-}$ with an elemental negative charge and a spin $S = ½$ on a PC$_{61}$BM is shown as well.

to overlap and create new bandgap energy bands that may merge with the VB and CB allowing freedom for extensive charge transfer. The analogous band structure is formed also in case of other PPP-like polymers (Lu 1988). The energy levels ΔE_1 and ΔE_2 of polarons depend on the polymer structure and the nature of dopant molecule.

Polaron quasi-particles can be also formed on polymer chains, e.g., upon illumination of appropriate polymer:fullerene composites (Sun and Sariciftci 2005, Poortmans and Arkhipov 2006, Pagliaro, Palmisano, and Ciriminna 2008, Krinichnyi 2009, Brabec, Scherf, and Dyakonov 2014, Krinichnyi 2016b, a). Fullerene molecules embedded into polymer matrix of such systems form so called "bulk heterojunctions" (BHJ) with polymer chains. In this case polymer matrix and adduct become as electron donor D (hole transporter, p-type material) and electron acceptor A (electron transporter, n-type material), respectively. Beyond photo induced charge exciting and separation, positive carriers are transported to electrodes by polarons diffusing in the polymer phase and electrons hopping between fullerene domains embedded into polymer matrix. A definitive advantage of BHJ is that it can be made by simply mixing these materials in an organic solvent, and casting with well-known solution deposition techniques, e.g., spin coating (Shaheen et al. 2001). The illumination of such a system by photons with energy hvph higher than the π-π* energy bandgap of polymer matrix leads to the fast formation of spinless excitons in its BHJ. These quasi-particles can geminate ultrafast dissociate forming Coulomb bound electron-hole pairs (charge-transfer states) of electrons on the acceptor moiety and holes on the donor moiety. Then electrons and holes can leave the donor:acceptor interface relaxing into more favorable energy levels (Lupton, McCamey, and Boehme 2010). With increasing distance from the material interface, the

Coulomb attraction becomes less, and finally, the electrons and holes become independent of each other, forming quasi-pairs or charge-separated states. Finally, charge separation is accompanied by the formation of unbound (free) positively charged polarons on polymer chains and negatively charged radical on fullerene globes. After this stage, charge carrier recombination can occur.

Polymer:fullerene BHJ are characterized by efficient light-excited charge generation at the interface between two organic materials with different electron affinity. Figure 3 illustrates the energy diagram of two exemplary intrinsic semiconductors, e.g., regioregular poly(3-alkylthiophene) (P3AT) shown in Figure 1 and fullerene derivative, [6,6]-phenyl-C_{61}-butanoic acid methyl ester ($PC_{61}BM$), widely used in polymer:fullerene photovoltaic devices, before making a contact between them. A heterojunction formed by these materials inserted between a high work-function electrode (El_1) matching the highest occupied molecular orbital level of the donor ($HOMO_D$) and a low work-function electrode (El_2) matching the lowest unoccupied molecular orbital level of the electron acceptor ($LUMO_A$) should in principle act as a diode with rectifying current–voltage characteristics. Under the forward bias (the low work-function electrode is biased negative in respect to the high work-function electrode) the electron injection into the $LUMO_A$ layer from the low work-function electrode as well as the electron extraction out of the $HOMO_D$ by the high work-function electrode is energetically possible and a high current may flow through the heterojunction. Under reverse bias (the low work-function electrode is biased positive in respect to the high work-function electrode), the electron removal from the electron donor and electron injection to the electron acceptor is energetically unfavorable. The formation of the polaron $P^{+\bullet}$ and fullerene $F_{61}^{-\bullet}$ charge carriers is shown in Figure 3 as well. This process occurs in the femtosecond time domain (Banerji et al. 2010), whereas the electron back transfer with charge annihilation is much slower possibly due to dynamics and relative slow structural relaxation in such a system of lowed dimensionality. The separation and recombination of free charge carriers can be considered as concurring opposite directed processes. Recombination of opposite charge carriers can either be geminate, between electrons and holes originating from the same photo-generated exciton, or non-geminate, between separated charge carriers. The geminate recombination of polaron-fullerene pairs is monomolecular and, therefore, a first order process. The non-geminate, bimolecular recombination of separated polaron-fullerene quasi-pairs following Langevin theory of a second order. The formation, separation and recombination of charge carriers are governed also by the energy of initiating photons (Krinichnyi 2016a).

1.3. Resonant Manipulation of Interacting Polarons in Polymer Nanocomposites

The efficiency of energy conversion by organic polymer:fullerene quantum structures is governed partly by charge localization phenomena, which underlie their optical transitions. These molecular systems in which a charge is transferred by weakly spin-orbit coupled carriers are gaining particular interest because the use of spin carriers with additional spin degree of freedom opens an undoubted imperative to develop a remarkable new generation of electronic, "spintronic," devices with spin-assisted and, therefore, handling electronic properties. Because these materials have weak spin-orbit and hyperfine interactions, they can be used for hosting spin-based classical and quantum information. They exhibit intriguing interfacial effects and can be used to realize flexible multifunctional spintronic devices. Chiral organic systems can also play the role of efficient spin injectors and detectors, which can potentially replace ferromagnetic materials in spin devices. It is well known that the data processing in conventional computers is limited by the transport of elemental charge carriers, electrons, through silicon semiconductors. Exploiting the orientation of electron spin rather than its charge takes a possibility to create spintronic devices which will be smaller, more versatile and more robust than those currently making up silicon chips and circuit elements. This can be also used, e.g., for encoding and transferring information more efficiency then using spinless ones (Dediu et al. 2009). Deposition of isolated tetracyano-*p*-quinodimethane (TCNQ) molecules each possessing a spin to a graphene monolayer leads to formation in the latter of spatially extended spin-split electronic bands (Garnica et al. 2013) that may be used in organic spintronic devices (Shen et al. 2014). The careful controllable preparation, preservation, and manipulation of quantum states should form the backbone of the use of quantum information processing to create organic molecular spin-controlled electronic devices. Orientation of spin in charge carriers survives for a relatively long time (nanoseconds, compared to tens of femtoseconds during which electron momentum decays), which makes spintronic devices particularly attractive for memory storage and magnetic sensors applications, and, potentially for quantum computing where electron spin would represent a bit (called qubit) of information (Lloyd 1993).

Spin feature of charge carriers stabilized or/and induced, e.g., in the above described polymer:fullerene BHJ can also be used for manipulation by electronic properties of such system. Excitons formed in a polymer:fullerene BHJ under illumination at the intermediate step can be conversed into polaron pairs or donor-acceptor complexes which then collapsing into spin pairs of polarons and fullerene anion radicals. Besides, the triplet state of fullerene characterizing by high electron spin polarization can also be easily photoexcited in such a nanocomposite. This can possibly be used as an active media for masers and other molecular devices (Blank, Kastner, and Lebanon 1998). Diffusing and meeting in a system bulk, the above charge carriers recombine. Most organic hydrocarbon materials show only radiative recombination from the singlet state, so that high triplet formation rates form a formidable loss channel in devices with light-emitting recombination (Wohlgenannt et al. 2001). However, this process itself is not a trivial phenomenon which microscopic details still remain unknown. For example, one would not necessarily expect the recombination of a charge carrier with just the first opposite carrier. In this case both the charge carriers exchangeable spin-flip, so that their further recombination become dependent on their dynamics, number, polarization and mutual distinct. For large separations, when the thermal energy exceeds the interaction potential the charges can be considered as non-interacting. However, once the carriers become nearer than the inverted Coulombic interaction potential, their wave functions begin to overlap so that exchange interactions become non-negligible. This can lead to the formation in organic semiconductors of exciton of either singlet or triplet configuration. It is important to note that an electron spin-flip arising from spin-orbit coupling may thus transform the singlet exciton into a triplet one with lower energy level. This process prevails in the most planar structures with particular molecular symmetries (Baunsgaard et al. 1997). So, depending on various conditions, a singlet ($\uparrow\downarrow-\uparrow\downarrow$ or S) and three triplet ($\downarrow\downarrow$, $\uparrow\downarrow+\uparrow\downarrow$, $\uparrow\uparrow$ or T_+, T_0, T_-) spin configurations can be realized in these molecular devices (Lupton, McCamey, and Boehme 2010).

Free charge carriers are generated through dissociation of either singlet or triplet polaron pairs in polymer:fullerene BHJ. The dissociation rates of these carrier pairs are spin dependent, so that sum number of charge carriers should depend on the initial spin state of the polaron pair. A spin configuration of such pairs transforms between singlet and triplet ($S \leftrightarrow T$) either randomly due to irreversible spin-lattice relaxation (with characteristic time T_1) or coherently by an external electromagnetic field which induces electron spin precession and resonant reorientation. Such a coherent manipulation of the spin state requires phase coherence of the initiated spin with the initiating microwave (MW) field. Loss of phase coherence arising, e.g., due to the intrinsic (homogeneous) properties of the spin system, or extrinsic (inhomogeneous) characteristics provoking, e.g., by the system disorder or anisotropic local field distribution is characterizing by spin-spin relaxation time T_2 for homogeneous and T_2^* for inhomogeneous processes. If the charge carriers recombine with higher probability than own dissociation, they form excitons with single or triple multiplicity. This recombination process is strongly spin dependent due to the energetic difference between singlet and triplet states and, therefore, can be characterized by appropriate transition rates. So formed excitons may then decay, radiatively or nonradiatively, to the molecular ground state. This process involves charge transfer through a polymer:fullerene BHJ in which two-step exciton dissociation is accompanied by the transfer of its energy from the donor to the acceptor following by a polaron transfer to the donor (Lloyd, Lim, and Malliaras 2008). In this photovoltaic system, even rarely distributed defects may initiate exciton dissociation and the formation of polaron pairs (Im et al. 2002). Undoubtedly, this process is also spin dependent because in such a system only singlet excitons can be photo induced as well as the triplet excitons are bound stronger than the singlet ones, so that having different energetics. This means that in most photovoltaic systems the triplet exciton dissociation rate can, therefore, be nulled and the contribution of singlet excitons should only be considered.

1.4. Electron Paramagnetic Resonance of Polarons in Polymer Systems

Different methods can be used for the study of charge carriers stabilized or/and exited in polymers and their nanocomposites (Rouxel, Thomas, and Ponnamma 2016). Physics processes carrying out in such objects may be identified and analyzed comparing results obtained by all possible methods at similar experimental conditions, e.g., optical (fluorescence and phosphorescence) and/or electrically detected magnetic resonance (Lupton, McCamey, and Boehme 2010). The former can, in principle, sense singlet and triplet excitons in organic semiconductors (List et al. 2002), however, polaronic charge carriers themselves introduce efficient subgap optical transitions that obstructs unambiguous identification of excitations in such systems. In some cases, incorporation of a heavy-metal atom into an organic complex can increase spin-orbit coupling, effectively mixing singlet and triplet levels. A strong modulation of the phosphorescence signal can occasionally be observed in the presence of an electric field, thus proving the presence of spin triplet polaron pair species, which are much more polarizable than the ultimate triplet exciton (Reufer et al. 2006).

Polaron formed in conjugated polymers possesses a spin $S = 1/2$. So, Electron Paramagnetic Resonance (EPR) was proved (Schlick 2006, Eaton et al. 2010, Lupton, McCamey, and Boehme 2010, Ranby and Rabek 2011, Misra 2011) to be one of the most widely used and productive physical direct methods for structural and dynamic studies of high-molecular systems with free radicals, ion-radicals, molecules in triplet states, transition metal complexes, other PC. The method is based on resonant absorption of MW radiation by a paramagnetic sample due to the splitting of the energy levels in an external magnetic field. It allowed getting more detailed information on processes carrying out in various solids containing PC. As the original properties of conjugated polymers are related to the existence of PC localized or/and delocalized along and between their chains, a great number of EPR experiments has been performed for the study of their magnetic, relaxation and dynamics properties (Bernier 1986, Mizoguchi and Kuroda 1997, Kuroda 2002). For more than fifty years EPR such investigations are predominantly carried out at convenient X-band (3-cm, 9.7 GHz) EPR, i.e., at registration frequency $\omega_e/2\pi = \nu_e \leq 10$ GHz, and external magnetic fields $B_0 \leq 330$ mT. However, at these wavebands the signals of organic free radicals with g-factor lying near g-factor of free electron ($g_e = 2.00232$) are registered in a narrow magnetic field range and, therefore, can overlap the lines with close g-factors. At this waveband polarons in conjugated polymers demonstrate uninformative single spectra, so the line shape and concentration of these quasi-particles can only be directly measured. Besides, strong cross-relaxation of PC still perceptible at low magnetic fields (Altshuler and Kozirev 1972) additionally complicates the registration and identification of spin packets. EPR spectroscopy was proved also to be powerful method for the study of the exiton collapse into a pair of spin charge carriers, charge separation, transfer and recombination in organic donor:acceptor systems. Indeed, the evidence for a successful charge transfer is based on the fact that excitons generated by light absorption have zero spin and therefore cannot be detected by EPR spectroscopy. If the exciton is split at the interface between donor and acceptor, polaron and anion radical are created which have half integer spin leading to the appropriate EPR signal. The amount of light induced charge carriers can be determined by comparing the EPR spectrum in the dark and after illumination.

Because polarons obtain a spin, this can be used for creation of spintronic devices with spin-assisted electronic properties. Electron spin can be oriented parallel and antiparallel to a direction of the applied magnetic field forming two-level quantum system. To probe resonance spin hopping between these pure quantum states, weak spin-orbit coupling is required, which limits the applicability of the convenient magnetic resonance methods for the study of spin-assisted charge transfer in inorganic spintronic materials (Wolf et al. 2001). In organic semiconductors, however, spin–orbit interaction is very weak due to low atomic order number, so that the information about spin orientation can be easier established by magnetic spectroscopic techniques (Cinchetti et al. 2009). It allows using cutting-edge spectroscopic methods for determination of spin polarization in spin-assisted processes and also for their controlled manipulation. EPR spectroscopy was shown (Lupton, McCamey, and Boehme 2010) to be powerful direct tool able to reveal about the underlying nature of spin carriers excited in quantum systems including multispin polymer systems with weak spin-orbit coupling.

The efficiency of the method increases with spin precession frequency. It was shown that EPR investigation at high spin precession frequencies, $\omega_e/2\pi \geq 100$ GHz corresponding to field strength $B_0 \geq 3$ T, of organic radicals in different solids (Grinberg, Dubinskii, and Lebedev 1983, Krinichnyi 1995), especially polarons stabilized and initiated in conjugated polymers (Krinichnyi 1995, 2006, 2009, 2014b, a, 2016b, a) enables to increase considerably the precision and descriptiveness of the method. At 2-mm waveband EPR it were investigated in details structure, relaxation, dynamics and other specific characteristics of radical centers and their local environment, elementary charge transfer processes in different solids, biopolymers, conjugated polymers and their nanocomposites. It should, however, be noted that the advantages of the high-frequency/field EPR spectroscopy are limited in practice by a concentration sensitivity that decreases with increasing MW frequency. Nevertheless, multifrequency EPR spectroscopy seems to be powerful direct method for the detailed study of conjugated polymers and their nanocomposites with interacting spin packets.

This chapter summarizes the main results obtained in the investigation of relaxation, magnetic and dynamics parameters of polarons stabilized and initiated in various conjugated polymers and their nanocomposites at wide range of wavebands EPR. It is organized as following. The second part is devoted to an original data obtained in multifrequency EPR study of the nature, relaxation and dynamics of paramagnetic centers delocalized on polaronic charge carriers as well as the mechanisms of charge transfer stabilized and initiated in some widely known conjugated polymers. The use of some conjugated polymers as electron donor in organic composites is described as well. The third part reveals the possibility to handle of charge transport in some multispin polymer composites by using the spin-spin exchange. The fourth part denotes the prospects of the study of organic polymer systems for the further construction of novel elements of molecular electronics. An Appendix containing experimental details of EPR study of the main magnetic resonance, relaxation and dynamics parameters of polaron quasi-particles in organic compounds and short theoretical background finalizes the present chapter.

2. MAGNETIC, RELAXATION AND DYNAMIC PARAMETERS OF POLARONS IN CONJUGATED POLYMERS

Polymer doping and/or illumination leads to the formation of polaron on its chain. It possesses spin, elemental charge and high Q1D mobility along polymer chain. The main properties of such charge carrier are governed by the structure and dynamics of its environment as well as by the external physical effect. The terms of anisotropic g-factor and linewidth determined from EPR spectra of polarons stabilized in some conjugated polymers are summarized in Table 1. Below are analyzed the main magnetic, relaxation and dynamics parameters of polaronic charge carriers initiated in various initial and modified conjugated polymers.

2.1. Poly(p-Phenylene)

Poly(p-Phenylene) and its nanocomposites are considered as suitable material for constructing of various molecular devices (Brédas and Chance 1990, Nalwa 2001). This is mainly to its simple synthesis and high conductivity. A number of unpaired electrons stabilized in this conjugated polymer reaches $10^{17} - 10^{19}$ cm^{-3}(Bernier 1986) strongly depending on the technique used for its polymerization process. The room temperature (RT) linewidth of PPP increases at doping with the increase of atomic number of an alkali metal dopant due to a strong interaction between molecule of a dopant and an unpaired electron. Such polymer demonstrates a Korringa spin relaxation mechanism, typical of disordered metals, and the Mott's variable range hopping (VRH) mechanism (Mott and Davis 2012) for carriers localized within weakly conjugated carbonized regions (Matthews et al. 1999). The charge in PPP is transferred mainly by polarons at low doping level and bipolaron at metallic state (Xie, Mei, and Lin 1994), similar to that, as it is observed in analogous highly conjugated polymers (Lu 1988, Lacaze, Aeiyach, and Lacroix 1997).

A series of PPP film-like samples of near 10 μm thickness synthesized electrochemically on a platinum electrode in a BuPyCl-AlCl$_3$ melt were studied at X- and D-bands EPR: as prepared and evacuated PPP: Cl$_3^-$ film (PPP-1), the same film after its storage for forth days (PPP-2), that exposed for a few seconds to air oxygen (PPP-3); after Cl$_3^-$ dopant removal from the PPP-1 sample (PPP-4); and after BF$_4^-$ redoping of the PPP-1 film (PPP-5) (Goldenberg et al. 1990, 1991).

At X-band EPR, PPP-1 – PPP-3 samples demonstrate well-pronounced asymmetric single Dysonian spectrum with effective $g = 2.0029$ (Figure 4,a). The analogous PPP line shape was also registered at the study of highly lithium-doped PPP (Dubois, Merlin, and Billaud 1999). The line asymmetry factor A/B is changed depending on the sample modification. As dopant Cl$_3^-$ is removed, i.e., at the transition from PPP-1 film to PPP-4 one, the above spectrum transforms to a two-component one with $g_\perp = 2.0034$ and $g_\parallel = 2.0020$, where $2g_\perp = g_{xx} + g_{yy}$ and $g_\parallel = g_{zz}$ (Figure 4,b). Such a transition is accompanied by a line broadening and by a drastic decrease in the concentration of spin charge carriers from 1.1×10^{19} down to 3.9×10^{17} cm^{-3}. It should be noted that in earlier studies of different PPP samples and other π-conjugated polymers at X-band EPR there was not registered such an axially symmetric EPR spectra (Bernier 1986, Lacaze, Aeiyach, and Lacroix 1997). With PPP doped by BF$_4^-$ anions (PPP-5), the spectrum shape retains, however, a slight decrease in the concentration of polarons, and a change in the sign of its dependency on temperature are observed. Using the difference $g_\perp - g_\parallel = 1.4 \times 10^{-3}$, the minimum excitation energy of unpaired electron $\Delta E_{\sigma\pi^*}$ is calculated from Eq.(A.2) to be equal to 5.2 eV. This energy is close to the first ionization potential of polycyclic aromatic hydrocarbons (Traven' 1989). Consequently, PC can be localized in PPP-4 and PPP-5 samples near cross-linkages which appear as polycyclic hydrocarbons as it was predicted earlier (Kuivalainen et al. 1983).

At D-band EPR the bell-like contribution attributed to the manifestation of the fast passage of inhomogeneously broadened line appears at $g_{\mathrm{eff}} = 2.00319$ in sum EPR spectra of the two latter films (Figure 4,c). This effect allows us to evaluate spin-spin relaxation time, $T_1 \approx 10^{-4}$ s for PC in dedoped and redoped samples.

From the analysis of the X- and D-bands EPR line shape the rate of spin-packets exchange $v_{ex} = 4 \times 10^7$ s^{-1} was estimated for the neutral and redoped films (Goldenberg et al. 1991). This value obtained for doped film increases up to 1.8×10^8 s^{-1} due to the increase in PC concentration and mobility. Effective, isotropic g-factor, $g_{\mathrm{iso}} = 1/3(g_\parallel + 2g_\perp)$ of a neutral sample is equal to g-factor of Cl$_3^-$-doped one. This fact shows that **g**-tensor components of PC are averaged in PPP: Cl$_3^-$ sample due to Q1D spin diffusion with coefficient $D_{1D}^0 \geq 4.3 \times 10^7$ rad/s. However, Q1D spin diffusion coefficient calculated from relaxation times of the samples was appeared to be considerably low (Goldenberg et al. 1991) due possible to disordered of polymer matrix. *AC* conductivity calculated for PPP-2, PPP-2 and PPP-3 samples from Eq.(A.35), Eq.(A.38), and Eq.(A.39) changes from 1.5×10^5 S/m up to 1.8×10^5 S/m and then decrease down to 5.3×10^5 S/m, respectively. This parameter obtained for PPP-1 depends on the temperature as $\sigma_{ac} \propto T^{0.23}$. Such a dependency seems to indicate the existence in doped PPP sample of some conductivity mechanisms, namely VRH and isoenergetic tunneling (Kivelson 1980) of charge carriers. The strong decrease in PC concentration as well as the slowdown of spin exchange and spin-spin relaxation processes at the sample dedoping, i.e., at transition from PPP-1 to PPP-4, evidence for the pairing of most polarons into spinless bipolarons, whose Q1D diffusion causes electron relaxation of the whole spin system.

The intrachain spin diffusion coefficients D_{1D} calculated from Eq.(A.25) as well as conductivity σ_{1D} due to Q1D spin diffusion calculated from Eq.(A.34) were appeared to be too small (around $10^5 - 10^6$ rad/s) for the appearing of a Dysonian component in the PPP spectra.

Thus, EPR data analysis allow to conclude that in highly doped PPP-1, the charge is transferred mainly by mobile bipolarons and only small part of mobile polarons take part in this process. In PPP-5 these polarons couple into spinless bipolarons being the predominant charge carriers in this conjugated polymer.

At the electrochemical substitution of Cl_3^- anion by BF_4^- one, the location of the latter may differ from that of the dopant in the initial PPP sample. However, the morphology of the BF_4^--redoped PPP may be close to that of a neutral PPP-4 film.

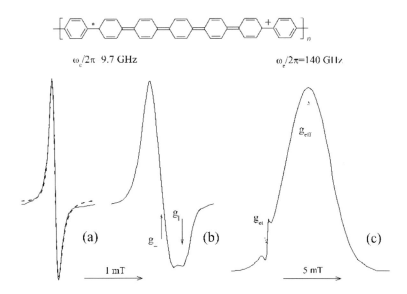

Figure 4. Typical X-band EPR in-phase modulation absorption spectrum of PPP-1 – PPP-3 (*a*) and PPP-4, PPP-5 (*b*) samples; (*c*) typical D-band EPR in-phase modulation dispersion spectrum of PPP-4 and PPP-5 films (see the text) registered at 300 K. The spectrum calculated from Eqs. (A.35), (A.38) and (A.39) with $D/A = 0.39$ and $\Delta B_{pp} = 0.109$ mT is shown at the left by dashed line. A narrow line in the right spectrum is attributed to the lateral standard, single crystal $(DBTTF)_3PtBr_6$ with $g = 2.00411$.

Table 1. The terms and trace of the g-tensor, g_{ii} and g_{iso}, respectively, as well as the linewidth ΔB_{pp}^{i} of the *i*-th spectral components (in mT) of polarons stabilized as electron donors in some conjugated polymers and measured at different wavebands (WB) and low (20-80 K) temperatures when all spin motions are considered to be frozen

Sample	g_{xx}	g_{yy}	g_{zz}	g_{iso}	ΔB_{pp}^{x}	ΔB_{pp}^{y}	ΔB_{pp}^{z}	ΔB_{pp}^{iso}	WB[2]	References
PPP	2.0034	2.0034	2.0020	2.0029	0.44	0.44	0.23	0.37	X	(Goldenberg et al. 1990)
PATAC	2.0433	2.00902	2.00243	2.01825	5.66	9.82	5.95	7.14	D	(Krinichnyi, Roth, and Schrödner 2002)
PPy	2.00380	2.00380	2.00235	2.00	0.57	0.57	0.57	0.57	D	(Pelekh, Goldenberg, and Krinichnyi 1991)
PANI-EB	2.00522	2.00401	2.00228	2.00384	1.25	1.25	2.27	1.59	D	(Krinichnyi, Chemerisov, and Lebedev 1997)
PANI-EB	2.00603	2.00382	2.00239	2.00408	0.45	0.45	3.02	1.31	D	(Krinichnyi et al. 2002)
PTTF-CH$_3$-C$_6$H$_4$	2.01191	2.00584	2.00185	2.00652	1.72	1.63	1.79	1.71	D	(Krinichnyi 2006)
PTTF-C$_2$H$_5$-C$_6$H$_4$	2.01405	2.00676	2.00235	2.00772	4.06	4.63	4.43	4.37	D	(Krinichnyi et al. 1993)
PTTF-THA	2.01292	2.00620	2.00251	2.00721	5.53	2.73	1.74	3.33	D	(Krinichnyi, Denisov, et al. 1998)
PT	2.00382	2.00266	2.00266	2.00305	1.45	1.79	1.79	1.68	D	(Krinichnyi et al. 1985)
P3HT	–	–	–	2.0030	–	–	–	0.66	X	(Krinichnyi, Troshin, and Denisov 2008)
P3HT	2.0030	2.0021	2.0011	2.0021	0.16	0.15	0.16	0.16	K	(Konkin et al. 2010)
P3HT	2.0028	2.0019	2.0009	2.0019	1.07	0.53	0.64	0.75	W	(Aguirre et al. 2008)
P3HT	2.00380	2.00230	2.00110	2.00240	0.85	0.88	0.76	0.83	D	(Poluektov et al. 2010)
P3OT	2.00409	2.00332	2.00232	2.00324	0.82	0.78	0.88	0.83	D	(Krinichnyi and Roth 2004)
P3DDT	2.0026	2.0017	2.0006	2.0016	0.25	0.14	0.15	0.18	X	(Krinichnyi, Yudanova, and Spitsina 2010)
PCDTBT	–	–	–	2.0022	–	–	–	0.14	X	(Krinichnyi, Yudanova, and Denisov 2014)
PCDTBT	2.00320	2.00240	2.00180	2.00247	0.44	0.49	0.38	0.44	D	(Niklas et al. 2013)

Notes: [1] the abbreviations mean the following polymers: PPP - poly(p-phenylene), PATAC - poly(*bis*-alkylthioacetylene), PPy - polypyrrole, PANI - polyaniline, PTTF – poly(tetrathiafulvalene), PT - polythiophene, PCDTBT - poly[N-9'-hepta-decanyl-2,7-carbazole-alt-5,5-(4',7'-di-2-thienyl-2',1',3'-benzothiadiazole)].

[2] the wavebands (WB) correspond to spin precession frequency $\omega_e/2\pi$ and resonant magnetic field B_0 of 9.7 GHz and 0.34 T (X), 24 GHz and 0.86 T (K), 38 GHz and 1.2 T (Q), 94 GHz and 3.4 T (W), 140 GHz and 4.9 T (D), respectively.

As it was established by Goldenberg et al. (Goldenberg et al. 1990, 1991), the PPP film synthesized in the BuPyCl-AlCl$_3$ melt is characterized by smaller number of benzoid monomers and by greater number of quinoid units. It leads to a more ordered structure and planar morphology of the polymer that in turn prevents the coupling of spin charge carriers to bipolaron in highly doped polymer. The anions are removed at dedoping, so then the packing density of the polymer chains grows. It can prevent an intrafibrillar implantation of BF_4^- anions and lead to the localization of dopant molecules in the intrafibrillar free volume of the polymer matrix. The change of charge transfer mechanism at the replace of dopants should be a result of such a morphological transition in PPP (Goldenberg et al. 1991).

2.2. Poly(*bis*-Alkylthioacetylene)

The derivative of *trans*-polyacetylene, poly(*bis*-alkylthioacetylene) (Richter et al. 1987) shown in Figure 2, is also an insulator in a neutral form. From the ^{13}C Nuclear Magnetic Resonance (NMR) study (Hempel et al. 1990) the conclusion was made that PATAC has sp^2/sp^3-hybridized carbon atom ratio typical for polyacetylene, however, in the contrast with the latter, pristine polymer has a more twisted backbone. The *dc* conductivity of PATAC increases under chemical doping from $\sigma_{dc} \approx 10^{-12}$ S/m up to $\sigma_{dc} \approx 10^{-8} - 10^{-2}$ S/m depending on the kind and/or concentration of an anion introduced into the polymer in a liquid or gas phase. Upon irradiated by argon laser this parameter increases up to $\sigma_{dc} \approx 1 \times 10^3 - 2 \times 10^4$ S/m depending on the absorbed dose (Roth, Gruber, et al. 1990). Hall coefficient measurement of the laser-modified PATAC (Roth, Gruber, et al. 1990) have shown that the charge carriers are of *p*-type and their mobility μ depends on the temperature as $\mu \propto T^n$ with $0.25 \leq n \leq 0.33$ at 80 $\leq T \leq 300$ K. The mobility was obtained at RT to be close to $\mu = 0.1 - 8$ cm^{2}V^{-1}s^{-1} that is close to $\mu = 2$ cm^{2}V^{-1}s^{-1} obtained for *trans*-PA (Bleier et al. 1988).

The X-band EPR study (Roth, Gruber, et al. 1990) has shown that π-like PC with different mobility exist in laser-modified PATAC, however, the conductivity of so treated polymer is mainly determined by the dynamics of diamagnetic bipolarons. It was proposed that as the temperature increases the bipolaron mobility decreases and their concentration in laser-modified PATAC increases, so then these processes should lead to the extremely low (close 10^{-3} K^{-1}) temperature coefficient of the PATAC *dc* conductivity. In laser-treated polymer the RT spin concentration, *g*-factor and peak-to-peak linewidth ΔB_{pp} change from $N \approx 2.7 \times 10^{14}$ cm^{-3}, *g*= 2.0056, and $\Delta B_{pp} = 0.72$ mT to $N \approx 4.7 \times 10^{17}$ cm^{-3}, *g*= 2.0039, and $\Delta B_{pp} = 0.65$ mT, respectively.

An initial, insulating, powder-like PATAC-1 sample and that irradiated by an argon ion laser with dose of 5 J/cm^3, PATAC-2, and with the photon energy/wavelength $\lambda_{ph}/h\nu_{ph} = 2.46$ eV/488 nm (here $h = 2\pi\hbar$ is the Planck constant) were studied at both the X- and D-bands EPR at RT (Roth et al. 1999, Krinichnyi, Roth, and Schrödner 2002).

At X-band the PC in the PATAC-2 sample demonstrates a slightly asymmetric Lorentzian spectrum with a weak component at low fields (Figure 5,a). The intensity of the latter component decreases during the PATAC laser modification and/or as the temperature increases. The spectrum simulation has shown that such line asymmetry arises rather due to the anisotropy of *g*-factor. This supposition was confirmed by a more detailed study of the samples at higher registration frequency. D-band EPR spectra of the PATAC-2 sample is also presented in Figure 5,b. Higher spectral resolution allowed to show the existing in PATAC of two types of PC, namely, polarons localized on the short π-conjugated polymer chain $P_{loc}^{+\bullet}$ with g_{xx} = 2.04331, g_{yy} = 2.00902, g_{zz} = 2.00243, and linewidth ΔB_{pp} = 6.1 mT (D-band), and polarons moving along the main π-conjugated polymer chain $P_{mob}^{+\bullet}$ with g_{xx} = 2.00551, g_{yy} = 2.00380, g_{zz} = 2.00232, $(2g_\parallel = g_{xx}+g_{yy}, g_\perp = g_{zz})$ and ΔB_{pp} = 2.7 mT (D-band). Simulated spectra of these charge carriers are shown in Figure 5 as well.

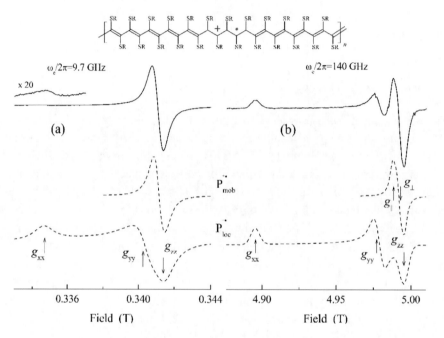

Figure 5. The absorption X- (*a*) and D- (*b*) bands EPR spectra of the laser modified PATAC-2 sample registered at 100 K. Dashed lines show the spectra of radical $P_{loc}^{+\bullet}$ calculated with g_{xx} = 2.0433, g_{yy} = 2.00902, g_{zz} = 2.00243, and radical $P_{mob}^{+\bullet}$ calculated with g_{xx} = 2.00551, g_{yy} = 2.00380, g_{zz} = 2.00232. Polaron initiated in this polymer with R ≡ −CH$_3$ by laser irradiation is shown schematically.

Since the isotropic *g*-factor of the laser-treated PATAC sample is considerably higher than that of most organic conjugated polymers, $g_{iso} \cong$ 2.003 (Krinichnyi 1995, Mizoguchi and Kuroda 1997, Krinichnyi 2016b), one can conclude that the unpaired electron in PATAC interacts with sulfur atoms. It is typical for other sulfur-containing compounds, e.g., poly(tetrathiafulvalenes) (Krinichnyi 1996, 2016b) and benzotrithioles (Cameron et al. 1991, 1992, Krinichnyi et al. 1997) in which sulfur atoms are involved into the conjugation. Taking into account that the overlapping integral I_{c-c}^p in such organic π-systems depends on the torsion (dihedral) angle θ between *p*-orbits of neighboring C-atoms as $I_{c-c}^p \propto \cos\theta$ (Traven' 1989, Masters et al. 1992), the shift of *g*-factor from the *g*-factor for the free electron in Eq.(A.2) should be multiplied by the factor $\lambda_s(1 - \cos\theta)/(1 + k_1\cos\theta)$, where λ_s = 0.047 eV (Carrington and McLachlan 1967) is the spin-orbit interaction constant for sulfur and k_1 is a constant. The *g*-factor of PC in sulfur-containing solids in which electrons are localized mainly on the sulfur atom lies in the region of $2.014 \leq g_{iso} \leq 2.020$ (Cameron et al. 1991, 1992, Krinichnyi et al. 1997, Bock et al. 1984). In tetrathiafulvalene (TTF) derivatives an unpaired electron is delocalized on 12 or more carbon atoms and four sulfur atoms leading to the decrease of both $\rho_s(0)$ and, therefore, g_{iso} values (Krinichnyi 2000a). An additional fast spin motion takes place in the PTTF (Krinichnyi 1995, Krinichnyi et al. 1997, Krinichnyi 2000a, 2016b) leading to a further decrease in the $\rho_s(0)$ value, and therefore in g_{iso} down to $2.007 - 2.014$ depending on the structure and effective polarity in PTTF samples. Due to the smaller g-factor in PATAC one can expect a higher spin delocalization in this polymer as compared with the above mentioned organic semi conjugated solids.

Assuming the energy of $n \rightarrow \pi^*$ transition, $\Delta E_{n\pi^*} \approx 2.6$ eV, typical for benzotrithioles and PTTF (Cameron et al. 1992, Bock et al. 1984), then the g-factor components of PC in the initial PATAC were obtained from Eq.(A.2) to be $\rho_s(0) \approx 1.1$ and $\Delta E_{n-\sigma^*} = 15.6$ eV. This means that in the initial polymer the spin is localized within one monomer unit.

It was shown (Krinichnyi 1995, Krinichnyi et al. 1997) that the storage of the sample lead to the increase g_{xx} and g_{yy} values of $P_{loc}^{+\bullet}$ up to 2.0451 and 2.00982, respectively. Besides, this originates also the line broadening of this PC up to $\Delta B_{pp} = 12.9$ mT. This means the increase of spin localization on the sulfur nucleus due to the shortness of polymer chains in amorphous regions during the PATAC storage. On the other hand, such destruction does not lead to the change in the magnetic parameters and concentration of mobile polarons.

Supposing the spin delocalization onto approximately five units (Devreux et al. 1987) also in PATAC, then the $\rho_s(0)$ value determined above should decrease down to 0.22. This fits very well the g-factor measured for the PATAC-2 sample. Higher spin delocalization upon laser treatment can be accompanied by the decrease of the θ value down to 21.3°. This should lead to an additional acceleration of spin diffusion along the polymer chains. The latter value is close to the change in θ ($\Delta\theta = 22 - 23°$) at transition from benzoid to quinoid form in PPP (Brédas et al. 1991), from emeraldine base to emeraldine salt form of polyaniline (PANI-EB and PANI-ES, respectively) (Krinichnyi, Chemerisov, and Lebedev 1997), and from the polytertathiafulvalene with phenyl bridges (PTTF-Ph) to that with tetrahydroanthracene (PTTF-THA) ones (Krinichnyi 2000a). This supports the assumption made by Roth et al. (Roth, Gruber, et al. 1990) that laser irradiation leads to a more planar morphology of polymer backbone and therefore to both the higher spin delocalization and conductivity of the sample.

The concentration ratio $[P_{mob}^{+\bullet}]/[P_{loc}^{+\bullet}] \approx 1:2$ obtained for the insulating sample PATAC-1, increases during its modification by laser up to 2:1 determined for the PATAC-2 sample. This means that such the treatment of conjugated polymer leads to a strong increase of the concentration of mobile polarons in laser-treaded system. On the other hand, the concentration of charge carriers determined from Hall and dc conductivity studies changes by more than 15 orders of magnitude reaching $N \sim 10^{19}$ cm^{-3} in relatively strong laser-irradiated polymer (Roth, Gruber, et al. 1990). This means that the charge, as in case of some other conjugated polymers, is predominantly transferred by paramagnetic polarons in an initial and slightly modified PATAC. The planarity of chains in highly irradiated polymers increases, so then the most polarons couple into diamagnetic bipolarons.

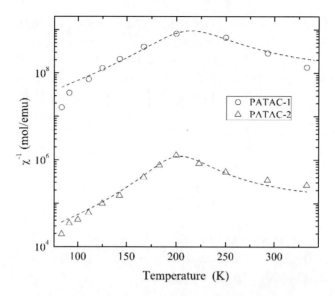

Figure 6. Temperature dependence of spin susceptibility of the PATAC samples. Top-to-bottom dashed lines show the dependencies calculated from Eq.(A.4) with $C = 2.4\times10^{-6}$ emu K/mol, $k_1 = 7.2\times10^{-4}$ emu K/mol, $J_{af} = 0.069$ eV and $C = 4.6\times10^{-3}$ emu K/mol, $k_1 = 0.48$ emu K/mol, $J_{af} = 0.050$ eV.

Spin susceptibility χ of PATAC depends not only on the laser irradiation dose absorbed but also on the temperature. Figure 6 demonstrates inversed temperature dependencies of this value obtained for the PATAC samples. The χ value decreases as the temperature decreases from maximum down to a critical temperature $T_c \approx 200 - 220$ K and then starts to increase at lower temperatures. The observed increase of the magnetic susceptibility at temperatures lower than T_c can result from the formation of clusters with collective localized Curie spins. At higher temperatures, as in case of polyaniline (Iida et al. 1992, Krinichnyi et al. 2002) and poly(3-dodecylthiophenes) (Barta et al. 1994, Cik et al. 1995, Kawai et al. 1996, Cik et al. 2002), the spin susceptibility of PATAC seems also to include a contribution due to singlet-triplet equilibrium described by the second term of Eq.(A.4). The data presented in Figure 6 evidence that $\chi(T)$ dependencies of the PATAC samples are fitted well by Eq.(A.4) with $J_{af} \approx 0.05 - 0.07$ eV.

It is obvious that the above processes should be accompanied by a reversible coupling of polarons into bipolarons and dissociation of bipolarons into polarons. However, Stafström et al. (Stafstrom and Brédas 1988, Stafström and Brédas 1988) have shown that the bipolaron state is not the favorable state in main conjugated polymers.

In order to study the change of spin dynamics in the PATAC backbone under its laser-modification, the temperature dependence of linewidth ΔB_{pp} of mobile polarons $P^{+\bullet}_{mob}$ in both the samples should be analyzed. These dependencies measured at X- and D-bands EPR are shown in Figure 7. It is seen from the Figure that the linewidth measured at X-band EPR is weakly dependent on temperature. However, this value becomes more temperature sensitive at higher spin precession frequency due to the decrease of an exchange interaction between spin packets. The D-band linewidth of the PATAC-1 sample increases with the temperature decrease. At the same time, this value of all laser-modified samples demonstrates the extremal temperature dependence with the critical temperature T_c close to $200 - 220$ K. These functions are similar to $\chi(T)$ presented in Figure 6, therefore, such behavior can be associated with the polaron-bipolaron transition at T_c and can probably not reflect the change of the mobility of charge carriers in the laser-treated sample. At the storage, the linewidth of the PATAC-2 sample becomes linearly dependent on the temperature (Figure 7). The RT linewidth of the sample PATAC-2 was measured at D-band EPR to be 1.45 mT. So, the linewidth of PC $P_{mob}^{+\bullet}$ increases by approximately five times at the transition from the X-band to the D-band mainly due to the effect of spin precession frequency on spin-spin relaxation. The dependences presented are reflect also an exchange spin-spin interaction described by E.(A.17). It is seen from the Figure that the data obtained experimentally are fitted well by this Equation. The energy ΔE_r characterizing the reorganization of polarons decreases more than threefold under the polymer modification by laser irradiation (see Figure 7). This value, however, increases back after the storage of the treated sample. This fact may indicate an increase in the planarity of the sample upon laser irradiation and a decrease in this parameter with time.

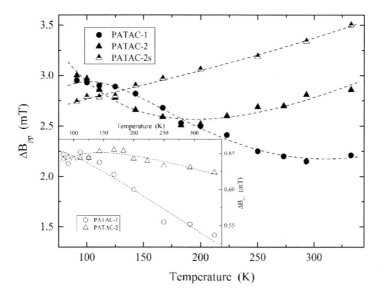

Figure 7. Temperature dependencies of the linewidth of mobile polarons R_2 in the PATAC-1 and PATAC-2 samples determined from their X- (insert) and D-band EPR absorption spectra. Semifilled points show such dependence obtained at D-band EPR for PATAC-2 sample stored for two years. Top-to-bottom dashed lines show the dependencies calculated from Eq.(A.17) with $\omega_{hop}^0 = 9.1 \times 10^8$ s^{-1}, $\Delta E_r = 0.023$ eV, $\omega_{hop}^0 = 7.8 \times 10^6$ s^{-1}, $\Delta E_r = 0.008$ eV, $\omega_{hop}^0 = 5.3 \times 10^7$ s^{-1}, $\Delta E_r = 0.029$ eV, as well as with $\omega_{hop}^0 = 8.7 \times 10^7$ s^{-1}, $\Delta E_r = 0.002$ eV, $\omega_{hop}^0 = 2.1 \times 10^8$ s^{-1}, $\Delta E_r = 0.001$ eV (insert).

As in case of other conjugated polymers, saturation effects were registered in dispersion spectra of the PATAC samples. This allowed investigating anisotropic torsion librations of the pinned polarons near the main x-axis of the polymer chains. Such dynamics in the PATAC-1 sample was analyzed within a model of activation motion of immobilized polarons and polymer chains with correlation time of $\tau_c^x = 6.3 \times 10^{-6} \exp(0.043 \text{ eV}/k_B T)$ s (here k_B is the Boltzmann constant) (Roth et al. 1999, Krinichnyi, Roth, and Schrödner 2002). These effects also made it possible to determine also effective spin-lattice and spin-spin relaxation times of the laser-modified PATAC-2 sample. Both relaxation times determined for the initial PATAC-2 polymer and the same sample stored for two years are presented in Figure 8,a as function of temperature. It is seen from the Figure that both relaxation times of polarons initiated in the sample decrease considerably during the storage. Besides, this process leads to the extremal temperature dependence with $T_c \approx 160$ K at which the semiconductor-metal transition occurs. The analogous phenomena was registered in polyaniline (Krinichnyi, Chemerisov, and Lebedev 1997, Krinichnyi et al. 2002), poly(p-phenylene vinylene) (Ahlskog et al. 1997), poly(3,4-ethylene-dioxy-thiophene) (Chang et al. 1999), and other conjugated polymers (Ahlskog, Reghu, and Heeger 1997).

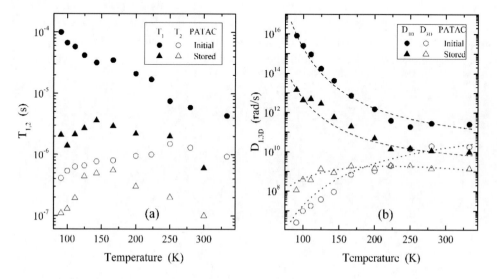

Figure 8. Temperature dependence of the effective spin-lattice T_1 and spin-spin T_2 relaxation times (a) intra- D_{1D} and interchain D_{3D} spin diffusion of mobile polarons determined from Eq.(A.25) and Eq.(A.26) (b) in as-modified and stored for two years PATAC-2 samples. The dashed lines show dependences calculated from Eq.(A.29) with $E_{ph} = 0.18$ eV and $E_{ph} = 0.15$ eV. The dotted lines depict dependences calculated from Eq.(A.30) with characteristic energy E_a equal to 0.061 and 0.051 eV, respectively.

The involvement of spectral components into the motional exchange and therefore their shift to the spectrum center at the laser modification of PATAC apparently indicates the appearance of polarons moving along the PATAC-2 polymer chains with the rate $D_{1D}^0 \geq 2 \times 10^{10}$ rad/s (Roth et al. 1999, Krinichnyi, Roth, and Schrödner 2002). Assuming the spin situation near sites of cubic backbone, the temperature dependencies of the D_{1D} and D_{3D} values were calculated for the initial and stored for two years PATAC-2 sample from Eq.(A.25) and Eq.(A.26) (see Figure 8,b). It is seen from the Figure that the rate of the spin intrachain diffusion in the samples remains invariable as the temperature decreases down to $T_c \cong 200$ K due probably to the compensation effect of the T_1 increase and the decrease in spin concentration, however, starts to increase at the further temperature decreases. At the same time, D_{3D} decreases monotonically with the temperature decrease of the as-modified PATAC-2 sample. With sample storage this value has a weak fall in temperature (Figure 8,b).

It was shown (Krinichnyi, Roth, and Schrödner 2002) that polaron diffusion along the PATAC chain is governed by the scattering of its spin on the lattice phonons of crystalline domains embedded into an amorphous polymer matrix. The energy of the lattice photons E_{ph}

of the initial and stored PATAC was determined to be 0.18 and 0.15 eV (see Figure 8,b). A difference in energy E_{ph} may indicate a decrease of polymer ordering during its storage. These values is close to that (0.13 eV) obtained for polyaniline doped by hydrochloric acid (Krinichnyi 2000b), however, exceeds by a factor 3 – 4 the activation energy of macromolecular librations in PATAC samples determined by saturation transfer EPR (ST-EPR) method. Spin diffusion between polymer chains of the initial and stored PATAC samples was described within activation mechanism by Eq.(A.30) with characteristic energy E_a equal to 0.061 and 0.051 eV, respectively. These values are close to the typical activation energy of the interchain spin hopping in some other compounds of low dimensionality. The decrease in E_a evidences the decrease of polymer ordering/crystallinity during storage. E_a values are close to those obtained from paramagnetic susceptibility and chain librations data of the samples. This fact leads to the conclusion about the interference of these processes in PATAC. From the average RT mobility $\mu \cong 0.5$ cm^{2}V^{-1}s^{-1} determined from the Hall study for charge carriers in highly irradiated polymer (Roth, Gruber, et al. 1990) one can evaluate an intrinsic conductivity of the treated PATAC to be close to 3.4×10^2 S/m. This value, however, significantly exceeds respective contributions to ac conductivity due to polaron Q1D and Q3D mobility calculated from Eq.(A.34), $\sigma_{1D} = 0.48$ S/m and $\sigma_{3D} = 8.5 \times 10^{-2}$ S/m, respectively. This means that the intrinsic conductivity in highly conjugated PATAC is determined mainly by dynamics of spinless bipolarons with concentration exceeding the number of mobile polarons by approximately two orders of magnitude.

So, multifrequency EPR spectroscopy allowed postulating the existence of two types of paramagnetic centers in laser-modified PATAC – polarons moving along polymer chains in highly ordered crystalline domains and polarons localized by spin traps in amorphous regions of polymer backbone. Assuming that the polaron is covered by electron and excited phonon clouds, one can propose that both spin relaxation and charge transfer should be accompanied by the phonon dispersion. The mobility of the polarons depends strongly on their interaction with other PC and with the lattice phonons. The charge transfer integral and therefore the intrinsic conductivity of the sample is modulated by macromolecular dynamics and such a dynamics reflects the effective crystallinity of PATAC with metal-like domains. The strong spin-spin interaction at high temperatures leads to the stripping of bipolarons into polaron pairs. The number of bipolarons exceeds the number of polarons at least by two orders of magnitude, so the total conductivity of PATAC is determined mainly by dynamics of diamagnetic charge carriers. Magnetic resonance, relaxation and dynamics parameters of PATAC are shown to change during its storage that can be explained by degradation of the polymer.

2.3. Polypyrrole

Polypyrrole is another conjugated polymer widely investigated as perspective molecular system for plastic electronics (Rodriguez, Grande, and Otero 1997, Scotheim and Reynolds 2007). The degree of local order varies for PPy dependent upon the preparation method, with the degree of crystallinity varying from nearly completely disordered up to ~50% crystalline (Epstein 2007). In contrast to polyaniline, the local order in the disordered regions of PPy does not resemble that in the ordered regions.

Neutral PPy exhibits a complex X-band EPR spectrum with a superposition of a narrow (0.04 mT) and a wide (0.28 mT) lines with $g \approx 2.0026$ (Bernier 1986, Saunders, Fleming, and Murray 1995), typical for radicals in polyene and aromatic π-systems (Rodriguez, Grande, and Otero 1997). The intensities of both lines correspond to one spin per a few hundred monomer units. In PPy was also registered single contribution with $g_{iso} = 2.0026$, $\Delta B_{pp} = 0.235$ mT (Wu, Chang, and Lin 2009) and $g_{iso} = 2.0026$, $\Delta B_{pp} = 0.022$ mT (Bartle et al. 1993) was also registered.

Paramagnetic susceptibility of the narrow line is thermally activated, while χ parameter of the broad one has a Curie behavior. These features, together with the temperature dependencies of the linewidth, were interpreted in terms of coexistence of two types of PC with different relaxation parameters in neutral PPy. Doped PPy sample exhibits only a strong and narrow (~0.03 mT) X-band EPR spectrum of polarons with $g = 2.0028$, which follows Curie law from 300 to 30 K (Bernier 1986). However, it was revealed that magnetic susceptibility of polarons in doped PPy is characterized by a Curie-like susceptibility and a weak Pauli contribution; however, the latter does not contribute to the conductivity mechanism (Schmeisser et al. 1998, Kanemoto and Yamauchi 2000b). It was obtained that the charge in PPy is transferred according to the VRH model (Taunk and Chand 2014). The linewidth of PC stabilized in partly stretch-oriented PPy were observed (Sakamoto et al. 1999) to be functions of temperature and the angle between the static magnetic field and the stretched direction. The linewidth of PC becomes higher as the oxygen molecules penetrate into PPy bulk due to the exchange and dipole-dipole interactions of polarons with oxygen biradicals (Kanemoto and Yamauchi 2001). Spin relaxation of PPy was shown (Kanemoto and Yamauchi 2000a, b) to depend on its doping level in relation to several physical properties such as the ratio of the Pauli susceptibility to the total one, the metallic behavior in the conductivity, and the polaron band observed in the optical spectrum. Most of the preceding results imply that EPR signal does not arise from the same species, which carries the charges, because of the absence of correlations of the susceptibility with concentration of charge carriers and the linewidth with the carrier mobility. This was interpreted in favor of the spinless bipolaron formation upon PPy doping (Scott et al. 1983, Chakrabarti et al. 1999). Thus, EPR signal of doped PPy is attributed mainly to neutral radicals and therefore reports little about the intrinsic conjugated processes.

In this case the method of spin probe seems to be more effective for the study of structural and electronic properties of PPy sample. Only a few papers reported the study of conjugated polymers by using spin label and probe at X-band (Audebert et al. 1987, Winter et al. 1990, Shchegolikhin, Yakovleva, and Motyakin 1995, Sersen, Cik, and Veis 2003). It is explained by the fact that main conjugated polymers in conjugated state are insoluble in convenient solvents that prevents an introduction of a stable radical into their bulk. A low spectral resolution at low-frequency wavebands did not allow the registration of all components of $\mathbf{g}$ and $\mathbf{A}$ tensors and therefore the separate determination of the magnetic susceptibility of both spin label and PC on the polymer chain, and the measurement of the dipole-dipole interaction between different PC. PPy modified during electrochemical synthesis by nitroxide doping anion, as a label covalently joined to the pyrrole cycle, was studied by Winter et al. (Winter et al. 1990). However, in spite of a high concentration of a spin probe introduced into PPy, effective X-band EPR spectrum of such sample did not contain lines of the probe.

D-band EPR method of spin probe becomes more effective at investigation of PPy synthesized electrochemically on a platinum electrode in an aqueous solution of 0.2 M pyrrole and 0.02 M 2,2,6,6-tetramethyl-1-oxypiperid-4-ylacetic acid (Pelekh, Goldenberg, and Krinichnyi 1991).

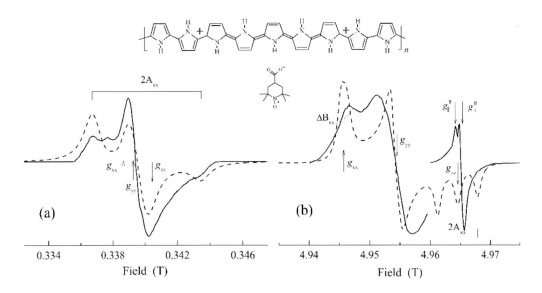

Figure 9. The X-band (*a*) and D-band (*b*) absorption spectra EPR of 4-carboxy-2,2,6,6-tetramethyl-1-oxypiperidiloxyl nitroxide radical introduced into frozen (120 K) toluene (dotted line) and conductive polypyrrole (solid line) as a spin probe. The anisotropic spectrum of localized paramagnetic centers marked by the symbol *R* and taken at a smaller amplification is also shown in the lower part of the Figure. The measured magnetic parameters of the probe and radical *R* are shown. Bipolaron on a PPy chain and nitroxide radical are also shown.

The absorption X- and D-bands EPR spectra of a nitroxide radical, introduced simultaneously as a probe and a counter-ion into PPy as well as a probe into a frozen nonpolar model system are shown in Figure 9. One can see that at X-band EPR the lines of nitroxide radical rotating with correlation time $\tau_c > 10^{-7}$ s overlap with the single line of PC (*R*) stabilized in PPy (Figure 9,a). Such an overlapping stipulated by a low spectral resolution, hinders the separate determination of magnetic resonance parameters of the probe and radical *R* in PPy together with the dipole-dipole broadening of its spectral components.

As it was expected, the spectra of both model and modified polymer systems become more informative at D-band EPR (Figure 9,b). At this waveband all canonic components of EPR spectra of the probe in PPy and toluene are completely resolved so then all the values of **g** and **A** tensors can be measured directly. Nevertheless, the asymmetric spectrum of radicals *R* with magnetic parameters $g_{\parallel}^R = 2.00380$, $g_{\perp}^R = 2.00235$ and $\Delta B_{pp} = 0.57$ mT is registered on the *z*-component of the probe spectrum. In non-polar toluene the probe is characterized by the following magnetic resonance parameters: $g_{xx} = 2.00987$, $g_{yy} = 2.00637$, $g_{zz} = 2.00233$; $A_{xx} = A_{yy} = 0.60$ mT and $A_{zz} = 3.31$ mT. The difference $\Delta g = g_{\parallel}^R - g_{\perp}^R = 1.45 \times 10^{-3}$ corresponds to an excited electron configuration in *R* with $\Delta E_{\sigma\pi^*} = 5.1$ eV lying near to an energy of electron excitation in neutral PPP. In conjugated PPy g_{xx} value of the probe decreases down to 2.00906 and the broadening of its *x*- and *y*-components, $\delta(\Delta B_{pp})$ is 4 mT

(Figure 9,b). In addition, the shape of the probe spectrum shows the localization of PC R on the polymer pocket of 1 nm size, i.e., the charge is transferred by spinless bipolarons in PPy, as it was proposed in the case of PPP: BF_4^- and PT: BF_4^-.

In neutral PPy the fragments with a considerable dipole moment are *a priori* absent. Besides, the dipole-dipole interactions between the radicals can be neglected due to low concentration of the probe and PC localized on the chain. Therefore, the above change in the probe magnetic resonance parameters taking place at transition from model non-polar system to the conjugated polymer matrix may be caused by Coulombic interaction of the probe active fragment with the extended spinless bipolarons, each carrying double elemental charge (see Figure 9). The effective electric dipole moment of such charge carriers moving near the probe was determined from the shift of g_{xx} component to be equal to dipole moment $\mu_v = 2.3$ D. The shift of **g** tensor component g_{xx} of the probe may be calculated within the frames of the electrostatic interaction of the probe and bipolaron dipoles. The potential of electric field induced by bipolaron in the place of the probe localization is determined by Eq.(A.27) taking μ_u as the dipole moment of the probe, ε is the dielectric constants for PPy, and r is the distance between an active fragment of the radical and bipolaron. By using the dependence of the growth of an isotropic hyperfine constant of the probe under microenvironment electrostatic field, $\Delta a = 7.3er_{NO}t_{cc}^{-1}$ (here r_{NO} is the distance between N and O atoms of the probe active fragment, t_{cc} is the resonant overlapping integral of C=C bond) and the relation $dg_{xx}/dA_{zz} = 2.3\times10^{-2}$ mT^{-1} for hexamerous unit ring nitroxide radical (Krinichnyi 1991b, a), one can write $\Delta g_{xx} = 6\times10^{-3}er_{NO}k_BT(x \coth x - 1)/(t_{cc}\mu_u)$. By using $\mu_u = 2.7$ D (Reddoch and Konishi 1979), $\mu_v = 2.3$ D and $r_{NO} = 0.13$ nm (Buchachenko and Vasserman 1973, Buchachenko, Turton, and Turton 1995), the value of $r = 0.92$ nm is obtained.

The rate of spin-spin relaxation which stipulates the radical spectrum broadening $T_2^{-1} = T_{2(D)}^{-1} + T_{2(0)}^{-1}$ consists of the relaxation rate of the radical non-interacting with the environment $T_{2(0)}^{-1}$ and the growth in the relaxation rate due to dipole-dipole interactions $T_{2(D)}^{-1} = \gamma_e\delta(\Delta B_{x,y})$. The characteristic time τ_c of such an interaction can be calculated from the broadening of the spectral lines using Eq.(A.26) with $J(\omega_e) = 2\tau_c/(1 + \omega_e^2\tau_c^2)$. The inequality $\omega_e\tau_c \gg 1$ is valid for most condensed systems of high viscosity, so then averaging the lattice sum over angles, $\Sigma\Sigma(1 - 3\cos^2\vartheta)^2 r_1^{-3}r_2^{-3} = 6.8r^{-6}$(Lebedev and Muromtsev 1972), and using $\gamma_e\delta(\Delta B_{pp}) = 7\times10^8$ s^{-1} (here γ_e is hyromagnetic ratio for electron) and $r = 0.92$ nm calculated above one can determine $T_{2(D)}^{-1} = 3<\omega^2>\tau_c$ or $\tau_c = 8.1\times10^{-11}$ s. This value lies near to the polaron interchain hopping time, $\tau_{3D}\cong 1.1\times10^{-10}$ s estimated for lightly doped PPy (Kanemoto and Yamauchi 2000b). Taking into account, that the average time between the translating jumps of charge carriers is defined by the diffusion coefficient D and by the average jump distance equal to a product of lattice constant d_{1D} on half width of charge carrier $N_p/2$, $\tau_c = 1.5 \langle d_{1D}^2 N_p^2 \rangle / D$, and by using then $D = 5\times10^{-7}$ m^2/s typical for conjugated polymers, one can determine $\langle d_{1D}N_p \rangle = 3$ nm equal approximately to four pyrrole rings. This value lies near to a width of the polaron in both polypyrrole and polyaniline, but, however, is smaller considerable then N_p obtained for polydithiophene (Devreux et al. 1987).

Thus, the shape of the probe spectrum reports about a very slow motion of the probe due probably to an enough high pack density of polymer chains in PPy. The interaction between spinless charge carriers with an active fragment of the probe results in the redistribution of the spin density between N and O nuclei in the probe and therefore in the change of its magnetic resonance parameters. This makes it possible to determine the distance between the radical and the chain along which the charge is transferred together with a typical bipolaron length in doped conjugated polymers. The method allows also to evaluate characteristic size of a cavity in which the probe is localized and, therefore, the morphology of the sample under study.

2.4. Polyaniline

Figure 10 shows X- and D-band EPR spectra of an initial, emeraldine base form of polyaniline (PANI-EB) (a), and X-, Q-, and D-band EPR spectra of emeraldine salt form of polyaniline (PANI-ES) highly doped with 2-acrylamido-2-methyl-1-propanesulphonic acid (AMPSA), PANI:AMPSA$_{0.6}$. At the X-band PANI-EB demonstrates a Lorentzian three-component EPR signal consisting of asymmetric (R_1) and symmetric (R_2) spectra of paramagnetic centers which should be attributed respectively to localized and delocalized PC. R_2PC keep line symmetry at higher doping levels. At the D-band the PANI-EB EPR spectra became Gaussian and broader compared with those obtained at X-band EPR (Figure 10), as is typical of PC in other conducting polymers (Krinichnyi 1996, 2000a, 2016b) At this waveband delocalized PC demonstrate an asymmetric EPR spectrum at all doping levels y. The analysis of EPR spectra obtained at both wavebands EPR showed that the line asymmetry of R_2PC in undoped and slightly doped PANI samples can be attributed to anisotropy of the g-factor which becomes more evident at the 140 GHz waveband EPR. The linewidth of these PC weakly depends on the temperature. Therefore, the R_1 with strongly asymmetric EPR spectrum can be attributed to a -(Ph-$\overset{\bullet +}{NH}$-Ph)- radical with $g_{xx}= 2.006032$, $g_{yy} = 2.003815$, $g_{zz} = 2.002390$, $A_{xx}=A_{yy}= 0.45$ mT, and $A_{zz} = 3.02$ mT, localized on a short polymer chain. The magnetic parameters of this radical differ weakly from those of the Ph-$\overset{\bullet +}{NH}$-Ph radical (Buchachenko and Vasserman 1973, Buchachenko, Turton, and Turton 1995), probably because of a smaller delocalization of an unpaired electron on the nitrogen atom ($\rho_N^\pi = 0.39$) and of the more planar morphology of the sample. Assuming a McConnell proportionality constant for the hyperfine interaction of the spin with nitrogen nucleus $Q = 2.37$ mT (Buchachenko and Vasserman 1973, Buchachenko, Turton, and Turton 1995), a spin density on the heteroatom nucleus of $\rho_N(0) = (A_{xx}+A_{yy}+A_{zz})/(3Q) = 0.55$is estimated. At the same time another radical R_2 is formed in the system with $g_\perp= 2.004394$ and $g_\parallel = 2.003763$ which can be attributed to PC R_1 delocalized on more polymer units of a longer chain. Indeed, the model spectra presented in Figure 10,a well fit both the PC with different mobility. The lowest excited states of the localized PC were determined from Eq.(A.2) with $\rho_N^\pi = 0.56$ (Long et al. 1994) to be $\Delta E_{n\pi*} = 2.9$ eV and $\Delta E_{\sigma\pi*} = 7.1$ eV.

The shape of EPR spectrum of PANI-ES depends on the nature of counter-ion and doping level y. Figure 10,b shows also EPR spectra of the PANI film highly doped with 2-acrylamido-2-methyl-1-propanesulfuric acid, PANI:AMPSA$_{0.6}$, registered at different frequencies of spin precession ω_e. In order to determine correctly all main magnetic resonance parameters, linewidth, paramagnetic susceptibility, g-factor, of PC with Dysonian contribution, all the effective spectra presented should be calculated using Eq.(A.35), Eq.(A.38) and Eq.(A.39) as described in (Kon'kin et al. 2002). From the analysis it was revealed that EPR spectra of the PANI:AMPSA$_{0.6}$ consist of two contributions due to different PC with Dysonian shape, namely narrow EPR spectrum of PC R_1 with $g = 2.0028$ localized in amorphous polymer backbone and broader EPR spectrum of PC R_2 with $g = 2.0020$ and higher mobility in crystalline phase of the polymer. The RT ΔB_{pp} value of PC R_2 decreases from 54 down to 20 and then down to 5.3 G at the increase of registration frequency $\omega_e/2\pi$ from 9.7 up to 36.7 and then up to 140 GHz (Figure 11), so one can express this value as $\Delta B_{pp}(\omega_e) = 0.15 + 2.2 \cdot 10^8 \omega_e^{-0.84}$ mT. Such extrapolation reveals the dependence of spin-spin relaxation time on the registration frequency and allows estimating correct linewidth at $\omega_e \to 0$ limit to be 0.15 mT. The cooling of the sample leads to the decrease in the relative concentration of PC R_2 and to the monotonous increase in its linewidth, as it is seen in Figure 11. In the same time, the linewidth of PC R_1 decreases monotonously and the sum spin concentration increases at the temperature decrease. Main magnetic resonance parameters of polarons stabilized in PANI are summarized in Table 1.

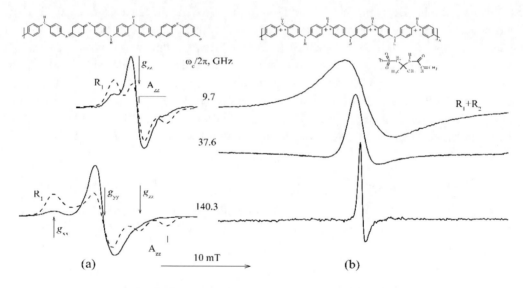

Figure 10. (*a*) Room temperature X- (9.7 GHz) and D- (140.3 GHz) bands EPR absorption spectra of emeraldine base form of polyaniline (PANI-EB) and these calculated with $g_{xx} = 2.006032$, g_{yy} 2.003815, $g_{zz} = 2.002390$, $A_{xx} = A_{yy} = 0.45$ mT, $A_{zz} = 3.02$ mT (R_1), and with $g_\perp = 2.004394$ and $g_\parallel = 2.003763$ (R_2) are shown by dashed lines. (*b*) Room temperature X- (9.7 GHz), Q- (37.6 GHz), and D- (140.3 GHz) bands EPR effective absorption spectra of emeraldine salt form of polyaniline (PANI-ES) highly doped by 2-acrylamido-2-methyl-1-propanesulphonic acid (AMPSA), PANI:AMPSA$_{0.6}$. Structures of PANI-EB, PaNI-ES and AMPSA are shown schematically.

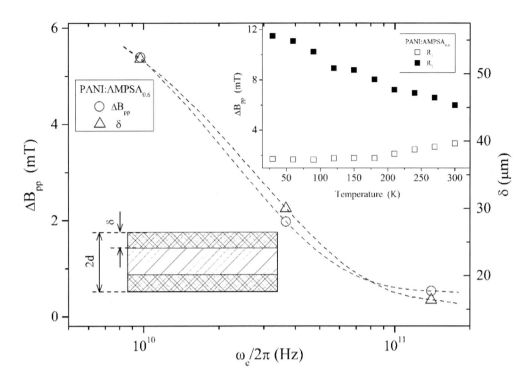

Figure 11. The peak-to peak linewidth ΔB_{pp} of the PANI:AMPSA$_{0.6}$ film as well as the thickness of skin-layer δ formed on its surface as function of spin precession frequency $\omega_c/2\pi$. In the inset is shown temperature dependence of the ΔB_{pp} value of the R_1 (open points) and R_2 (filled points) PC stabilized in this sample determined from its X-band EPR spectra with the Dyson contribution.

The narrowing of the line on raising the PANI temperature can be explained by averaging of the local magnetic field caused by hyperfine interaction between the localized spins whose energy levels lie near the Fermi level. The EPR line of the sample may also be broadened to some extent by relaxation due to the spin-orbital interaction responsible for linear dependence of T_1^{-1} on temperature (Sariciftci, Heeger, and Cao 1994), however, this such interaction can be neglected in case of PANI:AMPSA. It is significant that the linewidth of both types of PC is appreciably larger than that obtained previously for the fully oxidized powder-like and analogous films (Sariciftci, Heeger, and Cao 1994) which indicates a higher intrinsic conductivity of the PANI:AMPSA sample. Comparison of the ΔB_{pp} values suggested that a crystalline phase is formed in the amorphous phase of this PANI-ES beginning with the oxidation level $y = 0.3$, and that the PC in such phase exhibit a broader EPR spectrum. In the amorphous phase of the polymer, the PC R_1 are characterized by less temperature-dependent linewidth and are likely not involved in the charge transfer being, however, as probes for whole conductivity of the sample. At the same time, the magnetic resonance parameters of radicals of the R_2 type should reflect the charge transport in the crystalline domains of PANI:AMPSA.

Figure 12 depicts the inversed effective paramagnetic susceptibility χ and χT product (insert) of the R_1 and R_2 PC stabilized in the PANI:AMPSA$_{0.6}$ sample as function of temperature. It is seen that at low temperatures when $T \leq T_c \approx 100$ K the Pauli and Curie terms prevail in the total paramagnetic susceptibility χ of both type PC in this sample. At $T \geq T_c$, when the energy of phonons becomes comparable with the value $k_B T_c \approx 0.01$ eV, the spins start to interact that causes the increase in the third term of Eq.(A.4) of sum susceptibility as result of the equilibrium between the spins with triplet and singlet states in the system. It is evident that the R_1 signal susceptibility obeys mainly the Curie law typical for localized isolated PC, whereas the R_2 susceptibility consists of the Curie-like and Pauli-like contributions. The dependences calculated from Eq.(A.4) with respective $\chi_P = 2.2 \cdot 10^{-5}$ emu/mol, $C = 1.7 \cdot 10^{-2}$ emu K/mol, $J_{af} = 4$ meV and $\chi_P = 5.3 \cdot 10^{-3}$ emu/mol, $C = 6.5 \cdot 10^{-1}$ emu K/mol, $J_{af} = 6$ meV are fitted well experimental data obtained for PC R_1 and R_2, respectively. The J_{af} values obtained are much lower of the corresponding energy (0.078 eV) obtained for ammonia-doped PANI-ES (Iida et al. 1993).

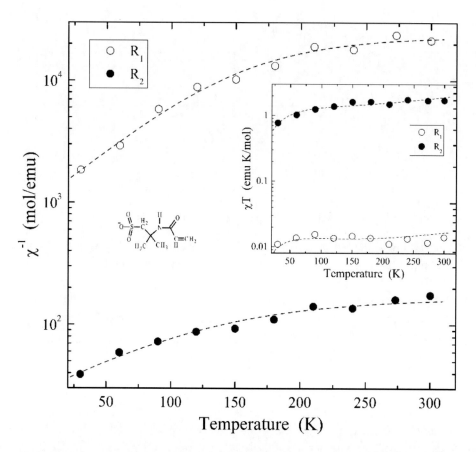

Figure 12. Temperature dependence of inversed effective paramagnetic susceptibility χ and χT product (insert) of the R_1 (open points) and R_2 (filled points) PC stabilized in the PANI:AMPSA$_{0.6}$ sample with different doping levels y. Dashed lines show the dependences calculated from Eq.(A.4) with respective $\chi_P = 2.2 \cdot 10^{-5}$ emu/mol, $C = 1.7 \cdot 10^{-2}$ emu K/mol, $k_1 = 1.7 \cdot 10^{-2}$ emu K/mol, $J_{af} = 4$ meV and $\chi_P = 5.3 \cdot 10^{-3}$ emu/mol, $C = 6.5 \cdot 10^{-1}$ emu K/mol, $k_1 = 1.2 \cdot 10^{-2}$ emu K/mol, $J_{af} = 6$ meV.

The density of states at the Fermi level ε_F, $n(\varepsilon_F) = 3.5$ states/eV Ph obtained for the charge carriers in the PANI:AMPSA$_{0.6}$ is in agreement with that determined in the optical (Lee, Heeger, and Cao 1995) and EPR (Sariciftci, Heeger, and Cao 1994) study of structurally close polymer system. The Fermi energy of the Pauli-spins, $\varepsilon_F \approx 0.2$ eV, is lower than that (0.4 - 0.5 eV) obtained for other highly doped PANI-ES (Lee, Heeger, and Cao 1993, Krinichnyi 2000b, Krinichnyi et al. 2002). Assuming that the mass of charge carrier in heavily doped polymer is equal to the mass of free electron ($m_c = m_e$), the number of charge carriers in such a quasi-metal (Blakemore 1985), $N_c \approx 4.3 \cdot 10^{21}$ cm^{-3}, can be determined. This is close to the spin concentration in this polymer; therefore, one can conclude that all delocalized PC are involved in the charge transfer in the PANI:AMPSA$_{0.6}$ sample. The velocity of charge carriers near the Fermi v_F level was calculated for PANI:AMPSA$_{0.6}$ to be $6.2 \cdot 10^5$ m/s that slightly excesses that, $(2.8 - 4.0) \times 10^5$ m/s, evaluated for other PANI-ES samples from their EPR magnetic susceptibility data (Beau, Travers, and Banka 1999, Beau et al. 1999, Krinichnyi, Konkin, and Monkman 2012).

AC conductivity of the highly doped PANI:AMPSA$_{0.6}$ film determined from Dysonian spectra of the R_2 PC and also *dc* conductivity of this polymer determined by the *dc* conductometry method (Kon'kin et al. 2002) are given in Figure 13 as function of temperature (Krinichnyi, Konkin, and Monkman 2012). The analysis of these data evidenced on the complex charge transfer in the sample. Charge carriers were shown to hop through amorphous part of the sample and then to diffuse through its crystalline domain, so then the *dc* term of the total conductivity of the samples should be determined by 1D VDH between metal-like domains and their scattering on the lattice phonons in these domains (Krinichnyi 2014a). These processes occur in parallel, so the effective conductivity should be explained in terms of both models. Indeed, Figure 13 shows that the $\sigma_{ac}(T)$ dependences obtained experimentally for charge carriers R_1 and R_2 can be fitted by Eq.(A.28), Eq.(A.29) with $h\nu_{ph} = 0.039$ and 0.020 eV, respectively, and Eq.(A.34) (Krinichnyi 2014a). The energy determined for phonons in this PANI-ES sample lies near that obtained for other conductive polymers (Krinichnyi 2016b) and evaluated (0.066 eV) from the data determined by Wang et al. for HCl-doped PANI (Wang, Li, et al. 1991, Wang et al. 1992). It is evident that E_a determined for the PANI:CSA$_{0.6}$ sample lie near. This means that protons situated in crystalline domains sense electron spin dynamics. The data obtained can be evidence of the contribution of the R_1 and R_2 PC in the charge transfer through respectively amorphous and crystalline parts of the polymer. RT σ_{ac} values determined from Dysonian spectra of R_2 PC lies near respective σ_{dc} values that is characteristic for classic metals. Besides, RT σ_{ac} values determined from Dysonian spectra of R_2 PC lies near respective σ_{dc} values that is characteristic for classic metals. The data obtained can be evidence of indirect contribution of the R_1 PC and direct contribution of the R_2 PC in the charge transfer through respectively amorphous and crystalline parts of the polymer matrix.

Thus, both pinned and delocalized PC are formed simultaneously in the PANI regions with different crystallinity. An anti-ferromagnetic interaction in crystalline domains is stronger than that in amorphous regions of PANI-ES. Charge transport between crystalline metal-like domains occurs through the disordered amorphous regions with more localized charge/spin carriers. The change of conductivity with temperature is consistent with a disordered metal close to the critical regime of the metal-insulator transition with the Fermi energy close to the mobility edge (Sariciftci, Heeger, and Cao 1994, Sariciftci et al. 1995).

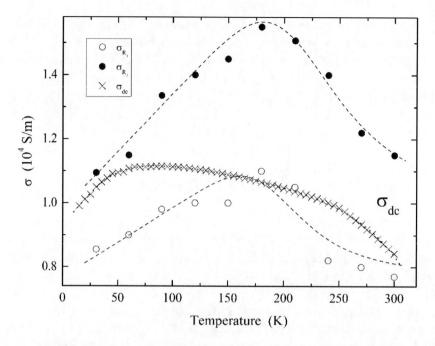

Figure 13. Temperature dependence of *ac* and *dc* conductivity of the PANI:AMPSA$_{0.6}$ film. Dashed lines show the dependences theoretically calculated in framework of VRH and polaron scattering on the lattice phonons from Eq.(A.28), Eq.(A.29) with $h\nu_{ph}$ = 0.039 and 0.020 eV, respectively, and Eq.(A.34).

The main polarons are localized in the highly doped PANI:AMPSA sample at $T \leq T_c$. This is the reason for the Curie type of susceptibility of the sample and should lead to the VRH charge transfer between the polymer chains. The spin-spin exchange appears at $T \geq T_c$ due likely to the activation librations of the polymer chains (Krinichnyi 2014a). The energy required for such dynamics lies within the energy range characteristic of other PANI-ES (Pratt et al. 1997, Krinichnyi, Chemerisov, and Lebedev 1997, Krinichnyi 2014a) and poly(tetrathiafulvalenes) (Krinichnyi 1996, Krinichnyi, Denisov, et al. 1998, Krinichnyi 2016b).This energy is governed by the effective rigidity and planarity of the polymer chains that are eventually responsible for the electronics properties of such polymer system. The results of the EPR study of other PANI-ES with various structure and number of dopants are described in (Krinichnyi 1995, 2014b, a, 2016b).

2.5. Poly(Tetrathiafulvalenes)

In recent decades, the electron donor TTF and its derivatives have been a subject of chemical and physical studies, due to the fact that many compounds of this group can form electrically conjugated charge transfer salts (Williams et al. 1992). In order to design TTF-based elements of molecular electronics, e.g., chemical sensors (Faridbod et al. 2008), powder-like PTTF matrices were synthesized in which TTF units with hydrogen, methyle and ethyle side substitutes *R* are linked *via* phenyl bridges (PTTF-R-C$_6$H$_4$) (Hinh, Schukat, and

Fanghänel 1979, Trinh et al. 1989) and *via* tetrahydroanthracene bridges (PTTF-THA) (Quang 1987) (Figure 1).

Iodine doped PTTF is a p-semiconductor with highest dc conductivity on the order of $\sigma_{dc} \approx 0.1 - 0.01$ S/m depending on the structure of monomer unit (Roth et al. 1988). The temperature dependency of dc conductivity was obtained to be explained in terms of a VRH and a thermally activated hopping at low and high temperature ranges, respectively (Gruber et al. 1990). EPR and Mössbauer measurements of doped PTTF indicate a polaron-bipolaron charge transfer mechanism governed by the doping level and temperature (Roth et al. 1989, Roth, Brunner, et al. 1990, Patzsch and Gruber 1992).

PTTF-CH$_3$-C$_6$H$_4$, PTTF-C$_2$H$_5$-C$_6$H$_4$ and PTTF-THA samples schematically shown in Figure 1 were studied at X-band EPR (Roth et al. 1988, Roth, Brunner, et al. 1990). EPR spectrum of a simplest PTTF-H-C$_6$H$_4$ sample was analyzed to be a superposition of a strongly asymmetric spectrum of immobilized PC with $g_{xx} = 2.0147$, $g_{yy} = 2.0067$, $g_{zz} = 2.0028$ and a symmetric spectrum caused by mobile polarons with $g = 2.0071$. A relatively high value of **g** tensor evidences for the interaction of an unpaired electron with sulfur atom having large spin-orbit coupling constant. Roth et al. shown (Roth et al. 1988, Roth, Brunner, et al. 1990) that the spin-lattice relaxation time T_1 of an undoped PTTF-C$_2$H$_5$-C$_6$H$_4$ sample depends on the temperature as $T_1 \propto T^{-\alpha}$ with α decreases from two at $100 < T < 150$ K temperature range down to one at higher temperatures. The addition of a dopant causes the change of a line shape of this sample due to the appearance of a larger number of mobile PC. Such a change in the magnetic and relaxation parameters was attributed to the conversion of spinless bipolarons into the paramagnetic polarons induced by the doping or/and heating of the polymer. In neutral and slightly doped polymers the charge is transferred by small polarons (Patzsch and Gruber 1992) whom dynamics is described by Eq.(A.32). However, it is difficult to carry out at X-band EPR the detailed investigation of doped PTTF samples with low concentration of the immobilized PC for the further analysis of spin effect in electron relaxation and dynamics.

The nature, composition and dynamics of PC in initial and iodine-doped PTTF samples above mentioned were studied by multifrequency EPR method more completely (Krinichnyi et al. 1993, Krinichnyi, Denisov, et al. 1998, Krinichnyi 2000a, 2006, 2016b).

Multifrequency EPR spectra of exemplary PTTF-CH$_3$-C$_6$H$_4$ sample iodine-doped up to $y = 0.08$ are presented in Figure 14. They allow one to determine more correctly all terms of the anisotropic **g**-tensor and to separate the lines attributed to polarons with different mobility. Computer simulation shows that the anisotropic EPR spectrum of the sample consists of localized PC $P_{loc}^{+\bullet}$ with slowly temperature dependent magnetic parameters $g_{xx} = 2.01191$, $g_{yy} = 2.00584$, $g_{zz} = 2.00185$, and more mobile PC $P_{mob}^{+\bullet}$ with $g_{iso} = 2.00655$ and $\Delta B_{pp}^{iso} = 5.6$ mT. The analogous spectrum of $P_{loc}^{+\bullet}$ in PTTF-C$_2$H$_5$-C$_6$H$_4$ is characterized by the magnetic parameters $g_{xx} = 2.01405$, $g_{yy} = 2.00676$, $g_{zz} = 2.00235$, whereas PC with nearly symmetric spectrum are registered at $g^p = 2.00774$ with $\Delta B_{pp}^{iso} = 13.4$ mT. The canonic components of **g** tensor of PC localized in PTTF-THA are $g_{xx} = 2.01292$, $g_{yy} = 2.00620$, $g_{zz} = 2.00251$, whereas more mobile PC with weakly asymmetric spectrum are characterized by the parameters $g_{\parallel}^p = 2.00961$ and $g_{\perp}^p = 2.00585$. These g-factors exceed the corresponding magnetic parameters of the polarons in PATAC evidencing of the larger interaction of an unpaired electron with sulfur nuclear in PTTF. The ratio of concentrations of the localized and mobile PC is 1:4.5 in neutral PTTF-CH$_3$-C$_6$H$_4$, 1:21 in PTTF-C$_2$H$_5$-C$_6$H$_4$, and 1:15.8 in neutral PTTF-THA. The existing in these systems of two polaron ensemble with different mobility was confirmed by their ^{1}H NMR study (Krinichnyi 2016b).

As effective $g^p = 1/3(g_{\parallel}^P + 2g_{\perp}^P)$ is close to the average g-factor of immobilized polarons, PC of two types with approximately equal magnetic parameters exist in PTTF, namely polarons moving along the polymer main axis with minimum rate of $D_{1D}^0 \geq 3 \times 10^{10}$ rad/s and polarons pinned on traps or/and on short polymer chains. The comparatively large iodine ions soften the polymer matrix at the doping of a polymer, so then the mobility of its chains increases. It seems just a reason for the growth in a polymer of a number of delocalized polarons (Figure 14). The main terms of **g**-tensor of some PC in PTTF-C$_2$H$_5$-C$_6$H$_4$ are averaged completely due to their mobility, whereas such an averaging takes place only partially in the case of other PTTF samples. This fact can be explained by a different structure and morphology of the polymers' matrix.

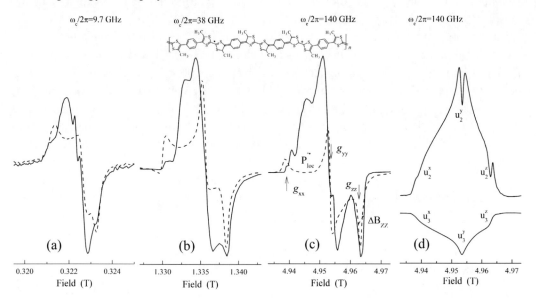

Figure 14. In-phase X- (*a*), Q- (*b*) and D- (*c*) bands absorption spectra EPR of the initial (dashed lines) and iodine-doped (solid lines) PTTF-CH$_3$-C$_6$H$_4$ samples registered at room temperature. (*d*) Typical in-phase (u_2) and $\pi/2$-out-of-phase (u_3) terms of D-band dispersion spectra of the samples are shown. The formation of the polaron on PTTF chain and measured magnetic parameters are shown as well.

So, the high spectral resolution at D-band EPR allows to determine separately all components of spectra of the polarons with different mobility and then to analyze their temperature dependence. Figure 15 shows the linewidth of polarons stabilized in the initial and iodine-doped PTTF-CH$_3$-C$_6$H$_4$ and PTTF-THA samples as function of temperature. It is seen from the Figure that the linewidth of localized polarons changes slightly in PTTF-CH$_3$-C$_6$H$_4$, however, this parameter doubles under doping of PTTF-THA. The increase of the spectrum linewidth of the mobile polarons in PTTF-CH$_3$-C$_6$H$_4$ sample indicates the acceleration of polaron diffusion in this system. Such a change in ΔB_{pp}^P of mobile polaron is analogous to the line narrowing of spin charge carriers in the PANI:AMPSA organic metal described above. Polaron motion along the polymer chain initiates interactions between electron spins and also between electron and proton spins. Such interactions depend also on the spin precession frequency. RT linewidth ΔB_{pp} of the EPR spectral components of polarons immobilized, e.g., in PTTF-C$_2$H$_5$-C$_6$H$_4$ increases from 0.28 to 0.38 and then to 3.9 mT while electron spin precession frequency $\omega_e/2\pi$ increases from 9.5 to 37 and then to 140 GHz, respectively. On the other hand, linewidth of mobile polarons increases from 1.02 to 1.15 and then to 17.5 mT, respectively at such a transition. The width of all NMR lines also increase by factor 1.5 – 2 with the increase of $\omega_p/2\pi$ from 300 MHz to 400 MHz (Krinichnyi 2016b). The fact, that the mobile PC has a broader line than the pinned ones, can be explained by the higher probability of its interaction with other electron and nuclear spins and with the dopant ions due mobility. This feature is typical for conjugated polymers (Krinichnyi 2000a, 2016b), however, disagrees with that obtained earlier for such system at X-band EPR (Roth et al. 1988, Roth, Brunner, et al. 1990).

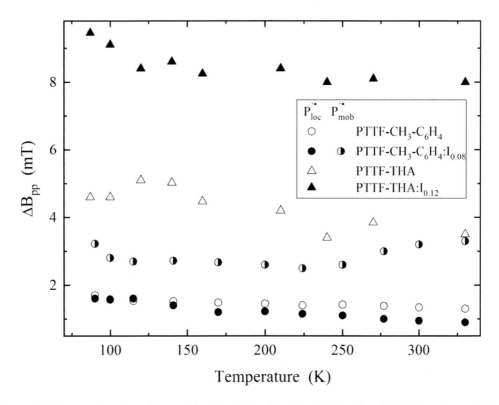

Figure 15. Temperature dependence of the linewidth of localized $P_{loc}^{+\bullet}$ and mobile $P_{mob}^{+\bullet}$ polarons in the initial (open points) and iodine-doped (filled points) PTTF-CH$_3$-C$_6$H$_4$ and PTTF-THA samples up to y = 0.08 and 0.12, respectively, determined from their D-band EPR spectra.

The concentration of PC in PTTF-CH$_3$-C$_6$H$_4$ slightly increases from 2×10^{17} cm^{-3} up to 3×10^{17} cm^{-3} at the polymer doping. Such process of PTTF-THA also changes slightly this parameter, from 3×10^{18} cm^{-3} up to 4×10^{18} cm^{-3}. PC concentration in doped PTTF-C$_2$H$_5$-C$_6$H$_4$ is 5×10^{17} cm^{-3}.

Paramagnetic susceptibility of the initial PTTF-CH$_3$-C$_6$H$_4$ and PTTF-THA samples as well as that reached upon their iodine doping up to $y = 0.08$ and 0.12, respectively, is presented in Figure 16 as function of temperature. This value changes monotonically within all temperature region The analysis of the data obtained shown that the susceptibility of the initial and iodine-doped (up to $y = 0.08$) PTTF-CH$_3$-C$_6$H$_4$ samples is determined mainly by the second term of Eq.(A.4) with $C = 1.9 \times 10^{-4}$ and 5.7×10^{-4} emu K/mol, respectively. This parameter of the initial and slightly iodine doped (up to $y = 0.12$) PTTF-THA samples is also governed by the Curie term of Eq.(A.4) with $C = 1.1 \times 10^{-2}$ and 4.2×10^{-3} emu K/mol, respectively. The results presented evidence that spin susceptibility of the PTTF-CH$_3$-C$_6$H$_4$ system increases with the doping, whereas such parameter of the PTTF-THA polymer decreases. This can indicates the charge transfer mainly by spinless bipolarons in the latter.

The increase of spectral linewidth within the $37 \leq \omega_e/2\pi \leq 140$ GHz registration range confirms a weak interaction between spin packets in this polymer. This provokes MW saturation of PC in PTTF at comparatively small B_1 values at D-band EPR. It causes the appearance of bell-like terms in both the in-phase and $\pi/2$-out-of-phase components of its dispersion signal due to manifestation of fast passage effects (see Figure 14,d). The u_i^x, u_i^y, and u_i^z terms of the dispersion signal U in Figure 14,d are attributed to PC with a strongly asymmetric distribution of spin density which are differently oriented in an external magnetic field. The simulation of the PTTF dispersion spectra allowed to identify them as a superposition of a spectrum, attributed to mobile polarons and a predominant asymmetric spectrum with $g_{xx} = 2.01189$, $g_{yy} = 2.00564$, $g_{zz} = 2.00185$ in undoped PTTF-CH$_3$-C$_6$H$_4$, $g_{xx} = 2.01356$, $g_{yy} = 2.00603$, $g_{zz} = 2.00215$ in undoped PTTF-C$_2$H$_5$-C$_6$H$_4$, and $g_{xx} = 2.01188$, $g_{yy} = 2.00571$, $g_{zz} = 2.00231$ in undoped PTTF-THA of immobile polarons.

The temperature dependencies of effective relaxation times of polarons in PTTF samples determined from their D-band EPR dispersion spectra using the method described earlier (Krinichnyi et al. 1993) are summarized in Figure 17. T_1 value of such charge carriers in all PTTF samples with phenyl bridges and in undoped PTTF with THA bridges was shown to change monotonically with temperature as $T^{-\alpha}$ with exponent α lying near 3 and 5 for mobile and pinned polarons, respectively. This value determined for PTTF at D-band EPR is larger than that measured for immobile radicals by using spin echo technique at X-band EPR (Roth et al. 1988). A difference between T_1^{mob} and T_1^{loc} can be caused, e.g., by a strong interaction between different PC.

Analyzing MW saturated spins stabilized on polymer chains one can use the ST-EPR method (Hyde and Dalton 1979) to determine super slow macromolecular librations in conjugated polymer systems (Krinichnyi 2006). Such dynamics in PTTF-CH$_3$-C$_6$H$_4$ and PTTF-THA systems was studied at D-band EPR (Krinichnyi et al. 1993, Krinichnyi, Denisov, et al. 1998, Krinichnyi 2006). The energy E_a required for activation of such dynamics in the first initial and doped system was determined from Eq.(A.30) to be 0.11 and 0.14 eV, respectively. This parameter determined for the in initial and doped latter sample is 0.19 and 0.07 eV, respectively. E_a values obtained at D-band EPR are comparable with that determined at lower registration frequency for interchain charge transfer in doped PTTF (Roth et al. 1988, Roth, Brunner, et al. 1990) that indicates the interaction of pinned and mobile polarons in this polymer matrix.

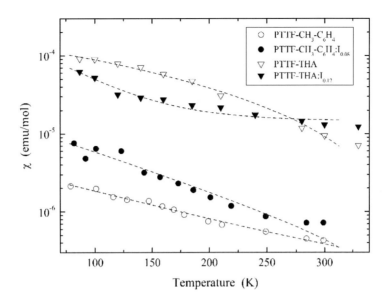

Figure 16. Temperature dependence of paramagnetic susceptibility of polarons stabilized on the initial PTTF-CH$_3$-C$_6$H$_4$ and PTTF-THA samples as well as on those iodine doped up to $y = 0.08$ and 0.12, respectively. Top-to-bottom dashed lines show dependences calculated from Eq.(A.4) with $C = 1.1\times10^{-2}$, 4.2×10^{-3}, 5.7×10^{-4} and 1.9×10^{-4} emu K/mol, respectively.

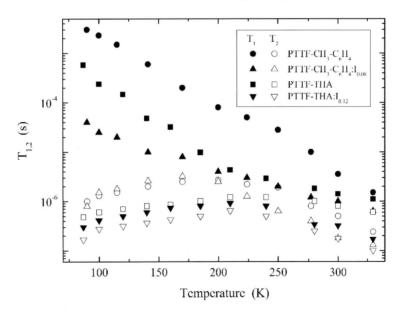

Figure 17. Spin-lattice T_1 (points) and spin-spin T_2 (points) relaxation times determined for polarons stabilized in the initial PTTF-CH$_3$-C$_6$H$_4$ and PTTF-THA samples as well as on those iodine doped up to $y = 0.08$ and 0.12, respectively, as function of temperature.

In order to compare experimental results with the polaron theory, Q1D diffusion motion of PC diffusing along and between PTTF's chains with the diffusion coefficients D_{1D} and D_{3D}, respectively, was also assumed. The temperature dependencies of effective D_{1D} and D_{3D} calculated for PC in different PTTF samples by using Eq.(A.25), Eq.(A.26) and the data presented in Figure 17 are shown in Figure 18. Assuming that the spin delocalization over the polaron in PTTF occupies approximately five monomer units (Devreux et al. 1987) the maximum value of D_{1D} does not exceed 2×10^{12} rad/s for PTTF samples at room temperature. This value is at least two orders of magnitude lower than that determined earlier by low-frequency magnetic resonance methods for polarons in polypyrrole (Devreux and Lecavelier 1987) and polyaniline (Devreux et al. 1987), but higher then D_{1D}^0 evaluated above. RT anisotropy of spin dynamics in PTTF is $A = D_{1D}/D_{3D} \geq 10$. Figure 18 shows that D_{3D} value increases at RT by ca. an order of magnitude at transition from PTTF-CH$_3$-C$_6$H$_4$ to PTTF-THA sample due probably to more intensive interaction between spins in the latter. The variation of g_{xx} value is $\Delta g_{xx} = 1.0\times10^{-3}$ at such a transition. Assuming, that the overlapping integral t_{cc} of macromolecules depends on dihedral angle θ (i.e., the angle between p-orbits of neighboring C-atoms) as $t_{cc} \propto \cos\theta$, and that spin density on sulfur atom ρ_s depends as $\rho_s \propto \sin\theta$(Traven' 1989), one can calculate from Eq.(A.2) the difference $\Delta\theta$ at such a transition to be $\Delta\theta = 22^0$. Note, that analogous change in θ takes place at transition from benzoid to quinoid form of PPP (Brédas 1986), at transition from EB to ES form of PANI (Krinichnyi, Nazarova, et al. 1998, Krinichnyi et al. 2002) and laser treatment of PATAC.

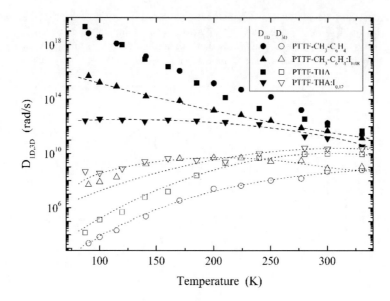

Figure 18. Temperature dependencies of the effective rates of polaron intrachain D_{1D} and interchain D_{3D} diffusion in the PTTF samples with phenyl and THA bridges determined from Eq.(A.25), Eq.(A.26) and the data presented in Figure 17. Top-to-bottom dashed lines show dependences calculated from Eq.(A.29) with E_{ph} = 0.11 and 0.025 eV, respectively. Top-to-bottom dotted lines show dependences calculated from Eq.(A.30) with E_a = 0.14, 0.07, 0.19, and 0.11 eV, respectively.

The D_{1D} and D_{3D} values of the PTTF samples calculated using Eq.(A.25) and Eq.(A.26) are presented in Figure 18 as function of temperature. It is seen that the D_{1D} values of the initial PTTF samples demonstrate stronger temperature dependence. Such dependencies cannot be described by tunnel charge transfer mechanisms. The most acceptable charge dynamic process can be suggested in terms the Kivelson-Heeger theory of polaron interaction with the lattice phonons. The data presented in Figure 18 evidence that the D_{1D} parameter of the iodine-doped PTTF-CH$_3$-C$_6$H$_4$ and PTTF-THA can indeed be fitted by Eq.(A.29) with E_{ph} = 0.11 and 0.025 eV, respectively. The interchain diffusion coefficient D_{3D} of the samples monotonically increase with the temperature (Figure 18). The data presented can be suggested within the thermally activated interchain polaron hopping in conduction band tails with activation energy E_a. Temperature dependencies of conductivity due to such spin motion calculated from Eq.(A.30) are also shown by dotted lines on Figure 18. The energy required for activation of charge hopping between the chains of the iodine-doped PTTF-CH$_3$-C$_6$H$_4$ systems was determines to be $E_a = 0.11$ and 0.14 eV, respectively. The respective values were obtained for PTTF-THA, $E_a = 0.19$ and 0.07 eV, respectively.

It is seen from Figure 18, that D_{3D} value increases at the replacement of phenyl bridges by tetrahydroanthracene ones in PTTF. This fact allows concluding that this parameter and therefore the anisotropy of charge carrier transfer is governed by the polymer structure. Indeed, the structure of a polymer becomes more planar at such transition, so then the polymer chains are situated closer in PTTF-THA and the probability of interchain transfer increases due to the increase of appropriate transfer integral. This obviously leads to the decrease in the anisotropy of charge transport in this polymer. In other words, such a transition increases the dimensionality (crystallinity) of the system, which is typical for highly ordered metal-like organic systems.

The Fermi velocity v_F was determined for polaron diffusion in PTTF samples to be near to 1.9×10^7 cm/s (Krinichnyi 2000a). So the mean free path l_i of a charge was determined to be $l_i = v_{1D} c_{1D|}^2 v_F^{-1} = 10^{-2} - 10^{-4}$ nm for the PTTF samples. The l_i is less then lattice constant d (Patzsch 1991) therefore the charge transfer is incoherent in this polymer which cannot be considered as Q1D metal. For such case the interchain charge transfer integral can be determined as (Wang et al. 1992) $t_\perp = v_F \hbar /2d = 0.05$ eV. This value lies near activation energy of chain librations and also near normalized activation energy of interchain polaron hopping in doped PTTF-CH$_3$-C$_6$H$_4$ and PTTF-THA samples. These facts allow us to conclude that charge transfer in PTTF is determined mainly by interchain phonon-assisted hopping of polarons which is stimulated by super slow macromolecular dynamics. This evidences for the correlation of molecular and charge dynamics in such polymer system. It confirms also the supposition (Madhukar and Post 1977), that the fluctuations of lattice oscillations, librations among them, can modulate the electron interchain transfer integral in conjugated compounds.

The rise of libron-exciton interactions at the polymer doping evidences for the formation of a complex quasi-particle, namely molecular-lattice polaron (Silinsh, Kurik, and Chapek 1988) in doped PTTF. According to this phenomenological model, molecular polaron is additionally covered by lattice polarization so then its mobility becomes as sum of mobilities of molecular and lattice polarons. The energy of formation of such molecular-lattice polaron E_p in PTTF was determined to be 0.19 eV (Krinichnyi et al. 1993). This value is smaller than that (0.1 eV) obtained for other disordered conjugated polymers (Zuppiroli, Paschen, and Bussac 1995). Therefore, the characteristic time, necessary for polarization of both atomic and molecular orbits of polymer, can be determined as $\tau_p \approx \hbar E_p = 3.5 \times 10^{-15}$ s. This value is sufficiently smaller than intra- and interchain hopping times for charge carriers in PTTF (Figure 18). This allow to conclude that the time τ_h required for the hopping of charge carriers in PTTF sufficiently exceeds the polarization time for charge carriers' microenvironment in the polymer, i.e., $\tau_h \gg \tau_p$. This inequality is a necessary and sufficient condition for electronic polarization of polymer chains by a charge carrier.

Thus, both the Q1D and Q3D spin dynamics are realized in PTTF affecting the charge transfer process. Q3D polaron diffusion dominates in the polymer conductivity; however,

38

Q1D diffusion of spin and spinless charge carriers also plays an important role in effective conductivity of the polymer.

2.6. Polythiophene

Polythiophene, its alkyl derivatives and their nanocomposites shown in Figure 1 and Figure 3 are also considered as perspective systems for design of various molecular devices (Samuelsen and Mardalen 1997, Zanardi et al. 2009, Zaumseil 2014). Pristine PT demonstrates at X-band EPR a single symmetric line with $g = 2.0026$ and $\Delta B_{pp} = 0.8$ mT, showing that the spins do not belong to a sulfur-containing moiety and are localized on the polymer chains (Bernier 1986, Mizoguchi and Kuroda 1997). The low concentration of PC ($n \cong 66$ ppm or 6.6×10^{-5} spin per a monomer unit) is consistent with a relatively high-purity of material, containing few chain defects. The doping of polymer leads to the formation on its chains of polarons which spins are delocalized along eight thiophene units (Springborg 1992). This process also provokes the appearance in the gap of asymmetric states with $\Delta E_1 = 0.32$ eV and $\Delta E_2 = 0.47$ eV (Stafström and Brédas 1988) (Figure 2). Stafström and Brédas (Stafström and Brédas 1988) found that these energies increase up to 0.57 and 0.73 eV, respectively, as pairs of polarons pair into diamagnetic bipolarons. Kaneto et al. (Kaneto et al. 1985) obtained from EPR study a maximum number of spins in PT doped with BF_4^- up to $y \approx 0.03$, suggesting a cross-over from polaron to bipolaron at this doping level. On the other hand, a fairly small number of spins in iodine-doped PT were reported (Moraes et al. 1985, Hayashi et al. 1986). Chen $et\ al.$ (Chen, Heeger, and Wudl 1986) found only a vanishingly small EPR signal in PT electrochemically doped with ClO_4^- up to $y = 0.14$, suggesting a bipolaron ground state.

Spin dynamics in PT electrochemically doped with ClO_4^- was studied at RT by EPR (Mizoguchi et al. 1994). It was shown that highly doped sample reveals temperature dependence of linewidth due to the Elliott mechanism (Elliott 1954), characteristic of metals. RT coefficients of spin diffusion along and between polymer chains as well as the conductivity terms due to such spin diffusion were obtained to be respectively $D_{1D} = 1.9 \times 10^{15}$ rad/s, $D_{3D} = 5.5 \times 10^9$ rad/s and $\sigma_{1D} = 1.2 \times 10^5$ S/m, $\sigma_{3D} = 3.7$ S/m at $n(\varepsilon_F) = 0.12$ states per eV per C atom. The conductivity of PT:ClO_4^- changes as T^{-2} at high temperatures, whereas a metal-insulator (or semiconductor) transition takes place in this sample at 30 K (Masubuchi et al. 1993). The temperature dependence of conductivity of PT:BF_4^- supports the Mott's VRH mechanism (Demirboga and Onal 2000). Temperature dependence of activation energy indicated that the charge carrier hopping is the dominating mechanism of charge transport in this sample.

Powder-like nanocomposites of polythiophene synthesized electrochemically from monothiophene (PT) and dithiophene (PdT) and BF_4^-, ClO_4^-, and I_3^- counter-ions as a dopant were studied at both the X- and D-bands EPR (Krinichnyi et al. 1985).

At the X- band EPR these samples demonstrate a symmetric single line with effective $g \cong g_e$ and the width, slightly changing in a wide temperature range. However, the spectrum of the PT:I_3^- sample is broadened significantly with the temperature increase. At this waveband

EPR spectrum of PdT:ClO_4^- appears as a single symmetric line, which width decreases smoothly monotonically from 0.70 down to 0.25 mT with the temperature decrease from RT down to 77 K. So, magnetic resonance parameters of polarons are expectable governed by the structure of counter-ions.

D-band EPR spectra of these conjugated polymers demonstrate a greater variety of line shape (Figure 19). The Figure shows that PC stabilized in PT:BF_4^- and PT:ClO_4^- samples are characterized by an axially symmetric spectrum typical for PC localized on a polymer backbone. The analogous situation seems to be realized also for PT:I_3^-, for which the broadening and overlapping of canonic components of EPR spectrum can take place due to a stronger spin-orbit interaction of PC with counter-ions. PdT:ClO_4^- sample also demonstrates a single EPR line at this waveband EPR in a wide temperature range, thus indicating the domination of delocalized PC in this polymer.

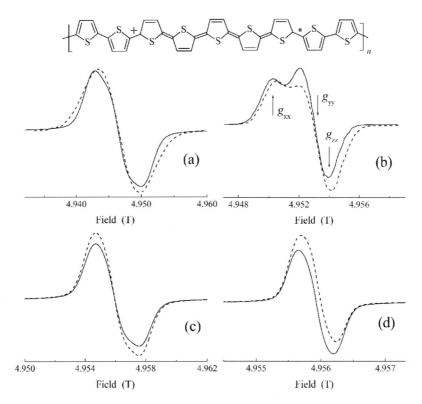

Figure 19. D-band absorption EPR spectra of electrochemically synthesized polythiophene modified by I_3^- (a), BF_4^- (b), ClO_4^- (c), (d) counter-ions and poly(di)thiophene (PdT) doped with ClO_4^- counter-ions (d) and registered at $T = 300$ (solid line) and 200 K (dashed line). The components of g tensor are shown. At the top the PT-based nanocomposite with polaron charge carrier is shown schematically.

Main magnetic resonance parameters, evaluated, e.g., for PT:BF_4^- sample from its D-band EPR spectra are also presented in Table 1. The terms of g-factor of PT doped with I_3^-, BF_4^-, ClO_4^- as well as of PdT doped with ClO_4^- were measured to be $g_\parallel = 2.00679$ and $g_\perp = 2.00232$, $g_\parallel = 2.00412$ and $g_\perp = 2.00266$, $g_\parallel = 2.00230$ and $g_\perp = 2.00239$, $g_\parallel = 2.00232$ and

$g_\perp$ = 2.00364, respectively. It derives from the analysis of the data, that the energy of an excited configuration $\Delta E_{\sigma\pi} \propto \Delta g^{-1}$ determined from Eq.(A.2), increases more than four times at the transition from I_3^- to BF_4^- and to ClO_4^- anions. The width of EPR spectral components of PT increase by factor of seven with the increase of registration frequency from 9.7 up to 140 GHz (Krinichnyi et al. 1985, Krinichnyi 2000a), indicating a strong spin-spin exchange in this system. Spin susceptibility of these nanocomposites changes within the series from 2.7×10^{-5} up to 6.8×10^{-5} down to 4.4×10^{-5} and up to 8.7×10^{-5} emu/mol, respectively. Such a transition leads also to the acceleration of spin dynamics and the growth of film conductivity. As the temperature decreases, a Dyson-like line is displayed in the region of a perpendicular component of the PT:BF_4^- EPR spectrum without a noticeable change of signal intensity (Figure 19,b). The further temperature decrease results in the increase of the line asymmetry factor A/B without reaching the extreme in the $100 - 300$ K temperature range, thus being the evidence for the growth of ac conductivity as it occurs in case of other semiconductors of lower dimensionality. Therefore, an intrinsic conductivity of PT samples can be determined from Eq.(A.35), Eq.(A.36) and Eq.(A.37) using the characteristic size of sample particles. The ac conductivity of these samples was estimated to change from 3.2×10^2 up to 1.2×10^3 then down to 6.8×10^2 S/m (Krinichnyi et al. 1985, Krinichnyi 2000a). Then applying relationship (A.34), one can determine the rate of Q1D diffusion D_{1D} of charge carriers in the samples. This value calculated for the above series changes from 1.2×10^{13} down to 1.1×10^{13} and then up to 1.0×10^{14} rad/s. This may be the evidence for the realization of charge transfer in PT both by polarons and bipolarons, whose concentrations depend on the origin of anion, introduced into a polymer.

With the temperature growth a linewidth of the PdT:ClO_4^- sample first increases and starts to decrease as T becomes lower than $T_c \approx 170$ K (Figure 20). This is accompanied by the resembling change in an inverted paramagnetic susceptibility χ^{-1} (Figure 20). Such a fact is evidence that the $\Delta B_{pp}(T)$ and $\chi(T)$ dependencies obtained for this sample are not symbasys. An extremal change in $\Delta B_{pp}(T)$ can be interpreted in terms of the Houzé-Nechtschein model (Houze and Nechtschein 1996) of the exchange interaction of spins localized on neighboring polymer chains in Q1D polymer system. Figure 20 shows the adaptability of this model for the PdT:ClO_4^- sample and evidences for the strong and weak interaction between spins below and above T_c, respectively. The energy of activation of such interaction obtained from Eq.(A.19), $E_r = 0.040$ eV, lies near that determined for macromolecular librations in other conjugated polymers (Krinichnyi 1995, 2000a, 2006). This leads to reversible pairing of polarons into bipolarons at $T \leq T_c$ and to bipolaron dissociation to polarons at higher temperatures. Figure 20 also shows that the temperature dependence of an effective magnetic susceptibility of the sample follows Eq.(A.4) with $\chi_P = 8.2\times10^{-5}$ emu/mol, $C = 2.1\times10^{-2}$ emu K/mol, and $J_{af} = 0.071$ eV. Assuming a linear dependency for the bipolaron decay rate and frequency of polymer chain librations, the activation energy of the latter process can be evaluated from $\chi(T)$ dependency to be equal to $E_a = 0.025$ eV at $T \geq 200$ K. Such a complex character of polaron-bipolaron transformation and spin-spin interaction seems to explain the above mentioned narrowing of X-band EPR spectrum of this sample with temperature.

RT intrinsic conductivity of the samples was also evaluated for the PdT:ClO_4^- sample from their Dysonian D-band EPR spectra to be 5.5×10^3 S/m. The rate of Q1D diffusion of charge carrier, $D_{1D} = 4.1 \times 10^{13}$ rad/s, obtained for this system is less considerably than $D_{1D} = 1.1 \times 10^{13}$ rad/s obtained above and 1.9×10^{15} rad/s determined by Mizoguchi et al. (Mizoguchi et al. 1994) for PT:ClO$_4^-$ sample, but close on the order of value to that evaluated for polarons in doped PANI-ES, in which interchain charge transfer dominates. The conductivity of PdT:ClO$_4^-$ sample reveals an extremal character with bending point $T_c \approx$ 170 K characteristic for the $\Delta B_{pp}(T)$ dependence (Figure 20). The charge transfer in the PdT:ClO$_4^-$ sample, as in case of other conjugated polymers, was analyzed in terms of Q1D Mott's hopping and electron scattering on the lattice phonons. As Figure 20 evidences, $\sigma_{ac}(T)$ is approximated well by Eq.(A.29) with E_{ph} = 0.13 eV. The addytivity of the ΔB_{pp} and σ_{ac} values supports the Houzé-Nechtschein model (Houze and Nechtschein 1996) predicted linear dependence of these values.

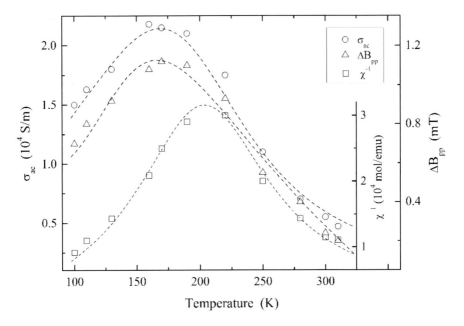

Figure 20. Temperature dependence of the intrinsic conductivity σ_{ac}, linewidth ΔB_{pp}, and inverted magnetic susceptibility χ of PdT:ClO$_4^-$ determined from its Dysonian D-band EPR spectra. Top-to-bottom dashed lines present the dependencies calculated from Eq.(A.29) with E_{ph} = 0.13 eV, Eq.(A.19) with E_r = 0.040 eV, and Eq.(A.4) with χ_P = 8.2×10^{-5} emu/mol, C = 2.1×10^{-2} emu K/mol, c_{af} = 10.2 emu K/mol, J_{af} = 0.071 eV.

2.7. Poly(3-Alkylthiophenes)

Poly(3-alkylthiophenes) (Figure 1) were shown (Kaeriyama 1997, Hotta 1997) to be suitable model system for understanding the electronic and optical properties of sulphur-based Q1D conjugated polymers with non-degenerate ground states (Gronowitz 1991, Kaeriyama 1997, Hotta 1997). Later, these polymers were appeared to be most effective active matrix for organic electronic and photonics devices (Sun and Sariciftci 2005, Poortmans and Arkhipov 2006, Jeffries-El and McCullough 2007, Pagliaro, Palmisano, and Ciriminna 2008, Brabec, Scherf, and Dyakonov 2014, Zaumseil 2014).

P3AT in undoped state are semiconductors, whom energy bandgap is determined by the presence of the π-orbital conjugation along the main polymer axis. For P3AT the gap amounts ca. 2 eV at ambient temperatures giving rise to its characteristic red color. This parameter, however, demonstrates temperature dependency, so that this color transforms to yellow at the temperature change ("thermochromism"). McCullough et al. (McCullough et al. 1993) have reported the maximum RT σ_{dc} of I_2-doped P3AT to be 6×10^3 S/m for poly(3-hexylthiophene) (P3HT, $m = 6$ in Figure 1), 2×10^4 S/m for poly(3-octylthiophene) (P3OT, $m = 8$ in Figure 1), and 1×10^5 S/m for poly(3-dodecylthiophene) (P3DDT, $m = 12$ in Figure 1). This shows the correlation of the P3AT dc conductivity with the alkyl group length and the morphology of the sample. It is well known that the transport properties of this class of materials are mainly governed by the presence of positively charged mobile polarons originating from the synthesis and the adsorption of oxygen from ambient atmosphere (spontaneous p-type doping) (Conwell 1997, Menon 1997). These charge carriers are subsequently partly trapped by the impurities, as opposed to inorganic semiconductors where, in the case of p-type doping, the hole is transferred from the impurity to the valence band. The presence of polaron in polythiophene and its derivatives was revealed by optical absorption measurements and EPR (Mizoguchi and Kuroda 1997). When the concentration of dopant y (oxygen, iodine, etc.) increases, the number of polarons also increases and, starting from some doping level, polarons combine into diamagnetic bipolarons. Kunugi et al. (Kunugi et al. 2000) have shown by the electrochemical study of charge transport that the RT carrier mobility μ in a regioregular P3OT film is 5×10^{-7} m^2 V^{-1}s^{-1} at $y = 1.4\times10^{-4}$. This value decreases down to $\mu = 5\times10^{-8}$ m^2 V^{-1}s^{-1} at $y = 1.0\times10^{-2}$ due to the scattering of polarons by ionized dopants and the formation of immobile π-dimers. Then it increases up to $\mu = 0.5$ cm^{2}V^{-1}s^{-1} at $y = 0.23$ due to the formation of bipolarons, followed by the evolution of the metal-like conduction.

Polarons can also be initiated by light illumination of P3AT with an electron acceptor introduced into its bulk as it is seen in Figure 3. This process is reversible, so both the spins photo induced in a polymer:fullerene BHJ tend to recombine after the light off.

Polarons stabilized in P3AT possess spin $S = \frac{1}{2}$ as well that also stipulates their wide investigation by different magnetic resonance methods. ^{1}H NMR proton spin-lattice relaxation time study of an initial and ClO$_4$-doped P3OT samples have shown (Masubuchi et al. 1999) that the molecular motion of the chains and side octyl groups occurs at different temperatures. At X-band EPR the spectrum of polaron in P3AT is characterized by a single line with the linewidth $\Delta B_{pp} \approx 0.6 - 0.8$ mT and the g-factor close to the g-factor of a free electron (Mizoguchi and Kuroda 1997). EPR signal of a slightly BF$_4$-doped poly(3-methylthiophene) (P3MT, $n = 1$ in Figure 1) was found to be a superposition of Gaussian line with $g_1 = 2.0035$ and $\Delta B_{pp} \cong 0.7$ mT attributed to the presence of localized PC and a Lorentzian one with $g_2 = 2.0029$ and $\Delta B_{pp} = 0.15$ mT due to delocalized PC (Scharli et al. 1987). In this sample the total concentration of PC amounts to about 3×10^{19} cm^{-3}, $i.e.$, about one spin per 300 thiophene rings. After doping, only one symmetric Lorentzian component of the former spectrum is observed. This line is symmetric at $y \leq 0.25$ and demonstrates a Dyson-like line at higher y (Tourillon et al. 1984). This process is accompanied by a sufficient decrease of electron both spin-lattice and spin-spin relaxation times (Scharli et al. 1987), which may indicate the growth of system dimensionality upon doping process. The analysis of $\chi(y)$ dependency shows that polarons are formed predominantly at low doping

EPR Spectroscopy of Polarons in Conjugated Polymers and Their Nanocomposites 43

level and then start to combine into bipolarons at higher y. Main magnetic resonance parameters, determined at various wavebands EPR for polarons stabilized or initiated in some P3AT are presented in Table 1.

Below are presented experimental data obtained in the study of polarons stabilized and initiated in some P3AT.

2.7.1. Poly(3-Hexylthiophene)

Detached regioregular P3HT modified by bis-PC$_{62}$BM shown in Figure 21 are characterized by the absence of both "dark" and light induced EPR (LEPR) signals over the entire range of temperatures studied. Their composite irradiated by visible light directly in a cavity of the EPR spectrometer, two overlapping LEPR lines appear reversibly at $T \leq 200$ K (Figure 21). Low- and high-field lines were attributed to positively charged polarons $P^{+\bullet}$ with isotropic g-factor, g_{iso}^{P}, and negatively charged methanofullerene ion radical with effective g_{iso}^{F} background photo induced in the P3HT:bis-PC$_{62}$BM BHJ. The effective values $g_{iso}^{P} = 2.0023$ and $g_{iso}^{F} = 2.0007$ are close to those obtained for charge carriers photo induced in other polymer:fullerene composites (Marumoto et al. 2003, Janssen, Moses, and Sariciftci 1994, Krinichnyi 2009), however, exceed those, $g_{iso}^{P} = 2.0017$ and $g_{iso}^{F} = 1.9996$, obtained for P3HT:PC$_{61}$BM (Krinichnyi, Yudanova, and Spitsina 2010). Note, that the g_{iso}^{P} value of the latter measured more accurately at W-band (94 GHz) EPR is equal to 2.00187 (Aguirre et al. 2008). These parameters were appeared to become weakly temperature dependent than those of the P3HT:PC$_{61}$BM composite (Krinichnyi, Yudanova, and Spitsina 2010). As in case of other polymer:fullerene systems (Krinichnyi and Balakai 2010, Krinichnyi, Yudanova, and Spitsina 2010, Krinichnyi 2016a), LEPR spectra of the P3HT:bis-PC$_{62}$BM BHJ consists of two Lorentzian contributions of mobile polarons and bis-methanofullerene anion radicals in quasi-pairs $P_{mob}^{+\bullet} \leftrightarrow bmF_{mob}^{-\bullet}$ as well as a contribution of localized polarons, $P_{loc}^{+\bullet}$, pinned in polymer traps (shown by dashed lines in Figure 21). It should be to note that, as compared with the P3HT:PC$_{61}$BM system, there is absent contribution of pinned fullerene radicals, $bmF_{loc}^{+\bullet}$. This implies that the number of deep traps able to capture a anion radical in the P3HT:bis-PC$_{62}$BM composite is sufficiently lower than that in the P3HT:PC$_{61}$BM one due to the better ordinary of the former. The best fit of the LEPR spectra of the sample was achieved using a convolution of Gaussian and Lorentzian line shapes, which means that electron excitation leads to inhomogeneous and homogeneous line broadening, respectively, due to unresolved hyperfine interaction of unpaired spin with neighboring protons and also to its different mobility.

Effective g-factor of polarons is strongly governed by the structure and conformation of a conjugated π-electron system. Our D-band EPR study of P3OT showed (Krinichnyi and Roth 2004) that an unpaired electron delocalized on polaron and extended over L lattice units weakly interacts with sulfur heteroatoms involved in the polymer backbone (see below). This provokes rhombic symmetry of spin density and, therefore, anisotropic g-factor and linewidth. Since the backbone of this and other P3AT polymers can be expected to lie preferably parallel to the film substrate (Kim et al. 2006), the lowest principal g-value is associated with the polymer backbone. The macromolecule can take any orientation relative to the z-axis, i.e., the polymer backbone direction as is derives from the presence of both the

g_{xx} and g_{yy} components in the spectra for all orientations of the film. Thus, the g-factor anisotropy is a result of inhomogeneous distribution of additional fields along the x and y directions within the plane of the polymer σ-skeleton rather than along its perpendicular z direction. This was confirmed later at millimeter waveband LEPR study of the P3HT:PC$_{61}$BM composite. Thus, it was shown that the spin of polarons photo induced in the P3HT chains is also characterized by rhombic symmetry originating anisotropic g-factor (see Table 1). Spin distribution in the $mC_{61}^{-\bullet}$ anion radical embedded into P3HT matrix is also characterized by rhombic symmetry that leads to respective anisotropy of its g-factor with g_{xx} = 2.00058, g_{yy} = 2.00045, g_{zz} = 1.99845 (Poluektov et al. 2010).

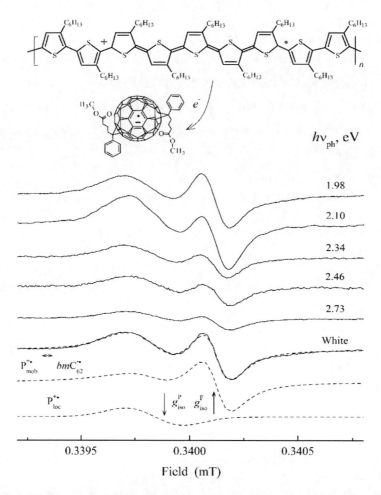

Figure 21. X-band LEPR spectra of charge carriers background photo induced at 77 K in BHJ formed by macromolecules of regioregular P3HT with globes of *bis*-PC$_{62}$BM by monochromic light with different photon energy $h\nu_{ph}$ and by polychomatic white light at T = 77 K. Sum spectrum as well as its Lorentzian contributions caused by mobile radical quasi-pairs, $P_{mob}^{+\bullet} \leftrightarrow bmF_{mob}^{-\bullet}$, and localized polarons, $P_{loc}^{+\bullet}$, are shown by dashed lines. The positions of LEPR spectra of localized and mobile charge carries are shown as well. At the top are schematically depicted P3HT with a polaron and *bis*-PC$_{62}$BM.

EPR Spectroscopy of Polarons in Conjugated Polymers and Their Nanocomposites 45

P3HT HOMO energy level E_{HOMO} depends on the overlap of adjacent thiophene MOs and, therefore, is expected to shift with ring angle (Conwell et al. 1988) similarly to the valence band involved in the π-π^* transition. The band gap, $E_{LUMO} - E_{HOMO}$, should depend on torsion angle θ between the planes of the neighboring thiophene rings (Harigaya 1998), being near 30° in regioregular P3HT (Łużny, Trznadel, and Proń 1996). A decrease in g_{iso}^P occurs at electron excitation from the unoccupied shell to the antibonding orbit, $\pi \rightarrow \sigma^*$ (Buchachenko, Turton, and Turton 1995), so one may conclude that the energy of antibonding orbits decreases as $PC_{61}BM$ globes is replaced by bis-$PC_{62}BM$ ones in this composite. This increases g_{iso}^P of the P3HT:bis-$PC_{62}BM$ composite and decreases the slope of its temperature dependency which is characteristic of more ordered system. Indeed, the changes in total energy with the torsion angle θ appear as effective steric potential energy. The angular dependence of this energy is an harmonic, with larger angles becoming more probable with the temperature increase. In this case the decrease of molecular regioregularity or a greater distortion of the thiophene rings out of coplanarity reduces charge mobility along the polymer chains (Lan and Huang 2008). This is usually attributed to a decrease in the effective conjugation lengths of the chain segments. The intrachain transfer integral t_{1D} is primarily governed by the degree of overlap between the p_z atomic orbitals of the carbon atoms forming polymer units and, therefore, should evolve a square-cosine function of the torsion angle θ (Van Vooren, Kim, and Cornil 2008). This allows one to evaluate the decrease in the θ value by ~10° at the replacement of the $PC_{61}BM$ by bis-$PC_{62}BM$ in the P3HT:methanofullerene system. This fact and the weaker temperature dependence of g-factors of both charge carriers in the P3HT:bis-$PC_{62}BM$ composite indicate the growing in planarity and ordering of polymer matrix occurring at such replacement.

The limiting number of polarons n_p and fullerene anion radicals n_f simultaneously formed per each polymer unit in the P3HT:bis-$PC_{62}BM$ BHJ can be determined, as in the case of analogous systems (Krinichnyi, Yudanova, and Spitsina 2010), by I_2-doping of the sample. Limiting paramagnetic susceptibility χ of iodine-treated composite was determined at 310 and 77 K to be 8.7×10^{-6} and 9.9×10^{-5} emu/mole, respectively. The analysis showed that the cooling of the sample leads to the appearance in its sum EPR spectra of anisotropic contribution attributed to trapped polarons. The ratio of a number of mobile to trapped polarons at 77 K is appeared to be near 7:1. Mobile polarons initiated in the sample by the I_2-doping demonstrate at 310 K single Lorentzian EPR spectra with peak-to-peak linewidth ΔB_{pp} of 0.56 mT, which are broader than that obtained for polarons stabilized in the P3HT:$PC_{61}BM$ (Krinichnyi, Yudanova, and Spitsina 2010) and other conjugated polymer matrices (Krinichnyi 2000a) Such broadening of the EPR line becomes most likely due to stronger dipole-dipole interaction between charged polarons. The contribution to linewidth due to such interaction can be estimated as $\Delta B_{dd} = \mu_B R_0^{-3} = 4/3\,\pi\mu_B n_p$, where μ_B is Bohr magneton, R_0 is distance between polarons proportional to their concentration n_p on the polymer chain. At 77 K, the ΔB_{pp} values of mobile and trapped polarons are equal to 0.18 and 0.19 mT and characterized by Lorentzian and Gaussian distribution of spin packets, respectively. Assuming intrinsic linewidth of polarons $\Delta B_{pp}^0 = 0.15$ mT in regioregular P3HT (Breiby et al. 2003), one can obtain $R_0 \approx 1.3$ nm from the line broadening due to dipole-dipole interaction in P3HT:bis-$PC_{62}BM$. The maximum density of polaron transport states was estimated (Westerling, Osterbacka, and Stubb 2002), e.g., for regioregular P3HT to be near 10^{21} cm^{-3}

assuming that one polaron occupies approximately five monomer units. Intrinsic concentration of doping-initiated polarons counting only upon polymer fraction in the P3HT:*bis*-PC$_{62}$BM composite was determined at 77 K to be 2.2×10^{19} cm^{-3}. This value lies near 2×10^{19} cm^{-3} obtained for concentration of acceptors in ZnO-treated P3HT (Marchant and Foot 1995). Effective concentration calculated for both the polymer and fullerene phases is 2.1×10^{18} cm^{-3} in this composite. This allows one to evaluate an effective number of both types of charge carriers per each polymer unit initiated in the polymer:fullerene composites by light irradiation (see Figure 22) and I$_2$-initiated polarons, n_p = 3.8×10^{-3}. The n_P values obtained are considerably lower than that, $n_p \approx 0.05$, estimated for polarons excited in doped polyaniline (Houze and Nechtschein 1996).

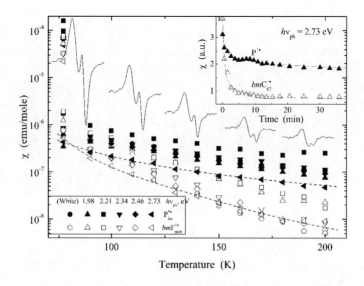

Figure 22. Temperature dependence of paramagnetic susceptibility of the $P_{loc}^{+\bullet}$ and $bmF_{mob}^{-\bullet}$ charge carriers photo induced in BHJ P3HT:*bis*-PC$_{62}$BM by polychromatic white and monochromatic light with different photon energy hv_{ph}. Respective limiting values evaluated for these carriers at 77 K by using iodine doping of the composite are shown as well. Exemplary dashed lines show dependences calculated from Eq.(A.12) with E_r = 0.018 and 0.041 eV. LEPR spectra of the sample registered at respective temperatures are shown at the top. In the insert are shown the decay of spin susceptibilities of pinned polarons $P_{loc}^{+\bullet}$ (filled points) and methanofullerene anion radicals $bmF_{mob}^{-\bullet}$ (open points) photo induced in the BHJ of P3HT:*bis*-PC$_{62}$BM composite by photons with hv_{ph} = 2.73 eV at 77 K. Dashed lines show the dependences calculated from Eq.(A.10) with E_0 = 0.011 and 0.023 eV.

The deconvolution of LEPR spectra of the P3HT:*bis*-PC$_{62}$BM composite allowed us to obtain separately all terms of its effective paramagnetic susceptibility χ as contributions of mobile and localized polarons χ_P and methanofullerene anion radicals χ_F. Figure 22 illustrates the changes in LEPR spectra of the P3HT:*bis*-PC$_{62}$BM composite with its heating and temperature dependences of all contributions into sum χ. Since concentration of main charge carriers decreases dramatically at T > 200 K, the precision of determination of their spin susceptibility falls significantly.

Fast initiation and consequent slow recombination of spin charge carriers in polymer:fullerene BHJ at background illumination allow one to register spin susceptibility as a differential result of these processes. The probability of the latter is mainly governed by polaron multistage diffusion along a polymer chain through an energetic barrier and further electron tunneling from a fullerene anion to a polymer chain (Yan et al. 2000). Motion of polarons is assumed to be described by the a multiple trapping in sites with respective energy E_t (Nelson 2003, Tachiya and Seki 2010). Since crystalline subsystems in a composite are characterized by different band gaps, their E_t energies are also different and distributed exponentially. If energy of a trapped polaron exceeds E_t, it is occasionally thermally detrapped to the free state. A positive charge of polaron is not required to be recombined with negative charge on first fullerene. Diffusing along polymer backbone with positive elemental charge it may collide with the nearest fullerene anion radical located between polymer chains and then to recombine with a charge on a subsequent counter-anion. Assuming that polaron motion is not disturbed by the presence of fullerene molecules, we can conclude that the collision duration is governed by polaron diffusion. As it walks randomly along the chain, it passes a given fullerene molecule with the frequency $v_c \approx \omega_{hop}/L^2$. The polaron that diffuses between initial i and final j sites spatially separated by distance R_{ij} spends the energy equal to the difference of their depths, ΔE_{ij}. Thus, spin susceptibility of the sample should become energy-dependent and follow Eq.(A.6). A positive charge of polaron is not required to be recombined with negative charge on first fullerene; it can to recombine with a charge on a subsequent counter-ion. The fullerene molecules can be considered as fast relaxing impurities, so the susceptibility of polarons in the composite should follow Eq.(A.12) with the spin flip-flop probability p expressed Eq.(A.11). As it is seen from Figure 22, the net electronic processes in the composites can be described in terms of the spin exchange mechanism. The **energy barrier tunneling by polaron** was determined from Eq.(A.12) at irradiation of the composite by photons with hv_{ph} = 1.98, 2.10, 2.34, 2.46, 2.73 eV and by the white light to be ΔE_{ij} = 0.001, 0.029, 0.014, 0.016, 0.018, and 0.003 eV, respectively.

It is evident that the energy required for polaron trapping in the polymer matrix is lower than that obtained for other charge carriers. ΔE_{ij} evaluated from $\chi(T)$ for these charge carriers increases considerably indicating higher energy required for their trapping in the system. The data presented evidence additionally that all spin-assisted processes are governed mainly by the structure of anion radicals as well as by the nature and dynamics of charge carriers photo induced in BHJ. It is seen that the χ value of both charge carriers becomes distinctly higher at characteristic energy $hv_{ph} \approx 2.1$ eV lying near the band gap of P3AT (Al Ibrahim et al. 2005). Such a dependence of spin concentration on photon energy can be explained either by the formation of spin pairs with different properties in homogeneous (higher ordered) composite fragments or by the excitation of identical charge carriers in heterogeneous domains (lower ordered) of the system under study. Different spin pairs can be photo induced as a result of the photon-assisted appearance of traps with different energy depths in a polymer matrix. However, the revealed difference in the parameters of radicals seems to be a result of their interaction with their microenvironment in domains in homogeneously distributed in polymer:fullerene composite. Different ordering of these domains can be a reason for variation in their band gap energy leading, hence, to their sensitivity to photons with definite but different energies. This can give rise to the change in the interaction of charge carriers with a lattice and other spins. Effective spin susceptibility of the sample discussed somewhat exceeds that obtained for P3HT:PC$_{61}$BM (Krinichnyi, Yudanova, and Spitsina 2010). This effect and the absence of trapped anion radicals in the former allowed to conclude additionally more ordered BHJ in the P3HT:*bis*-PC$_{62}$BM composite which interfere in the formation of traps.

If one includes Coulomb interactions, this should affect the activation energy for either defrosting or thermally assisted tunneling by an amount $U_c = e^2/4\pi\varepsilon\varepsilon_0 r$, where e is elemental charge, ε is a dielectric constant, and r is charge pair separation. Assuming $\varepsilon = 3.4$ for P3HT (Deibel et al. 2010), minimum separation of charge carriers is equal to the radius a of π electrons on the C atoms which two times longer than the Bohr radius, i.e., 0.106 nm, r equal to interchain separation, 0.38 nm (Chen, Wu, and Rieke 1995), one obtains the decrease in U_c from ~0.4 eV down to 0.02 eV during dissociation of an initial radical pair. Therefore, both the photo induced polaron and anion radical should indeed be considered as noninteracting that prolongs their life.

When initiating background illumination is switched off, photo initiation of charge carriers in BHJ stops and the concentration of spin charge carriers excited starts to decrease as shown in the insert of Figure 22. Live time of charge carriers seems to be much longer than $t \sim 0.1$ μs obtained by optical absorption spectroscopy for relevant recombination times of mobile photoexcitations in organic solar cells (So, Kido, and Burrows 2008, Sharma 2010). So, the data presented are mainly pertinent to carriers trapped in polymer matrix. Generally, charge recombination is described as a thermally activated bimolecular recombination (Brabec et al. 2003) which consists of temperature-independent fast and exponentially temperature-dependent slow steps (Westerling, Osterbacka, and Stubb 2002). At the latter step, polaronic charge carrier can either be retrapped by vacant trap sites or recombine with electron on fullerene anion radical. Trapping and retrapping of a polaron reduces its energy that results in its localization into deeper trap and in the increase in number of localized polarons with time. So, the decay curves presented can be interpreted in terms of bulk recombination of holes and electrons during their repeated trapping into and detrapping from trap sites with different depths in energetically disordered semiconductor (Tachiya and Seki 2010). The traps in such a system are characterized by different energy depths and energy distribution E_0. Polarons fast translative diffuse along a polymer backbone, and fullerene anion radicals can be considered to be immobilized between polymer chains. The dependences calculated in frames of this approach are shown in Figure 22. Trap energy distribution in the composite was determined upon its irradiation by photons with $h\nu_{ph} = 1.98$, 2.10, 2.34, 2.46, 2.73 eV and by the white light to be $E_0 = 0.041$, 0.027, 0.018, 0.014, 0.011, and 0.031 eV, respectively. Therefore, the decay of long-lived polaron excitations originated from initial excitons and spin quasi-pairs photo induced in the polymer:fullerene composite can be described in terms of the above model in which the low-temperature recombination rate is strongly governed by temperature and the width of energy distribution of trap sites. Non-linear change $E_0(h\nu_{ph})$ dependence should indicate that the local structure and ordering govern the depth of spin traps and their distribution in this composite. This allows one to conclude the crucial role of the photon energy on the formation and energy properties of the traps in a BHJ of disordered systems.

Effective EPR linewidth of both charge carriers photo induced in the P3HT:bis-PC$_{62}$BM composite is presented in Figure 23 as a function of temperature and photon energy. It is seen that this value obtained for polarons changes extremely with the temperature with characteristic temperature $T_c \approx 130$ K, whereas linewidth of the methanofullerene anion radicals demonstrates monotonic temperature dependence and decreases with the system heating. This indicates echange interaction of polaron spins which broadens their lines. Besides, sulphur and hydrogen atoms in the composite possess a nuclear magnetic moment initiating the hyperfine interaction between the electrons and the nuclei. Polaron translational and fullerene pseudorotational diffusion in a polymer:fullerene composite should also be taken into account. In this case a polaron should interact with the other PC with the collision probability p expressed by Eq.(A.10), so then its absorption line should additionally be broaden by the value expressed by Eq.(A.19). Successful fitting of experimental data with t_{1D} = 1.18 eV (Cheung, McMahon, and Troisi 2009) are presented in Figure 23 should evidence the applicability of such approach for interpretation of electronic processes realized in the P3HT:*bis*-PC$_{62}$BM composite. The energy E_r obtained lies near that evaluated for regioregular P3HT from *ac* conductometric (0.080 eV) (Obrzut and Page 2009) and ^{13}C NMR (0.067 – 0.085 eV at T< 250 K) (Yazawa et al. 2010) data.

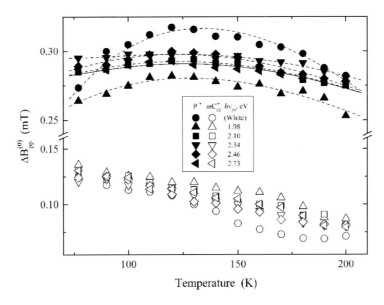

Figure 23. Linewidth of charge carriers photo induced in the P3HT:*bis*-PC$_{62}$BM composite as a function of temperature and photon energy $h\nu_{ph}$. Top-to-bottom dashed lines show the dependences calculated from Eq.(A.19) with E_r = 0.045, 0.047, 0.051, 0.050, 0.050, 0.048 eV. The symbol (0) in $\Delta B_{pp}^{(0)}$ implies that the LEPR spectra were measured far from MW saturation, when $B_1 \to 0$.

The main relaxation parameters of the pined and mobile polarons determined using steady-state saturation method are presented in Figure 24,a as a function of temperature and photon energy. It is seen from the Figure that the interaction of most charge carriers with the lattice is characterized by monotonic temperature dependence. All relaxation parameters are governed by structure and morphology of the polymer:fullerene composite as well as by the energy of initiating phonons. Polarons induce an additional magnetic field in the BHJ accelerating electron relaxation of all spin ensembles. As their relaxation in conjugated polymers are defined mainly by their dipole-dipole interaction (Krinichnyi et al. 1992, Krinichnyi 2000a) the respective coefficients of spin intrachain D_{1D} and interchain D_{3D} diffusion can be calculated from Eq.(A.25), Eq.(A.26) and the data presented in Figure 24,a. Figure 24,b depicts both these parameters obtained for polarons in the composite as a function of the energy of initiating phonons and temperature. The Figure shows that both the coefficients of polaron diffusion are governed sufficiently by the energy of initiated photons $h\nu_{ph}$. To account for the LEPR mobility data obtained, different theoretical models can be used. Intrachain polaron dynamics in the P3HT:*bis*-PC$_{62}$BM composite is characterized by strong temperature dependence (Figure 24,b). Typically, such a behavior can be associated with the scattering of polarons on the lattice phonons of crystalline domains immersed into an amorphous polymer matrix. It is seen from Figure 24,b that the experimental polaron intrachain dynamics data obtained at irradiation of the composite by photons with $h\nu_{ph}$ = 1.98, 2.10, 2.34, 2.46, 2.73 eV and by the white light can successfully be fitted by Eq.(A.31) with E_{ph} = 0.077, 0.075, 0.091, 0.088, 0.065, and 0.062 eV, respectively. This parameter lies near that determined for PANI-ES (0.09 – 0.32 eV) (Krinichnyi 2000b) and laser modified PATAC (0.15 – 0.18 eV) (Krinichnyi, Roth, and Schrödner 2002).

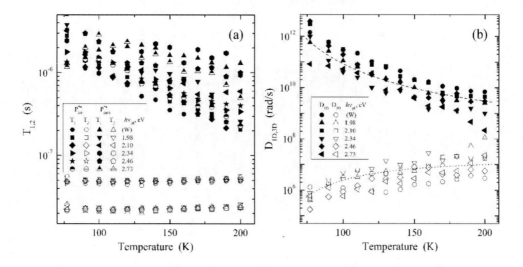

Figure 24. Temperature dependence of spin-lattice T_1 (filled points) and spin-spin T_2 (open points) relaxation times (*a*) as well as intrachain (D_{1D}, filled points) and interchain (D_{3D}, open points) diffusion coefficients of localized $P^{+\bullet}_{loc}$ and mobile $P^{+\bullet}_{mob}$ polarons photo induced in the P3HT:*bis*-PC$_{62}$BM composite by the polychromatic white and monochromatic light with different photon energy $h\nu_{ph}$. Above and below exemplary dashed lines show dependences calculated from Eq.(A.29) with E_{ph} = 0.077 eV and from Eq.(A.31) with E_t = 0.128eV, respectively.

The interchain spin hopping dynamics can be analyzed in terms of the Hoesterey-Letson formalism modified for amorphous low-dimensional systems (Fishchuk et al. 2002). According to this model, spin hoping between polymer chains should be controlled by the traps with concentration n_t and depth E_t. Figure 24,b shows also the temperature dependences calculated from Eq.(A.31) with T_{cr} = 127 – 140 K. The Figure evidences that interchain polaron dynamics in the composite irradiated by photons with $h\nu_{ph}$ = 1.98, 2.10, 2.34, 2.46, 2.73 eV and by the white light can be described in the frames of the above mentioned theory with E_t = 0.128, 0.129, 0.128, 0.119, 0.125, and 0.123 eV, respectively. It should be noted that the E_t values obtained for P3HT:PC$_{61}$BM composite prevail those characteristic of P3HT:*bis*-PC$_{62}$BM (Krinichnyi 2016a) that is additional evidence of deeper traps formed in the former polymer matrix. This fact, probably, indicates the decrease in trap concentration due the increase in effective crystallinity of the polymer matrix.

So, the main magnetic resonance, relaxation and dynamic parameters of polarons photo initiated in the P3HT:bis-PC$_{62}$BM composite are governed by the structure, conformation and ordering of BHJ as well as by the energy of excited photons. This can be partly as a result of structural inhomogeneity of the polymer:fullerene composite, conditioning reversible photon-initiated traps with different depth and distribution. The data obtained allowed us to suggest the importance of the ring-torsion and ring-librative motions on the charge initiation, separation and diffusion in this disordered organic system. It was shown that a polaron diffusing along a polymer backbone exchanges with the spin of a counter methanofullerene anion radical in terms of the modified Marcus theory. Charge recombination occurs in terms of spin multitrapping in energetically disordered semiconductor whose local structure and ordering govern the number, depth and distribution of charge traps. The interaction of most charge carriers with the lattice is characterized by monotonic temperature dependence. Electron relaxation of charge carriers was shown to be also governed by dynamics, structure and conformation of their microenvironment as well as by photon energy. Selectivity to photon energy is governed by properties of donor and acceptor forming BHJ and can be used, for example, in plastic sensoric photovoltaics. The energetic barrier required for polaron interchain hopping predominantly prevails that of its intrachain diffusion in the composite.

2.7.2. Poly(3-Octylthiophene)

P3OT is also expected to be a suitable and perspective material for molecular electronics (Carter 1982, Brédas and Chance 1990, Salaneck, Clark, and Samuelsen 1991, Ashwell 1992, Scrosati 1993, Wong 1993), e.g., polymer sensors (Bobacka, Ivaska, and Lewenstam 1999) and polymer:fullerene solar cells (Sariciftci and Heeger 1995, Lee et al. 1993, Lee et al. 1994, Gebeyehu et al. 2000, Gebeyehu et al. 2001). The piezoelectric effect has been also registered in P3OT (Taka et al. 1993). DC conductivity of an as synthesized P3OT is about 10^{-4} S/m (Chen, Wu, and Rieke 1995). A constriction of the P3OT bandgap was observed (Kaniowski et al. 1998) due to the decrease of the torsion angle between its adjacent thiophene rings and the enhancement of interchain interactions between parallel polymer planes. More detail information on the magnetic and electronic properties of the initial P3OT sample and treated by an annealing (P3OT-A) and by both recrystallization and annealing (P3OT-R) was obtained at X- and D-bands EPR (Roth and Krinichnyi 2003, Krinichnyi and Roth 2004). The transition temperature of P3OT is close to 450 K, so respective treatment was made at this temperature.

Figure 25 shows EPR spectra of polarons stabilized in the initial and treated P3OT samples obtained at both wavebands EPR. At the X-band the samples show a single nearly Lorentzian EPR line with g_{iso} = 2.0019. The asymmetry factor of the lines is A/B =1.1, and their peak-to-peak linewidth was determined to be ΔB_{pp} = 0.27 (P3OT), 0.27 (P3OT-A), and 0.26 mT (P3OT-R). The latter value is smaller than $\Delta B_{pp} \approx 0.6 - 0.8$ mT obtained for PT and P3MT (Mizoguchi and Kuroda 1997), however, is close to 0.32 mT registered for polarons in regioregular P3OT (Marumoto et al. 2002). The X-band EPR spectrum of the P3OT sample stored for two years is also drawn in Figure 25,a by the dotted line. Its computer modeling has shown that it consists of anisotropic and isotropic spectra. The effective g-factors obtained from these spectra are close; therefore one can conclude that localized PC with more anisotropic magnetic parameters appear at the polymer storage.

The RT total spin concentration of P3OT increases during the polymer treating from 3.9×10^{19} cm^{-3} in P3OT up to 3.1×10^{20} cm^{-3} in P3OT-A and 3.5×10^{20} cm^{-3} in P3OT-R or from 0.013 to 0.11 and to 0.11 spin per a monomer unit, respectively. Their effective paramagnetic susceptibility χ and the χT value are presented in Figure 26 as function of temperature.

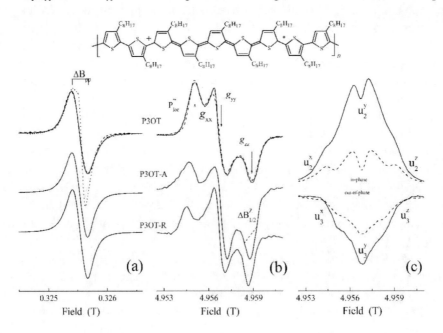

Figure 25. RT X-band (*a*) and D-band (*b*) absorption EPR spectra of the P3OT, P3OT-A, and P3OT-R samples. The X-band EPR spectrum of the P3OT stored for two years is shown on the left by the dotted line. The magnetic resonance parameters measured are shown. The spectra calculated using $g_{xx} = 2.004089$, $g_{yy} = 2.003322$, $g_{zz} = 2.002322$, $\Delta B_{pp}^x = \Delta B_{pp}^y = \Delta B_{pp}^z = 0.25$ mT, and Lorentzian/Gaussian = 0.9 line shape ratio (*a*), and $\Delta B_{pp}^x = 0.82$ mT, $\Delta B_{pp}^y = 0.78$ mT, $\Delta B_{pp}^z = 0.88$ mT, and Lorentzian/Gaussian = 0.4 line shape ratio (*b*) are shown by the dashed lines. The sum spectrum of the lines calculated with $g_\parallel = 2.00387$, $g_\perp = 2.00275$, $\Delta B_{pp} = 0.11$ mT and $g_{iso} = 2.00312$ and $\Delta B_{pp} = 0.35$ mT with amplitude ratio of 2.5:1 is shown by dotted line on the right as well. (*c*) In-phase (u_2) and $\pi/2$-out-of-phase (u_3) terms of D-band dispersion spectra of the samples registered at $T = 100$ K (solid lines) and 145 K (dashed lines) are shown. The formation of the polaron on P3OT chain and measured magnetic parameters are shown as well.

In the high-temperature region the spin susceptibility of the P3OT and P3OT-R samples seems to include a contribution due to a strong spin-spin interaction as it was revealed in PANI (Iida et al. 1992, Krinichnyi et al. 2002), PATAC (Krinichnyi, Roth, and Schrödner 2002) and P3DDT (Barta et al. 1994, Cik et al. 1995, Kawai et al. 1996, Cik et al. 2002). This contribution disappears at low temperatures due to the phase transition opening an energy gap at the Fermi level (Mizoguchi and Kuroda 1997), so then the susceptibility demonstrates the Curie behavior. Figure 26 shows that the total spin susceptibility of all P3OT samples follows Eq.(A.4). As in the case of sulfonated polyaniline (Kahol et al. 1999), these results evidence that the annealing of the P3OT polymer affects an effective exchange coupling between spins which results in the change of an effective number of the Bohr magnetons per a monomer unit, however, the recrystallization neglects this effect.

EPR Spectroscopy of Polarons in Conjugated Polymers and Their Nanocomposites 53

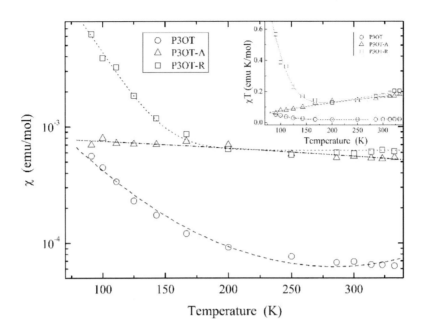

Figure 26. The $\chi(T)$ and $\chi T(T)$ (insert) dependencies of paramagnetic susceptibility of the P3OT, P3OT-A, and P3OT-R samples. The dependencies calculated from Eq.(A.4) with $\chi_P = 4.5 \times 10^{-4}$ emu/mol, $C = 0.044$ emu K/mol, $c_{af} = 0$ (dotted line), $\chi_P = 0$, $C = 4.1$ emu K/mol, $c_{af} = 52.1$ emu K/mol, and $J_{af} = 0.011$ eV (dash-dotted line), and $\chi_P = 0$, $C = 0.72$ emu K/mol, $c_{af} = 10.6$ emu K/mol, and $J_{af} = 0.0062$ eV (dashed line) are shown as well.

At the D-band EPR the samples demonstrate the superposition of more broadened convoluted Gaussian and Lorentzian lines (with the Lorentzian/Gaussian line shape ratio ca. 0.4) with the anisotropic *g*-factor (Figure 25,b) as it is typical for PC in some other conjugated polymers with heteroatoms above described. The linewidth of the spectra increases by a factor of ca. 3 at the increase of the registration frequency (Figure 27). It is seen that the linewidths of the PC in the samples depend on the polymer treatment and temperature. The treatment leads to the decrease in the averaged linewidth $<\Delta B_{pp}>$ from 0.82 mT in P3OT down to 0.78 mT in P3OT-A and then to 0.67 mT in P3OT-R confirming the above supposition of the growth of the system crystallinity. The linewidth of the P3OT increases with the temperature increase from 90 K up to the phase transition characteristic temperature $T_c = 200$ K and then decreases at the further temperature growth. The analogous tendency demonstrates P3OT-R, however, its T_c value decreases down to ca. 170 K (Figure 27). This effect of the linewidth decrease below T_c was detected also in the study of doped polyaniline (Krinichnyi, Chemerisov, and Lebedev 1997, Krinichnyi et al. 2002) and was not registered in other conjugated polymers; it can be interpreted, e.g., as the manifestation of defrosting of molecular motion and/or acceleration of relaxation processes at low temperatures. On the other hand, the ΔB_{pp}^x value of P3OT-A decreases with the temperature increase of up to $T_c = 200 - 250$ K and starts to increase at $T \geq T_c$, whereas the other spectral components are broadened linearly with the temperature growth from 100 K (Figure 27). If one supposes that molecular dynamics and/or electron relaxation should stimulate activated broadening of the *i*-th line, $\Delta B_{pp}^i = \Delta B_{pp}^{i(0)} \exp(E_a/k_B T)$, from the slopes of these dependencies it is possible to determine separately the parameters of electron relaxation and molecular dynamics near the principal macromolecular axes. The energies E_a^x, E_a^y, and E_a^z required for activation of molecular motion near the principal x-, y- and z-axes in the samples P3OT, P3OT-A, and P3OT-R were determined to be 3.6, 4.9, 4.3 meV, 7.2, 2.0, 2.1 meV, and 2.4, 1.9, 1.8 meV, respectively.

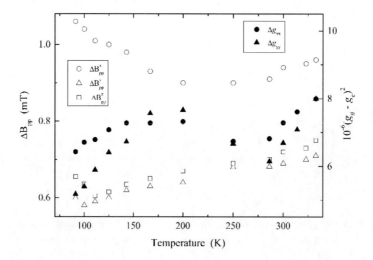

Figure 27. The D-band spectral component linewidth (*a*) and squared shifts of the g_{xx} and g_{yy} values from $g_e = 2.00232$ (*b*) of the P3OT as function of temperature.

From the RT spectra of P3OT the main components of its **g** tensor were also determined to be $g_{xx} = 2.00409$, $g_{yy} = 2.00332$, and $g_{zz} = 2.00235$. The treatment of the samples leads to the change of these parameters to $g_{xx} = 2.00404$, $g_{yy} = 2.00315$, and $g_{zz} = 2.00231$ for P3OT-A and to $g_{xx} = 2.00402$, $g_{yy} = 2.00313$, and $g_{zz} = 2.00234$ for P3OT-R. The structure of a polymer should affect the distribution of an unpaired electron in polaron changing the principal values of its PC **g**-tensor and hyperfine structure. The shift of the g_{xx} and g_{yy} values from the g_e-factor for a free electron can be compared with that calculated from Eq.(A.2) with the constant of spin-orbit interaction of the electron spin with the sulfur nuclear λ_s equal to 0.047 eV.

The effective g-factor of the P3AT is higher than that of most hydrocarbonic conjugated polymers, therefore one can conclude that in P3AT the unpaired electron interacts with sulfur atoms. This is typical for other sulfur-containing compounds, e.g., PTTF (Krinichnyi 2000a), PATAC (Krinichnyi, Roth, and Schrödner 2002) and benzo-trithioles (Krinichnyi et al. 1997) in which sulfur atoms are involved into the conjugation. In such organic solids an unpaired electron is localized mainly on the sulfur atom so its an effective g-factor is 2.014 $<g_{iso}<$ 2.020 (Williams et al. 1992, Cameron et al. 1991, 1992, Krinichnyi et al. 1997, Bock et al. 1984). The g-factor of PC in P3OT is much smaller so one can expect a higher spin delocalization in the monomer units. Indeed, assuming $\Delta E_{n\pi^*} \approx 2.6$ eV as typical for sulfuric solids, then the g-factor components of the initial P3OT yields the decrease in $\rho_s(0)$ by a factor of 1.8 as compared with PATAC (Krinichnyi, Roth, and Schrödner 2002) and by a factor of 2.6 – 3.9 as compared with poly(tetrathiafulvalenes) (Krinichnyi 2000a). One can evaluate the lower limit of RT probability (in frequency units) for the duration of the stay of spin at the sulfur site in the P3OT samples by the total shift of the spectral components registered at the above mentioned positions g_{ii}^{P3OT} relative to the g_{ii}^{S} value typical for the sulfuric radical to be $D_{1D}^{0} \geq 3.4 \times 10^{9}$ rad/s.

As in the case of other organic radicals, the g_{zz} values of the P3OT samples are close to the g-factor for a free electron so they feel the change in the system properties weakly. In contrast to X-band EPR, high spectral resolution achieved at D-band EPR allows one to register separately the structure and/or dynamic changes in all spectral components. Figure 27 evidences that the g_{xx} and g_{yy} values reflect more efficiently the properties of the radical microenvironment. These values of the initial P3OT decrease with the temperature decrease from 333 down to 280 K possibly due to the transition to the more planar morphology of the polymer chains. Below 280 K, these values increase at the sample freezing down to 160 – 220 K and then also decrease at the further temperature decrease. The decrease of g_{xx} and g_{yy} values at low temperatures can be explain by a harmonic vibration of macromolecules which evokes the crystal field modulation and is characterized by the $g(T) \propto T$ dependence (Owens 1977). The temperature polymer treatment weakly affects these parameters. This fact can be interpreted by the growth of the system crystallinity due to the higher chain packing in the treated polymers. The analysis of the data presented in Figure 27 shows that the linewidth can also correlate with the spin-orbit coupling in the framework of the Elliott mechanism playing an important role in the charge transfer in organic ion-radical salts (Williams et al. 1992) and conductive polymers with pentamerous unit rings (Mizoguchi and Kuroda 1997). Indeed, $\Delta B_{pp}^{x} \propto \Delta g_{xx}^{2}$ and $\Delta B_{pp}^{y} \propto \Delta g_{yy}^{2}$ dependencies are valid for, e.g., P3OT at least at $T \leq T_{c}$. This means that different mechanisms can affect the individual components of the P3OT spectrum and that the scattering of charge carriers should be governed by the potential of the polymer backbone.

Figure 25,c exhibits also in-phase and $\pi/2$-out-of-phase terms of D-band EPR dispersion spectra of P3OT registered at different temperatures. It is seen that the bell-like contribution with Gaussian spin packet distributions appears in both dispersion terms. This is attributed to the adiabatically fast passage of saturated spin packets by a modulating magnetic field at this waveband EPR, as in case of PTTF (see Figure 14,d) and other conjugated polymers (Krinichnyi 2006). The intensities of the $\pi/2$-out-of-phase spectral components change with the temperature. This effect evidences the appearance of the saturation transfer over the quadrature spectrum term due to super slow macromolecular dynamics (Krinichnyi 2008b, a). The analysis of spectral u_{3}^{i} components allowed determining the correlation time τ_{c}^{x} of libration motion of the chain macromolecular segments near the principal molecular x axis in the initial and treated P3OT samples. The energy required for activation of such libration dynamics in the P3OT, P3OT-A and P3OT-R samples was also determined to be $E_{a} = 0.069$, 0.054 and 0.073 eV, respectively.

The linear compressibility of an initial P3OT with planar chains is strongly anisotropic, being 2.5 times higher for the direction along molecular a-axis than along the b-axis (Mardalen et al. 1998). It was proved that the low- and high-frequency modes exist in polythiophenes (Sauvajol et al. 1997). These modes differently superposed in P3OT originating "successive" macromolecular dynamics in P3OT and P3OT-A and "parallel" molecular librations in P3OT-R (Krinichnyi 2008b, a). Osterbacka et al. (Österbacka et al. 2001) have found that the interchain coupling existing in self-assembled lamellae in P3AT drastically changes the properties of polaron excitations. So that the polaron, normally Q1D delocalized in conjugated polymers, can be delocalized in two dimensions in P3OT that results in a much reduced relaxation energy and multiple absorption bands. The upper limit for the correlation time of anisotropic molecular motion in the P3OT registered by the ST-EPR method is $\tau_c^x \leq 4.4\times10^{-4}$ s at 66 K.

It is evident that the polymer structure and the polymer treatment should influence on spin charge relaxation and dynamics. At $T \geq 200$ K the inequality $\omega_m T_1 < 1$ holds for polarons in all P3OT samples. The opposite inequality is fulfilled at lower temperatures when the dispersion spectrum is determined by the two last terms presented in Figure 25,c. The semilogarithmic temperature dependence of the relaxation times determined from these saturated EPR dispersion spectra terms are shown in Figure 28,a. One can conclude from these data that the polymer treatment leads to the acceleration of the PC effective relaxation possibly due to the increase of the interaction of polaron charge carriers with the lattice phonons. The relaxation times of the samples increase simultaneously at the temperature decreases from 333 down to ca. 250 K (P3OT, P3OT-R) and 150 K (P3OT-A) (Figure 28,a). The spin-spin relaxation is accelerated below this point leading to the appropriate change of the spectral component linewidth (Figure 27).

The spin-spin relaxation time of the P3OT samples determined from their X-band EPR absorption and D-band EPR dispersion spectra increases approximately by a factor of six. This means that the experimental data can rather be explained in terms of a modulation of spin relaxation by the polaron intra- and interchain motion in the polymers. In the framework of such approach, the relaxation time of the electron or proton spins in the sample should vary as $T_{1,2} \propto \omega_e^{1/2}$ (Butler, Walker, and Soos 1976, Nechtschein 1997) that increase relaxation times approximately by a factor of four. Note, that Q2D spin motion should lead to $T_{1,2} \propto \ln(\omega_e)$ dependency (Butler, Walker, and Soos 1976, Nechtschein 1997) that increase these parameters by factor of about two at the same change in spin precession frequency.

In Figure 28,b are shown the temperature dependencies of the effective dynamic parameters D_{1D} and D_{3D} calculated for PC in the initial and treated P3OT samples from the data presented in Figure 28,a with Eq.(A.25) and Eq.(A.26) at $L \approx 5$ (Devreux et al. 1987). The RT D_{3D} value obtained is about $D \approx 2.1\times10^{10}$ s^{-1} evaluated from the charge carrier mobility in slightly doped P3OT (Kunugi et al. 2000) and the D_{1D} value exceeds by $1 - 2$ orders of magnitude the lower limit of the spin motion D_{1D}^0. The RT anisotropy of spin dynamics D_{1D}/D_{3D} increases from 6 in P3OT up to 18 in P3OT-A and decreases down to 2 in P3OT-R. At the temperature decreases down to 200 K this value increases up to 2.5×10^3, 1.4×10^2, and 3.9×10^2, respectively, and then up to 3.8×10^8, 6.5×10^7, and 3.4×10^{10}, respectively, as the temperature decreases down to 100 K (Figure 28,b).

The $D_{1D}(T)$ dependence calculated from the ^{1}H 50 MHz spin-lattice relaxation data obtained by Masubuchi et al. (Masubuchi et al. 1999) for the initial P3OT with the respective equation and spectral density function is also presented in Figure 28,b. It is seen from the Figure that the D_{1D} value calculated from the NMR data is changed weaker with temperature. Besides, this value considerably exceeds the D_{1D} value obtained by EPR at high temperatures and is close to that determined for the low-temperature region. Such discrepancy occurs probably because NMR is not a direct method for studying electron spin dynamics in this and other conjugated polymers.

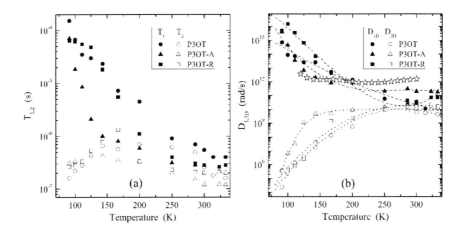

Figure 28. Temperature dependencies of the relaxation times T_1 (filled symbols), T_2 (open symbols) (a) and spin diffusion coefficients D_{1D} (filled symbols), D_{3D} (open symbols) (b) of the P3OT, P3OT-A, and P3OT-R samples. Open stars show the $D_{1D}(T)$ dependence calculated from the data obtained from ^{1}H NMR study of P3OT (Masubuchi et al. 1999). The dependencies calculated from Eq.(A.29) with E_{ph} = 0.18, 0.20, 0.13 eV (dashed lines), and Eq.(A.30) with E_a = 0.18, 0.19, 0.19 eV (dotted lines) are shown.

The temperature dependencies calculated from Eq.(A.29) are also presented in Figure 28,b. The Figure evidences that the intramolecular spin dynamics follows well Eq.(A.29) with the energy of optical phonons of 0.13 eV for P3OT, 0.20 eV for P3OT-A, and 0.18 eV for P3OT-R samples. This value is close to the respective energy determined for PANI-ES and laser-modified PATAC (see above).

The polaron-phonon interaction seems play an important role also in the interchain charge transfer at $T \geq T_c$. However, another mechanism should prevail at lower temperatures. Figure 28,b shows that spin interchain dynamics increases with the temperature increase at the low-temperature region and then slightly decreases at $T \geq T_c$. Such dependence is typical for the systems with a strong coupling of the charge with the lattice phonons. The analogous dependence with the characteristic temperature $T_c \approx 150$ K was obtained, e.g., for dc conductivity of P3HT (Yamauchi et al. 1997). In this case the strong temperature dependence of the hopping conductivity is more evidently displayed in spin interchain dynamics. This interaction should lead to the narrow energy gap E_a much higher than the RT thermal energy ($k_B T \approx 0.026$ eV). The strong temperature dependence for D_{3D} at low temperatures can be due to the thermal activation of spin charge carriers from widely separated localized states in the gap to closely localized states in the tails of the polymer VB and CB. Indeed, the data obtained experimentally for the spin interchain dynamics can be approximated by Eq.(A.30) with respective E_a values. The increase of the dimensionality of the polymer system should lead to the decrease in E_a. Charge transfer through bulk of P3MT was analyzed by Parneix, et al. (Parneix and El Kadiri 1987) in terms of such approach with E_a = 0.18 eV. So, we can also to explain the data obtained for Q3D spin dynamics in the P3OT samples in terms of this theory. The temperature dependencies calculated from Eq.(A.30) for the P3OT samples are also shown in Figure 28,b for comparison. The activation energies required for spin interchain diffusion in the P3OT, P3OT-A, and P3OT-R samples were obtained to be E_a = 0.18, 0.19, 0.19 eV, respectively. These values are close to those determined for the charge transfer in nanomodified PATAC, PTTF-C_2H_5-C_6H_4, PANI, P3HT, and poly[N-9'-hepta-decanyl-2,7-carbazole-alt-5,5-(4',7'-di-2-thienyl-2',1',3'-benzothiadiazole)] (PCDTBT) conjugated polymers (Krinichnyi 2016b). *DC* conductivity of P3OT is close to 10^{-4} S/m and is determined mainly by the activation hopping of a charge carrier between high-conductive crystalline domains (Chen, Wu, and Rieke 1995). Because an intrinsic conductivity of this polymer is defined by the σ_{1D} and σ_{3D} values determined from Eq.A.34), so the $\sigma_{1,3D}>>\sigma_{dc}$ relation should hold for all the samples.

Thus, the interaction of the spin charge carrier with the heteroatom in sulfurous organic polymer semiconductors P3OT leads to the appearance of the *g*-factor anisotropy in their D-band EPR spectra. Spin relaxation and dynamics are determined by the interaction of the mobile polaron with optical phonons of the polymer lattice. Polymer modification leads to a distinct change in magnetic, relaxation and dynamics properties of the polaron and its microenvironment. The close values of the energies of spin interchain transport, dipole-dipole interaction and the optical lattice phonons indicates the correlation of charge transport and macromolecular dynamics in P3OT. These energies increase at the polymer treatment indicating the increase of its effective crystallinity. It can be concluded that the two types of charge transport mechanism can be associated with change in the lattice geometry at the characteristic (the polymer-glass transition) temperature T_c, e.g., with its thermochromic effect.

3. SPIN-ASSISTED POLARON DYNAMICS IN POLYMER:DOPANT/POLYMER:FULLERENE NANOCOMPOSITES

An unique capability of spin orientation in external magnetic field and its very weak interaction with own environment in semiconductors open an undoubted imperative to developing new spin electronic (magneto-electronic, spintronic) devices simultaneously exploiting both charge and spin of electrons in the same device and appropriate cutting-edge spectroscopic methods suitable for registration of spin polarization and its controlled manipulation (Maekawa 2006, Dediu et al. 2009). EPR spectroscopy was proved to be the most effective direct tool able to register spin excitations, study their nature, properties and manipulate in such systems (Lupton, McCamey, and Boehme 2010).

The interaction between spin charge carriers should also affect electronic properties of organic polymers with spin charge carriers. PANI-ES seems to be one of the most suitable systems for the study of spin-assisted charge transfer in organic semiconductors. Above it was demonstrated the existence in PANI-ES of polarons trapped on chains in amorphous polymer phase and polarons diffusing along and between polymer chains. Polarons diffusing along polymer chains appeared to be accessible for triplet excitations injected into the polymer bulk. Spin exchange interaction leads to collision of domestic and guest spins dramatically changing their magnetic, relaxation and electron dynamics parameters. Such effect was registered only in PANI-ES highly doped with *para*-toluenesulfuric acid (PANI:*p*TSA) (Krinichnyi, Roth, et al. 2006, Krinichnyi, Tokarev, et al. 2006). PANI:*p*TSA becomes Fermi glass with high density of states near the Fermi energy level ε_F (Kahol 2000, Wessling et al. 2000), *dc* conductivity of PANI-ES follows the Q3D Mott's VRH model (Kapil, Taunk, and Chand 2010). Other PANI-ES doped with, e.g., sulfuric acid (PANI:SA) can also be

EPR Spectroscopy of Polarons in Conjugated Polymers and Their Nanocomposites   59

characterized as Fermi glass but with localized electronic states at the Fermi level ε_F due to disorder. This predestined the use of both these PANI-ES as reservoir of stabilized spins. P3DDT:PC$_{61}$BM was used as a second spin reservoir in comparative experiments (Krinichnyi, Yudanova, and Wessling 2013, Yudanova, Bogatyrenko, and Krinichnyi 2016). EPR study of PANI-ES/P3DDT:PC$_{61}$BM composites and their ingredients is expected to provide a good framework for understanding the underlying nature of exchange interactions between different spins in multispin system. Below are presented the first results of the LEPR study of main magnetic resonance parameters of polarons stabilized in highly doped PANI:pTSA, polarons and fullerene radical anions photo induced in the P3DDT:PC$_{61}$BM composite under background illumination by white light, as well as these charge carriers stabilized in the PANI:pTSA/P3DDT:PC$_{61}$BM composite in a wide temperature range. To emphasize these results, the analogous investigation with PANI:SA and PANI:SA/P3DDT:PC$_{61}$BM composite was also made.

To analyze the nature of all paramagnetic centers in both the PANI:SA/P3DDT:PC$_{61}$BM and PANI:pTSA/P3DDT:PC$_{61}$BM composites, first related study of spin properties of their ingredients was done (Krinichnyi, Yudanova, and Wessling 2013, Krinichnyi 2014b, a, 2016a).

Initial PANI:SA and PANI:pTSA samples exhibit single X-band EPR spectra presented in Figure 29,a and Figure 29,b attributed to polarons $P_1^{+\bullet}$ with effective g-factor equal to 2.0031 and 2.0028 stabilized in backbone of PANI-SA and PANI:pTSA, respectively. These values remain almost unchanged within all temperature range used during long time that is characteristic of paramagnetic centers in crystalline high-conductive solids (Krinichnyi 1996, 2000b, a). As-prepared P3DDT:PC$_{61}$BM sample does not demonstrate any EPR spectrum without light irradiation (Figure 29,c). When illuminated by visible light, positively charged polaron is formed on a polymer backbone due to electron transfer to methanofullerene (see Figure 3). Thus, two partly overlapping LEPR lines are observed at low temperatures (Figure 29,c). As in case of other polymer nanocomposites (Krinichnyi et al. 2007, Krinichnyi, Yudanova, and Spitsina 2010, Krinichnyi and Balakai 2010, Krinichnyi 2016a), this spectrum was attributed to radical pairs of positively charged diffusing polarons $P_2^{+\bullet}$ with isotropic (effective) $g_{iso} = 2.0018$ and negatively charged radical anions $C_{61}^{-\bullet}$ with effective $g_{iso} = 1.9997$ rotating about the main axis. These values are close to those obtained for spin charge carriers photo induced in other polymer/fullerene BHJ (Janssen, Moses, and Sariciftci 1994, Marumoto et al. 2003, Krinichnyi 2016a). The EPR line shape due to dipole or hyperfine broadening is normally Gaussian. For spin Q3D motion in metal-like crystallites of PANI-ES or exchange of different spin packets, the line shape becomes close to Lorentzian shape, corresponding to an exponential decay of transverse magnetization. The effective LEPR spectrum presented in Figure 29,c exhibits mainly a Lorentzian doublet of mobile radical pairs $P_2^{+\bullet} - mC_{61}^{-\bullet}$, shown in Figure 29,d as a sum of equally contributed mobile polarons $P_2^{+\bullet}$ and methanofullerene radical anions $mC_{61}^{-\bullet}$ as well as a Gaussian contribution of the former charge carriers pinned by deep traps appeared in a polymer matrix under its illumination (shown in Figure 2,e). Both mobile charge carriers recombine with the probability increasing with temperature, and their effective spectrum shown in Figure 29,d becomes too weak to be registered at $T \geq 200$ K at a reasonable signal-to-noise ratio. Captured polarons $P_2^{+\bullet}$ are characterized by higher stability, and their spectrum can be observed for several hours even at high temperatures. Nevertheless, these carriers indirectly participate in collective charge transfer through BHJ.

It is evident that LEPR spectra of both PANI:SA/P3DDT:PC$_{61}$BM and PANI:pTSA/P3DDT:PC$_{61}$BM composites presented in Figure 29,f and Figure 29,g, respectively, can be considered as a sum of P3DDT:PC$_{61}$BM and appropriate PANI-ES contributions. As an illumination is turned off, the spectra originated from the polarons $P_1^{+\bullet}$ stabilized in PANI-ES (shown in Figure 29,a and Figure 29,b) and polarons $P_2^{+\bullet}$ pinned in P3DDT:PC$_{61}$BM (shown in Figure 2,e) can only be registered (see Figure 29,f and Figure 29,g). In order to study charge-separated states and spin-spin interactions in these systems more precisely, their spectra should be tentatively deconvoluted, as it was successfully done for analogous spin-modified systems (Takeda et al. 1998, Yanilkin et al. 2007, Poluektov et al. 2010, Krinichnyi 2016a). This allowed to obtain separately magnetic resonance parameters of all paramagnetic centers stabilizing in initial polymers and appropriate composites and analyze their interaction in BHJ.

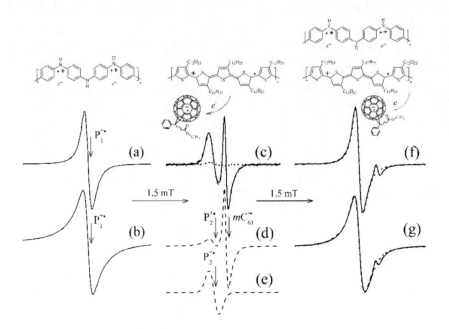

Figure 29. X-Band EPR spectra of polarons $P_1^{\bullet+}$ stabilized in fully doped with sulfuric acid, PANI:SA (*a*) and *para*-toluenesulfuric acid, PANI:pTSA (*b*), P3DDT:PC$_{61}$BM BHJ illuminated by achromatic white light with color temperature 5500 K at $T = 77$ K (*c*) with contributions due to mobile polaron-methanofullerene radical quasi-pairs $P_2^{+\bullet}$— $mC_{61}^{-\bullet}$ (*d*) and localized polarons $P_2^{+\bullet}$ pinned by deep traps (*e*) as well as PANI:SA/P3DDT:PC$_{61}$BM (*f*) and PANI:pTSA/P3DDT:PC$_{61}$BM (*g*) composites illuminated by achromatic white light with color temperature 5500 K at $T = 77$ K. Dashed line show sum EPR spectrum (*c*) and deconvoluted its contributions (*d*, *e*) calculated using $\Delta B_{pp}^P = 0.27$ mT, $\Delta B_{pp}^F = 0.12$ mT, and $[P_2^{+\bullet}]/[mC_{61}^{-\bullet}] = 2.0$. In (*c*, *f*, *g*), dotted lines exhibit "dark" EPR spectra of PC stabilized in appropriate composites, whereas the spectra calculated at their illumination by white light at 77 K are shown by dashed lines. The positions of paramagnetic centers are shown as well. Structure of PANI-ES fully doped with sulfuric (A$^-$ ≡ SA, *a*) and (A$^-$ ≡ pTSA, *b*) acids, regioregular P3DDT (*c*), and PC$_{61}$BM (*d*). The appearance of positive charged polaron in the P3DDT backbone due to photo initiated charge separation and transfer to the fullerene globe is shown schematically.

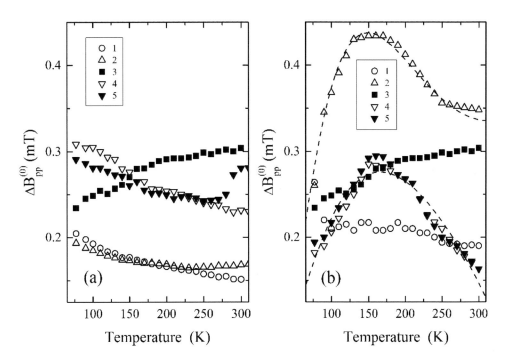

Figure 30. Temperature dependence of peak-to-peak linewidth $\Delta B_{pp}^{(0)}$ determined for domestic polarons $P_1^{+\bullet}$ stabilized in the initial PANI-ES backbones (1), PANI-ES/P3DDT:PC$_{61}$BM composites (2), polarons $P_2^{+\bullet}$ background photo initiated by white light in the P3DDT:PC$_{61}$BM BHJ (3), as well as polarons $P_2^{+\bullet}$ stabilized in the darkened (4) and irradiated by white light (5) PANI-ES/P3DDT:PC$_{61}$BM composites with A$^-$ ≡ SA (a) and A$^-$ ≡ pTSA (b) counter-ions. The upper (0) symbol in $\Delta B_{pp}^{(0)}$ implies this parameter to be measured far from the spectrum microwave saturation. Dashed lines show the dependences calculated from Eq.(A.19) with ω_{hop}^0 = 1.2×10^9 s^{-1}, E_r = 0.006 eV (above line), ω_{hop}^0 = 1.3×10^9 s^{-1}, E_r = 0.12 eV (below line), J_{ex} = 0.110 eV, and n_g = 1.2×10^{-4}.

Figure 30,a shows temperature dependences of effective absorption peak-to-peak linewidth ΔB_{pp} of polarons $P_1^{+\bullet}$ stabilized in the PANI:SA, $P_2^{+\bullet}$ photo initiated in P3DDT:PC$_{61}$BM BHJ, and these values obtained for darkened and illuminated PANI:SA/P3DDT:PC$_{61}$BM composite. It is seen that the EPR linewidth for both polarons stabilized in these systems depends on structure of a polymer matrix. Indeed, the heating of the initial PANI:SA samples is accompanied by a monotonic decrease in ΔB_{pp} of polarons $P_1^{+\bullet}$ stabilized on their chains. However, this parameter for polarons $P_2^{+\bullet}$ photo initiated in the P3DDT:PC$_{61}$BM BHJ evidences an opposite temperature dependence as compared with that for polarons $P_1^{+\bullet}$ (Figure 30). This effect can be explained by different interaction of these polarons with appropriate polymer lattice (see below). The formation of the PANI:SA/P3DDT:PC$_{61}$BM composite does not noticeably changes the linewidth for paramagnetic centers $P_1^{+\bullet}$. However, this originates the change in the temperature dependence of $P_2^{+\bullet}$ charge carriers photo initiated in the P3DDT matrix (see Figure 30,a).

Spin properties of both polaronic reservoirs in the PANI-ES/P3DDT:PC$_{61}$BM composites are strongly governed by the conformation of PANI-ES chains which determines its main electronic properties (Wessling 2010). As the PANI:SA matrix is replaced by the PANI:pTSA one, both polarons $P_1^{+\bullet}$ and $P_2^{+\bullet}$ start to demonstrate extreme temperature dependent linewidths characterized by appropriate critical point $T_{ex} \approx 150$ K. A similar effect was observed in the EPR study of exchange interaction of polarons with guest oxygen biradicals $^\bullet O - O^\bullet$ in PANI-ES highly doped with hydrochloric acid (Houze and Nechtschein 1996) and pTSA (Krinichnyi, Roth, et al. 2006, Krinichnyi, Tokarev, et al. 2006, Krinichnyi 2014a). This effect was identified as exchange interaction in quasi-pairs formed by guest spins with domestic mobile polarons hopping between sites of polymer chain with rate ω_{hop} across energy barrier E_r. Thus, the data, presented in Figure 30,b can perfectly be described in terms of the spin-spin exchange interaction of polarons $P_1^{+\bullet}$ and $P_2^{+\bullet}$ hopping in the nearly located solitary polymer chains. The collision of both type spins should additionally broaden the absorption term of EPR line by the value described by Eq.(A.19). In this case the extreme temperature dependence should indicate strong and weak spin interaction limits at $T \leq T_c$ and $T \geq T_c$, respectively ($T_c = 150 - 160$ K). In particular, it is an evidence of a stronger interaction of polarons with own microenvironment in P3DDT as compared with PANI-ES.

Assuming activation character of polaron motion in both polymer matrices and $n_P = 1.2 \times 10^{-4}$ obtained for P3DDT:PC$_{61}$BM BHJ (Krinichnyi, Yudanova, and Spitsina 2010), the linewidth of the polarons $P_1^{+\bullet}$ and $P_2^{+\bullet}$ can be fitted by Eq.(A.19) with $E_r = 0.006$ and 0.012 eV, respectively, at $J_{ex} = 0.110$ eV (see Figure 30,b). J_{ex} is less than that (0.360 eV) determined for air-filled PANI:pTSA (Krinichnyi, Roth, et al. 2006, Krinichnyi, Tokarev, et al. 2006) due probably to less number of guest radical and higher interpolaron distance.

Figure 31,a and Figure 31,b show the temperature dependence of spin susceptibility χ with contributions due to polarons $P_1^{+\bullet}$, $P_2^{+\bullet}$ and methanofullerene radical anions $mC_{61}^{-\bullet}$ forming spin pairs in the P3DDT:PC$_{61}$BM BHJ stabilized in the darkened and background illuminated PANI:SA/P3DDT:PC$_{61}$BM and PANI:pTSA/P3DDT:PC$_{61}$BM composites. The dependences obtained are important to reveal mobile or localized character of spins and their possible interaction. Indeed, for non-interacting and localized (or slightly delocalized) electrons in disordered phase, susceptibility normally follows the Curie law $\chi_C \propto 1/T$, whereas polarons delocalized in the conduction band of ordered crystallites cause temperature-independent Pauli behavior, χ_P. However, such simple picture has been questioned especially for polyaniline and other conducting polymers because most of the spins are localized (Mizoguchi and Kuroda 1997, Raghunathan et al. 1998). Disorder localizes electron spins and conducting polymer systems exhibit significant disorder. Therefore, we have to interpret the results obtained for the composites in terms of the model of exchange of $N_s/2$ coupled spin pairs with an uniform distribution originating the last term in Eq.(A.4). It is seen that the curves calculated from Eq.(A.4) with C, a_d, and J values determined for polarons carrying a charge in the initial PANI-ES and respective PANI-ES/P3DDT:PC$_{61}$BM composites (Krinichnyi, Yudanova, and Wessling 2013) provide excellent fits to all the experimental data sets within all temperature range used. Spin susceptibility obtained for $P_1^{+\bullet}$ is close to that obtained for PANI highly doped by sulfuric (Kahol 2000) and hydrochloric (Wang, Ray, et al. 1991) acids. The latter parameter is normally a function of distance. When polymer chains vibrate, J for polarons diffusing along neighboring chains would oscillate and should be described by a stochastic process (Adrain and Monchick 1979). However, such effect appears at low temperatures, when $k_B T \ll J$. Thus, it can be neglected within all temperature range used. Nevertheless, this constant increases as polarons $P_1^{+\bullet}$ start to interact with polarons $P_2^{+\bullet}$ and, on the side, SA counter ions are replaced by pTSA ones (Krinichnyi, Yudanova, and Wessling 2013). This is additional evidence of strong interaction of polarons stabilized in both PANI-ES and P3DDT matrices. When the Fermi energy ε_F is close to the mobility edge, the temperature dependence of spin susceptibility gradually changes from Curie-law behavior $\chi_C \propto 1/T$ to temperature-independent Pauli-type behavior with increasing temperature. Corresponding density of states $n(\varepsilon_F)$ for both spin directions per monomer unit at ε_F can be determined from the analysis of the $\chi(T)T$ dependence for all polarons stabilized in both polymers (see inserts in Figure 31). It is seen that the pTSA-treated system is characterized by higher $n(\varepsilon_F)$ as compared with PANI:SA. This can be explained by the difference in their above mentioned metallic properties and also by on-site electron-electron interaction (Ginder et al. 1987).

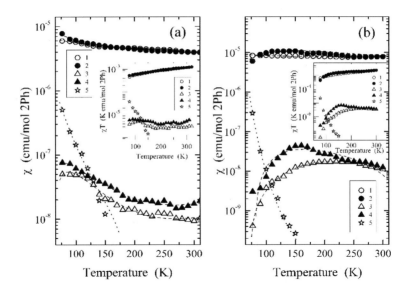

Figure 31. Temperature dependences of spin susceptibility χ and χT product (inserts) obtained for domestic polarons $P_1^{+\bullet}$ stabilized in the initial PANI-ES network (1) and respective PANI-ES/P3DDT:PC$_{61}$BM composite (2), polarons $P_2^{+\bullet}$ stabilized in the darkened (3) and illuminated by white light (4) PANI-ES/P3DDT/PCBM composites, as well as methanofullerene radical anions $mC_{61}^{-\bullet}$ (5) photo initiated in these composites with SA (a) and pTSA (b) counter-ions. Dashed lines show the dependences calculated from last term of Eq.(A.4) with respective C, a_d, and J constants. The dotted lines show the dependences calculated from Eq.(A.12) with $E_r = 0.024$ (a) and 0.050 (b) eV.

Spin susceptibility obtained for methanofullerene radical anions $mC_{61}^{-\bullet}$ in both composites demonstrates sharper temperature dependence (Figure 31). This can be explained by fast recombination of the polaron-methanofullerene radical pairs $P_2^{+\bullet} - mC_{61}^{-\bullet}$ whom spectrum is shown in Figure. 29,d. During background illumination of the P3DDT:PC$_{61}$BM BHJ, two processes are realized simultaneously, namely, photo initiation and recombination of spin pairs. As a result, we detected only net (effective) spin concentration. Therefore, effective paramagnetic susceptibility of both charge carriers photo initiated in P3DDT:PC$_{61}$BM BHJ should inversely depend on the probability of their recombination. Such process is also governed by multi-stage activation Q1D polaron hopping between polymer units (Yan et al. 2000). Positive charge on a polaron is not required to be recombined with the first negatively charged methanofullerene radical anion. Activation traveling of a polaron near such a center localized near a polymer chain should interact with its unpaired electron with the above probability p_f expressed by Eq.(A.11). In this case, effective spin susceptibility of such interacting spin sub-pairs can finally be described by Eq.(A.12). The dependences calculated from Eq.(A.12) with E_r = 0.024 and 0.050 eV are also presented in Figure 31. The latter value is higher than E_r = 0.012 eV obtained from Eq.(A.19) for polaron diffusion in the respective system. It can probably be explained by more complex exchange interaction of methanofullerene radical anion with polarons in both the PANI:pTSA and P3DDT backbones. Equation (A.12) fits well the experimental data presented in Figure 31. Therefore, the decay of long-lived charge carriers originated from initial spin pairs photo induced in the PANI-ES/P3DDT:PC$_{61}$BM composites can successfully be described in terms of the above model in which the low-temperature recombination rate is strongly governed by temperature and the width of energy distribution of trap sites.

It is seen from Figure 31 that spin susceptibility of polarons $P_1^{+\bullet}$ stabilized in both initial PANI-ES samples is characterized by weak temperature dependence without any anomaly. This also holds for polarons $P_2^{+\bullet}$ photo initiated in the PANI:SA/P3DDT:PC$_{61}$BM composite. The shape of $\chi(T)$ changes dramatically as SA counter ions are replaced by pTSA ones. Such a replacement provokes extremal χ vs. T dependence obtained for polarons $P_1^{+\bullet}$ and $P_2^{+\bullet}$ (see Figure 31,b). This is evidence of the above mentioned exchange interaction between these polarons formed on neighboring PANI and P3DDT chains. Such interaction increases the overlapping of their wave functions (which, however, slightly decreases at further light flashing) and the energy barrier which overcomes the polaron crossing a BHJ. This affects the polaron intrachain mobility and, therefore, probability of its recombination with fullerene anion. However, the character of the $mC_{61}^{-\bullet}$ quasi-rotation changes weakly under such a replacement (see Figure 31).

The analysis of electron spin relaxation of charge carriers can expand the information about spin localization, matrix dimensionality and spin-assisted electronic processes carrying out in the polymer systems. The EPR linewidth is determined mainly by spin-spin relaxation time. Spin-lattice relaxation also shortens the lifetime of a spin state and broadens the line. So, the effective linewidth can be expressed by Eq.(A.16) taking into account all contributions due to the structural, conformational and electronic properties of polarons' microenvironment. So, it would be important to analyze also how spin exchange affect spin-lattice relaxation. The interaction of polarons $P_1^{+\bullet}$ stabilized in the PANI:SA matrix was appeared to be depending weakly on the presence of guest spins. Their dominantly contribute in an effective spin susceptibility of the composites under study. So, a spin-lattice relaxation of these charge carriers in the PANI:pTSA and PANI:pTSA/P3DDT:PC$_{61}$BM BHJ can be analyzed with more degree of certainty.

Figure 32 exhibits temperature dependencies of both the T_1 and T_2 values of polarons $P_1^{+\bullet}$ stabilized in shadowed PANI:pTSA and PANI:pTSA/P3DDT:PC$_{61}$BM composites. Spin-spin relaxation was shown above to be governed by the spin-spin exchange interaction. RT spin-lattice relaxation time of the samples was measured to be 0.45×10^{-7} and 0.33×10^{-7} s, respectively. These values are in good agreement with T_1 = 0.98×10^{-7} s obtained by Wang at all. (Wang et al. 1992) for polyaniline highly doped by hydrochloric acid. It is seen, that spin-lattice relaxation of $P_1^{+\bullet}$ stabilized in the initial polymer changes weakly as the temperature increases up to ~180 K that is typical for organic ordered systems. This process accelerating suddenly near T ~ 210 K possibly to a phase transition and then plateaus at higher temperatures. The latter value differs from T_c mentioned above because they characterize different processes. As $P_1^{+\bullet}$ start to interact with other paramagnetic centers in the PANI:pTSA/P3DDT:PC$_{61}$BM composite, their spin-lattice relaxation strongly accelerates and becomes more temperature-dependent (Figure 32). It is once more evidence of the exchange between polarons stabilized in neighboring polymer chains. Figure 32 demonstrates then T_1 tends to T_2 at high temperatures. This is typical for organic systems of lowed dimensionality and can be explained by the defrosting of macromolecular dynamics.

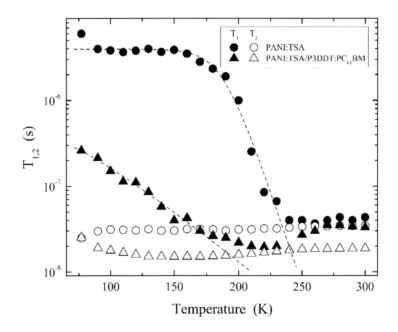

Figure 32. Temperature dependence of spin-lattice, T_1, and spin-spin, T_2, relaxation times determined for polarons $P_1^{+\bullet}$ stabilized in the PANI:pTSA backbone (*1*) and respective PANI:pTSA/P3DDT:PC$_{61}$BM composite (*2*) without light illumination.

So, light excitation of P3DDT:PC$_{61}$BM BHJ in the PANI-ES/P3DDT:PC$_{61}$BM composites leads to charge separation and transfer from a P3DDT chain to a methanofullerene molecule. This is accompanied by the excitation of two paramagnetic centers with clearly resolved LEPR spectra, namely, the positively charged polaron $P_2^{+\bullet}$ on the polymer backbone and the negatively charged radical anion $mC_{61}^{-\bullet}$ located between polymer chains. Both radicals are spatially separated due to fast Q1D diffusion of the former, so that they become non-interacting. After the forming and separation, spin charge carriers tend to recombine in polymer backbone. This process is governed by structure and morphology of polymer matrix and anion radical inserted. Polarons $P_2^{+\bullet}$ moving in P3DDT solitary chains interact with $P_1^{+\bullet}$ stabilized on neighboring PANI-ES chains due to overlapping of their wave functions. Such interaction is governed mainly by nanomorphology of PANI-ES subdomains and defines insulating and conducting forms of PANI:SA and PANI:pTSA, respectively. Spin exchange and transverse relaxation of polarons is governed by Q1D activation hopping $P_2^{+\bullet}$ along P3DDT chains and strongly increase as PANI:SA polymer is replaced by PANI:pTSA one in the triple composite. Spin-lattice relaxation of polarons stabilized in PANI:pTSA is also accelerated by their exchange interaction with guest spin ensemble. Paramagnetic susceptibility of these polarons is realized according the model of exchange coupled spin pairs differently distributed in appropriate polymer matrices. This deepens overlapping of wave functions of these charge carriers and leads to the increase in the energy barrier which overcomes the polaron under its crossing through a BHJ. It is evident that separate EPR investigation of spin properties of domestic and photo excited paramagnetic centers in the polymer:dopant/polymer:fullerene composite and its ingredients may give a possibility to control its texture and other structural properties over the entire range of temperatures studied. The data obtained for such model systems can contribute to open new horizon in creation of flexible and scalable organic molecular devices with spin-assisted electronic properties. The results described suggest an important role played by interchain coupling of different spin charge carriers on a handling of charge transfer in BHJ of PANI-ES/P3DDT:PC$_{61}$BM composite. Photo initiation of additional spins allows making such handling more delicate that is a critical strategy in creating systems with spin-assisted charge transfer. The correlations established between dynamics, electronic and structural parameters of these systems can be used for controllable synthesis of various organic spintronic devices with optimal properties. Since coherent spin dynamics in organic semiconductors is anisotropic, our strategy seems to make it possible obtaining complex correlations of anisotropic electron transport and spin dynamics for the further design of progressive molecular electronics. Electronic properties of such devices seem to be improved by the use of more ordered composites. The use, for example, of PCDTBT modified by C$_{70}$-based methanofullerene instead of the P3DDT:PC$_{60}$BM system described should facilitate the excitation to reach the polymer:fullerene interface for charge separation before it becomes spatially self-localized and bound within an exciton (Moon, Jo, and Heeger 2012). Therefore, the main properties of an exciton are irrelevant to ultrafast charge transfer and do not limit effective charge transfer in such composite.

It was shown that charge transfer in the PCDTBT:PC$_{61}$BM composite depends not only on the polymer structure and morphology but also on the number and nature of guest spin ensemble. In order to obtain all parameters of such system in details, the main magnetic resonance and dynamics parameters of PC in its ingredients should be first studied. Figure 33,a shows X-band EPR dark spectrum of the PANI:pTSA measured at $T = 77$ K. This spectrum consists of two components contributed by polarons $P_{11}^{+\bullet}$ and $P_{12}^{+\bullet}$ stabilized in different phases of polymer backbone. Linewidth and concentration of these PC differ by three and eight times, respectively. X-band LESR sum spectrum and its contributions due to polarons pinned by deep traps $P_2^{+\bullet}$ and mobile polaron-methanofullerene radical quasi-pairs $P_2^{+\bullet} - mC_{61}^{-\bullet}$ photo initiated by white light with color temperature 5500 K in the PCDTBT:PC$_{61}$BM composite are shown in the same Figure. Sum LEPR spectrum and the

EPR Spectroscopy of Polarons in Conjugated Polymers and Their Nanocomposites 67

contributions of these PC photo initiated in the PANI:pTSA/PCDTBT:PC$_{61}$BM composite are depicted in Figure 33,c. It should be noted some conclusions from the analysis of the data presented. First of them is the increase of relative number of polarons $P_{12}^{+\bullet}$ stabilized in crystalline, more ordered PANI phase of such multispin composite. An interaction of mobile fullerene anion radicals with other spin ensembles decreases in this complex system. These effects can be due to the increase in exchange interaction between all spin ensembles stabilized and photo initiated in the polymer composite. They can be used for the handling of charge transport in polymer composites by using of spin interactions. Electronic properties of multispin composite are also governed by the energy of initiated photons $h\nu_{ph}$.

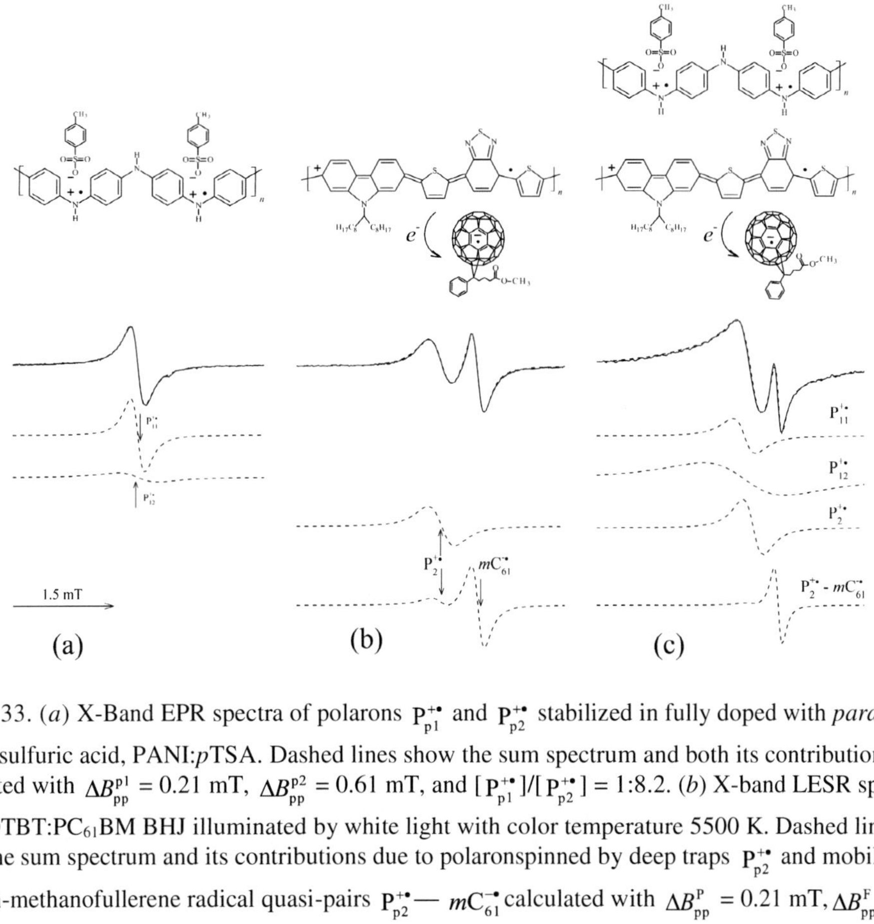

Figure 33. (*a*) X-Band EPR spectra of polarons $P_{p1}^{+\bullet}$ and $P_{p2}^{+\bullet}$ stabilized in fully doped with *para*-toluenesulfuric acid, PANI:pTSA. Dashed lines show the sum spectrum and both its contributions calculated with $\Delta B_{pp}^{p1} = 0.21$ mT, $\Delta B_{pp}^{p2} = 0.61$ mT, and $[P_{p1}^{+\bullet}]/[P_{p2}^{+\bullet}] = 1:8.2$. (*b*) X-band LESR spectrum of PCDTBT:PC$_{61}$BM BHJ illuminated by white light with color temperature 5500 K. Dashed lines show the sum spectrum and its contributions due to polaronspinned by deep traps $P_{p2}^{+\bullet}$ and mobile polaron-methanofullerene radical quasi-pairs $P_{p2}^{+\bullet}$— $mC_{61}^{-\bullet}$ calculated with $\Delta B_{pp}^{P} = 0.21$ mT, $\Delta B_{pp}^{F} = 0.13$ mT, $[P_{p2}^{+\bullet}]/[mC_{61}^{-\bullet}] = 1.1:1.0$, and respective g-factors presented in Table 1. (*c*) X-band LESR spectrum of PANI:pTSA/PCDTBT:PC$_{61}$BM multispin composite illuminated by achromatic white light with color temperature 5500 K at $T = 77$ K. Dashed lines show the sum spectrum and it's the above contributions with $\Delta B_{pp}^{p1} = 0.33$ mT, $\Delta B_{pp}^{p2} = 1.22$ mT, $\Delta B_{pp}^{P} = 0.28$ mT, $\Delta B_{pp}^{F} = 0.12$ mT, and $[P_{p1}^{+\bullet}]:[P_{p2}^{+\bullet}]:[P_{p2}^{+\bullet}]:[mC_{61}^{-\bullet}] = 51.4:8556:5.4:1.0$, and respective g-factors presented in Table 1. The positions of paramagnetic centers are shown as well. Structures of the PANI:pTSA (*a*), PCDTBT:PC$_{61}$BM BHJ (*b*), as well as multispin composite PANI:pTSA/PCDTBT:PC$_{61}$BM (*c*) are shown schematically. The formation of positive charged polaron in the PCDTBT backbone due to photoinitiated charge separation and transfer to the fullerene globe is also shown schematically (*b,c*).

Figure 34 demonstrates the dependence of the main properties of polaronic charge carriers in the PANI:pTSA/PCDTBT:PC$_{61}$BM composite on the photon energy. One can to note non-linear dependence of the number of mobile and pinned polarons photoinitiated in the PCDTBT:PC$_{61}$BM BHJ on hv_{ph}. This parameter of both charge carriers demonstrates dependences with explicit extremes lying near 1.8 and 2.7 eV. The former value lies near the PCDTBT bandgap, $2\Delta = 1.87$ eV (Park et al. 2009, Kim et al. 2014). Such a peculiarity can probably be as result of specific morphology and band structure of the samples with inhomogeneously distributed spin traps. It is quite clear that the higher concentration ratio $[P_{mob}^{+\bullet}]/[P_{loc}^{+\bullet}]$, the better efficiency of energy conversion should be expected for respective system. This ratio weakly depends on the light photon energy. Once PANI:pTSA is introduced into the composite, this ratio becomes more sensitive to the photon energy with extremum at $hv_{ph} \approx 1.8$ eV. Besides, the insert of the guest spin ensemble into the initial composite increases the selectivity of its polarons to the photon energy. This can be also used in novel elements of polymer electronics for spin-assisted energy conversion.

Figure 34 demonstrates how polaron dynamics in both composites is governed by photon energy and changes due to spin-spin interaction. Both the constants of spin intrachain and interchain diffusion in the PCDTBT:PC$_{61}$BM composite are shown in Figure 34,c as function of hv_{ph}. These values as well as the anisotropy of spin dynamics (D_{1D}/D_{3D}) demonstrate weak dependence on the photon energy. Dynamics of all spin packets becomes sensitive to the photon energy in the PANI:pTSA/PCDTBT:PC$_{61}$BM BHJ (Figure 34,d). It is seen that all the constants of spin diffusion change extremely with characteristic energy $hv_{ph} \approx 1.8 - 2.0$ eV. The system ordering or dimensionality proportional to the ratio D_{3D}/D_{1D} changes extremely in this point as well. Such effect can be used for constructing of molecular device with spin-assisted charge transport properties. It should be noted that the main change in spin concentration and dynamics occurs at $hv_{ph} \approx 1.8$ eV. According the Eq.(A.34), the effective charge transport is determined by the product of the number of charge carriers by their mobility. Therefore, the photons with such energy should mainly influence spin-controlled charge transport in the multispin composite described.

So, the results described show that the main part of these carriers transfers the charge through PCDTBT:PC$_{61}$BM BHJ, whereas some quantity of polarons is captured by deep spin traps reversibly initiated in polymer backbone. The spatial distribution, number, and energy depth of such traps depend on a structure, morphology of polymer matrix as well as on the energy of initiating photons. Magnetic resonance, relaxation and dynamics parameters of mobile and fixed charge carriers are governed by their exchange interaction and, therefore, all the process carrying out in the copolymer composites become spin-assisted. Besides, these parameters were shown to be governed by the number and energy of initiating photons. Concentration of both charge carriers in the systems studied shows extreme dependence on the photon energy with explicit extremes around 1.8 and 2.8 eV. Such a peculiarity can appears as a result of specific morphology and band structure of the composite matrix with spin traps inhomogeneously distributed in its bulk. Dynamics of polarons photoinitiated in the PCDTBT:PC$_{61}$BM composite weakly depends on the photon energy. At the inserting of guest spin ensemble into the PCDTBT:PC$_{61}$BM composite, dynamics parameters become stronger dependent on the photon energy demonstrating extremes around 1.8 eV. The sensitivity of polaron properties to the energy of photons can be used for creation of perspective molecular electronic elements with spin-light-assisted magnetic and electronic characteristics. The methodology described can be used also for the study of electronic properties of other organic multispin polymer composites.

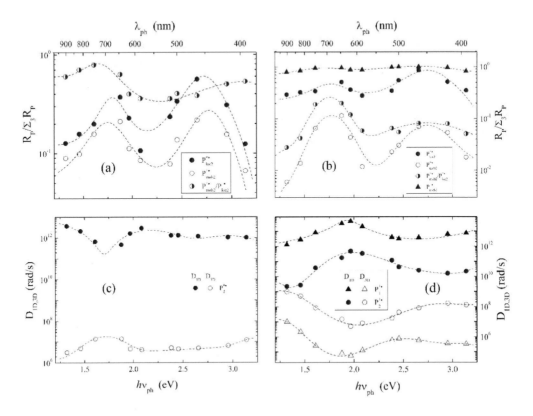

Figure 34. Relative concentrations of polarons, $[P^{+\bullet}_{1,2}]/\Sigma[P^{+\bullet}_i]$ and $[P^{+\bullet}_{mob2}]/[P^{+\bullet}_{loc2}]$ (a,b) as well as intrachain, D_{1D}, and interchain, D_{3D}, diffusion coefficients (c,d) of polarons $P^{+\bullet}_1$ and $P^{+\bullet}_2$ photoinduced in the PCDTBT:PC$_{61}$BM (a,c) and PANI:TSA/PCDTBT:PC$_{61}$BM (b,d) composites as a function of photon energy $h\nu_{ph}$. The values obtained were normalized to the luminous emittance of the light sourses. The dashed lines are drawn arbitrarily only for illustration to guide the eye.

CONCLUSION

The results presented show principal advantage of the multifrequency EPR spectroscopy in obtaining qualitatively new information on molecular dynamics, transfer mechanism of polarons in conjugated polymers and their nanocomposites. The method enables to detect such charge carriers with close magnetic resonant characteristics, to provide their reliable identification and to establish the correlation of their structural, dynamics and magnetic parameters in such systems. Besides, the interaction between different spin-packets is waned significantly in high magnetic fields, so then they may be considered as noninteracting and their parameters can be determined more precessionaly. It takes qualitatively new information on metrology, relaxation and mobility of polarons as well as matrix's morphology, polarity, etc.

Polarons are stabilized in conjugated polymers as dimerization distortion of their chains. They possess a spin, elemental charge and are characterized by high mobility along polymer chains. The main properties of such charge carriers depend on various factors, e.g., method of synthesis, structure and morphology of an initial and modified polymer. The pairing of polarons into diamagnetic bipolarons with slower dynamics and higher effective mass can occurs with the increase of doping level of a polymer. Besides, the doping process leads to the change in charge transfer mechanism. Spin and charge dynamics in initial and weakly nanomodified conjugated polymer is realized mainly by interchain tunneling of the "small polarons," which highly interact with phonons of polymer lattice. Once the doping level of a polymer increases and its matrix becomes more ordered, these mechanisms ceases to dominate and the charge can be transferred in result of its thermal activation from widely separated localized states in the bandgap to close localized states in the tails of both the valence and conjugated bands. This stipulates the formation in some matrices of complex quasi-particles, namely the molecular-lattice polarons, due to libron-phonon interactions analogously to that as it is realized in organic molecular crystals. Polaron dynamics in conjugated polymers are *a priori* highly anisotropic due their lower dimensionality. In heavily doped samples crytalline Q3D clasters are formed in bulk of polymer matrix. Polaron transport in such systems is realized in terms of Q1D spin hopping between Q3D clasters characterizing by its strong interaction with lattice phonons. Polarons diffusing along a polymer chains interact with other spin charge carriers stabilized on neighboring chains and/or with radicals embedded into polymer matrix as counter anions. Such exchange interaction complicates charge transport in polymer:fullerene and polymer:dopant/polymer:fullerene multispin nanocomposites. Charge dynamics in these systems is also realized in the above mentioned mechanisms which tangled by spin exchange interaction. These spin-assisted processes are governed mainly by the structure of ingredients of a composite as well as by the nature and dynamics of photoinduced charge carriers. The specific morphology of polymer nanocomposites changes their energy levels and shifts the competition between excited states in their BHJ. Correlations obtained for polymer:fullerene and polymer:dopant/polymer:fullerene composites and their ingredients by direct EPR method can be used for a further development and optimization of nanomodified polymer photovoltaic devices.

The spin of a polaron plays a key role in coupling of different polarons stabilized in close polymer chains. This can be used for a handling of charge transfer through BHJ of multispin nanocomposite. Such spin-assisted charge transfer can be more delicate under initiation of polarons by different photons. The correlations established between dynamics, electronic and structural parameters of these systems at wide region of spin precession frequency can be used for the further controllable synthesis of various conjugated polymers with optimal properties. This will open new horizon in creation of flexible and scalable organic devices with field- and spin-assisted electronic properties. Such organic systems can also play the role of efficient spin injectors and detectors, which can potentially eliminate ferromagnetic materials in spin devices. Solitary spin carriers trapped in bulk of darkened polymer matrix can in principle be used as elemental dots for quantum computing. Since coherent spin dynamics in such organic BHJ is anisotropic, our strategy seems to make it possible obtaining complex correlations of anisotropic electron transport and spin dynamics for the further design of progressive molecular electronics and spintronics.

APPENDIX

EPR method is based on the excitation of unpaired electron spin to the higher energy level upon an external magnetic field and MW radiation. Electrons possess a property called "spin," resulting in an angular momentum. Because the electron is charged, there is associated with the angular momentum a magnetic moment which points in the opposite direction to the angular momentum vector. In an external magnetic field, the spin precesses around the field direction at the Larmor frequency and thus a component of the magnetic moment is either parallel or anti-parallel to the field direction. If a microwave field of this frequency is applied to a spin containing sample, then the spins can change their direction relative to the magnetic field. This results in absorption of the microwave field, which may be measured. Spin reorientation is also affected by microenvironment. Microwave absorption depends on fundamental properties of spin reservoir described above. Thus, EPR spectra can yield detail information not only about spin properties of a sample but also about its structure and composition.

A.1. Theoretical Backgrounds of Electron Paramagnetic Resonance of Polarons in Conjugated Polymers

A.1.1. Landé Factor

The main magnetic resonance parameters directly obtained by the continuous wave (CW) EPR spectroscopy for paramagnetic centers in condensed systems are the Landé factor (or g-factor that is the ratio of electron mechanic momentum to a magnetic moment), spin susceptibility, as well as line shape and width. The first of them is characterized by the Zeeman interaction of an unpaired electron with external magnetic field. If the fundamental resonance condition (Assenheim 2014).

$$\hbar\,\omega_e = \gamma_e\,h\,B_0 = g\mu_B B_0 \tag{A.1}$$

is fulfilled (here μ_B is the Bohr magneton), an unpaired electron absorbs an energy quantum and is transferred to a higher excited state. It can be seen that the higher B_0 (or ω_e) value, the higher excited state an electron can reach and the higher spectral resolution can therefore be realized. It is stipulated by the distribution of spin density in polymer unit, the energy of exited configurations and its interaction with nearest nuclear. If spin of polaron weakly interacts with own environments, its Landé-factor lies near g-factor of free electron, $g_e = 2.00232$. At higher interaction, environmental nuclei induce an additional magnetic field resulting tensoric character of its Landé-factor (Buchachenko, Turton, and Turton 1995, Misra 2011, 2014),

$$\mathbf{g} = \begin{vmatrix} g_{xx} & & \\ & g_{yy} & \\ & & g_{zz} \end{vmatrix} = \begin{vmatrix} 2\left(1+\dfrac{\lambda\rho(0)}{\Delta E_{n\pi^*}}\right) & & \\ & 2\left(1+\dfrac{\lambda\rho(0)}{\Delta E_{\sigma\pi^*}}\right) & \\ & & 2 \end{vmatrix}, \tag{A.2}$$

where λ is the spin-orbit coupling constant, $\rho(0)$ is the spin density, $\Delta E_{n\pi^*}$ and $\Delta E_{\sigma\pi^*}$ are the energies of the unpaired electron $n{\to}\pi^*$ and $\sigma{\to}\pi^*$ transitions, respectively. Normally, polarons in organic conjugated polymers require a small energy of $n{\to}\pi^*$ transition. This leads to deviation of its g_{xx} and g_{yy} values from g_e, so then the inequality $g_{xx} > g_{yy} > g_{zz} \approx g_e$ holds for these PC.

Weak interaction of an unpaired electron delocalized on polaron over L lattice units with heteroatoms involved in a polymer backbone provokes rhombic symmetry of spin density and, therefore, anisotropy of its magnetic resonance parameters. Since the backbone of a polymer can be expected to lie preferably parallel to the film substrate (Kim et al. 2006), the lowest principal g-value is associated with the polymer backbone. The macromolecule can take any orientation relative to the z-axis, i.e., the polymer backbone direction as is derives from the presence of both the g_{xx} and g_{yy} components in the spectra for all orientations of the film. Thus, the g-factor anisotropy is the result of inhomogeneous distribution of additional fields in such systems along the x and y directions within the plane of their σ-skeleton rather than along its perpendicular z direction. Multifrequency EPR spectroscopy allows to resolve some paramagnetic centers with near g-factors or spectral components of paramagnetic centers with anisotropic g-factor (Krinichnyi 1995, 2000a, 2006, Misra 2011, 2014). Harmonic librations of polymer chains with localized polarons can modulate the charge transfer integrals in polymer composites as it is typical for organic molecular ordered systems (Silinsh, Kurik, and Chapek 1988). This should change effective g-factor as

$$g = g_0 + \frac{A}{\hbar\omega_l}\coth\left(\frac{\hbar\omega_l}{2k_B T}\right), \tag{A.3}$$

where g_0 and A are constants, $\omega_l = \omega_0\exp(-E_l/k_B T)$ is librational frequency, E_l is the energy required for activation of such a motion, and T is the temperature.

A.1.2. Spin Susceptibility

A static paramagnetic susceptibility χ is also important characteristic of a paramagnetic system. Generally, this parameter of N spins consists of temperature independent Pauli susceptibility of the Fermi gas χ_P and temperature dependent contributions of localized Curie paramagnetic centers χ_C (Vonsovskii 1974). However, such simple picture has been questioned especially for conjugated polymers and their composites. These systems are characterized by significant disorder which localizes spins (Mizoguchi and Kuroda 1997, Raghunathan et al. 1998). This originates the appearance in effective χ of additional contribution χ_{ST} coming due to a possible singlet-triplet spin equilibrium in the system

(Vonsovskii 1974), contribution χ_{ECP} described in terms of an exchange coupled pairs (ECP) model of spin exchange interaction in pairs randomly distributed in a polymer matrix (Kahol and Mehring 1986, Clark and Tippie 1979) and contribution χ_m coming due to polaron Q1D mobility characterizing by mid-gap energy $E_g = 2E_a$ near the Fermi level ε_F (Jonston 1984, Barnes 1981). Finally, one can write the equation for sum χ as

$$\chi=\chi_P+\chi_C+\chi_{ST}+\chi_m+\chi_{ECP}= N_A\mu_{eff}^2 n(\varepsilon_F) + \frac{C}{3k_BT} + \frac{k_1}{T}\left[\frac{\exp(-J_{af}/k_BT)}{1+3\exp(-J_{af}/k_BT)}\right]^2 +$$

$$\frac{Ca_d}{3k_BT}\left[3+\exp\left(-\frac{2J}{k_BT}\right)\right]^{-1} + C(1-a_d)\left\{\frac{J}{3k_BT} + \ln\left[3+\exp\left(-\frac{2J}{k_BT}\right)\right]\right\} + k_2\sqrt{\frac{E_a}{k_BT}}\exp\left(1-\frac{E_a}{k_BT}\right), \quad (A.4)$$

where N_A is the Avogadro's number, $n(\varepsilon_F)$ is the density of states per unit energy (in eV) for both spin orientations per monomer unit at ε_F, $\mu_{eff} = \mu_B g\sqrt{S(S+1)}$ is the effective magneton, S is a spin normally equal to ½ for paramagnetic centers in organic polymers, $C = N\mu_B^2 g^2 S(S+1)$ is the Curie constant per mole-C/mole-monomer, k_1 and k_2 are constants, J is the exchange coupling constant, and a_d is a fraction of spin pairs interacting in disordered polymer regions.. The contributions of these terms to the total paramagnetic susceptibility depend on various factors, for example, on the nature and mobility of charge carriers can vary at the system modification. A small value of J corresponds to spin localization in a strongly disordered matrix and it increases at overlapping of wave functions of interacting spins in more ordered regions.

In most polymer semiconductors, polarons are formed as very stable quasi-particles in result of their doping and/or treatment, e.g., by annealing or irradiation. Such charge carriers can also be excited on polymer chains quite reversibly, and such effect is used for conversion of solar light (Sun and Sariciftci 2005, Brabec, Scherf, and Dyakonov 2014). The treatment of polymer semiconductors modified by some electron acceptor normally leads to the transfer of electron from their chains to the acceptor that is accompanied by the formation of polarons on polymer chains and anion-radicals on acceptors. Paramagnetic carriers charged positively and negatively recombine after an irradiation down. Therefore, an effective spin susceptibility of such system is sum of these two alternating processes (Krebs 2012).

In polymer:fullerene composites, both initiated charges diffusing to the opposite electrodes must reach them prior to recombination. If these chargers after their transfer are still bound by the Coulomb potential, which is typical for the compounds with low-mobile charge carriers described here, they cannot escape from each other's attraction and will finally recombine. When the carrier dissipation distance is longer than the Coulomb radius, the excitons initiated, e.g., by light in their heterojunctions can be split into positive and negative charge carriers. To fulfill this condition, the Coulomb field must be shielded or charge carrier hopping distance must exceed the Coulomb radius. In this case charges are transferred to the electrodes either by the diffusion of appropriate carriers or by the drift induced by the electric field. In order to excite a radical pair by each photon, charge carrier transit time t_{tr} should be shorter considerably than the lifetime of a radical pair τ, i.e., $t_{tr} \ll \tau$. The former value is determined by charge carrier mobility μ, sample thickness d, and electric field E inside the film, $t_{tr} = d/\mu E$. If photocurrent is governed by the carrier drift in the applied electric field, the drift distance $l_{dr} = \mu\tau E$. If this current is governed by carrier diffusion, the diffusion distance

$l_{diff} = (D\tau)^{1/2} = (\mu\tau k_B T/e)^{1/2}$, where D is the diffusion coefficient, and e is the elemental electron charge. Thus, the $\mu\tau$ product governs the average distance passed by the charge carrier before recombination and, therefore, is an important parameter determining whether the efficiency of solar cells is limited by charge transport and recombination. The latter, generally, is described as a thermally activated bimolecular recombination (2003) which consists of temperature-independent fast and exponentially temperature-dependent slow steps (Westerling, Osterbacka, and Stubb 2002).

Let a polaron possessing a positive charge multihops along a polymer chain from one initial site i to other available site j close to a position occupied by a negatively charged fullerene globule. A charge hops easier between fullerenes than from polaron and fullerene, and an effective charge recombination is still limited by the transport of polarons towards fullerene molecules. The recombination is mainly stipulates by sequential charge transfer by polaron along a polymer chain and its transfer from polymer chain to a site occupied by a fullerene. Polaronic dynamics in undoped and slightly doped conjugated polymers is highly anisotropic (Krinichnyi 2000a). Therefore, the probability of a charge transfer along a polymer chain exceeds considerably that of its transfer between polymer macromolecules.

According to the tunneling model (Nelson 2003), positive charge on a polaron can tunnel from this carrier toward a fullerene and recombine with its negative charge during the time

$$\tau(R_{ij}^|) = \frac{\ln X}{\nu_{pn}} \exp\left(\frac{2R_{ij}^|}{a_0}\right),\tag{A.5}$$

where $R_{ij}^|$ is the spatial separation of sites i and j, a_0 is the effective localization (Bohr) radius, X isa random number between 0 and 1, and ν_{pn} is the attempt to jump frequency for positive charge tunneling from polymer chain to fullerene. The charge can also be transferred by the polaron thermally assisted multistep tunneling through energy barrier $\Delta E_{ij} = E_j - E_i$, so then

$$\chi(R_{ij}, E_{ij}) = \chi_0 \frac{\ln X}{\nu_{pp}} \exp\left(\frac{2R_{ij}}{a_0}\right) \exp\left(\frac{\Delta E_{ij}}{k_B T}\right),\tag{A.6}$$

where ν_{pp} is the attempt frequency for a hole tunneling between the polymer chains. The values in the couples ν_{pn}, ν_{pp} and $R_{ij}^|$, R_{ij} may be different due, for instance, to the different electronic orbits.

Undoubtedly, both charge carriers have different localization radii. The localization radius for a negative charged carrier should be on the order of the radius of the fullerene globule. The distance R_0 should depend, e.g., on the length of a side alkyl chain substitute in a polymer:fullerene matrix (Tanaka et al. 2007). Polaron stabilized in conjugated polymers is normally covers 3 - 5 monomer units (Devreux et al. 1987, Westerling, Osterbacka, and Stubb 2002, Niklas et al. 2013). The nearest-neighbor distance of spin pair with the typical radiative lifetime τ_0 changes with time t as

$$R_0(t) = \frac{a}{2} \ln\left(\frac{t}{\tau_0}\right).\tag{A.7}$$

EPR Spectroscopy of Polarons in Conjugated Polymers and Their Nanocomposites 75

Assuming that photoexcitation is turned off at some initial time $t_0 = 0$ at a charge carrier concentration n_0 and taking into account a time period of geminate recombination $t_1 - t_0$, one can write for concentration of charge carriers

$$n(R) = \frac{n}{1 + \dfrac{4\pi}{3} n_1 \left(R_0^3 - R_1^3 \right)}, \tag{A.8}$$

where R_0 is specified by Eq.(A.8), $R_1 = R(t_1)$ describes the distance between the nearest-neighbor charge carriers at time t_1 after which solely non-geminate recombination is assumed, and n_1 is the charge carrier concentration at time t_1. It follows from Eq.(A.8) that the time dependence of residual carrier concentration does not follow a simple exponential decay but shows a more logarithmic time behavior. After very long times, i.e., at large R_0, one obtains $n(R_0) = (3/4\pi) R_0^{-3}$ which is independent of the initial carrier density n_1 and also n_0. It follows from Eq.(A.5) that photoexcited charge carriers have comparable long lifetimes which are solely ascribed to the large distances between the remaining trapped charge carriers. The excited carrier concentration n_1 follows directly from light induced EPR (LEPR) measurements, whereas the a and τ_0 values can be guessed in a physically reasonable range. Finally, the concentration of spin pairs should follow the relation (Schultz et al. 2001)

$$\frac{n(t)}{n_0} = \frac{\dfrac{n_1}{n_0}}{1 + \left(\dfrac{n_1}{n_0} \right) \dfrac{\pi}{6} n_0 a^3 \left[\ln^3 \left(\dfrac{t}{\tau_0} \right) - \ln^3 \left(\dfrac{t_1}{\tau_0} \right) \right]}. \tag{A.9}$$

The analysis showed that the spin concentration initially photoexcited at $t = 0$ is governed by some factors. One of them is the number and distribution of spin traps inversely formed in polymer matrix under irradiation. A number and a depth of such traps depend on the photon energy $h\nu_{ph}$ (Krinichnyi 2009, Krinichnyi, Yudanova, and Spitsina 2010, Krinichnyi 2016b, a). At the latter step, polaronic charge carrier can either be retrapped by vacant trap site or recombine with electron on fullerene anion radical. Trapping and retrapping of a polaron reduces its energy that results in its localization into deeper trap and in the increase in number of localized polarons with time. So, the decay curves presented can be interpreted in terms of bulk recombination between holes and electrons during their repeated trapping into and detrapping from trap sites with different depths in energetically disordered semiconductor (Tachiya and Seki 2010). Analysing LEPR spectra, it becomes possible to separate the decay of mobile and pinned spin charge carriers excited in the composite. The traps in such a system should be characterized by different energy depths and energy distribution E_0. Polarons fast translative diffuse along a polymer backbone, and fullerene anion radicals can be considered to be immobilized between polymer chains. This approach predicts the following law for decay of charge carriers (Tachiya and Seki 2010):

$$\frac{n(t)}{n_0} = \frac{\pi \alpha \delta (1 + \alpha) v_d}{\sin(\pi \alpha)} \, t^{-\alpha}, \tag{A.10}$$

76   *Victor I. Krinichnyi*

where n_0 is the initial number of polarons at $t = 0$, δ is the gamma function, $\alpha = k_B T/E_0$, ν_d is the attemptjump frequency for polaron detrapping.

Positive charge on polaron is not required to be recombined with the first negative charge on subsequent acceptor. Thus, the probability of annihilation of charges can differ from the unit. Q1D hopping of positively charged polaron from site i to site j with the frequency ω_{hop} may collide with the acceptor located near the polymer matrix. While polaron is mobile, the molecule of acceptor can be considered as a translative fixed, but librating near own main molecular axis. In this case the spin flip-flop probability p_{ff} during a collision should depend on the amplitude of exchange and ω_{hop} value as (Molin, Salikhov, and Zamaraev 1980, Houze and Nechtschein 1996)

$$p_{ff} = k_1 \cdot \frac{\alpha^2}{1+\alpha^2},$$ (A.11)

where k_1 is constant equal to ½ and 16/27 for $S = \frac{1}{2}$ and $S = 1$, respectively, $\alpha = (3/2) \, 2\pi J_{ex}/\hbar \, \omega_{hop}$ and J_{ex} is the constant of exchange interaction of spins in a radical pair. In the polymer composites weak and strong exchange limits can be realized when the increase of ω_{hop} may result in decrease or increase in exchange frequency, respectively. If the ratio $J_{ex}/\hbar$ exceeds the frequency of collision of both types of spins, the condition of strong interaction is realized in the system leading to the direct relation of spin-spin interaction and polaron diffusion frequencies, so then $\lim(p) = 1/2$. In the opposite case $\lim(p) = 9/2$ $(\pi/\hbar)^2 (J_{ex}/\omega_{hop})^2$. It is evident that the longer both the above tunneling times or/and the lesser the probability p_{ff}, the smaller the number of ion-radical pairs possible to recombine and, therefore, higher spin susceptibility should be reached. A combination of Eq.(A.6) and Eq.(A.11) gives

$$\chi_p = \chi_{pn} + \chi_P^0 \frac{\hbar}{J_{ex}} \left(\alpha + \frac{1}{\alpha} \right) \exp\left(\frac{\Delta E_r}{k_B T} \right).$$ (A.12)

Assuming the above introduced activation character for polaron multistep hopping with the frequency $\omega_{hop} = \omega_{hop}^0 \exp(-\Delta E_r/k_B T)$ and the absence of dipole-dipole interaction between fullerene anion-radicals, one can determine ΔE_r from temperature dependences of paramagnetic susceptibility.

A.1.3. Spectral Line Shape and Width

In contrast with a solitary and isolated spin characterized by δ-function absorption spectrum, the spin interaction with own environment in a real system leads typically to the change in line shape and increase of linewidth. Analyzing the shape and intensity of experimental spectrum it is possible to obtain direct information on electronic processes in polymer systems. It is known that an electron spin is affected by local magnetic fields, induced by another nuclear and electron n r_{ij}-distanced spins (Roth, Keller, and Schneider 1984):

$$B_{\text{loc}}^2 = \frac{1}{4n}\gamma_e^2 h^2 S(S+1)\sum_{i,j}\frac{\left(1-3\cos^2\theta_{ij}\right)}{r_{ij}^6} = \frac{M_2}{3\gamma_e^2},$$ (A.13)

where M_2 is the second moment of a spectral line. If a line broadening is stipulated by local magnetic field fluctuating faster than the rate of interaction of a spin with nearest environment, the first derivative of the Lorentzian resonant line with a distance between positive and negative peaks $\Delta B_{\text{pp}}^{\text{L}}$ and maximum intensity between these peaks $I_{\text{L}}^{'(0)}$ is registered at resonance frequency $\omega_e^{(0)}$ (Blumenfeld, Voevodski, and Semenov 1962, Weil, Bolton, and Wertz 2007)

$$I_{\text{L}}' = \frac{16}{9}I_{\text{L}}^{'(0)}\frac{(B-B_0)}{\Delta B_{\text{pp}}^{\text{L}}}\left[1+\frac{4}{3}\frac{(B-B_0)^2}{\left(\Delta B_{\text{pp}}^{\text{L}}\right)^2}\right]^{-2},$$ (A.14)

whereas at slower fluctuation of an additional local magnetic field the spectrum is defined by Gaussian function of distribution of spin packets

$$I_{\text{G}}' = \sqrt{e}\; I_{\text{G}}^{'(0)}\frac{(B-B_0)}{\Delta B_{\text{pp}}^{\text{G}}}\exp\left[-\frac{2(B-B_0)^2}{\left(\Delta B_{\text{pp}}^{\text{G}}\right)^2}\right].$$ (A.15)

The EPR line shape due to dipole or hyperfine broadening is normally Gaussian. An exchange interaction between the spins in real system may result in the appearance of more complicated line shape, described by a convolution of Lorentzian and Gaussian distribution function. This takes a possibility from the analysis of such a line shape to define the distribution, composition and local concentrations of spins in such a system. For example, if equivalent paramagnetic centers with concentration n are arranged chaotically or regularly in the system their line shape is described by the Lorentzian and Gaussian distribution function, respectively, with the width $\Delta B_{\text{pp}}^{\text{L}} = \Delta B_{\text{pp}}^{\text{G}} = 4\gamma_e\hbar\,n$ (Lebedev and Muromtsev 1972). In the mixed cases the line shape transforms to Lorentzian at a distance from the center $\delta B \leq 4\gamma_e\hbar/r^3$ (here r is a distance between magnetic dipoles) with the width $\Delta B_{\text{pp}}^{\text{L}} = 4\gamma_e\hbar\,n$ in the center and becomes Gaussian type on the tails at $\delta B \geq \gamma_e\hbar/r^3$ with the width $\Delta B_{\text{pp}}^{\text{G}} = \gamma_e\hbar\sqrt{n/r^3}$.

Linewidth is mainly determined by transverse (spin-spin) relaxation time T_2. However, there are several relaxation processes in a polymer composites which cause the shortening of T_2 and hence the broadening of the EPR line. One of them is spin longitudinal (spin-lattice) relaxation on the lattice phonons with time T_1, which shortens the lifetime of a spin state and therefore broadens the line. Representing all other possible relaxation processes by the time $T_2^|$, one can write for effective peak-to-peak linewidth ΔB_{pp} as (Wertz and Bolton 2013)

$$\Delta B_{\text{pp}} = \Delta B_{\text{pp}}^0 + \frac{2}{\sqrt{3}\gamma_e}\cdot\frac{1}{T_2} = \Delta B_{\text{pp}}^0 + \frac{2}{\sqrt{3}\gamma_e}\cdot\left(\frac{1}{T_2^|}+\frac{1}{2T_1}\right),$$ (A.16)

where ΔB_{pp}^0 is the linewidth at the absence of spin dynamics and interaction. The collision of these paramagnetic centers should to broad EPR spectrum by (Molin, Salikhov, and Zamaraev 1980, Houze and Nechtschein 1996)

$$\delta(\Delta B_{pp}) = p_{ff}\omega_{hop}n_g = k_1\omega_{hop}n_g\left(\frac{\alpha^2}{1+\alpha^2}\right), \tag{A.17}$$

where p_{ff} is the flip-flop probability inserted above and n_g is the number of guest paramagnetic centers per each polymer unit. In this case the guest spin acts as a nanoscopic probe of the polaron dynamics. Note, that the n_g parameter is temperature dependent that should be taken into account when calculating the effective linewidth. According to the spin exchange fundamental concepts (Molin, Salikhov, and Zamaraev 1980), if exchange interaction changes between weak and strong exchange limits (see above), an appropriate $\delta(\Delta\omega)(T)$ dependency may demonstrate extremal dependence with characteristic temperature T_c. This should evidence the realization of high and low of spin-spin interaction at $T \leq T_c$ and $T \geq T_c$, respectively, realized, e.g., in highly doped polyaniline samples (Houze and Nechtschein 1996, Krinichnyi, Roth, et al. 2006, Krinichnyi, Tokarev, et al. 2006, Krinichnyi, Yudanova, and Wessling 2013).

The rate of charge hopping between two adjacent polymer units can be estimated to a good approximation using a semiclassical Marcus theory adopted for conjugated polymers (Van Vooren, Kim, and Cornil 2008, Lan and Huang 2008)

$$\omega_{hop} = \frac{4\pi^2}{\hbar}\frac{t_{1D}^2}{\sqrt{4\pi E_r k_B T}}\exp\left(-\frac{E_r}{4k_B T}\right), \tag{A.18}$$

where t_{1D} is electronic coupling between initial and final states (intrachain transfer integral) and E_r is both the inner- and outer-sphere reorganization energy of charge carriers due to their interaction with the lattice phonons. The t_{1D} value decreases slightly with temperature, whereas its distribution broadens a line due to thermal motion of polymer units (Cheung, McMahon, and Troisi 2009) similar to that happening in organic crystals (Troisi and Orlandi 2006, Kirkpatrick et al. 2007). Note, that the n_g parameter is temperature dependent that should be included into finalized equation. Combination of Eq.(A.17) and Eq.(A.18) yields

$$\delta(\Delta\omega) = \frac{\pi t_{1D}^2 n_g(T)}{\hbar\sqrt{\dfrac{E_r k_B T}{\pi}}} \cdot \frac{\exp\left(-\dfrac{E_r}{4k_B T}\right)}{1+\left[\dfrac{3J_{ex}}{2t_{1D}^2}\sqrt{\dfrac{E_r k_B T}{\pi}}\exp\left(\dfrac{E_r}{4k_B T}\right)\right]^{-2}}. \tag{A.19}$$

Except fast electron spin diffusion, EPR line can also be broadened by the acceleration of molecular dynamics processes, for example oscillations or slow torsion librations of the polymer macromolecules. The approach of random walk treatment provides (Butler, Walker, and Soos 1976), that such Q1D, Q2D, and Q3D spin diffusion with respective diffusion coefficients D_{1D}, D_{2D} and D_{3D} in the motionally narrowed regime changes the respective linewidth of a spin-packet as (Krasicky, Silsbee, and Scott 1982)

EPR Spectroscopy of Polarons in Conjugated Polymers and Their Nanocomposites   79

$$\Delta B_{pp} \approx \frac{\gamma_e^{1/3}(\Delta B_{pp}^0)^{4/3}}{D_{1D}^{1/3}},$$
(A.20)

$$\Delta B_{pp} \approx \frac{\gamma_e(\Delta B_{pp}^0)^2}{\sqrt{D_{1D}D_{3D}}},$$
(A.21)

$$\Delta B_{pp} \approx \frac{\gamma_e(\Delta B_{pp}^0)^2}{D_{3D}}.$$
(A.22)

This theory postulates that at the transition from Q1D to Q2D and then to Q3D spin motion the shape of the EPR line should transform from Gaussian to Lorentzian. This approach allows evaluating an effective dimension of the system under study, say from an analysis of temperature dependence of its EPR spectrum linewidth. For spin Q3D motion or exchange, the line shape becomes close to Lorentzian shape, corresponding to an exponential decay of transverse magnetization with time t, proportional to $\exp(-\eta t)$; for a Q1D spin motion, this value is proportional to $\exp(-\rho t)$ (here η and ρ are constants (Hennessy, McElwee, and Richards 1973). In order to determine the type of spin dynamics in Q1D system appropriate anamorphoses $I_0^|/I(B)$ vs. $[(B - B_0)/\Delta B_{1/2}]^2$ and $I_0^|/I(B)$ vs. $[(B - B_0)/\Delta B_{pp}]^2$ (here $\Delta B_{1/2}$ is the half-width of an integral line) (Hennessy, McElwee, and Richards 1973) should be analyzed.

A.1.4. Electron Relaxation and Spin Dynamics

As the magnetic term B_1 of the steady-state microwave field increases, the linewidth ΔB_{pp} of a LEPR spectrum broadens and its intensity I_L first increases linearly, plateaus starting from some B_1 value and then decreases. This occurs due to manifestation of the microwave steady-state saturation effect in the LEPR spectrum of composite. Polaron and fullerene anion radical are non-interacting and, therefore, independent of one another. This allows us to use such effects for separate estimation of their spin-lattice T_1 and spin-spin T_2 relaxation times from relations (Poole 1983)

$$\Delta B_{pp} = \Delta B_{pp}^{(0)} \sqrt{1 + \gamma_e^2 B_1^2 T_1 T_2}$$
(A.23)

and

$$I_L = I_L^{(0)} B_1 \left(1 + \gamma_e^2 B_1^2 T_1 T_2\right)^{-3/2},$$
(A.24)

where $I_L^{(0)}$ is intensity of non-saturated spectrum and $T_2 = 2/\sqrt{3}\gamma_e \Delta B_{pp}^{(0)}$. Normally, the inflection point characteristic for polarons' saturation curve is distinct from that obtained for fullerene anion radicals. This is evidence of different relaxation parameters of these paramagnetic centers and confirms additionally their mutual independence.

The mechanism and the rate of electron relaxation depend on the structure and conformation of an initial and modified polymer:fullerene composites in which radical pairs are photoinduced in differently ordered domains with respective band-gaps. Various spin-assisted dynamic processes occur in polymer:fullerene composites, e.g., polaron diffusion along and between polymer chains with coefficients D_{1D} and D_{3D}, respectively, and librative rotational motion of fullerene anion radical near own main molecular axis with coefficient D_{rot}. These processes induce additional magnetic fields in the whereabouts of electron and nuclear spins which, in turn, accelerates relaxation of both spin ensembles. Relaxation of the whole spin reservoir in organic conjugated polymers is defined mainly by dipole-dipole interaction between electron spins (Krinichnyi et al. 1992), so then these coefficients can be determined from the following equations (Carrington and McLachlan 1967):

$$T_1^{-1}(\omega_e) = \langle \omega^2 \rangle [2J(\omega_e) + 8J(2\omega_e)],$$ (A.25)

$$T_2^{-1}(\omega_e) = \langle \omega^2 \rangle [3J(0) + 5J(\omega_e) + 2J(2\omega_e)],$$ (A.26)

where $\langle \omega^2 \rangle = 1/10\,\gamma_e^4 \hbar^2 S(S+1)n\Sigma_{ij}$ is the constant of dipole-dipole interaction for powder, n is a number of polarons per each monomer, Σ_{ij} is the lattice sum for powder-like sample, $J(\omega_e) = (2D_{1D}^{|}\omega_e)^{-1/2}$ (at $D_{1D}^{|}>>\omega_e>>D_{3D}$), $J(0)=(2D_{1D}^{|}D_{3D})^{-1/2}$ (at $D_{3D}>>\omega_e$) are the spectral density functions for polaron longitudinal diffusion, and $J(\omega_e) = \tau_c/(1+\tau_c^2\omega_e^2)$ is the spectral density function for fullerene rotational libration with correlation time τ_c, $D_{1D}^{|}=4D_{1D}/L^2$, and L is a factor of spin delocalization over a polaron equal approximately to five monomer units in P3AT (Devreux et al. 1987, Westerling, Osterbacka, and Stubb 2002).

Spinless charge carrier induces at a distinct point with a guest charge an electric field with a potential (Buchachenko 1984)

$$E_d = \frac{k_B T (x\coth x - 1)}{\mu_u},$$ (A.27)

where $x = 2\mu_u\mu_v(\pi\varepsilon\varepsilon_0 k_B T r^3)^{-1}$, μ_v and μ_u are the dipole moments of the domestic and guest charge carriers, respectively, ε and ε_0 are the dielectric constants for a carrier and vacuum respectively, and r is the distance between both charge carriers. Such a field may change magnetic parameters of guest spin, and such variation can also be registered by EPR method.

A.1.5. Mechanism of Charge Transport in Polymer Systems

Different theoretical models can be used for explanation of spin dynamics in condensed systems.

To describe electron transfer in amorphous metals Mott developed the variable range hopping (VRH) model (Mott and Davis 2012). This theory balances the likelihood of tunneling between random energy electron potential wells with the likelihood of gaining enough thermal energy to move to a nearby site. In this hopping process of a charge transfer each state can have only one electron of one spin direction. If the localization of a charge carrier is very strong, it will hop to the nearest state with the probability proportional to $\exp(-\Delta E_{ij}/k_B T)$ (see the respective term of Eq.(A.6)). As the temperature is decreased, fewer states fall within the allowed energy range and the average hopping distance increases. This results in the following temperature and frequency dependence of spin diffusion coefficient in such system (Austin and Mott 1969)

$$D_{ac}(\omega,T) = D_0 \omega_e T^2 \left(\ln \frac{\omega_0}{\omega_e} \right)^4, \tag{A.28}$$

where ω_0 is the hopping attempt frequency.

Polaron dynamics in some polymer composites can be characterized by strong temperature dependence. This can probably be due to the scattering of polarons on the lattice phonons of crystalline domains embedded into an amorphous matrix. According to the model proposed for charge dynamics in crystalline domains of doped conjugated polymers, such scattering should affect polaron intrachain diffusion with an appropriate coefficient (Pietronero 1983, Kivelson and Heeger 1988)

$$D_{3D}(T) = \frac{\pi^2 M t_0^2 k_B^2 T^2}{h^3 \alpha_{eph}^2} \cdot \left[\sinh\left(\frac{E_{ph}}{k_B T} \right) - 1 \right] = D_{3D}^{(0)} T^2 \cdot \left[\sinh\left(\frac{E_{ph}}{k_B T} \right) - 1 \right], \tag{A.29}$$

where M is the mass of a polymer unit, t_0 is the transfer integral equal for π-electron to ~2.5 – 3 eV, α_{eph} is a constant of electron-phonon interaction, and E_{ph} is phonon energy.

Spin dynamics in less ordered polymer matrix of composites can be realized in the frames of the Elliot model based on spin hopping over energetic barrier E_b (Long and Balkan 1980). This model predicts the following temperature dependencies for diffusion coefficients of a charge carrier at alternating current

$$D_{ac}(\omega,T) = D_{ac}^0 T^2 \omega_e^s \exp\left(\frac{E_b}{k_B T} \right), \tag{A.30}$$

where the exponent $s = 1 - \alpha k_B T / E_b$ reflects system dimensionality and α is a constant. Comparison of spin dynamic parameters obtained at direct current and at different spin precession frequencies ω_e allowed one to determine more precisely details of charge transfer in organic polymer systems (Krinichnyi 1995, 2000a, 2006).

If spin traps are initiated in polymer matrix, the dynamics of spin charge carriers can be explained in terms of the Hoesterey-Letson formalism modified for amorphous low-dimensional systems containing spin traps with concentration n_t and depth E_t (Fishchuk et al. 2002). In the frames of such approach, the coefficient of trap-assisted spin diffusion in the case of low trap concentration limit can be

$$D_t(T) = v_0 \left(\frac{R_{ij}}{d} \right)^2 \exp\left(-\frac{2R_{ij}}{r} \right) \exp\left(\frac{E_t}{2k_B T_{cr}} \right) \exp\left[-\frac{E_t}{2k_B T} \left(\frac{\sigma_0}{k_B T} \right)^2 \right], \tag{A.31}$$

where v_0 is hopping attempt frequency, d is the lattice constant, $T_{cr} = E_t/2k_B \ln(n_t)$ is critical temperature at which the transition from trap-controlled to trap-to-trap hopping transport regimes occurs, and σ_0 is the width of intrinsic energetic distributions of hopping states in the absence of traps.

Diffusion of small polaron may correlate with the dynamics of polymer matrix. According to the Friedman-Holstein model (Friedman and Holstein 1963), in an adiabatic regime this should lead to the following temperature dependence of such charge carrier diffusion coefficient (Nagels 1980)

$$D_{sp} = \frac{\omega_0}{4\pi} \, f(T) \, \exp\left(-\frac{E_h - t_h}{3k_B T}\right), \tag{A.32}$$

where d_i is the distance between neighboring hopping sites, $f(T) = 1$ for $t_h \ll k_B T$, and $f(T) = k_B T / t_h$ for $t_h \gg k_B T$, E_h is the hopping activation energy, and t_h is the hopping transfer integral. In a non-adiabatic regime, small polaron cannot follow the lattice oscillations and the time required for its hopping from one site to another is large compared to the duration of a coincidence even between two polaron sites. For this case the latter Equation should be written as (Nagels 1980)

$$D_{sp} = \frac{t_h}{2h} \left(\frac{\pi k_B T}{4E_h}\right)^{1/2} \exp\left(-\frac{E_h}{3k_B T}\right). \tag{A.33}$$

The conductivity of a conjugated polymer due to dynamics of N_i polarons can be calculated from the modified Einstein relation

$$\sigma_{1,3D} = N_i e^2 \mu = \frac{N_i e^2 D_{1,3D} d_{1,3D}^2}{k_B T}, \tag{A.34}$$

where μ is the charge carrier mobility and $c_{1,3D}$ is the respective lattice constant.

A.1.6. Dysonian EPR Spectral Contribution

EPR line of paramagnetic centers in conducting composites can be complicated by the fact that the magnetic term B_1 of MW field used to excite resonance sets up eddy currents in the material bulk. These currents effectively confine the magnetic flux to a surface layer of thickness of order of the "skin depth". The appearance of such skin layer on various surfaces is shown in inserts of Figure A.1. This phenomenon affects the absorption of microwave energy incident upon a sample and results in the less intensity of electron absorption per unit volume of material for large particles than for small ones. This also leads to the appearance of asymmetric Dyson-like contribution (Dyson 1955) in EPR spectra of some polymer systems containing ordered domains embedded into their amorphous phase (Krinichnyi 1995, 2006, 2009, 2014b, 2016b, a, 2017, Konkin et al. 2014) (see Figure A.1). Such effect appears when the skin-layer thickness δ becomes comparable or thinner than a characteristic size of a sample, e.g., due to the increase of conductivity. In this case the time of charge carrier diffusion through the skin-layer becomes essentially less than a spin relaxation time and the Dysonian line with characteristic asymmetry factor A/B (the ratio of intensities of the spectral positive peak to negative one) is registered. Such line shape distortion is accompanied by the line shift into higher magnetic fields and the drop of sensitivity of EPR technique.

EPR Spectroscopy of Polarons in Conjugated Polymers and Their Nanocomposites

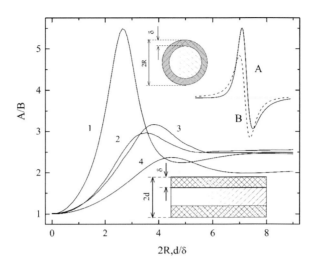

Figure A.1. The dependence of the spectrum asymmetry factor A/B on the thickness of skin-layer δ formed on the conjugated plate (*1*), shank with square (*2*) and circular (*3*) section, and sphere (*4*). In the insert are shown EPR spectra of PC in insulating (dashed line, $D/A = 0$, $A/B = 1$) and conducting (solid line, $D/A = 0.8$, $A/B = 2.2$) materials with characteristic diameter and thickness size.

Generally, the Dysonian line consists of dispersion $\chi^{!}$ and absorption $\chi^{\parallel}$ terms, therefore one can write for its first derivative the following relation (Chapman, Rhodes, and Seymour 1957):

$$\frac{d\chi}{dB} = A\frac{2x}{(1+x^2)^2} + D\frac{1-x^2}{(1+x^2)^2}, \tag{A.35}$$

where $x = 2(B-B_0)/\sqrt{3}\Delta B_{pp}^L$. The line asymmetry parameter A/B is correlated with the above coefficients A and D simply as $A/B = 1 + 1.5\, D/A$ independently on the EPR linewidth. Organic polymers are usually studied as powder and film. Appropriate coefficients of absorption A and dispersion D in Eq.(A.35) for skin-layer on the surface of a spherical powder particle with radius R and intrinsic *ac* conductivity σ_{ac} can be calculated from equations:

$$\frac{4A}{9} = \frac{8}{p^4} - \frac{8\,(\sinh p + \sin p)}{p^3(\cosh p - \cos p)} + \frac{8\,\sinh p\,\sin p}{p^2(\cosh p - \cos p)^2} + \frac{(\sinh p - \sin p)}{p\,(\cosh p - \cos p)} - \frac{(\sinh^2 p - \sin^2 p)}{(\cosh p - \cos p)^2} + 1, \tag{A.36}$$

$$\frac{4D}{9} = \frac{8\,(\sinh p - \sin p)}{p^3(\cosh p - \cos p)} - \frac{4\,(\sinh^2 p - \sin^2 p)}{p^2(\cosh p - \cos p)^2} + \frac{(\sinh p + \sin p)}{p\,(\cosh p - \cos p)} - \frac{2\sinh p\,\sin p}{(\cosh p - \cos p)^2}, \tag{A.37}$$

where $p = 2R/\delta$, $\delta = \sqrt{2/\mu_0 \omega_e \sigma_{ac}}$, and μ_0 is the magnetic permeability for vacuum. In case of the formation of skin-layer on the flat plate with a thickness of $2d$ the above coefficients can be determined from relations

$$A = \frac{\sinh p + \sin p}{2p \ (\cosh p + \cos p)} + \frac{1 + \cosh p \ \cos p}{(\cosh p + \cos p)^2}, \qquad (A.38)$$

$$D = \frac{\sinh p - \sin p}{2p \ (\cosh p + \cos p)} + \frac{\sinh p \ \sin p}{(\cosh p + \cos p)^2}, \qquad (A.39)$$

where $p = 2d/\delta$.

Figure A.1 depicts the dependence of the spectrum asymmetry factor A/B on the thickness of skin-layer δ formed in different conducting bulks. The analysis of multifrequncy EPR spectra with Dysonian term allows determining directly *ac* conductivity of crystalline domains embedded into amorphous polymer matrix at spin precession frequency ω_e.

A.2. EPR Techniques

The study of spin properties of spin topological distortions in organic substances is mainly carried out by using X-band EPR technique operating at $\omega_e/2\pi \approx 9.7$ GHz and $B_0 \approx$ 0.33 T. This is due mainly to the availability starting from the 1940s of an appropriate element base and a relative simplicity of the experiments at this waveband. Typical X-band CW EPR spectrometer is drawn schematically in Figure A2. The main parts of this device are MW power oscillator *1*, magnet *9*, and MW reflecting cavity (resonator) *11*. The EPR spectrum is usually measured keeping constant MW frequency during sweeping the magnetic field strength. If the energy of the MW quants equals the energy splitting of the spin states, the resonance condition (A.1) is fulfilled and MW radiation is absorbed. Typically, the EPR spectrum is given as the first derivative of the absorption peak, resulting from magnetic field modulation necessary for the use of a lock-in amplifier. The field modulation determines among others the time resolution of the EPR spectrometer. It is limited to the inverse of the field modulation frequency. Usually a modulation frequency of 100 kHz is used which would result in a maximum time resolution of 10 ps. MW power is generated by oscillator *1* previously based on vacuum klystron or backward wave oscillator and then on solid-state Gunn diode. This part contains also appropriate automatic frequency control circuit. MW power is splitted and transmitted to the two branches of the MW bridge. The referent branch of the bridge consists of MW attenuator *2* and MW phase shifter *3*, whereas the signal measuring part contains more precession MW attenuator *6*, one-way MW circulator *7* and MW reflecting cavity (resonator) *11* situated between the magnet poles *9*. The sample under study in quartz ampoule is placed at the centre of the MW cavity *11* and both the branches are tuned to reach balanced zero signal at the output of the MW diode *4*. Once the Eq.(A.1) is fulfilled for PC analyzed, they absorb a little part of MW energy in the measuring branch and are excited to the higher energy level. This causes imbalance in the MW bridge and appearance at its output of resonant *ac* signal under a scanning of a sample in an external magnetic field B_0. This signal is transformed into a resulting *dc* EPR response by phase detector *5*. Normally, the cavity and magnet of the spectrometer are provided with a hole through which the PC can be initiated in the sample under its direct irradiation or illumination by appropriated source *8*. The more spins are stabilized or/and initiated in the sample the higher integral signal is registered on the device output. One of the simple exemplary devices, portable X-band EPR spectrometer PS 100X operating at 9.72 GHz is characterized by a sensitivity of 4×10^{11} spin/mT, resonance instability of 3×10^{-5} per a hour, maximal MW power 150 mW inducing $B_1 = 48$ μT in center of the cavity with an unloaded quality factor Q ≈ 5,000 and mode H_{102}.

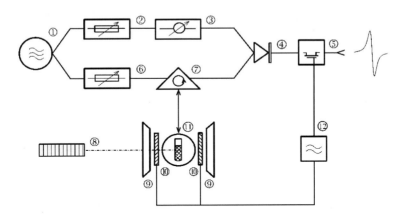

Figure A2. The sketch of X-band EPR spectrometer: *1* - MW oscillator, *2* - MW attenuator, *3* - MW phase shifter, *4* - MW detector, *5* – ac phase detector, *6* – MW attenuator, *7* – MW circulator, *8* – light source (laser), *9* - magnet pole, *10* - ac modulation coils, *11* - MW cavity with a sample, *12* - ac oscillator.

If some spin ensembles with near magnetic parameters are stabilized in a sample, their lines can overlap. The width of individual EPR lines of most organic radicals usually amounts to 0.1-1.0 mT, and *g*-factor value differs from g_e by $(1-10) \times 10^{-4}$. Therefore the spectral resolution or splitting of lines of different radicals, $\delta B \cong \delta g B_0 / g \cong 10 - 100$ mT, is smaller than their width at X-band, resulting in the overlapping of their EPR spectra. This complicates the identification of such radials in solids and the analysis of structural and dynamic properties of microenvironment by using their *g*-factors (Misra 2011). This means that in order to attain the satisfactory resolution of free radicals' EPR spectra the condition $\delta B/B_0 < 2 \times 10^{-5}$ should be valid. If the linewidth of PC do not changes with registration frequency ω_e, such condition is fulfilled at millimeter wavebands EPR. The separation in field of two lines with different *g*-factors will scale with the field, so if the linewidths remain constant, then the lines will eventually be resolved as the field is increased. Besides, the susceptibility of the method also increases at high spin precession frequency ω_e (Krinichnyi 2006, Misra 2011, Krinichnyi 2016b). The operating frequency was defined in most cases by the investigation purposes and physical properties of the objects under study. Except X-band (of frequences extending from 8 up to 12 GHz), such investigations are also carried out at K- (18 – 26.5 GHz), Q- (30 – 50 GHz), W- (75 – 110 GHz), and D- (110 – 170 GHz) bands EPR. The main advantages of the high-frequency/field CW EPR spectroscopy may be compiled as following:

- Increase in spectral resolution due to the higher magnetic field
- Increase in detection sensitivity for samples of limited quantity due to higher resonator filling factor
- Increase in orientation selectivity in the investigation of disordered systems

- Accessibility of spin systems with larger zero-field splitting due to the larger microwave quantum energy
- Implification of spectra due to the reduction of second-order effects at high fields.

The basic block-scheme of the CW D-band EPR spectrometer (Galkin et al. 1977, Krinichnyi 1991a, b) is drawn in Figure A.3. This device is assembled as a direct amplification circuit with H_{011} type reflecting cavity, a double MW T-bridge and n-SbIn bolometer operating as MW detector at temperature of liquid helium (4.2 K). The main part of the spectrometer includes the MW klystron or Gunn diode oscillator *11* with appropriate elements of the waveguide section, a cryostat *1* with a superconducting solenoid *2* and MW bolometric n-SbIn sensor *14*. In the center of the cryostat is placed the tunable MW cavity *4* with a sample *5* encircled by modulating *3* and temperature-sensitive *6* copper coils. A sample filled into a quartz capillary with the external diameter and length of 0.6 and 7 mm, respectively, is placed into the center of the cavity between two cylindrical pistons. The temperature of the sample can be controlled within 6 – 380 K region by helium or nitrogen gaseous flow. The quality factor Q of the cavity with the inner diameter of 3.5 mm and operation length of 1.5 mm lies near 2,000. The value of microwave field magnetic component B_1 in the center of the cavity is equal to 20 µT. The magnetic field inhomogeneity in the point of sample arrangement does not exceed 10 µT/mm. The absolute point-sample sensitivity of the spectrometer is 5×10^8 spin/mT at room temperature that is unique for EPR spectroscopy. The latter value is three orders of magnitude lower than N_{min} calculated (Krinichnyi 2016b) due to the smaller P and Q_0 values as compared with X-band. The concentration sensitivity for aqueous samples is 6×10^{13} spin/mT cm^3.

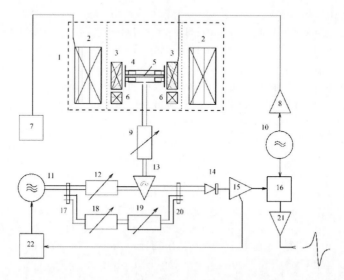

Figure A3. The sketch of D-band EPR spectrometer: *1* - helium cryostat, *2* - superconducting solenoid, *3* – *ac* modulation coil, *4* – MW cavity with two cylindrical side tuning and fixed pistons, *5* – quartz capillary with a sample, *6* - temperature-sensitive coil, *7* - solenoid current supply, *8* – *ac* modulator amplifier, *9* and *19* – MW phase shifter, *10* - *ac* oscillator, *11* - MW oscillator, *12* and *18* - MW attenuators, *13* - MW circulator, *14* – MW superlow temperature (4.2 K) barretter, *15* - *21* - *ac* preamplifiers, *16* - phase detector, *17* and *20* - directional MW couplers, *22* – MW oscillator power supply with a section of MW frequency auto adjustment.

The first CW D-band EPR investigation of PC stabilized in different biopolymers and organic conjugated polymers showed the growth of sensitivity and resolution of the method at this waveband (Krinichnyi 1991a, b, 1995). It was demonstrated that the study of these systems at higher registration frequencies allows increasing sufficiently the efficiency of the method, to obtain qualitative new information on these and other objects and to solve various scientific problems. During the last decade, multi frequency EPR spectroscopy allowed us to study the main properties of polarons and other PC in conjugated polymers and their nanocomposites (Krinichnyi 2006, 2009, 2014b, a, 2016b, a). During this time, the variety of EPR technique, pulse, combined, has been expanded (Misra 2011).

REFERENCES

Adrain, F.J., and L. Monchick. 1979. "Theory of chemically induced magnetic polarization. Effects of $S–T_{\pm 1}$ mixing in strong magnetic fields." *Journal of Chemical Physics* 71 (6):2600-2610.

Aguirre, A., P. Gast, S. Orlinskii, I. Akimoto, E. J. J. Groenen, H. El Mkami, E. Goovaerts, and S. Van Doorslaer. 2008. "Multifrequency EPR analysis of the positive polaron in I_2-doped poly(3-hexylthiophene) and in poly[2-methoxy-5-(3,7-dimethyloctyloxy)]-1,4-phenylenevinylene." *Physical Chemistry Chemical Physics* 10 (47):7129-7138.

Ahlskog, M., R. Menon, A. J. Heeger, T. Noguchi, and T. Ohnishi. 1997. "Metal-insulator transition in oriented poly(p-phenylenevinylene)." *Physical Review B* 55 (11):6777-6787.

Ahlskog, M., M. Reghu, and A. J. Heeger. 1997. "The temperature dependence of the conductivity in the critical regime of the metal-insulator transition in conducting polymers." *Journal of Physics -Condensed Matter* 9 (20):4145-4156.

Al Ibrahim, M., H. K. Roth, M. Schrödner, A. Konkin, U. Zhokhavets, G. Gobsch, P. Scharff, and S. Sensfuss. 2005. "The influence of the optoelectronic properties of poly(3-alkylthiophenes) on the device parameters in flexible polymer solar cells." *Organic Electronics* 6 (2):65-77.

Altshuler, S. A., and B. M. Kozirev. 1972. Электронный парамагнитный резонанс соединений элементов промежуточных групп [*Electron paramagnetic resonance of compounds of elements of intermediate groups*]. Academic Press. Original edition, M.: Наука, 1972, 670 с.

Ashwell, G. J., ed. 1992. *Molecular Electronics*. Edited by G. J. Ashwell. New York: John Wiley.

Assenheim, H.M. 2014. *Introduction to Electron Spin Resonance*: Springer.

Audebert, P., G. Binan, M. Lapkowski, and Limosin. 1987. "Grafting, ionomer composites, and auto-doping of conductive polymers." In *Electronic Properties of Conjugated Polymers, Springer Series in Solid State Sciences*, edited by H. Kuzmany, M. Mehring and S. Roth, 366. Berlin: Springer-Verlag.

Austin, I. G., and N. F. Mott. 1969. "Polarons in crystalline and non-crystalline materials." *Advances in Physics* 18 (71):41-102.

Banerji, N., S. Cowan, M. Leclerc, E. Vauthey, and A. J. Heeger. 2010. "Exciton formation, relaxation, and decay in PCDTBT." *Journal of the American Chemical Society* 132 (49):17459-17470.

Barnes, S. E. 1981. "Theory of electron spin resonance of magnetic ions in metals." *Advances in Physics* 30 (6):801-938.

Barta, P., S. Niziol, P. Leguennec, and A. Pron. 1994. "Doping-Induced magnetic phase-transition in poly(3-alkylthiophenes)." *Physical Review B* 50 (5):3016-3024.

Bartle, A., L. Dunsch, H. Naarmann, D. Schmeisser, and W. Gopel. 1993. "ESR studies of polypyrrole films with a two-dimensional microstructure." *Synthetic Metals* 61(1-2):167-170.

Baunsgaard, D., M. Larsen, N. Harrit, J. Frederiksen, R. Wilbrandt, and H. Stapelfeldt. 1997. "Photophysical properties of 2,3,6,7,10,11-hexakis(n-hexylsulfanyl)triphenylene and 2,3,6,7,10,11-hexakis(n-hexylsulfonyl)triphenylene insolution." *Journal of the Chemical Society, Faraday Transactions* 93 (10):1893-1901.

Beau, B., J. P. Travers, and E. Banka. 1999. "NMR evidence for heterogeneous disorder and quasi-1D metallic state in polyaniline CSA." *Synthetic Metals* 101 (1-2):772-775.

Beau, B., J. P. Travers, F. Genoud, and P. Rannou. 1999. "NMR study of aging effects in polyaniline CSA." *Synthetic Metals* 101 (1-2):778-779.

Bernier, P. 1986. "The magnetic properties of conjugated polymers: ESR studies of undoped and doped systems." In *Handbook of Conducting Polymers*, edited by T. E. Scotheim, 1099-1125. New York: Marcel Dekker, Inc.

Blakemore, J. S. 1985. *Solid State Physics*. 2 ed. Cambridge: Cambridge University Press.

Blank, A., R. Kastner, and H. Lebanon. 1998. "Exploring new active materials for low noise, room temperature, microwave amplifiers and other devices." *IEEE Transactions of Microwave Theory and Techniques* 46 (12):2137-2144.

Bleier, H., S. Roth, Y. Q. Shen, and D. Schafer-Siebert. 1988. "Photoconductivity in trans-polyacetylene: Transport and recombination of photogenerated charged excitations." *Physical Review B* 38:6031-6040.

Blumenfeld, L. A., V. V. Voevodski, and A. G. Semenov. 1962. *Application of Electron Paramagnetic Resonance in Chemistry (Russ)*. Novosibirsk: Izdat. SO AN SSSR. Original edition, [Л.А. Блюменфельд, В.В. Воеводский, А.Г. Семёнов, Применение электронного парамагнитного резонанса в химии], Новосибирск, Издат. СА АН СССР, 1962, с.

Bobacka, J., A. Ivaska, and A. Lewenstam. 1999. "Plasticizer-free all-solid-state potassium-selective electrode based on poly(3-octylthiophene) and valinomycin." *Analytica Chimica Acta* 385 (1-3):195-202.

Bock, H., P. Rittmeyer, A. Krebs, K. Schultz, J. Voss, and B. Kopke. 1984. "Radical ions. 56. One-electron oxidation of 1,2-dithiete derivatives." *Phosphorus Sulfur and Silicon and the Related Elements* 19:131-136.

Brabec, C., V. Dyakonov, J. Parisi, and N. S. Sariciftci. 2003. *Organic Photovoltaic: Concepts and Realization, Springer Series in Materials Science*. Berlin: Springer.

Brabec, C., U. Scherf, and V. Dyakonov, eds. 2014. *Organic Photovoltaics: Materials, Device Physics, and Manufacturing Technologies*. 2- ed. Vol. 1. Weinheim: Wiley-VCH.

Brédas, J. L. 1986. "Electronic structure of highly conducting polymers." In *Handbook of Conducting Polymers*, edited by T. E. Scotheim, 859-913. New York: Marcel Dekker, Inc.

Brédas, J. L., and R. R. Chance, eds. 1990. *Conjugated Polymeric Materials: Opportunities in Electronics, Optoelectronics, and Molecular Electronics*. Edited by J. L. Brédas and R. R. Chance, *NATO Advanced Study Series*. Dordrecht: Kluwer Academic Publishers.

Brédas, J. L., R. R. Chance, R. H. Baughman, and R. Silbey. 1982. "Abinitio effective Hamiltonian study of the electronic properties of conjugated polymers." *Journal of Chemical Physics* 76 (7):3673-3678.

Brédas, J. L., C. Quattrocchi, J. Libert, A. G. MacDiarmid, J. M. Ginder, and A. J. Epstein. 1991. "Influence of ring-torsion dimerization on the band-gap of aromatic conjugated polymers." *Physical Review B* 44 (12):6002-6010.

Brédas, J. L., and G. B. Street. 1985. "Polarons, bipolarons, and solitons in conducting polymers." *Accounts of Chemical Research* 18 (10):309-315.

Breiby, D. W., S. Sato, E. J. Samuelsen, and K. Mizoguchi. 2003. "Electron spin resonance studies of anisotropy in semiconducting polymeric films." *Journal of Polymer Science Part B-Polymer Physics* 41 (23):3011-3025.

Buchachenko, A. L. 1984. *Complexes of Radicals and Molecular Oxygen with Organic Molecules (Russ)*. Moscow: Nauka. Original edition, [Бучаченко, А. Л., Комплексы радикалов и молекулярного кслорода с органическими молекулами], М.: Химия, 1984.

Buchachenko, A. L., C.N. Turton, and T.I. Turton. 1995. *Stable Radicals*. New York: Consultants Bureau.

Buchachenko, A. L., and A. M. Vasserman. 1973. *Stable Radicals (Russ)*. Moscow: Khimija. Original edition, [Бучаченко, А.Л., Вассерман, А.М., Стабильные радикалы, М.: Химия], 1973.

Butler, M. A., L. R. Walker, and Z. G. Soos. 1976. "Dimensionality of spin fluctuations in highly anisotropic TCNQ salts." *Journal of Chemical Physics* 64 (9):3592-3601.

Cameron, T. S., R. C. Haddon, S. M. Mattar, S. Parsons, J. Passmore, and A. P. Ramirez. 1991. "The synthsis, characterization, X-ray crystal-structure and solution ESR-spectrum of the paramagnetic solid, 4,5-bis(trifluoromethyl)-1,2,3-trithiolium hexafluorarsenate - Implication for the identity of 1,2-dithiete cations." *Journal of the Chemical Society-Chemical Communications* (6):358-360.

Cameron, T. S., R. C. Haddon, S. M. Mattar, S. Parsons, J. Passmore, and A. P. Ramirez. 1992. "Preparation and crystal-structure of the paramagnetic solid $F_3CCSSSCCF_3ASF_6$ - Implication for the identity of RCSSCR-bullet[+]." *Journal of the Chemical Society-Dalton Transactions* (9):1563-1572.

Carrington, F., and A. D. McLachlan. 1967. *Introduction to Magnetic Resonance with Application to Chemistry and Chemical Physics*. New York, Evanston, London: Harrer & Row, Publishers.

Carter, F. L., ed. 1982. *Molecular Electronic Devices II*. Edited by F. L. Carter. Vol. 2. New York: Marcel Dekker.

Chakrabarti, S., B. Das, P. Banerji, D. Banerjee, and R. Bhattacharya. 1999. "Bipolaron saturation in polypyrrole." *Physical Review B* 60 (11):7691-7694.

Chang, Y. H., K. Lee, R. Kiebooms, A. Aleshin, and A. J. Heeger. 1999. "Reflectance of conducting poly(3,4-ethylenedioxythiophene)." *Synthetic Metals* 105 (3):203-206.

Chapman, A. C., P. Rhodes, and E. F. W. Seymour. 1957. "The effect of eddy currents on nuclear magnetic resonance in metals." *Proceedings of the Physical Society* B 70 (4):345-360.

Chatterjee, A., and S. Mukhopadhyay, eds. 2018. *Polarons and Bipolarons: An Introduction, Chapman & Hall Pure and Applied Mathematics*. Boca Raton: CRC Press.

Chen, J., A. J. Heeger, and F. Wudl. 1986. "Confined soliton pairs (bipolarons) in polythiophene - *insitu* magnetic-resonance measurements." *Solid State Communications* 58 (4):251-257.

Chen, T. A., X. M. Wu, and R. D. Rieke. 1995. "Regiocontrolled synthesis of poly(3-alkylthiophenes) mediated by Rieke zink - their characterization and solid-state properties." *Journal of the American Chemical Society* 117 (1):233-244.

Cheung, D. L., D. P. McMahon, and A. Troisi. 2009. "Computational study of the structure and charge-transfer parameters in low-molecular-mass P3HT." *Journal of Physical Chemistry B* 113 (28):9393–9401.

Cik, G., F. Sersen, L. Dlhan, I. Cerven, A. Stasko, and D. Vegh. 2002. "Study of magnetic properties of copolymer of 3-dodecylthiophene and 2,3-R,R-thieno[3,4-b]pyrazine." *Synthetic Metals* 130 (2):213-220.

Cik, G., F. Sersen, L. Dlhan, L. Szabo, and J. Bartus. 1995. "Anomaly in magnetic properties of poly(3-alkylthiophene)s depending on alkyl chain length." *Synthetic Metals* 75 (1):43-48.

Cinchetti, M., K. Heimer, J. P. Wustenberg, O. Andreyev, M. Bauer, S. Lach, C. Ziegler, Y. L. Gao, and M. Aeschlimann. 2009. "Determination of spin injection and transport in a ferromagnet/organic semiconductor heterojunction by two-photon photoemission." *Nature Materials* 8:115-119.

Clark, W. G., and L. C. Tippie. 1979. "Exchange-coupled pair model for the random-exchange Heisenberg antiferromagnetic chain." *Physical Review B* 20 (7):2914-2923.

Conwell, E. M. 1997. "Transport in conducting polymers." In *Handbook of Organic Conductive Molecules and Polymers*, edited by H. S. Nalwa, 1-45. Chichester: John Wiley & Sons.

Conwell, E. M., C. B. Duke, A. Paton, and S. Jeyadev. 1988. "Molecular conformation of polyaniline oligomers: optical absorption and photoemission of three-phenyl molecules." *Journal of Chemical Physics* 88 (5):3331-3337.

Dediu, V. A., L. E. Hueso, I. Bergenti, and C. Taliani. 2009. "Spin routes in organic semiconductors." *Nature Materials* 8:707 - 716.

Deibel, C., D. Mack, J.Gorenflot, A.Schöll, S. Krause, F.Reinert, D.Rauh, and V. Dyakonov. 2010. "Energetics of excited states in the conjugated polymer poly(3-hexylthiophene)." *Physical Review B* 81 (8):085202 - 085206.

Demirboga, B., and A. M. Onal. 2000. "ESR and conductivity investigations on electrochemically synthesized polyfuran and polythiophene." *Journal of Physics and Chemistry of Solids* 61 (6):907-913.

Devreux, F., F. Genoud, M. Nechtschein, and B. Villeret. 1987. "On polaron and bipolaron formation in conducting polymers." In *Electronic Properties of Conjugated Polymers*, edited by H. Kuzmany, M. Mehring and S. Roth, 270-276. Berlin: Springer-Verlag.

Devreux, F., and H. Lecavelier. 1987. "Evidence for anomalous diffusion in a conducting polymer." *Physical Review Letters* 59 (22):2585-2587.

Dubois, M., A. Merlin, and D. Billaud. 1999. "Electron spin resonance in lithium and sodium electrochemically intercalated poly(paraphenylene)." *Solid State Communications* 111 (10):571-576.

Dyson, F. J. 1955. "Electron spin resonance absorbtion in metals. II. Theory of electron diffusion and the skin effect." *Physical Review B* 98 (2):349-359.

Eaton, G.R., S.S. Eaton, D.P. Barr, and R.T. Weber. 2010. *Quantitative EPR*. Wien, New York: Springer.

Elliott, R. J. 1954. "Effect of spin-orbit coupling on paramagnetic resonance in semi-conductors." *Physical Review* 96:266-279.

Elsenbaumer, R. L., and L. W. Shacklette. 1986. "Phenylene-Based conducting polymers." In *Handbook of Conducting Polymers*, edited by T. E. Scotheim, 213-263. New York: Marcel Dekker, Inc.

Emin, D. 2013. *Polarons*. Cambridge: Cambridge University Press.

Epstein, A. J. 2007. "Conducting polymers: Electrical conductivity." In *Physical Properties of Polymers. Handbook*, edited by J. E. Mark, 725-744. Springer.

Faridbod, F., M. R. Ganjali, R. Dinarvand, and P. Norouzi. 2008. "Developments in the field of conducting and non-conducting polymer based potentiometric membrane sensors for ions over the past decade." *Sensors* 8 (4):2331-2412.

Fishchuk, I. I., A. K. Kadashchuk, H. Bässler, and D. S. Weiss. 2002. "Nondispersive charge-carrier transport in disordered organic materials containing traps." *Physical Review B* 66 (20):205208/01 - 205208/12.

Friedman, L., and T. Holstein. 1963. "Studies of polaron motion. Part III: The Hall mobility of the small polaron." *Annals of Physics* 21 (3):494-549.

Galkin, A. A., O. Y. Grinberg, A. A. Dubinskii, N. N. Kabdin, V. N. Krymov, V. I. Kurochkin, Y. S. Lebedev, L. G. Oransky, and V. F. Shuvalov. 1977. "EPR spectrometer in 2-mm range for chemical research." *Instruments and Experimental Techniques* 20 (4):1229-1229.

Garnica, M., D. Stradi, S. Barja, F. Calleja, C. Diaz, M. Alcami, N. Martin, de Parga Vazquez, L. Amadeo, F. Martin, and R. Miranda. 2013. "Long-range magnetic order in a purely organic 2D layer adsorbed on epitaxial graphene." *Nature Physics* 9 (6):368-374.

Gebeyehu, D., C. J. Brabec, F. Padinger, T. Fromherz, J. C. Hummelen, D. Badt, H. Schindler, and N. S. Sariciftci. 2001. "The interplay of efficiency and morphology in photovoltaic devices based on interpenetrating networks of conjugated polymers with fullerenes." *Synthetic Metals* 118 (1-3):1-9.

Gebeyehu, D., F. Padinger, T. Fromherz, J. C. Hummelen, and N. S. Sariciftci. 2000. "Photovoltaic properties of conjugated polymer/fullerene composites on large area flexible substrates." *Bulletin of the Chemical Society of Ethiopia* 14 (1):57-68.

Ginder, J. M., A. F. Richter, A. G. MacDiarmid, and A. J. Epstein. 1987. "Insulator-to-metal transition in polyaniline." *Solid State Communications* 63 (2):97-101.

Goldenberg, L. M., A. E. Pelekh, V. I. Krinichnyi, O. S. Roshchupkina, A. F. Zueva, R. N. Lyubovskaja, and O. N. Efimiv. 1990. "Investigation of poly(p-phenylene) obtained by electrochemical oxidation of benzene in a BuPyCl-AlCl$_3$melt." *Synthetic Metals* 36 (2):217-228.

Goldenberg, L. M., A. E. Pelekh, V. I. Krinichnyi, O. S. Roshchupkina, A. F. Zueva, R. N. Lyubovskaja, and O. N. Efimiv. 1991. "Investigation of poly(*para*-phenylene) obtained in electrochemical oxidation of benzene in the BuPyCl-AlCl$_3$ melt and in organic solvents with oleum." *Synthetic Metals* 43 (1-2):3071-3074.

Grinberg, O. Y., A. A. Dubinskii, and Y. S. Lebedev. 1983. "Electron-Paramagnetic Resonance of free radicals in a 2-millimeter range of wave-length." *Russian Chemical Reviews* 52 (9):1490-1513.

92   *Victor I. Krinichnyi*

Gronowitz, S., ed. 1991. *Thiophene and its Derivatives*. Edited by S. Gronowitz. Vol. 44, *The Series of Heterocyclic Compounds*. New York: Wiley.

Gruber, H., H. K. Roth, J. Patzsch, and E. Fanghänel. 1990. "Electrical-Properties of poly(tetrathiafulvalenes)." *Makromolekulare Chemie-Macromolecular Symposia* 37:99-113.

Harigaya, K. 1998. "Long-range excitons in conjugated polymers with ring torsions: poly(para-phenylene) and polyaniline." *Journal of Physics -Condensed Matter* 10 (34):7679-7690.

Hayashi, S., K. Kaneto, K. Yoshino, R. Matsushita, and T. Matsuyama. 1986. "Electrical-Conductivity and Electron-Spin-Resonance studies in iodine-doped polythiophene from semiconductor to metallic regime." *Journal of the Physical Society of Japan* 55 (6):1971-1980.

Heeger, A. J., S. Kivelson, J. R. Schrieffer, and W. P. Su. 1988. "Solitons in conducting polymers." *Reviews of Modern Physics* 60 (3):781-850.

Heeger, A. J., N. S. Sariciftci, and E. B. Namdas. 2010. *Semiconducting and Metallic Polymers*. London: Oxford Univertsity Press.

Hempel, G., A. M. Richter, E. Fanghänel, and H. Schneider. 1990. "NMR-Untersuchungen zu Struktur und Beweglichkeit von substituierten Polyacetylenen." *Acta Polymerica* 41 (10):522-527.

Hennessy, Michael J., Carl D. McElwee, and Peter M. Richards. 1973. "Effect of interchain coupling on electron-spin resonance in nearly one-dimensional systems." *Physical Review B* 7 (3):930-947.

Hinh, L. V., G. Schukat, and E. Fanghänel. 1979. "Tetrathiafulvalene. VII [1]. Arylenverbrückte polymere Tetrathiafulvalene." *Journal für Praktische Chemie* 321 (2):299-307.

Hotta, S. 1997. "Molecular conductive materials: polythiophenes and oligothiophenes." In *Handbook of Organic Conductive Molecules and Polymers*, edited by H. S. Nalwa, 309-387. Chichester, New York: John Wiley & Sons.

Houze, E., and M. Nechtschein. 1996. "ESR in conducting polymers: Oxygen-induced contribution to the linewidth." *Physical Review B* 53 (21):14309-14318.

Hyde, J. S., and L. R. Dalton. 1979. "Saturation-transfer spectroscopy." In *Spin Labeling II: Theory and Application*, edited by L. J. Berliner, 1-70. New York: Academic.

Iida, M., T. Asaji, R. Ikeda, M. B. Inoue, M. Inoue, and D. Nakamura. 1992. "Electron-Paramagnetic Resonance study of intrinsic paramagnetism in poly(aniline trifluoromethanesulfonate), $[(—C_6H_4NH—)(CF_3SO_3)_{0.5}\cdot0.5H_2O]_x$" *Journal of Materials Chemistry* 2 (3):357-360.

Iida, M., T. Asaji, M. B. Inoue, and M. Inoue. 1993. "EPR study of polyaniline perchlorates - Spin species-related to charge-transport." *Synthetic Metals* 55 (1):607-612.

Im, C., J. M. Lupton, P. Schouwink, S. Heun, H.Becker, and H. Bassler. 2002. "Fluorescence dynamics of phenyl-substituted polyphenylenevinylene–trinitrofluorenone blend systems." *Journal of Chemical Physics* 117:1395-1402.

Janssen, R. A. J., D. Moses, and N. S. Sariciftci. 1994. "Electron and energy transfer processes of photoexcited oligothiophenes onto tetracyanoethylene and C60." *Journal of Chemical Physics* 101 (11):9519-9527.

Jeffries-El, M., and R. D. McCullough. 2007. "Regioregular Polythiophenes." In *Handbook of Conducting Polymers*, edited by T. E. Scotheim and J. R. Reynolds, 9-1 - 9-49. Boca Raton: CRC Press.

Jonston, D. C. 1984. "Thermodynamics of charge-density waves in quasi one-dimensional conductors." *Physical Review Letters* 52 (23):2049-2052.

Kaeriyama, K. 1997. "Synthesis and properties of processable polythiophenes." In *Handbook of Organic Conductive Molecules and Polymers*, edited by H. S. Nalwa, 271-308. Chichester, New York: John Wiley & Sons.

Kahol, P. K. 2000. "Magnetic susceptibility of polyaniline and polyaniline-polymethylmethacrylate blends." *Physical Review B* 62 (21):13803-13804.

Kahol, P. K., and M. Mehring. 1986. "Exchange-coupled pair model for the non-curie-like susceptibility in conducting polymers." *Synthetic Metals* 16 (2):257-264.

Kahol, P. K., A. Raghunathan, B. J. McCormick, and A. J. Epstein. 1999. "High temperature magnetic susceptibility studies of sulfonated polyanilines." *Synthetic Metals* 101 (1-3):815-816.

Kanemoto, K., and J. Yamauchi. 2000a. "Doping-induced variation of electron spin relaxation behavior in polypyrroles." *Synthetic Metals* 114 (1):79-84.

Kanemoto, K., and J. Yamauchi. 2000b. "Electron-spin dynamics of polarons in lightly doped polypyrroles." *Physical Review B* 61 (2):1075-1082.

Kanemoto, K., and J. Yamauchi. 2001. "ESR broadening in conducting polypyrrole because of oxygen: Application to the study of oxygen adsorption." *Journal of Physical Chemistry B* 105 (11):2117-2121.

Kaneto, K., S. Hayashi, S. Ura, and K. Yoshino. 1985. "Electron-Spin-Resonance and transport studies in electrochemically doped polythiophene film." *Journal of the Physical Society of Japan* 54 (3):1146-1153.

Kaniowski, T., S. Niziol, J. Sanetra, M. Trznadel, and A. Proń. 1998. "Optical studies of regioregular poly(3-octylthiophene)s under pressure." *Synthetic Metals* 94 (1):111-114.

Kapil, A., M. Taunk, and S. Chand. 2010. "Preparation and charge transport studies of chemically synthesized polyaniline." *Journal of Materials Science-Materials in Electronics* 21 (4):399-404.

Kawai, T., H. Mizobuchi, S. Okazaki, H. Araki, and K. Yoshino. 1996. "Ferromagnetic tendency of TDAE-fullerene-conducting polymer system." *Japanese Journal of Applied Physics Part 2-Letters* 35 (5B):L640-L643.

Kim, D. H., H. J. Song, S. W. Heo, K. W. Song, and D. K. Moon. 2014. "Enhanced photocurrent generation by high molecular weight random copolymer consisting of benzothiadiazole and quinoxaline as donor materials." *Solar Energy Materials and Solar Cells* 120:94-101.

Kim, Y., S. Cook, S. M. Tuladhar, S. A. Choulis, J. Nelson, J. R. Durrant, D. D. C. Bradley, M. Giles, I. McCulloch, C. S. Ha, and M. Ree. 2006. "A strong regioregularity effect in self-organizing conjugated polymer films and high-efficiency polythiophene: fullerene solar cells." *Nature Materials* 5 (3):197-203.

Kirkpatrick, J., V. Marcon, J. Nelson, K. Kremer, and D. Andrienko. 2007. "Charge mobility of discotic mesophases: A multiscale quantum and classical study." *Physical Review Letters* 98 (22):227402/01–227402/04.

Kivelson, S. 1980. "Hopping conduction and the continuous-time random-walk model." *Physical Review B* 21 (12):5755-5767.

Kivelson, S., and A. J. Heeger. 1988. "Intrinsic conductivity of conducting polymers." *Synthetic Metals* 22 (4):371-384.

Kobayashi, S., and K. Müllen, eds. 2015. *Encyclopedia of Polymeric Nanomaterials*: Springer.

Kon'kin, A. L., V. G. Shtyrlin, R. R. Garipov, A. V. Aganov, A. V. Zakharov, V. I. Krinichnyi, P. N. Adams, and A. P. Monkman. 2002. "EPR, charge transport, and spin dynamics in doped polyanilines." *Physical Review B* 66 (7):075203/01-075203/11.

Kondawar, Subhash B., and Hemlata J. Sharma, eds. 2017. *Conducting Polymer Nanocomposites: Synthesis, Characterizations and Applications*: Studera Press.

Konkin, A., U. Ritter, P. Scharff, G. Mamin, A. Aganov, S. Orlinskii, V. I. Krinichnyi, D. A. M. Egbe, G. Ecke, and H. Romanus. 2014. "Multifrequency X-, W-band ESR study on photo-induced ion radical formation in solid films of mono- and di-fullerenes embedded in conjugated polymers." *Carbon* 77:11-17.

Konkin, A., U. Ritter, P. Scharff, H.-K. Roth, A. Aganov, N. S. Sariciftci, and D. A. M. Egbe. 2010. "Photo-induced charge separation process in (PCBM-C_{120}O)/(M3EH-PPV) blend solid film studied by means of X and K-bands ESR at 77 and 120 K" *Synthetic Metals* 160 (5-6):485-489.

Krasicky, P. D., R. H. Silsbee, and J. C. Scott. 1982. "Studies of a polymeric chromium phosphinate - Electron-Spin Resonance and spin dynamics." *Physical Review B* 25 (9):5607-5626.

Krebs, F. C. 2012. *Stability and Degradation of Organic and Polymer Solar Cells*. Chichester, UK: John Wiley & Sons.

Krinichnyi, V. I. 1991a. "Investigation of biological systems by high-resolution 2-mm wave band ESR" *Journal of Biochemical and Biophysical Methods* 23 (1):1-30.

Krinichnyi, V. I. 1991b. "Investigation of biological systems by high resolution 2-mm wave band EPR." *Applied Magnetic Resonance* 2 (1):29-60.

Krinichnyi, V. I. 1995. *2-mm Wave Band EPR Spectroscopy of Condensed Systems*. Boca Raton, FL: CRC Press. Book.

Krinichnyi, V. I. 1996. "The nature and dynamics of nonlinear excitations in conducting polymers. Heteroaromatic polymers." *Russian Chemical Reviews* 65 (6):521-536.

Krinichnyi, V. I. 2000a. "2-mm Waveband electron paramagnetic resonance spectroscopy of conducting polymers (Review)." *Synthetic Metals* 108 (3):173-222.

Krinichnyi, V. I. 2000b. "The nature and dynamics of nonlinear excitations in conducting polymers. Polyaniline." *Russian Chemical Bulletin* 49 (2):207-233.

Krinichnyi, V. I. 2006. "High Field ESR Spectroscopy of Conductive Polymers." In *Advanced ESR Methods in Polymer Research*, edited by S. Schlick, 307-338. Hoboken, NJ: Wiley.

Krinichnyi, V. I. 2008a. "2-mm Waveband saturation transfer electron paramagnetic resonance of conducting polymers." *Journal of Chemical Physics* 129 (13):134510-134518.

Krinichnyi, V. I. 2008b. "The 140-GHz (D-band) saturation transfer electron paramagnetic resonance studies of macromolecular dynamics in conducting polymers." *Journal of Physical Chemistry B* 112 (32):9746-9752.

Krinichnyi, V. I. 2009. "LEPR spectroscopy of charge carriers photoinduced in polymer/fullerene composites." In *Encyclopedia of Polymer Composites: Properties,*

Performance and Applications, edited by M. Lechkov and S. Prandzheva, 417-446. Hauppauge, New York: Nova Science Publishers.

Krinichnyi, V. I. 2014a. "Dynamics of spin charge carriers in polyaniline." *Applied Physics Reviews* 1 (2):021305/01 – 021305/40.

Krinichnyi, V. I. 2014b. "Relaxation and dynamics of spin charge carriers in polyaniline." In *Advances in Materials Science Research*, edited by M. C. Wythers, 109-160. Hauppauge, New York: Nova Science Publishers.

Krinichnyi, V. I. 2016a. "EPR spectroscopy of polymer:fullerene nanocomposites." In *Spectroscopy of Polymer Nanocomposites*, edited by S. Thomas, D. Rouxel and D. Ponnamma, 202-275. Amsterdam: Elsevier.

Krinichnyi, V. I. 2016b. *Multi Frequency EPR Spectroscopy of Conjugated Polymers and Their Nanocomposites*. Boca Raton, FL: CRC Press Taylor & Francis Group. Book.

Krinichnyi, V. I., and A. A. Balakai. 2010. "Light-induced spin localization in poly(3-dodecylthiophen)/PCBM composite." *Applied Magnetic Resonance* 39 (3):319-328.

Krinichnyi, V. I., S. D. Chemerisov, and Y. S. Lebedev. 1997. "EPR and charge-transport studies of polyaniline." *Physical Review B* 55 (24):16233-16244.

Krinichnyi, V. I., N. N. Denisov, H. K. Roth, E. Fanghänel, and K. Lüders. 1998. "Dynamics of paramagnetic charge carriers in poly(tetrathiafulvalene)." *Polymer Science, Series A* 40 (12):1259-1269.

Krinichnyi, V. I., O. Y. Grinberg, I. B. Nazarova, G. I. Kozub, L. I. Tkachenko, M. L. Khidekel, and Y. S. Lebedev. 1985. "Polythiophene and polyacetylene conductors in the 3-cm and 2-mm electron spin resonance bands." *Bulletin of the Academy of Sciences of the USSR, Division of Chemical Science* 34 (2):425-427.

Krinichnyi, V. I., R. Herrmann, E. Fanghänel, W. Mörke, and K. Lüders. 1997. "Spin relaxation and magnetic properties of benzo-1,2,3- trithiolium radical cations." *Applied Magnetic Resonance* 12 (2-3):317-327.

Krinichnyi, V. I., A. L. Konkin, and A. Monkman. 2012. "Electron paramagnetic resonance study of spin centers related to charge transport in metallic polyaniline." *Synthetic Metals* 162 (13-14):1147–1155.

Krinichnyi, V. I., I. B. Nazarova, L. M. Goldenberg, and H. K. Roth. 1998. "Spin dynamics in conducting poly(aniline)." *Polymer Science, Series A* 40 (8):835-843.

Krinichnyi, V. I., A. E. Pelekh, H. K. Roth, and K. Lüders. 1993. "Spin relaxation studies on conducting poly(tetrathiafulvalene)." *Applied Magnetic Resonance* 4 (3):345-356.

Krinichnyi, V. I., A. E. Pelekh, L. I. Tkachenko, and G. I. Kozub. 1992. "Study of spin dynamics intrans -polyacetylene at 2-mm waveband EPR." *Synthetic Metals* 46 (1):1-12.

Krinichnyi, V. I., and H. K. Roth. 2004. "EPR study of spin and charge dynamics in slightly doped poly(3-octylthiophene)" *Applied Magnetic Resonance* 26:395-415.

Krinichnyi, V. I., H. K. Roth, G. Hinrichsen, F. Lux, and K. Lüders. 2002. "EPR and charge transfer in H_2SO_4-doped polyaniline." *Physical Review B* 65 (15):155205/01-155205/14.

Krinichnyi, V. I., H. K. Roth, and M. Schrödner. 2002. "Spin charge carrier dynamics in poly(*bis*-alkylthioacetylene)" *Applied Magnetic Resonance* 23:1-17.

Krinichnyi, V. I., H. K. Roth, M. Schrödner, and B. Wessling. 2006. "EPR study of polyaniline highly doped by*p*-toluenesulfonic acid" *Polymer* 47 (21):7460-7468.

Krinichnyi, V. I., H. K. Roth, S. Sensfuss, M. Schrödner, and M. Al Ibrahim. 2007. "Dynamics of photoinduced radical pairs in poly(3-dodecylthiophene)/fullerene composite." *Physica E* 36 (1):98-101.

Krinichnyi, V. I., S. V. Tokarev, H. K. Roth, M. Schrödner, and B. Wessling. 2006. "EPR study of charge transfer in polyaniline highly doped by *p*-toluenesulfonic acid." *Synthetic Metals* 156 (21-24):1368-1377.

Krinichnyi, V. I., P. A. Troshin, and N. N. Denisov. 2008. "The effect of fullerene derivative on polaronic charge transfer in poly(3-hexylthiophene/fullerene compound." *Journal of Chemical Physics* 128 (16):164715/01 - 164715/7.

Krinichnyi, V. I., E. I. Yudanova, and N. N. Denisov. 2014. "The role of spin exchange in charge transfer in low-bandgap polymer:fullerene bulk heterojunctions." *Journal of Chemical Physics* 141 (4):044906/01 - 044906/11.

Krinichnyi, V. I., E. I. Yudanova, and N. G. Spitsina. 2010. "Light-induced EPR study of poly(3-alkylthiophene)/fullerene composites." *Journal of Physical Chemistry C* 114 (39):16756–16766.

Krinichnyi, V. I., E. I. Yudanova, and B. Wessling. 2013. "Influence of spin–spin exchange on charge transfer min PANI-ES/P3DDT/PCBM composite." *Synthetic Metals* 179:67–73.

Krinichnyi, V. I. 2017. *2-mm Wave Band EPR Spectroscopy of Condensed Systems*. Boca Raton, FL: CRC Press Taylor & Francis Group. Book.

Kuivalainen, P., H. Stubb, P. Raatikainen, and C. Holmstrom. 1983. "Electron Spin Resonance studies of magnetic defects in $FeCl_3$-doped polyparaphenylene $(p-C_6H_4)_x$." *Journal de Physique Colloques* 44:C3-757-C3-760.

Kunugi, Y., Y. Harima, K. Yamashita, N. Ohta, and S. Ito. 2000. "Charge transport in a regioregular poly(3-octylthiophene) film." *Journal of Materials Chemistry* 10 (12):2673-2677.

Kuroda, S. 2002. "ESR and ENDOR spectroscopy of solitons and polarons in conjugated polymers." In *EPR in the 21st Century: Basics and Applications to Material, Life and Earth Sciences*, edited by A. Kawamori, J. Yamauchi and H. Ohta, 113-124. Amsterdam, Tokio: Elsevier Science B.V.

Lacaze, P. C., S. Aeiyach, and J. C. Lacroix. 1997. "Poly(*p*-phenylenes): preparation, techniques and properties." In *Handbook of Organic Conductive Molecules and Polymers*, edited by H. S. Nalwa, 205-270. Chichester, New York: Wiley.

Lan, Y. K., and C. I. Huang. 2008. "A theoretical study of the charge transfer behavior of the highly regioregular poly-3-hexylthiophene in the ordered state." *Journal of Physical Chemistry B* 112 (47):14857-14862.

Launay, J. P., and M. Verdaguer. 2013. *Electrons in Molecules: From Basic Principles to Molecular Electronics*. Oxford, New York: Oxford University Press.

Lebedev, Y. S., and V. I. Muromtsev. 1972. *EPR and Relaxation of Stabilized Radicals (Russ)*. Moscow: Khimija. Original edition, [Я.С. Лебедев, В.И. Муромцев, ЭПР и релаксация стабилизированных радикалов, М.: Химия], 1972, 256 с.

Lee, C. H., G. Yu, D. Moses, K. Pakbaz, C. Zhang, N. S. Sariciftci, A. J. Heeger, and F. Wudl. 1993. "Sensitization of the photoconductivity of conducting polymers by C-60 - photoinduced electron-transfer." *Physical Review B* 48 (20):15425-15433.

Lee, K. H., A. J. Heeger, and Y. Cao. 1993. "Reflectance of polyaniline protonated with camphor sulfonic acid - Disordered metal on the metal-insulator boundary."*Physical Review B* 48 (20):14884-14891.

Lee, K. H., R. A. J. Janssen, N. S. Sariciftci, and A. J. Heeger. 1994. "Direct evidence of photoinduced electron-transfer in conducting-polymer-C60 composites Infrared photoexcitation spectroscopy." *Physical Review B* 49 (8):5781-5784.

Lee, K., A. J. Heeger, and Y. Cao. 1995. "Reflectance spectra of polyaniline." *Synthetic Metals* 72 (1):25-34.

List, E. J. W., U. Scherf, K. Mullen, W. Graupner, C. H. Kim, and J. Shinar. 2002. "Direct evidence for singlet-triplet exciton annihilation in π-conjugated polymers." *Physical Review B* 66 (23):235203-235207.

Little, W. A. 1964. "Possibility of synthesizing an organic superconductor." *Physical Review* 134 (6A):1416-1424.

Lloyd, M. T., Y. F. Lim, and G. G. Malliaras. 2008. "Two-step exciton dissociation in poly(3-hexylthiophene)/fullerene heterojunctions." *Applied Physics Letters* 92 (14):143308-143310.

Lloyd, S. 1993. "A potentially realizable quantum computer." *Science* 261:1569-1571.

Long, A. R., and N. Balkan. 1980. "AC loss in amorphous germanium" *Philosophical Magazine B* 41(3):287-305.

Long, S. M., K. R. Cromack, A. J. Epstein, Y. Sun, and A. G. MacDiarmid. 1994. "ESR of pernigraniline base solutions revisited." *Synthetic Metals* 62 (3):287-289.

Lu, Y., ed. 1988. *Solitons and Polarons in Conducting Polymers*. River Edge, NJ, Singapore: World Scientific.

Lupton, J. M., D. R. McCamey, and C. Boehme. 2010. "Coherent spin manipulation in molecular semiconductors: Getting a handle on organic spintronics." *Chemical Physics Chemistry* 11 (14):3040-3058.

Łużny, W., M. Trznadel, and A. Proń. 1996. "X-Ray diffraction study of reoregular poly(3-alkylthiophenes)" *Synthetic Metals* 8171-74.

Madhukar, A., and W. Post. 1977. "Exact solution for the diffusion of a particle in a medium with site diagonal and off-diagonal dynamic disorder." *Physical Review Letters* 39 (22):1424-1427.

Maekawa, S., ed. 2006. *Concepts in Spin Electronics*. Oxford: Oxford University Press.

Marchant, S., and P. J. S. Foot. 1995. "Poly(3-hexylthiophene)-zinc oxide rectifying junctions."*Journal of Materials Science-Materials in Electronics* 6 (3):144-148.

Mardalen, J., E. J. Samuelsen, O. R. Konestabo, M. Hanfland, and M. Lorenzen. 1998. "Conducting polymers under pressure: synchrotron x-ray determined structure and structure related properties of two forms of poly(octyl-thiophene)." *Journal of Physics - Condensed Matter* 10:7145-7154.

Marumoto, K., Y. Muramatsu, N. Takeuchi, and S. Kuroda. 2003. "Light-induced ESR studies of polarons in regioregular poly(3-alkylthiophene)-fullerene composites." *Synthetic Metals* 135 (1-3):433-434.

Marumoto, K., N. Takeuchi, T. Ozaki, and S. Kuroda. 2002. "ESR studies of photogenerated polarons in regioregular poly(3-alkylthiophene)-fullerene composite."*Synthetic Metals* 129:239-247.

Masters, J. G., J. M. Ginder, A. G. MacDiarmid, and A. J. Epstein. 1992. "Thermochromism in the insulating forms of polyaniline - role of ring-torsional conformation." *Journal of Chemical Physics* 96 (6):4768-4778.

Masubuchi, S., R. Imai, K. Yamazaki, S. Kazama, J. Takada, and T. Matsuyama. 1999. "Structure and electrical transport property of poly(3-octylthiophene)." *Synthetic Metals* 101 (1-3):594-595.

Masubuchi, S., S. Kazama, K. Mizoguchi, M. Honda, K. Kume, R. Matsushita, and T. Matsuyama. 1993. "Metallic transport-properties in electrochemically as-grown and heavily-doped polythiophene and poly(3-methylthiophene)." *Synthetic Metals* 57 (2-3):4962-4967.

Matthews, M. J., M. S. Dresselhaus, N. Kobayashi, T. Enoki, M. Endo, and K. Nishimura. 1999. "Localized spins in partially carbonized polyparaphenylene." *Physical Review B* 60 (7):4749-4757.

McCullough, R. D., R. D. Lowe, M. Jayaraman, P. C. Ewbank, and D. L. Anderson. 1993. "Synthesis and physical-properties of regiochemically well-defined, head-to-tail coupled poly(3-alkylthiophenes)." *Synthetic Metals* 55 (2-3):1198-1203.

Menon, R. 1997. "Charge Transport in Conducting Polymers." In *Handbook of Organic Conductive Molecules and Polymers*, edited by H. S. Nalwa, 47-145. Chichester: John Wiley & Sons.

Menon, R., C. O. Yoon, D. Moses, and A. J. Heeger. 1997. "Metal-Insulator transition in doped conducting polymers." In *Handbook of Conducting Polymers*, edited by T. A. Skotheim, R. L. Elsenbaumer and J. R. Reynolds, 27-84. New York: Marcel Dekker.

Misra, S. K., ed. 2011. *Multifrequency Electron Paramagnetic Resonance. Theory and Applications*. 2 vols. Vol. 2. Weinheim: Wiley-VCH.

Misra, S. K., ed. 2014. *Handbook of Multifrequency Electron Paramagnetic Resonance: Data and Techniques*. Weinheim, Germany: Wiley-VCH.

Mizoguchi, K., M. Honda, S. Masubuchi, S. Kazama, and K. Kume. 1994. "Study of spin dynamics and electronic-structure in polythiophene heavily-doped with ClO_4." *Japanese Journal of Applied Physics Part 2-Letters* 33 (9A):L1239-L1241.

Mizoguchi, K., and S. Kuroda. 1997. "Magnetic properties of conducting polymers." In *Handbook of Organic Conductive Molecules and Polymers*, edited by H. S. Nalwa, 251-317. Chichester, New York: John Wiley & Sons.

Molin, Y. N., K. M. Salikhov, and K. I. Zamaraev. 1980. *Spin Exchange*. Berlin: Springer. [К. И. Замараев, Ю. Н. Молин, К. М. Салихов, Спиновыйобмен. Теория и физико-химические приложения, Наука СО: Новосибирск, 1977, 320 с.].

Moon, J. S., J. Jo, and A. J. Heeger. 2012. "Nanomorphology of PCDTBT:PC_{70}BM bulk heterojunction solar cells." *Advanced Energy Materials* 2 (3):304-308.

Moraes, F., D. Davidov, M. Kobayashi, T. C. Chung, J. Chen, A. J. Heeger, and F. Wudl. 1985. "Doped poly(thiophene) - Electron-Spin Resonance determination of the magnetic susceptibility." *Synthetic Metals* 10 (3):169-179.

Mott, N. F., and E. A. Davis. 2012. *Electronic Processes in Non-Crystalline Materials, Oxford Classic Texts in the Physical Sciences*. Oxford: Clarendon Press.

Nagels, P. 1980. *The Hall Effect and its Application*. New York: Plenum.

Nalwa, H. S., ed. 1997. *Handbook of Organic Conductive Molecules and Polymers*. Edited by H. S. Nalwa. 1-4 vols. Vol. 1-4. Chichester, New York: John Wiley & Sons.

Nalwa, H. S., ed. 2001. *Advanced Functional Molecules and Polymers: Electronic and Photonic Properties*. Edited by H. S. Nalwa. Vol. 3. Boca Raton: CRC Press.

Nechtschein, M. 1997. "Electron spin dynamics." In *Handbook of Conducting Polymers*, edited by T. A. Skotheim, R. L. Elsenbaumer and J. R. Reynolds, 141-163. New York: Marcel Dekker.

Nelson, J. 2003. "Diffusion-limited recombination in polymer-fullerene blends and its influence on photocurrent collection." *Physical Review B* 67 (15):155209/01-155209/10.

Niklas, Jens, Kristy L. Mardis, Brian P. Banks, Gregory M. Grooms, Andreas Sperlich, Vladimir Dyakonov, Serge Beaupře, Mario Leclerc, Tao Xu, Luping Yue, and Oleg G. Poluektov. 2013. "Highly-efficient charge separation and polaron delocalization in polymer–fullerene bulk-heterojunctions: a comparative multi-frequency EPR and DFT study." *Physical Chemistry Chemical Physics* 15 (24):9562-9574.

Obrzut, J., and K. A. Page. 2009. "Electrical conductivity and relaxation in poly(3-hexylthiophene)." *Physical Review B* 80 (19):195211/01 - 195211/07.

Österbacka, R., C. P. An, X. M. Jiang, and Z. V. Vardeny. 2001. "Delocalized polarons in self-assembled poly(3-hexylthiophene) nanocrystals." *Synthetic Metals* 116 (1-3):317-320.

Owens, J. 1977. "Evidence for zero-field fluctuations in Cr^{3+} near the phase transition in $NH_4Al(SO_4)_2 \cdot 12H_2O$." *Physica Status Solidi B* 79 (2):623-628.

Pagliaro, M., G. Palmisano, and R. Ciriminna. 2008. *Flexible Solar Cells*: Wiley-WCH.

Park, S. H., A. Roy, S. Beaupre, S. Cho, N. Coates, J. S. Moon, D. Moses, M. Leclerc, K. Lee, and A. J. Heeger. 2009. "Bulk heterojunction solar cells with internal quantum efficiency approaching 100%." *Nature Photonics* 3:297-302.

Parneix, J. P., and M. El Kadiri. 1987. "Frequency- and temperature-dependent dielectric losses in lightly doped conducting polymers." In *Electronic Properties of Conjugated Polymers*, edited by H. Kuzmany, M. Mehring and S. Roth, 23-26. Berlin: Springer-Verlag.

Patzsch, J. 1991. "Elektrische Eigenschaften von polymeren Tetrathiafulvalen (PTTF)." Ph.D., Technische Hochschule Leipzig.

Patzsch, J., and H. Gruber. 1992. "Small Polarons in Polymeric Polytetrathiafulvalenes (PTTF)." In *Electronic Properties of Polymers*, edited by H. Kuzmany, M. Mehring and S. Roth, 121-124. Berlin: Springer-Verlag.

Peierls, R. E. 1996. *Quantum Theory of Solids*. London, U.K.: Oxford University Press.

Pelekh, A. E., L. M. Goldenberg, and V. I. Krinichnyi. 1991. "Study of dopped polypyrrole by the spin probe method at 3-cm and 2-mm waveband EPR." *Synthetic Metals* 44 (2):205-211.

Pietronero, L. 1983. "Ideal conductivity of carbon π-polymers and intercalation compounds." *Synthetic Metals* 8:225-231.

Poluektov, O. G., S. Filippone, N. Martín, A. Sperlich, C. Deibel, and V. Dyakonov. 2010. "Spin signatures of photogenerated radical anions in polymer-[70]fullerene bulk heterojunctions: High frequency pulsed EPR spectroscopy." *Journal of Physical Chemistry B* 114 (45):14426-14429.

Poole, Ch P. 1983. *Electron Spin Resonance, A Comprehensive Treatise on Experimental Techniques*. New York: John Wiley & Sons.

Poortmans, J., and V. Arkhipov, eds. 2006. *Thin Film Solar Cells: Fabrication, Characterization and Applications*Edited by J. Poortmans and V. Arkhipov. West Sussex: Wiley.

Pratt, F. L., S. J. Blundell, W. Hayes, K. Nagamine, K. Ishida, and A. P. Monkman. 1997. "Anisotropic polaron motion in polyaniline studied by muon spin relaxation." *Physical Review Letters* 79 (15):2855-2858.

Quang, T. 1987. PhD, *Chemical*, Carl Schorlemmer Technical University Merseburg.

Raghunathan, A., P. K. Kahol, J. C. Ho, Y. Y. Chen, Y. D. Yao, Y. S. Lin, and B. Wessling. 1998. "Low-temperature heat capacities of polyaniline and polyaniline polymethylmethacrylate blends." *Physical Review B* 58 (24):R15955-R15958.

Ranby, B., and J. F. Rabek. 2011. *ESR Spectroscopy in Polymer Research Series: Polymers - Properties and Applications (Book 1)*: Springer.

Reddoch, A., and S. Konishi. 1979. "The solvent effect on di-tert- butyl nitroxide. A dipole-dipole model for polar solutes in polar solvents." *Journal of Chemical Physics* 70:2121.

Reufer, M., P. G. Lagoudakis, M. J. Walter, J. M. Lupton, J. Feldmann, and U. Scherf. 2006. "Evidence for temperature-independent triplet diffusion in a ladder-type conjugated polymer." *Phys. Rev. B* 74 (24):241201-241204.

Richter, A. M., J. M. Richter, N. Beye, and E. Fanghänel. 1987. "Organische Elektronenleiter und Vorstufen. V. Synthese von Poly(organylthio-acetylenen)." *Journal Praktische Chemie* 329 (5):811-816.

Rodriguez, J., H. J. Grande, and T. F. Otero. 1997. "Polypyrroles: From basic research to technological applications." In *Handbook of Organic Conductive Molecules and Polymers*, edited by H. S. Nalwa, 415-468. Chichester, New York: Wiley.

Roth, H. K., W. Brunner, G. Volkel, M. Schrödner, and H. Gruber. 1990. "ESR and ESE studies on polymer semiconductors of weakly doped poly(tetrathiafulvalene)." *Makromolekulare Chemie-Macromolecular Symposia* 34:293-307.

Roth, H. K., H. Gruber, E. Fanghänel, and Trinh vu Quang. 1988. "ESR on polymer semiconductors of poly(tetrathiafulvalene)." *Progress in Colloid Polymer Science* 78:75-78.

Roth, H. K., H. Gruber, E. Fanghänel, A. M. Richter, and W. Horig. 1990. "Laser-Induced generation of highly conducting areas in poly(*bis*-alkylthioacetylenes)." *Synthetic Metals* 37:151-164.

Roth, H. K., H. Gruber, G. Voelkel, W. Brunner, and E. Fanghänel. 1989. "Electron spin resonance and relaxation studies on conducting poly(tetrathiafulvalenes)." *Progress in Colloid and Polymer Science* 80 (1):254-263.

Roth, H. K., F. Keller, and H. Schneider. 1984. *Hochfrequenzspectroskopie in der Polymerforschung*. Berlin: Academie Verlag.

Roth, H. K., and V. I. Krinichnyi. 2003. "Spin and charge transport in poly(3-octylthiophene)." *Synthetic Metals* 137 (1-3):1431-1432.

Roth, H. K., V. I. Krinichnyi, M. Schrödner, and R. I. Stohn. 1999. "Electronic properties of laser modified poly(*bis*-alkylthio-acetylene)"*Synthetic Metals* 101 (1-3):832-833.

Rouxel, Didier, Sabu Thomas, and Deepalekshmi Ponnamma, eds. 2016. *Spectroscopy of Polymer Nanocomposites*. 1 ed. Oxford, Cambridge, Amsterdam: Elsevier. Original edition, Spectroscopy of Polymer Nanocomposites.

Sakamoto, H., N. Kachi, K. Mizoguchi, K. Yoshioka, S. Masubuchi, and S. Kazama. 1999. "Origin of ESR linewidth for polypyrrole." *Synthetic Metals* 101 (1-3):481-481.

Salamone, J. C., ed. 1996. *Polymeric Material Encyclopedia*. Edited by J. C. Salamone. 12 vols. Boca Raton, Fl.: CRC Press.

Salaneck, W. R., D. T. Clark, and E. J. Samuelsen, eds. 1991. *Science and Applications of Conducting Polymers, Papers from the Sixth European Industrial Workshop*. Edited by W. R. Salaneck, D. T. Clark and E. J. Samuelsen. New York (Bristol): Adam Hilger.

Samuelsen, E. J., and J. Mardalen. 1997. "Structure of polythiophenes." In *Handbook of Organic Conductive Molecules and Polymers*, edited by H. S. Nalwa, 87-120. Chichester, New York: John Wiley.

Sariciftci, N. S., and A. J. Heeger. 1995. "Photophysics of semiconducting polymer C-60 composites - A comparative-study." *Synthetic Metals* 70 (1-3):1349-1352.

Sariciftci, N. S., A. J. Heeger, and Y. Cao. 1994. "Paramagnetic susceptibility of highly conducting polyaniline - Disordered metal with weak electron-electron interactions (Fermi glass)." *Physical Review B* 49 (9):5988-5992.

Sariciftci, N. S., A. C. Kolbert, Y. Cao, A. J. Heeger, and A. Pines. 1995. "Magnetic resonance evidence for metallic state in highly conducting polyaniline." *Synthetic Metals* 69 (1-3):243-244.

Saunders, B. R., R. J. Fleming, and K. S. Murray. 1995. "Recent advances in the physical and spectroscopic properties of polypyrrole films, particularly those containing transition-metal complexes as counteranions." *Chemistry of Materials* 7 (6):1082-1094.

Sauvajol, J. L., D. Bormann, M. Palpacuer, J. P. Lere-Porte, J. J. E. Moreau, and A. J. Dianoux. 1997. "Low and high-frequency vibrational dynamics as a function of structural order in polythiophene." *Synthetic Metals* 84 (1-3):569-570.

Scharli, M., H. Kiess, G. Harbeke, W. Berlinger, K. W. Blazey, and K. A. Muller. 1987. "ESR of BF_4-doped poly(3-methylthiophene)." In *Electronic Properties of Conjugated Polymers*, edited by H. Kuzmany, M. Mehring and S. Roth, 277-280. Berlin: Springer-Verlag.

Schlick, S., ed. 2006. *Advanced ESR Methods in Polymer Research*. Edited by S. Schlick. New York: John Wiley & Sons Inc.

Schmeisser, D., A. Bartl, L. Dunsch, H. Naarmann, and W. Gopel. 1998. "Electronic and magnetic properties of polypyrrole films depending on their one-dimensional and two-dimensional microstructures." *Synthetic Metals* 93 (1):43-58.

Schultz, N. A., M. C. Scharber, C. J. Brabec, and N. S. Sariciftci. 2001. "Low-temperature recombination kinetics of photoexcited persistent charge carriers in conjugated polymer/fullerene composite films." *Physical Review B* 64 (24):245210/01-245210/07.

Scotheim, T. E., R. L. Elsenbaumer, and J. R. Reynolds, eds. 1997. *Handbook of Conducting Polymers*. Edited by T. E. Scotheim, R. L. Elsenbaumer and J. R. Reynolds. New York: Marcel Dekker.

Scotheim, T. E., and J. R. Reynolds, eds. 2007. *Handbook of Conducting Polymers*. Edited by T. E. Scotheim and J. R. Reynolds. Vol. 3d. Boca Raton: CRC Press.

Scott, J. C., P. Pfluger, M. T. Kroumbi, and G. B. Street. 1983. "Electron-spin-resonanse studies of pyrrole polymers: Evidence for bipolarons" *Physical Review B* 28 (4):2140-2145.

Scrosati, B., ed. 1993. *Application of Electroactive Polymers*. Edited by B. Scrosati. London: Chapman & Hall.

Sersen, F., G. Cik, and P. Veis. 2003. "Study of conformational changes in poly(3-dodecylthiophene) dependent on backbone stereoregularity using a spin-probe technique." *Journal of Applied Polymer Science* 88 (9):2215-2223.

Shaheen, S. E., C. J. Brabec, N. S. Sariciftci, F. Padinger, T. Fromherz, and J. C. Hummelen. 2001. "2.5% Efficient organic plastic solar cells." *Applied Physics Letters* 78 (6):841-843.

Sharma, G. D. 2010. "Advances in nano-structured organic solar cells." In *Physics of Nanostructured Solar Cells*, edited by V. Badescu and M. Paulescu, 361−460. New York: Nova Science Publishers.

Shchegolikhin, A. N., I. V. Yakovleva, and M. Motyakin. 1995. "Optical-Properties of poly(diacetylene) block-copoly(ether-urethanes), containing covalently bound nitroxyl spin labels in the main-chain." *Synthetic Metals* 71 (1-3):2091-2092.

Shen, L., M. Zeng, Q. Wu, Z. Bai, and Y. P. Feng. 2014. "Graphene spintronics: Spin generation and manipulation in graphene." In *Graphene Optoelectronics: Synthesis, Characterization, Properties, and Applications*, edited by Rashid bin Mohd Yusoff, 167-188. Weinheim: Wiley.

Silinsh, E. A., M. V. Kurik, and V. Chapek. 1988. *Electronic Processes in Organic Molecular Crystals: The Phenomena of Localization and Polarization (Russian).* Riga: Zinatne. Original edition, [Силиньш, Э.А., Курик, М.В., Чапек, В., Электронные процессы в органических молекулярных кристаллах: Явления локализации и поляризации, Рига: Зинатне], 1988, 329 с.

Singleton, J. 2001. *Band Theory and Electronic Properties of Solids, Oxford Master Series in Condensed Matter Physics.* Oxford: Oxford University Press.

So, F., J. Kido, and P. Burrows. 2008. "Organic light-emitting devices for solid-state lighting." *Materials Research Society Bulletin* 33 (7):663–669.

Springborg, M. 1992. "The electronic-properties of polythiophene." *Journal of Physics - Condensed Matter* 4 (1):101-120.

Stafstrom, S., and J. L. Brédas. 1988. "Band-Structure calculations for the polaron lattice in the highly doped regime of polyacetylene, polythiophene, and polyaniline." *Molecular Crystals and Liquid Crystals* 160:405-420.

Stafström, S., and J. L. Brédas. 1988. "Evolution of the electronic-structure of polyacetylene and polythiophene as a function of doping level and lattice conformation." *Physical Review B* 38 (6):4180-4191.

Sun, S. S., and N. S. Sariciftci, eds. 2005. *Organic Photovoltaics: Mechanisms, Materials, and Devices (Optical Engineering).* Edited by S. S. Sun and N. S. Sariciftci, *Optical Science and Engineering Series.* Boca Raton: CRC Press.

Tachiya, M., and K. Seki. 2010. "Theory of bulk electron-hole recombination in a medium with energetic disorder."*Physical Review B* 82 (8):085201/01 - 085201/08.

Taka, T., O. Jylha, A. Root, E. Silvasti, and H. Osterholm. 1993. "Characterization of undoped poly(3-octylthiophene)." *Synthetic Metals* 55 (1):414-419.

Takeda, K., H. Hikita, Y. Kimura, H. Yokomichi, and K. Morigaki. 1998. "Electron spin resonance study of light-induced annealing of dangling bonds in glow discharge hydrogenated amorphous silicon: Deconvolution of electron spin resonance spectra." *Japanese Journal of Applied Physics Part 1-Regular Papers Short Notes & Review Papers* 37 (12A):6309-6317.

Tanaka, H., N. Hasegawa, T. Sakamoto, K. Marumoto, and S. I. Kuroda. 2007. "Light-induced ESR studies of quadrimolecular recombination kinetics of photogenerated charge carriers in regioregular poly(3-alkylthiophene)/C-60 composites: Alkyl chain dependence." *Japanese Journal of Applied Physics Part 1-Regular Papers Brief Communications & Review Papers* 46 (8A):5187-5192.

Taunk, M., and S. Chand. 2014. "Variable range hopping transport in polypyrrole composite films." In *Physics of Semiconductor Devices*, edited by V. K. Jain and A. Verma, 903-904. Heidelberg New York Dordrecht London: Springer.

Tourillon, G., D. Gourier, F. Garnier, and D. Vivien. 1984. "Electron spin resonance study of electrochemically generated poly(thiophene) and derivatives." *Journal of Physical Chemistry* 88:1049.

Traven', V. F. 1989. *Electronic Structure and Properties of Organic Molecules (Russian)*. Moscow: Khimija. Original edition, [Травень, В. Ф. "Электронная структура и свойства органических молекул." Москва: Химия]. 1989.

Trinh, V. Q., L. V. Hinh, G. Shukat, and E. Fanghänel. 1989. "Tetrathiafulvalene. XXV [1]. Konjugativ verknüpfte polymere Tetrathiafulvalene (TTF)." *Journal für Praktische Chemie* 331 (5):826-834.

Troisi, A., and G. J. Orlandi. 2006. "Dynamics of the intermolecular transfer integral in crystalline organic semiconductors." *Journal of Physical Chemistry A* 110 (11):4065–4070.

Van Vooren, A., J.-S. Kim, and J. Cornil. 2008. "Intrachain versus interchain electron transport in poly(fluorene-alt-benzothiadiazole): A quantum-chemical insight." *Chem Phys Chem* 9 (7):989-993.

Vonsovskii, S. V. 1974. *Magnetism. Magnetic Properties of Dia-, Para-, Ferro-, Antiferro-, and Ferrimagnetics*. 2 vols. Vol. 1,2. New York: John Wiley & Sons. Original edition, [Вонсовский, С.В., Магнетизм, Магнитные свойства диа-, пара-, ферро-, антиферро- и ферримагнетиков, М.: Наука], 1971, 1032 с.

Wan, M. X. 2008. *Conducting Polymers with Micro or Nanometer Structure*. Berlin, Heidelberg, New York: Springer.

Wang, Z. H., C. Li, E. M. Scherr, A. G. MacDiarmid, and A. J. Epstein. 1991. "3 Dimensionality of metallic states in conducting polymers - Polyaniline." *Physical Review Letters* 66 (13):1745-1748.

Wang, Z. H., A. Ray, A. G. MacDiarmid, and A. J. Epstein. 1991. "Electron localization and charge transport inpoly(*o*-toluidine) - A model polyaniline derivative." *Physical Review B* 43 (5):4373-4384.

Wang, Z. H., E. M. Scherr, A. G. MacDiarmid, and A. J. Epstein. 1992. "Transport and EPR studies of polyaniline - A quasi-one-dimensional conductor with 3-dimensional metallic states." *Physical Review B* 45 (8):4190-4202.

Weil, J. A., J. R. Bolton, and J. E. Wertz. 2007. *Electron Paramagnetic Resonance: Elementary Theory and Practical Applications*. Vol. 2d. New York: Wiley-Interscience.

Wertz, J. E., and J. R.Bolton. 2013. *Electron Spin Resonance: Elementary Theory and Practical Applications*: Springer Verlag.

Wessling, B. 1997. "Metallic properties of conductive polymers due to dispersion." In *Handbook of Organic Conductive Molecules and Polymers*, edited by H. S. Nalwa, 497-632. Chichester: John Wiley & Sons.

Wessling, B. 2010. "New insight into organic metal polyaniline morphology and structure." *Polymers* 2:786-798.

Wessling, B., D. Srinivasan, G. Rangarajan, T. Mietzner, and W. Lennartz. 2000. "Dispersion-induced insulator-to-metal transition in polyaniline." *European Physical Journal E* 2 (3):207-210.

Westerling, M., R. Osterbacka, and H. Stubb. 2002. "Recombination of long-lived photoexcitations in regioregular polyalkylthiophenes." *Physical Review B* 66 (16):165220/01 - 165220/07.

Williams, J. M., J. R. Ferraro, R. J. Thorn, K. D. Carlson, U. Geiser, H. H. Wang, A. M. Kini, and M. H. Whangboo. 1992. *Organic Superconductors (Including Fullerenes): Synthesis, Structure, Properties, and Theory*. New Jersey: Prentice-Hall, Inc., Englewood Cliffs.

Winter, H., G. Sachs, E. Dormann, R. Cosmo, and H. Naarmann. 1990. "Magnetic-Properties of spin-labeled polyacetylene." *Synthetic Metals* 36 (3):353-365.

Wohlgenannt, M., K. Tandon, S. Mazumdar, S. Ramasesha, and Z. V. Vardeny. 2001. "Formation cross-sections of singlet and triplet excitons in p-conjugated polymers." *Nature* 409:494-497.

Wolf, S. A., D. D. Awschalom, R. A. Buhrman, J. M. Daughton, S. von Molnar, M. L. Roukes, A. Y. Chtchelkanova, and D. M. Treger. 2001. "Spintronics: A spin-based electronics vision for the future." *Science* 294 (5546):1488-1495.

Wong, C. P., ed. 1993. *Polymers for Electronic & Photonic Application*. Boston, USA: Elsevier - Academic Press Inc.

Wu, T. M., H. L. Chang, and Y. W. Lin. 2009. "Synthesis and characterization of conductive polypyrrole/multi-walled carbon nanotubes composites with improved solubility and conductivity." *Composites Science and Technology* 69 (5):639-644.

Xie, S. J., L. M. Mei, and D. L. Lin. 1994. "Transitio between bipolaron and polaron states in doped heterocycle polymers." *Physical Review B* 50 (18):13364-13370.

Yamauchi, T., H. M. Najib, Y. W. Liu, M. Shimomura, and S. Miyauchi. 1997. "Positive temperature coefficient characteristics of poly(3-alkylthiophene)s." *Synthetic Metals* 84 (1-3):581-582.

Yan, B., N. A. Schultz, A. L. Efros, and P. C. Taylor. 2000. "Universal distribution of residual carriers in tetrahedrally coordinated amorphous semiconductors." *Physical Review Letters* 84 (18):4180-4183.

Yanilkin, V. V., N. V. Nastapova, V. I. Morozov, V. P. Gubskaya, F. G. Sibgatullina, L. S. Berezhnaya, and I. A. Nuretdinov. 2007. "Competitive conversions of carbonyl-containing methanofullerenes induced by electron transfer." *Russian Journal of Electrochemistry* 43 (2):184-203.

Yazawa, K., Y. Inoue, T. Shimizu, M. Tansho, and N. Asakawa. 2010. "Molecular dynamics of regioregular poly(3-hexylthiophene) investigated by NMR relaxation and an interpretation of temperature dependent optical absorption." *Journal of Physical Chemistry B* 114 (3):1241-1248.

Yudanova, E. I., V.R. Bogatyrenko, and V. I. Krinichnyi. 2016. "EPR Study of spin interactions in poly(3-dodecylthiophene):fullerene/polyaniline:*p*-toluenesulfonic acid composite." *High Energy Chemistry* 50 (2):132-138.

Zanardi, C., F. Terzi, L. Pigani, and R. Seeber. 2009. Electrode coatings consisting of polythiophene-based composites containing metal centres. In *Encyclopedia of Polymer Composites: Properties, Performance and Applications*, edited by M. Lechkov and S. Prandzheva. Hauppauge, NY: Nova Science Publishiers.

Zaumseil, J. 2014. "P3HT and other polythiophene field-effect transistors." In *P3HT Revisited: From Molecular Scale to Solar Cell Devices*, edited by S. Ludwigs, 107-137.

Zuppiroli, L., S. Paschen, and M. N. Bussac. 1995. "Role of the dopant counterions in the transport and magnetic-properties of disordered conducting polymers." *Synthetic Metals* 69 (1-3):621-624.

ABOUT THE AUTHOR

Victor I. Krinichnyi

Institute of Problems of Chemical Physics,
Chernogolovka, Russian Federation
Email: kivirus@gmail.com

Victor I. Krinichnyi was born in Kazan, USSR, in 1953. He received the diploma in radiophysics and electronics from the Kazan State University, USSR, in 1975. He was employed as an engineer (1975-1979), a principal engineer (1979-1982), a young scientific researcher (1982-1987), a scientific researcher (1987-1992), a senior scientific researcher (1992-1996) and a leading scientific researcher (1997-present) in the former Institute of Chemical Physics in Chernogolovka RAS, currently the Institute of Problems of Chemical Physics RAS, Chernogolovka, Moscow Region, Russia. He received the Ph.D. (in 1986) and Sci.D. (in 1993) degrees in physics and mathematics from the Institute of Problems of Chemical Physics RAS. His research interests resulting from practical application of multifrequency EPR spectroscopy include the relaxation and dynamics of non-linear charge carriers, solitons and polarons, in conjugated polymers and their nanocomposites, mechanism of charge transport in molecular crystals, spin phenomena in condensed systems as well as organic molecular electronics, photonics and spintronics. He collaborated as invited summer researcher in the Center of Atomic Energy (1994), Grenoble, France, Merseburg (1994), Jena (1997), Stuttgart (2001) and Ilmenau (2010, 2017) Universities, Institute for Physical High Technology (2004), Jena, and Polymer Research Institute (1998, 2000, 2002, 2003, 2004, 2005), Rudolstadt, Germany. He is the author of two monographs, "2-mm Wave Band EPR Spectroscopy of Condensed Systems" (Boca Raton, Fl.: CRC Press, 1995) and "Multi Frequency EPR Spectroscopy of Conjugated Polymers and Their Nanocomposites" (Boca Raton, Fl.: CRC Press, 2016), five contributions in edited books, eleven reviews and more than 100 articles. Dr.Sci. Krinichnyi is a member of the International EPR (ESR) Society since 1992.

In: Polarons: Recent Progress and Perspectives
Editor: Amel Laref

ISBN: 978-1-53613-935-8
© 2018 Nova Science Publishers, Inc.

Chapter 2

TIGHT-BINDING MODELS FOR THE CHARGE TRANSPORT IN ORGANIC SEMICONDUCTORS

Luiz Antônio Ribeiro Junior[1,2],
Antonio Luciano de Almeida Fonseca[2],
Jonathan Fernando Teixeira[3], Wiliam Ferreira da Cunha[4]
and Geraldo Magela e Silva[4]

[1]Department of Physics, Chemistry and Biology (IFM),
Linkoping University, Linkoping, Sweden
[2]International Center for Condensed Matter Physics,
University of Brasília, Brasília, DF, Brazil
[3]Institute Federal of Brasilia, Brasilia, Brazil
[4]Institute of Physics, University of Brasília, Brasília, Brazil

ABSTRACT

This chapter is dedicated to an exposition of the most fundamental aspects concerning the Tight-Binding treatment of charge transport in organic semiconductors. We begin with a short section containing the historical background regarding our systems of interest, namely, conducting polymers. Following, we dedicate a section to the presentation of the most important theoretical methods associated to the Tight-Binding approach. The modified Su-Schrieffer-Heeger (SSH) model, the applied variant of the Tight-Binding Hamiltonian, is detailed in its own section together with a description of the tools used to solve the systems' equations. The chapter is concluded with a brief summary of some results obtained by following this methodology.

1. INTRODUCTION

In this section, a brief historical background regarding the discovery of conducting polymers is presented. An introduction about π-conjugated molecular systems, particularly polyacetylene, is provided. Furthermore, we briefly discuss the physical properties of some important applications to the recombination process between quasi-particles in organic-based optoelectronic devices.

1.1. Historical Background

Since the discovery of conductivity in organic molecular crystals in the 60s, a huge amount of studies has been performed using this class of materials to figure out the mechanisms involved in the charge transport phenomena and in order to obtain new species of organic semiconductors [1-3]. Devoted to this research field, the group of Prof. Hideki Shirakawa was working on the synthesis of polymers from acetylene gas since the late 1960s [1]. During an experiment, in 1974, a PhD student under his guidance accidentally added thousands times more catalyst than it was usually required. As a result, it was obtained a silvery and shining film with elasticity similar to a plastic thin-sheet, named *trans*-polyacetylene [4-8].

Albeit the material obtained by Shirakawa and his student seemed a metallic-like film, they have found that it was not electrically conductive. In 1977, Shirakawa started a very fruitful collaboration with Alan Heeger and Alan MacDiarmid at the University of Pennsylvania that was aimed to obtain polyacetylene films with greater conductivity [9]. In the same year, they increased significantly the electrical conductivity of trans-polyacetylene by doping it with bromine [10]. Moreover, using well oriented samples of the resulting film, conductivity levels nearly as high as that to cooper were reached [11].

The results derived from the above-mentioned collaboration gave rise to a new generation of organic semiconductors, termed conjugated polymers. The interest about these materials increased dramatically since that decade. Thereafter, many other conjugated polymers have been both theoretically and experimentally studied. Particularly, after the discovery of the electroluminescence properties in poly(*p*-phenylene vinylene) films [12], conjugated polymers have shown a bright future as active materials in several technological applications such as Organic Light-Emitting Diodes (OLEDs) [13, 14], Organic Solar Cells (OSCs) [15, 16], and Organic Thin-Film Transistors (OTFTs) [17, 18]. Furthermore, the electronic and optical properties of conjugate polymers, together with their mechanical properties, processing advantages, versatility of chemical synthesis, and low cost make them attractive materials for the electronics industry, mainly when it comes to the promising development of a new display technology [19]. Due to the discoveries and advances made in the field of conducting polymers Heeger, MacDiarmid, and Shirakawa were awarded the Nobel Prize in Chemistry in the year 2000.

1.2. The Polyacetylene

Polyacetylene is one of the simplest conjugated polymers with an one-dimensional structure being formed by a long chain of carbon atoms [2, 3]. The term *conjugation* represents the switching of double and single bonds between the carbon atoms that constitute the chain, where each carbon atom is bonded to a hydrogen atom and laterally connected to other two carbons in a linear arrangement [2, 3]. During the 30s and 40s, this polymer was subject of several controversial theoretical discussions regarding the role played by π-electrons. From a particular point of view, it was thought, by symmetry considerations, that if the polymer chain were conjugated, there would be a separation (*gap*) between the last occupied electronic state and the first empty one. Therefore, the material would have semiconducting properties. It was also thought that, if the electron cloud were delocalized throughout the system, the bond lengths would tend to posses the same size. Thus, the symmetry of the system would lead to a metallic-like band structure, i.e., with a solely half-filled band.

From the theoretical point of view, the controversy was solved by understanding that, due to its polymeric structure, polyacetylene would have one-dimensional properties, thus being subjected to the theorem enunciated in 1953 by Rudolph Peierls [20]: any *one-dimensional conductor is unstable, subject to structural changes that may make it semiconducting*. Therefore, this theorem was favorable to the point of view which pointed out that the conjugated polymer chain would be the more stable structure. In this kind of structure, polyacetylene is weakly coupled to the neighboring chain; consequently, the resulting material is flexible. The bond conjugation produces two different distances between the CH groups: the carbons bonded by a π-type bond are in a smaller distance between them than those bonded by a σ-type bond. This process is termed dimerization. Experimental studies showed the existence of two isomers for polyacetylene: the *trans*-polyacetylene (tPA) Figure 1(a), more stable from the thermodynamic point of view, and the *cis*-polyacetylene (cPA) Figure 1(b) [5-7]. For both configurations, experiments showed the existence of unpaired electrons whose mobility was higher in the tPA isomer. In addition, samples of the tPA isomer exhibited metallic levels for the conductivity when they were exposed to high concentrations of dopants [11]. On the other hand, samples of cPA, for similar dopant concentrations, presented quite lower conductivity [11]. Moreover, it was found that the doping process is reversible, i.e., the polyacetylene film regained its original properties by decreasing the impurities concentration [5]. This fact allowed the conductivity degree of the sample to be precisely controlled [12, 13].

Thereafter, many conducting polymers, with more complex structures, were synthesized. However, one fundamental feature was common to all: the alternating between single and double bonds, i.e., these complex structures are also conjugated polymers. The poly(*p*-phenylene vinylene) (PPV), which its structure is represented in Figure 1(c), for example, is a conjugate polymer widely used in the fabrication of OLEDs recently, due to the high photoluminescence yield it possesses.

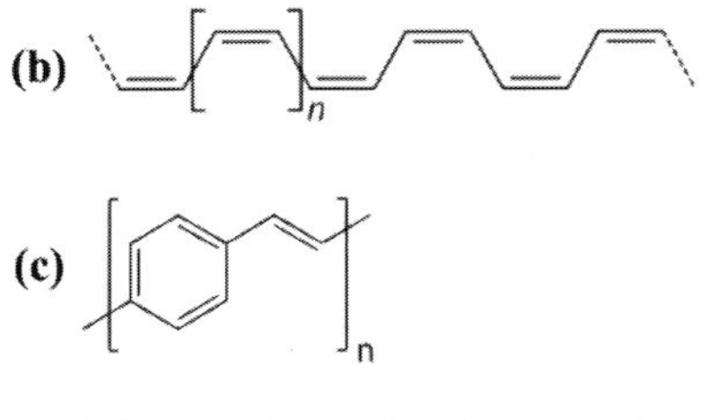

Figure 1. Schematic representation of three conjugated polymers: (a) trans-polyacetylene, (b) cis-polyacetylene and (c) poly(p-phenylene vinylene.

2. THEORETICAL BACKGROUND

The present section is intended to serve as an introduction to the theoretical methods referring to the Tight-Binding approach discussed in this chapter. We begin by considering the many-body problem. A brief overview about the second quantization formalism as well as the tight-binding model is followed in order to present the most convenient treatment of the system. Finally, for the sake of clarity, the Langevin formalism is presented to show how the temperature effects are included in our model.

2.1. The Many-Body Problem

Conducting polymers are molecules formed from smaller structural units (monomers), such as a CH group [21] in the case of the polyacetylene molecule already discussed. These monomers, in their turn, are composed by electrons and cores, so characterizing a many-body system [21, 22]. In this way, the quantum treatment of this system is performed using the time-dependent Schrödinger equation,

$$i\hbar \frac{\partial}{\partial t} |\Psi\rangle = \mathcal{H} |\Psi\rangle \quad (1)$$

The system may be represented as shown in Figure 2, where the distance between two electrons, i and j, is given by $r_{i,j} = |r_i - r_j|$, whereas the distance between two cores, A and B, can be defined as $R_{AB} = |R_A - R_B|$ [23]. In its turn, the distance between an electron and a core is represented, for example, as $r_{iA} = |r_i - R_A|$ [23]. Thereby, the electronic and nuclear degrees of freedom are described by a many-body Hamiltonian $\mathcal{H}$

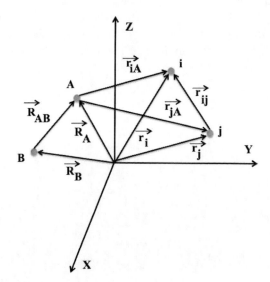

Figure 2. Molecular coordinate system for the many-body problem.

$$\mathcal{H} = \mathcal{H}_{c-c}(\{R\}) + \mathcal{H}_{e-e}(\{R\}) + \mathcal{H}_{e-c}(\{r, R\}) \tag{2}$$

$\mathcal{H}_{c-c}$ denotes the interaction energy between the cores, i.e., the core kinetic energy and its potential energy from the Coulomb interactions. The term $\mathcal{H}_{e-e}$ establishes the energy from the electron-electron interactions, including also its kinetic energy. Finally, the term $\mathcal{H}_{e-c}$ describes the potential energy due to the Coulomb interactions between electrons and cores. Thus, $\mathcal{H}$ can be placed, using atomic units, in the following form

$$\mathcal{H} = -\sum_{i=1}^{N} \frac{1}{2} \nabla_i^2 - \sum_{A=1}^{M} \frac{1}{2M_A} \nabla_A^2 - \sum_{i=1}^{N} \sum_{A=1}^{M} \frac{Z_A}{r_{iA}} + \sum_{i=1}^{N} \sum_{j>i}^{N} \frac{1}{r_{ij}} + \sum_{A=1}^{M} \sum_{B>A}^{M} \frac{Z_A Z_B}{R_{AB}} \tag{3}$$

where M_A is the mass for the core A and Z_A is its atomic number [23].

The first of these operators represents a sum of one-particle operators, which has the form

$$O_1 = \sum_{i=1}^{N} h(i), \tag{4}$$

where $h(i)$ is any operator involving only variables of the *i-th* particle. In this case, this operator can be placed in the form

$$\mathcal{H}_1 = -\sum_{i=1}^{N} \frac{1}{2} \nabla_i^2 - \sum_{A=1}^{M} \frac{1}{2M_A} \nabla_A^2, \tag{5}$$

In other words, this kind of operator contains dynamical variables that depend only on the position or momentum of the particle [23]. The first term in this operator describes the kinetic energy of the electrons whereas the the second one refers to the kinetic energy for the cores.

The second kind of operator considered in the scope of the many-body theory can be obtained by summing of two-particles operators as

$$O_2 = \sum_{i=1}^{N} \sum_{j>i} v(i,j), \tag{6}$$

which $v(i,j)$ represents any operator that depends on the position or momentum of the *i-th* and *j-th* particles [23]. Naturally, this class of operators can be represented using the equation

$$\mathcal{H}_2 = \sum_{i=1}^{N} \sum_{A=1}^{M} \frac{Z_A}{r_{iA}} + \sum_{i=1}^{N} \sum_{j>i}^{N} \frac{1}{r_{ij}} + \sum_{A=1}^{M} \sum_{B>A}^{M} \frac{Z_A Z_B}{R_{AB}}, \tag{7}$$

which is composed by terms that establishes the Coulomb interactions. The first part in Eq. 7 refers to the electron-core attraction whereas the two last terms describe the electron-electron and core-core repulsions, respectively.

2.2. Second Quantization Formalism

In the second quantization formalism, the antisymmetry property of the wavefunction is related to algebraic features of certain operators, thus being unnecessary to use the explicit form of the determinants [23]. Such formalism is frequently applied to fermionic systems and constitutes a convenient way to treat many-body systems.

Initially, within this approach it is defined a reference state in the Hilbert space, termed "*vacuum state*" $| \rangle$. This state represents a system without electrons. Therefore, the annihilation operator a_i can be defined with respect to its acting on the vacuum state,

$$a_i| \rangle = 0. \tag{8}$$

The creation operator, in its turn, which is the adjoint operator of a_i, is defined by its action on an arbitrary Slater Determinant as

$$a_i^\dagger |\chi_j\chi_k \cdots \chi_l\rangle = |\chi_i\chi_j\chi_k \cdots \chi_l\rangle. \tag{9}$$

Thus, $a_i^\dagger$ creates an electron in the orbital χ_i. It is important to note that the order which the creation and annihilation operators are applied is fundamental, because

$$a_i^\dagger a_j^\dagger |\chi_k \cdots \chi_l\rangle = a_i^\dagger |\chi_j\chi_k \cdots \chi_l\rangle = |\chi_i\chi_j\chi_k \cdots \chi_l\rangle, \tag{10}$$

and differently,

$$a_j^\dagger a_i^\dagger |\chi_k \cdots \chi_l\rangle = a_j^\dagger |\chi_i\chi_k \cdots \chi_l\rangle = -|\chi_i\chi_j\chi_k \cdots \chi_l\rangle. \tag{11}$$

For which was employed the antisymmetry principle of the Slater determinant itself. Considering now a sum of the Eqs. 10 and 11, it is obtained that

$$\left(a_j^\dagger a_i^\dagger + a_i^\dagger a_j^\dagger\right)|\chi_k \cdots \chi_l\rangle = 0. \tag{12}$$

By definition, the Slater Determinant is arbitrary. Therefore,

$$\{a_j^\dagger, a_i^\dagger\}a_j^\dagger a_i^\dagger + a_i^\dagger a_j^\dagger = 0 \tag{13}$$

i.e., the anticommutator of any two creation operators is null. Thus,

$$a_j^\dagger a_i^\dagger = -a_i^\dagger a_j^\dagger, \tag{14}$$

where to exchange the application order of the operators, it is only necessary to change the signal of the operator $a_i^\dagger a_j^\dagger$. Particularly, if the indexes are equal, then

$$a_i^\dagger a_i^\dagger = -a_i^\dagger a_i^\dagger = 0. \tag{15}$$

This shows that it is not possible to create two electrons in the same orbital. Eq. 15 is the form of the Pauli Exclusion Principle in the second quantization formalism [23].

Consider now any state $|Z\rangle$, so that

$$|Z\rangle = |\chi_i\chi_j\rangle = a_i^\dagger||\chi_j\rangle. \tag{16}$$

The adjoint of this state must be

$$(|Z\rangle)^\dagger = \left(a_i^\dagger|\chi_j\rangle\right)^\dagger = \langle\chi_j|\left(a_i^\dagger\right)^\dagger \equiv \langle\chi_j|a_i = \langle Z|, \tag{17}$$

where $|Z\rangle$ is a normalized state

$$\langle Z|Z\rangle = \langle\chi_j|a_i|\chi_i\chi_j\rangle = 1. \tag{18}$$

As $\langle\chi_j|\chi_j\rangle = \mathbf{1},$ in order to keep the formulation coherent, it is necessary to define here

$$a_i|\chi_i\chi_j\rangle = |\chi_j\rangle. \tag{19}$$

Thus, we have just the annihilation operator a_i as the adjoint of the creation operator $\left(a_i^\dagger\right)^\dagger$. In this way, the annihilation operator annihilates an electron in the orbital χ_i.

We shall now analyze the situation in which the two operators act as an applied product in the same state, a Slater Determinant without the orbital χ_i. Initially, it is considered a state $|\chi_k \ldots \chi_l\rangle$ which suffers the action of the operator $a_i a_i^\dagger + a_i^\dagger a_i$ as follows:

$$\left(a_i a_i^\dagger + a_i^\dagger a_i\right)|\chi_k \ldots \chi_l\rangle = a_i a_i^\dagger|\chi_k \ldots \chi_l\rangle = a_i|\chi_i\chi_k \ldots \chi_l\rangle = |\chi_k \ldots \chi_l\rangle. \tag{20}$$

On the other hand, if the state $|\chi_i\rangle$ is already occupied, then

$$\begin{aligned}
\left(a_i a_i^\dagger + a_i^\dagger a_i\right)|\chi_k \ldots \chi_i \ldots \chi_l\rangle &= a_i^\dagger a_i|\chi_k \ldots \chi_i \ldots \chi_l\rangle \\
&= -a_i^\dagger a_i|\chi_i \ldots \chi_k \ldots \chi_l\rangle \\
&= |\chi_k \ldots \chi_i \ldots \chi_l\rangle.
\end{aligned} \tag{21}$$

In order that, in the Eqs. 20 and 21 there were obtained the same determinants. Thus, it is possible to write

$$a_i a_i^\dagger + a_i^\dagger a_i \equiv \{a_i, a_i^\dagger\} = 1 \tag{22}$$

Generally, it is considered the case in which the operator $a_j a_i^\dagger + a_j^\dagger a_i$ is applied for $i \neq j$. Unlike the above-mentioned case, the action of this operator on a generic Slater Determinant can only be not null if the χ_i orbital is occupied whereas the χ_j orbital remains empty. In this way,

$$\left(a_i a_j^\dagger + a_j^\dagger a_i\right) |\chi_k \cdots \chi_i \cdots \chi_l\rangle = -(a_i a_j^\dagger + a_j^\dagger a_i)|\chi_i \cdots \chi_k \cdots \chi_l\rangle$$
$$= -a_i|\chi_j \chi_i \cdots \chi_k \cdots \chi_l\rangle - a_j^\dagger| \cdots \chi_k \cdots \chi_l\rangle$$
$$= a_i|\chi_i \chi_j \cdots \chi_k \cdots \chi_l\rangle - |\chi_j \cdots \chi_k \cdots \chi_l\rangle$$
$$= |\chi_j \cdots \chi_k \cdots \chi_l\rangle - |\chi_j \cdots \chi_k \cdots \chi_l\rangle$$
$$= 0 \tag{23}$$

Therefore, it is possible to conclude that

$$a_i a_j^\dagger + a_j a_i^\dagger \equiv \{a_i, a_j^\dagger\} = 0 \tag{24}$$

if $i \neq j$. From the Eqs. 24 and 25, it is obtained that

$$a_i a_j^\dagger + a_j a_i^\dagger \equiv \{a_i, a_j^\dagger\} = \delta_{ij}. \tag{25}$$

According to what was discussed before, the Slater Determinants can be represented using the creation and annihilation operators. Moreover, the anticommutation relations provide a representation by which the many-body wavefunction satisfies the antisymmetry principle. However, to a correct approach of the many-body theory without the use the determinants, the one- and two-electron operators (O_1 and O_2 respectively) should be addressed by means of the creation and annihilation operators. Thereby, it is possible to express the matrix elements of the operators O_1 and O_2 using only the algebraic properties of the creation and annihilation operators, so that O_1 and O_2 can be placed in the forms

$$O_1 = \sum_{ij}\langle i|h|j\rangle a_i^\dagger a_j \tag{26}$$

and

$$O_2 = \sum_{ijkl}\langle ij|h|kl\rangle a_i^\dagger a_j^\dagger a_k a_l \tag{27}$$

where

$$\langle ij|h|kl\rangle = \left\langle ij\left|\frac{1}{r_{ij}}\right|kl\right\rangle \tag{28}$$

The operators O_1 and O_2 are independent of the number of electrons [23]. The advantage to use the second quantization formalism is due to the fact that this approach can treat, in the same way, systems with different number of electrons, which is absolutely required when systems with a large number of particles, such as solids, are taken into account.

2.3. The Tight-Binding Approach

The translational invariance is a kind of discrete symmetry operation. Such symmetry possesses important applications in the solid state physics, i.e., for systems with spacial periodicity. Thereby, this consideration has proven to be fundamental to study conjugated polymers. The tight-binding approximation consists in considering a combination of the site atomic potentials as the potential of a crystalline lattice, i.e., it is assumes that these potentials are weakly superimposed [24]. In other words, this approximation adopts interactions between neighboring sites only. Considering one-dimensional conductors, the *tight-binding* approximation may be performed through the use of the periodicity of the lattice. Such periodicity is, in fact, a discrete symmetry of the system named translational symmetry.

Generally, by considering a lattice with an one-dimensional periodic potential a, which $V(x \pm a) = V(x)$, it is possible to analyze the electronic motion in a lattice with equally spaced ions[24]. The translational operator, represented by $\tau(l)$, in which l is arbitrary, has the following property

$$\tau^{\dagger}(l) x \tau(l) = x + l \tag{29}$$

However, when l coincides with the lattice spacing a, i.e., with the period of the potential, it should be noted that

$$\tau^{\dagger}(l) V(x) \tau(l) = V(x + a) = V(x) \tag{30}$$

Thus, due to the invariance under translations of the kinetic energy, for any displacement, and also to τ being a unitary operator, the Hamiltonian satisfies the equation

$$\tau^{\dagger}(a) \mathcal{H} \tau(a) = \mathcal{H} \tag{31}$$

that can be written as,

$$[\mathcal{H}, \tau(a)] = 0. \tag{32}$$

Therefore, the Hamiltonian and the translational operator can be diagonalized simultaneously [24]. Albeit the translational operator is unitary, it is not necessarily hermitian, and therefore it is expected its eigenvalue to be a complex number with unitary module.

It should now be discussed the case which the periodic potential is described by an infinity potential barrier between neighboring sites. Considering a state $|n\rangle$ given by $C_n^{\dagger}| \, \rangle$, where $|n\rangle$ represents an electron localized in the *i-th* lattice site, it is possible to note that the wavefunction $\langle x|n \rangle$ is finite only inside the site n. This fact shows that $|n\rangle$ is an eigenstate with eigenvalue E_0, i.e., $\mathcal{H}|n\rangle = E_0|n\rangle$. Moreover, one can see that similar states, at some other site of the lattice, may possess the same energy E_0. Thus, a linear combination of eigenstates $|n\rangle$ is also an eigenstate of the Hamiltonian.

It is possible to note that $|n\rangle$ is not an eigenstate of the translational operator, due to the fact that when this operator is applied in $|n\rangle$ it is obtained that

$$\tau(a)|n\rangle = |n + 1\rangle \tag{33}$$

Thereby, it becomes necessary to define an eigenstate that is, simultaneously, an eigenstate of $\mathcal{H}$ and $\tau(a)$. In this way, it is defined an eigenstate $|\theta\rangle$ that is a linear combination of eigenstates $|n\rangle$, which presents the following form

$$|\theta\rangle = \sum_n e^{i\theta n} |n\rangle, \tag{34}$$

where θ is a real parameter raging in the interval $[-\pi, \pi]$[24]. Now, applying the translational operator $\tau(a)$ on $|\theta\rangle$, we arrive at

$$\tau(a)|\theta\rangle = \sum_n e^{i\theta n} \tau(a)|n\rangle = \sum_n e^{i\theta n}|n+1\rangle = \sum_n e^{i\theta(n-1)} |n\rangle = e^{-i\theta}|\theta\rangle \tag{35}$$

Thus, considering a one-dimensional lattice with an infinity potential barrier between neighboring sites, the state $|\theta\rangle$ is an eigenstate of the translational operator $\tau(a)$ with eigenvalue $e^{-i\theta}$ and also eigenstate of the Hamiltonian operator having as eigenvalue E_0.

For real cases, i.e., when a finite potential barrier between neighboring sites is taken into account, the wavefunction $\langle x|\theta\rangle$ is not localized. In other words, the wavefunction $\langle x|\theta\rangle$ has its extremity extended through the other sites until the n-th lattice site. The diagonal elements of $\mathcal{H}$ on the basis $|n\rangle$ are all the same due to the translational invariance, which means that

$$\langle n| \mathcal{H}|n\rangle = E_0 \tag{36}$$

for any n. Assuming, for this moment, that the potential barriers between neighboring sites are high but not infinity, it is expected that the matrix elements of $\mathcal{H}$ between distant sites are null. Thus, it is assumed that only the outer elements of the diagonal are important, as shown in Eq. 37. In solid state physics, this affirmation is named Tight-Binding Approximation [24, 25], where

$$\langle n|\mathcal{H}|n'\rangle = \begin{cases} E_0 \to n = n' \\ -\Delta \to n = n' \pm 1 \\ 0 \to \forall n \neq n', \forall n \neq n' \pm 1. \end{cases} \tag{37}$$

Then, it is possible to write $\mathcal{H}$ as follows:

$$\mathcal{H} = \sum_{n',n''} \langle n'|\mathcal{H}|n''\rangle |n'\rangle\langle n''|. \tag{38}$$

Therefore, the acting of $\mathcal{H}$ on the basis $\{|n\rangle\}$ results in

$$\mathcal{H}|n\rangle = \sum_{n',n''} (\langle n'| \mathcal{H}|n''\rangle|n'\rangle\langle n''|n\rangle = \sum_{n'} (\langle n'| \mathcal{H}|n\rangle|n'\rangle$$
$$= E_0|n\rangle - \Delta|n+1\rangle - \Delta|n-1\rangle \tag{39}$$

One can see that $|n\rangle$ is not an eigenstate of $\mathcal{H}$. Now, acting $\mathcal{H}$ on $|\theta\rangle$, it is obtained that

$$\mathcal{H}|\theta\rangle = (E_0 - 2\Delta \cos\theta)|\theta\rangle \tag{40}$$

Thus, in summary, considering a lattice with finite potential barriers between the lattice sites, the basis set states $\{|\theta\rangle\}$ are degenerate and are eigenstates of the Hamiltonian operator with the eigenvalue ranging from $E_0 - 2\Delta$ to $E_0 + 2\Delta$ [24].

2.4. The Langevin Equation

The phenomenon that characterizes the motion of particles immersed in a certain fluid, moving under influence of random forces resulting from the collisions which are induced by thermal fluctuations is known as Brownian Motion [26]. In statistical mechanics, the Langevin Equation is defined as a stochastic differential equation that describes the brownian motion in a certain potential [27].

The simplest Langevin Equation is that which presents a constant potential. Thus, the brownian particle acceleration can be expressed is terms of its mass m subjected to a viscous force, that is considered to be proportional to the velocity (Stokes Law), a noise term $\zeta(t)$ which is used in stochastic processes as a time-dependent random variable that denotes the effect of several continuous collisions, and a force field $F(x)$ derived from intra and inter-molecular interactions. Thus, the Langevin Equation which describes this kind of system can be placed in the form [28]

$$m\frac{dv(t)}{dt} = -m\gamma v(t) + F(x) + \zeta(t). \tag{41}$$

For a generic potential, the usual form of the Langevin Equation is written as

$$m\frac{dv(t)}{dt} = -m\gamma v(t) + \zeta(t). \tag{42}$$

The random force $\zeta(t)$ has the following preperties: $\zeta(t)$ is a gaussian and stochastic noise with $\langle\zeta(t)\rangle = 0$; the time-correlation between $\zeta(t)$ and $\zeta(0)$ is infinitely small, in order that $\langle\zeta(t)\zeta(0)\rangle = \beta\delta(t)$; and, finally, the particle motion is occasioned due to thermal bath fluctuations where $\langle v(t)\zeta(t)\rangle = 0$ [28].

The generic solution to Eq. 42 may be obtained defining $v(t) = u(t)e^{-\gamma t}$, which $u(t)$ is a function to be determined. Thereby,

$$\frac{du}{dt} = e^{\gamma t}\zeta(t) \tag{43}$$

which the solution is

$$u = u_0 + \int_0^t e^{\gamma t'}\zeta(t')dt'. \tag{44}$$

Thus,

$$v = v_0 e^{\gamma t'} + e^{\gamma t'}\int_0^t e^{\gamma t'}\zeta(t')dt' \tag{45}$$

where v_0 is the particle's velocity at $t = 0$. Such solution is valid for any function $\zeta(t)$. Using the above-mentioned noise and $\zeta(t)$ properties, it is possible to calculate the variance for the velocity

$$\langle v \rangle = v_0 e^{-\gamma t} \tag{46}$$

and

$$v - \langle v \rangle = e^{-\gamma t} \int_0^t e^{\gamma t'} \zeta(t') dt'. \tag{47}$$

Considering that $\langle v(0)\zeta(t)\rangle = 0$, the Eq. 45 can be multiplied by $v(0)$, taking the average of all ensemble. Thus,

$$\langle v(0)v(t) \rangle = \langle v^2(0) \rangle e^{-\gamma t} \tag{48}$$

shows the loss of memory of the system regarding to its initial conditions and is consistent with the fact of considering the system to be markovian, i.e., a system in which the future event depends only on the immediately previous event [29]. Taking the square of the Eq. 45 and calculating, again, the average of all ensemble, considering the conditon $\langle \zeta(t)\zeta(0) \rangle = \beta \delta(t)$, we obtain

$$\langle v^2(t) \rangle = \langle v^2(0) \rangle e^{-2\gamma t} + \frac{e^{-2\gamma t}}{m^2} \int_0^t \int_0^{t'} e^{\gamma(t'+t'')} \langle \zeta(t')\zeta(t'') \rangle dt' dt''$$

$$= \langle v^2(0) \rangle e^{-2\gamma t} + \frac{\beta}{2\gamma m^2} (1 - e^{-2\gamma t}) \tag{49}$$

For longer periods of time, i.e., in the stationary regime where $\langle v \rangle = 0$, it is possible to obtain the β value. Therefore, taking $t \to \infty$, we obtain

$$\lim_{t\to\infty} \langle v^2(t) \rangle = \frac{\beta}{2\gamma m^2} \tag{50}$$

As it is known the energy equipartition theorem, where $(1/2)m\langle v^2(t)\rangle = (1/2)k_B T$. Thus, using this theorem, we can arrive at the following relation

$$\beta = 2\gamma m K_B T. \tag{51}$$

Considering that the velocities distribution satisfies the Fokker-Plank equation [30],

$$\frac{\partial}{\partial t} P(v,t) = \frac{\partial}{\partial v} \left(\gamma v + \frac{D_v}{2} \frac{\partial}{\partial v} \right) P(v,t), \tag{52}$$

in which D_v is related with the noise as follows

$$D_v = \frac{1}{m^2} \int_0^\infty \langle \zeta(t)\zeta(0) \rangle dt = \frac{\beta}{m^2}. \tag{53}$$

Assuming, by hypothesis that $v(t)$ is a gaussian process, the stationary solution of the Eq. 52 assumes the form,

$$P(v) = \left(\frac{m}{2\pi K_B T}\right)^{\frac{1}{2}} e^{-\frac{mv^2}{2K_B T}}. \tag{54}$$

Placing this equation in Eq. 52, is it possible to obtain that,

$$D_v = \frac{2K_B T}{m}\gamma = \frac{\beta}{m^2}, \tag{55}$$

and

$$\gamma = \frac{1}{2mK_B T}\int_0^\infty \langle \zeta(t)\zeta(0)\rangle dt, \tag{56}$$

where

$$\langle \zeta(t)\zeta(0)\rangle = 2m\gamma K_B T\delta(t), \tag{57}$$

which is the form of the Fluctuation-Dissipation theorem used in this work in order to include the temperature effects.

So far, we have introduced the theoretical formalism which is enough to discuss in detail the Su-Schrieffer-Heeger model together with important modifications implemented. The next section presents this model, highlighting its fundamentals aspects.

3. THE SU-SCHRIEFFER-HEEGER MODEL

In this section, it is presented a version of the Su-Schrieffer-Heeger (SSH) model modified to include an external electric field, electron-electron interactions, a symmetry-breaking term, impurities and temperature effects. Particularly, a detailed treatment regarding the quasi-particle dynamics in organic semiconductors, from the framework of a *tight-binding* approach, is provided in order to better analyze the mechanisms involved on the generation and recombination dynamics of excited states in these materials. Moreover, an overview of the quasi-particles (as agents as well as products) included in the aforementioned mechanisms is provided in order to clarify their understanding.

3.1. The SSH-Type Hamiltonian

An extremely convenient treatment of the conductivity in polyacetylene molecules was first presented in pioneering papers by Su, Schrieffer, and Heeger in 1979 [31-33]. The proposed model showed that the conductivity in this material is related to topological defects in its structure, which could easily arise when a sample of the material is exposed to doping agents [10, 34, 35]. Furthermore, it was shown that these charged defects could move through the lattice when an external electric field was applied.

Concerning the π-electrons theory, it is known that for an ion (a CH group for the polyacetylene) there exists six degrees of freedom per unity cell [36]. For the treatment of a polyacetylene molecule, in the scope of the SSH model, an important simplification is introduced: it is considered (from the electronic point of view) only the normal vibration modes that couple, predominantly, the π-electrons [33]. For the polyacetylene, this degree of freedom denotes the bond between the carbon atoms which can stretch and shrink alternately. Thus, projecting the ionic coordinates in a horizontal axis that extends through the lattice, the problem becomes one-dimensional, according represented in Figure 3.

In the polyacetylene molecule, the σ-electrons represent the strong covalent bonds between carbon-carbon and carbon-hydrogen atoms, of approximately 3 eV [37, 38]. This type of bond is the main responsible for the lattice backbone structure of polyacetylene. The π-electrons, in their turn, tend to form less localized bonds, of about 1 eV [38]. These bonds are formed by the overlap of two adjacent atomic orbitals $(2P_z)$ which are perpendicular to the plane of the polyacetylene molecule. Moreover, the π-bonds are considered weak if compared to the σ-bonds. The π-bonds are responsible for the alternation between the single and double bonds, i.e., these bonds are responsible by the dimers formation (the lattice dimerization) [37].

The u_n coordinates, as shown in Figure 3, are the displacements of the CH groups projected in the x-axis for the case where the lattice is not dimerized [31]. Considering the ground state, which is dimerized, if $u_n > 0$ then, necessarily, $u_{n+1} < 0$. It occurs due to the consideration of the tigh-binding approach (where only the first next neighbors can interact), as mentioned in the previous section, and also by considering that the system is one-dimensional, i.e., the coordinate u is the only necessary to describe the system [32]. For a dimerized lattice, $|u_{n+1} - u_n| \cong 0.08$Å and the lattice parameter $a = 1.40 \times \sqrt{3}/2$ Å $\cong$ 1.22 Å [32].

In summary, the SSH model is an extension of the tight-binding approach. Thus, the interchain interaction is neglected (one-dimensional system). Moreover, the coupling between the σ- and π-electrons is also neglected. The CH group, which has six degrees of freedom, is described by only its translations in the lattice direction by means of the u coordinate [36]. The other five degrees of freedom are ignored, since that in a first order approximation they are not related to the lattice dimerization [36]. Finally, the coupling between the π-electrons is considered in the scope of a mean-field approximation, i.e., the electronic correlation effects are not considered. In the basis of the atomic orbitals, for the π-electrons, the matrix elements are unknown, except for the first neighbors [33].

From the introduction about some important features discussed just above, we now can define the SSH Hamiltonian. In order to do so, it is initially considered that the displacements u_n are smaller than the bond between the carbon atoms, $u_n \cong 0.04$ Å. The potential energy of the σ-electrons can be expanded in a Taylor series truncated in second order as follows

$$E_\sigma = E_\sigma(0) + \sum_n \frac{\partial E_\sigma}{\partial (u_{n+1} - u_n)}(u_{n+1} - u_n) + \sum_n \frac{1}{2!}\frac{\partial^2 E_\sigma}{\partial (u_{n+1} - u_n)^2}(u_{n+1} - u_n)^2 + \ldots \qquad (58)$$

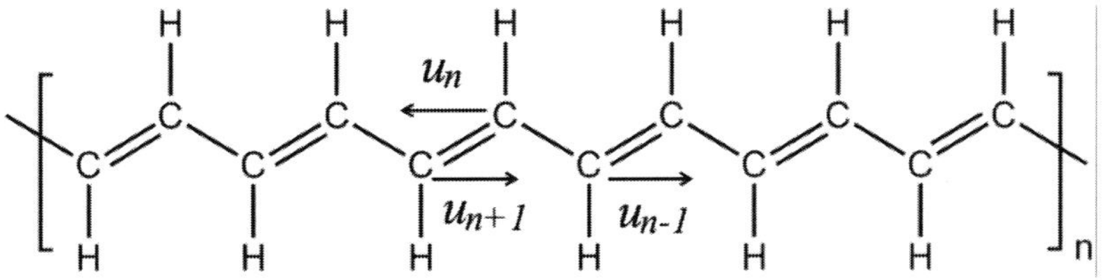

Tight-Binding Models for the Charge Transport in Organic Semiconductors 121

Figure 3. Schematic representation of a dimerized trans-polyacetylene molecule with its displacement coordinate u, which denotes the unique degree of freedom.

One can see that the first term of the expansion is a constant. Thereby, it can be defined as zero. The second term is also null, once the expansion is carried out considering a point that has vanishing first derivative. Due to the symmetry shown by the coordinates u_n, the coefficient of the second order terms could be defined as equal to a constant K [36] Thus, approximating this potential to a harmonic oscillator potential, it can be written as

$$\frac{1}{2}\sum_n K\,(u_{n+1} - u_n)^2. \tag{59}$$

The π-electrons, in their turn, are treated by the hopping term, in an expansion of the *tight-binding* approximation which considers coupling between the electronic and lattice parts of the system. They are described by a hopping integral according to the following equation

$$t_{n+1,n} = t_0 - \alpha(u_{n+1} - u_n), \tag{60}$$

where t_0 denotes the hopping integral of a π-electron between neighboring sites in an evenly spaced lattice and α is the electron-phonon coupling constant [33, 39-41]. Due to the fact that the bond lengths can not suffer larger variations, the first order expansion to the hopping integral is a reasonable approximation.

The kinetic energy of the sites, i.e., CH groups, is written as

$$E = \frac{1}{2}\sum_n M\dot{u}_n^2, \tag{61}$$

where M is their masses. Therefore, the SSH Hamiltonian, written in the second quantization formalism, assumes the form

$$H = -\sum_{n,s}\left(t_{n,n+1}C_{n+1,s}^\dagger C_{n,s} + h.c\right) + \frac{1}{2}\sum_n K(u_{n+1} - u_n^2) + \sum_n \frac{p_n^2}{2M\prime} \tag{62}$$

where n indexes the sites of the chain [31, 32]. The operator $C_{n,s}^\dagger (C_{n,s})$ creates (annihilates) a π-electron state at the nth site with spin s [31].

3.2. The Dimerized Lattice

The dimerized lattice is the simplest model to a semiconducting polymer and, particularly, characterizes the *trans*-polyacetylene structure. A semiconducting polymer chain is usually extensive and may present around 3000 sites. In the SSH model, such system can be represented by a N-size finite lattice with periodic boundary conditions. The stationary solution for a perfectly dimerized lattice was firstly investigated by Su, Schrieffer, and Heeger in the pioneering work which describes the SSH model [32]. In this solution, the term that describes the kinetic energy of the CH groups is naturally null. In order to describe this system, initially it is considered that a dimerized lattice can be treated in the framework of the Born-Oppenheimer approximation [32], where the u_n coordinates has the form

$$u_n = (-1)^n u, \tag{63}$$

which make the lattice perfectly dimerized by considering u as a constant. Rewriting the hopping term, given by the Eq. 60, it is obtained that

$$t_{n,n+1} = t_0 + 2\alpha u(-1)^n. \tag{64}$$

Thus, the Hamiltonian can be written as

$$H(u) = -\sum (t_0 + 2\alpha u(-1)^n) \left[C_{n+1,s}^\dagger C_{n,s} + C_{n,s}^\dagger C_{n+1,s} \right] + 2KNu^2. \tag{65}$$

Is it possible to diagonalize analytically such Hamiltonian to obtain the electronic spectrum. In order to do so, the following transforms are used [32]

$$C_{k,s}^{c\dagger} = \frac{-i}{\sqrt{N}} \sum_n (-1)^n e^{-ikna} C_{n,s'}^\dagger$$
$$C_{k,s}^{v\dagger} = \frac{1}{\sqrt{N}} \sum_n e^{-ikna} C_{n,s}^\dagger. \tag{66}$$

Such transforms describe the electronic spectrum at the first Brillouin zone, $\pi/2a \leq k \leq -\pi/2a$. Thus,

$$C_{n,s} = \frac{1}{\sqrt{N}} \sum_k \left[e^{-ikan} \left(C_{k,s}^v + i(-1)^n C_{k,s}^C \right) \right]. \tag{67}$$

Performing the sum for the index n, taking into consideration that

$$\frac{1}{N} \sum_n e^{ian(k-k\prime)} = \delta_{kk\prime}, \tag{68}$$

and

$$\sum_n (-1)^n e^{ian(k-k\prime)} = \sum_n e^{ian(k-k\prime+\pi/a)}, \tag{69}$$

the Hamiltonian can assume the form

$$H^d(u) = \sum_{k,s}\left[2t_0\cos(ka)\left(C_{k,s}^{c\dagger}C_{k,s}^{c} - C_{k,s}^{v\dagger}C_{k,s}^{v}\right) + 4asen(ka)\left(C_{k,s}^{c\dagger}C_{k,s}^{v} + C_{k,s}^{v\dagger}C_{k,s}^{c}\right)\right]$$
$$+ 2NKu^2. \tag{70}$$

Finally, in order to represent $H^d(u)$ in the diagonal form, it is important to define the new operators a_{ks}^c and a_{ks}^v that can be written as follows [32]

$$\begin{pmatrix} a_{ks}^v \\ a_{ks}^c \end{pmatrix} = \begin{pmatrix} \alpha_k & -\beta_k \\ \beta_k^* & \alpha_k^* \end{pmatrix}\begin{pmatrix} C_{ks}^v \\ C_{ks}^c \end{pmatrix}, \tag{71}$$

where

$$|\alpha_k|^2 + |\beta_k|^2 = 1. \tag{72}$$

By calculating the parameters α_k and β_k which diagonalize the Hamiltonian, we obtain

$$\alpha_k = \left[\frac{1}{2}\left(1 + \frac{\epsilon_k}{E_k}\right)\right]^{\frac{1}{2}} \tag{73}$$

and

$$\beta_k = \left[\frac{1}{2}\left(1 - \frac{\epsilon_k}{E_k}\right)\right]^{\frac{1}{2}} signal\,(k), \tag{74}$$

where

$$\epsilon_k = 2t_0\cos(ka), \tag{75}$$

$$\Delta_k = 4\alpha u\sin(ka), \tag{76}$$

and
$$E_k = \sqrt{\epsilon_k^2 + \Delta_k^2} \tag{77}$$

Thus, the Hamiltonian becomes diagonal in the representation of the operators a_{ks}^c and a_{ks}^v

$$H^d(u) = \sum_{k,s} E_k\left(a_{k,s}^{c\dagger}a_{k,s}^{c} - a_{k,s}^{v\dagger}a_{k,s}^{v}\right) + 2NKu^2. \tag{78}$$

In this equation, $E_k > 0$ represents the energy values for the conduction band, whereas $E_k < 0$ represents the valence band energies. Therefore, the gap is given by the difference between the energies of the conduction and valence bands presenting the value of $8\alpha u$ for the Fermi wave vector $\vec{k}_F = \pi/2a$. The ground state energy is given by the sum over all the occupied states

$$E_0(u) = \sum_{k,s}{}' E_k + 2NKu^2$$

$$= \sum_{k,s}{}' \sqrt{(2t_0\cos(ka))^2 + \left(4\alpha u\, sen(ka)\right)^2} + 2NKu^2,$$

(79)

where the prime represents the sum over occupied states only. Considering that the system has a large number of particles, it is possible to approximate the discrete energy in a continuous form as

$$E_0(u) = \frac{-2L}{\pi}\int_0^{\frac{\pi}{2a}} \sqrt{(2t_0\cos(ka))^2 + \left(4\alpha u\, sen(ka)\right)^2}\, dk + 2NKu^2$$

$$= \frac{4Nt_0}{\pi} E(1 - z^2) + \frac{NKt_0^2 z^2}{2\alpha^2}.$$

(80)

In this equation, $E(1 - z^2)$ is an elliptic integral of second specie, with $z = 2\alpha u/t_0$, and $L = Na$, the lattice length. The expansion of the elliptic integral for a small value for z can be written as,

$$E_0(z) = \frac{-4Nt_0}{\pi}\left[1 + \frac{1}{2}\left(\frac{\ln(4)}{|z|} - \frac{1}{2}\right)z^2 + \dots\right] + \frac{NKt_0^2 z^2}{2\alpha^2}.$$

(81)

The energy $E_0(u)$ has a local maximum for $u = 0$ (which denotes an undimerized lattice). Truncating the expansion of the elliptic integral, for a small value of z, we can obtain an u_0 value which minimizes the energy. It is also possible to write the density of states per spin to the perfectly dimerized lattice as,

$$\rho(E) = \frac{L}{2\pi\left|\frac{dE_k}{dk}\right|} = \begin{cases} \dfrac{N}{\pi}\dfrac{|E|}{\sqrt{(4t_0 - E^2)(E^2 - \Delta_0^2)}} & se\ \Delta_0 \leq |E| \leq 2t_0 \\ \qquad\qquad 0 & for\ the\ other\ cases. \end{cases}$$

(82)

Related to the ground state degeneracy, there exists an elementary excitation which corresponds to a domain wall (termed soliton) [31-33]. The equations presented so far in this section summarize the theory of the SSH model. The original work presents a stationary solution for a perfectly dimerized lattice and, also, the formulation of the solitons theory for polyacetylene lattices. This model is the most simple to describe the conductivity in polyacetylene lattices. However, this model should be extended in order to analyze the charge carrier dynamics in these materials under influence of an external electric field, Coulomb interactions, impurities, and temperature effects. The following sections are devoted to explain how such effects are taken into account in our model.

3.3. The Electric Field and the Symmetry Breaking Term

The first modification implemented in the standard SSH model was made in order to include the effects of an external electric field. The field is included in the Hamiltonian as a vector potential $\mathbf{A}$ modifying the Eq. 60 as follows,

$$t_{n,n+1} = e^{-i\gamma A(t)}[t_0 - \alpha y_n], \tag{83}$$

where y_n is defined as $y_n = u_{n+1} - u_n$. $\gamma = ea/\hbar c$, with e being the absolute value of the electronic charge, a is the lattice constant, and c is the speed of light [42-45]. The relation between the time-dependent vector potential $\mathbf{A}$ and the uniform electric field $\mathbf{E}$ is given by

$$E(t) = -\frac{1}{c}\dot{\mathbf{A}}(t). \tag{84}$$

The implementation of the electric field through the potential vector is convenient due to the fact that the system has periodic boundary conditions. Another point that is worthy of attention is that the abrupt inclusion of the electric field in the simulations may cause numerical errors. In many cases the charge carriers could be artificially annihilated in a few femtoseconds. Such problem is solved by turning on the field adiabatically according to the following equations

$$A(t) = \begin{cases} 0 & if\ t < 0, \\ -\frac{1}{2}cE\left[t - \frac{\tau}{\pi}sen\left(\frac{\pi t}{\tau}\right)\right] & if\ 0 \le t < \tau, \\ -c\left(t - \frac{\tau}{2}\right) & if\ \tau \le t < t_f, \\ -\frac{1}{2}cE\left[t + t_f - \tau + \frac{\tau}{\pi}sen\left(\frac{t}{\tau}(t - t_f + \pi)\right)\right] & if\ t_f \le t < t_f + \tau, \\ -cEt_f & if\ t \ge \tau. \end{cases} \tag{85}$$

where the parameters τ represents the time period for which the field acts on the system, whereas t_f denotes the time where the field is at its maximum.

In order to consider cis-polyacetylene chains, it should be included a symmetry breaking term in the hopping integral $t_{n+1,n}$, which is named Brazovskii-Kirova symmetry breaking term, represented here as δ_0. To include this term in the model, Eq. 83 should be rewritten as follows [44, 46, 47],

$$t_{n,n+1} = e^{-i\gamma A(t)}[(1 + (-1)^n\delta_0)t_0 - \alpha y_n]. \tag{86}$$

3.4. Electron-Electron Interactions and Impurity Effects

The doping processes in conjugated polymers are usually not performed by replacing atoms in the materials for metals or semimetals dopants, as in the case of inorganic semiconductors. In organic semiconductors, such processes are carried out by means sorption, i.e., the impurity is localized closer to the conjugated polymer lattice and the interaction between them affects only the eigen energies of the lattice site where it is adjacent. Consequently, this kind of doping process does not affect the nature of the bonds. In order to taken into account the impurity effects, a new term H_{imp} should be added to the SSH Hamiltonian presented in Eq. 62, henceforward named H_{SSH}. The new overall Hamiltonian (H_{total}) assumes the form $H_{total} = H_{SSH} + H_{imp}$, with the last term denoting the inclusion of impurity effects in a particular site of the lattice. This term is written as

$$H_{imp} = I_j C_{j,s}^\dagger C_{j,s}. \tag{87}$$

The electron-electron interactions are descried by the extended Hubbard Model. In this way, the overall Hamiltonian has now the form $H_{total} = H_{SSH} + H_{imp} + H_{e-e}$, where the last term is written as [48-50]

$$H_{ee} = U \sum_i \left(C_{i,\uparrow}^\dagger C_{i,\uparrow} - \frac{1}{2} \right) \left(C_{i,\downarrow}^\dagger C_{i,\downarrow} - \frac{1}{2} \right) + V \sum_i (n_i - 1)(n_{i+1} - 1), \tag{88}$$

where $n_i = C_{i,\uparrow}^\dagger C_{i,\uparrow} + C_{i,\downarrow}^\dagger C_{i,\downarrow}$, U is the onsite electron-electron coulombian interaction, and V is the neighboring sites electron-electron interactions. It should be noted that the present form of the Hubbard interactions maintain the electron-hole symmetry of the system.

3.5. The System Dynamics

In order to perform the time evolution of the system, initially it is considered a self-consistent state regarding to the degrees of freedom for electrons and phonons. This point will be further discussed latter. After that, the system dynamics is performed in the framework of the Ehrenfest Molecular Dynamics [44]. The lattice backbone dynamics is carried out in a classical approach by means of the Euler-Lagrange equations

$$\frac{d}{dt} \left(\frac{\partial \langle L \rangle}{\partial \dot{u}_n} \right) - \frac{\partial \langle L \rangle}{\partial u_n} = 0. \tag{89}$$

The expectation value for the Lagrangian can be obtained from the terms of the Hamiltonian Eq. 62. Thus,

$$\begin{aligned}
\langle L \rangle &= \langle T \rangle - \langle V \rangle \\
&= \sum_n \frac{M}{2} \dot{u}_n^2 \langle \psi | \psi \rangle - \sum_n \frac{K}{2} (u_{n+1} - u_n) \langle \psi | \psi \rangle \\
&\quad - \sum_{n,s} [t_0 - \alpha(u_{n+1} - u_n)] \langle \psi | C_{n+1}^\dagger C_{n,s} + C_{n,s}^\dagger C_{n+1,s} | \psi \rangle
\end{aligned} \tag{90}$$

where the last term represents the expectation value to the electronic Hamiltonian.

The representation of the expectation value of the electronic Hamiltonian can be written in terms of

$$B_{n,n'} \equiv \sum_{k,s} {}' \psi_{k,s}^*(n, t) \psi_{k,s}(n', t), \tag{91}$$

as

$$\langle H_e \rangle = -\sum_n \left[t_0 - \alpha(u_{n+1} - u_n) \right] \left(B_{n,n+1} + B^*_{n,n+1} \right). \tag{92}$$

This permits to rewrite the expected value to the Lagrangian as,

$$\langle L \rangle = \sum_n \frac{M}{2} \dot{u}_n^2 - \sum_n \frac{K}{2} (u_{n+1} - u_n)$$
$$+ \sum_n [t_0 - \alpha(u_{n+1} - u_n)] \left(B_{n,n+1} + B^*_{n,n+1} \right). \tag{93}$$

The lattice dynamics is given by the solution of Eq. 89 for the expectation value of the Lagrangian $\langle L \rangle$ expressed in Eq. 93. From this, we obtain [39]

$$M\ddot{u}_n = F_n(t), \tag{94}$$

where

$$F_n(t) = -K[2u_n(t) - u_{n+1})(t) - u_{n-1}(t)]$$
$$+ \alpha \left[(B_{n,n+1} + B_{n-1,n}) + (B_{n+1,n} + B_{n,n-1}) \right]. \tag{95}$$

Using the definition of the time-derivative for u_n, i.e.,

$$\dot{u} = \frac{u_n(t+dt) - u_n(t)}{dt} \tag{96}$$

we can integrate the equations of motion using the Verlet Algorithm [44, 45]

$$u_n(t + dt) = u_n(t) + \dot{u}_n(t)dt, \tag{97}$$

and

$$\dot{u}_n(t + dt) = \dot{u}_n(t) + \frac{F_n(t)}{M} dt. \tag{98}$$

In its turn, the electronic dynamics is governed by the time-dependent Schrödinger equation,

$$i\hbar \frac{\partial \psi_k}{\partial t} = H_e \, \psi_k \tag{99}$$

which can be formally solved using

$$\psi_k(t) = exp \left[\int_0^t \frac{H_e(t')}{\hbar} dt' \right] \psi_k(0), \tag{100}$$

where $H_e(t')$ is the electronic Hamiltonian at a given time t'. Especifically,

$$\psi_k(t + dt) = exp\left[\frac{-i}{\hbar}H_e(t)dt\right]\psi_k(t),$$ (101)

We can expand $\psi_k(t)$ as follows,

$$\psi_k(t) = \Sigma_l C_{lk}\phi_l(t).$$ (102)

where $C_{lk} = \langle\phi_l|\psi_k\rangle$, with $\{\phi_l\}$ and $\{\varepsilon_l\}$ being the eigenfunctions and eigenvalues of the electronic Hamiltonian at a given time t. In this way, placing ψ_k in Eq. 101, we obtain [39]

$$\psi_{n,k}(t + dt) = \Sigma_l[\Sigma_m \phi^*(m, t)\psi_k(m, t)]exp\left[\frac{-i}{\hbar}\varepsilon_l dt\right]\phi_l(n, t).$$ (103)

Thus, by knowing the set of eigenstates $\{\psi_k\}$ at a given time t, it is possible to calculate $\{\psi_k\}$ at $t+dt$.

We now turn our attention to the preparation of the initial self-consistent state regarding the degrees of freedom of electrons and phonons. For the construction of the initial state (stationary state), where $du_n/dt = 0$, the Lagrangian can be written as

$$\langle L \rangle = -\sum_n \frac{K}{2}(u_{n+1} - u_n)^2\langle\psi|\psi\rangle +$$
$$+ \Sigma_{n,s}[t_0 - \alpha(u_{n+1} - u_n)]\langle\psi|C_{n+1,s}^\dagger C_{n,s} + C_{n,s}^\dagger C_{n+1,s}|\psi\rangle.$$ (104)

Using the definition $y_n \equiv u_{n+1} - u_n$, and Eq. 91, we have

$$\langle L \rangle = -\Sigma_n \frac{K}{2}y_n^2 + \Sigma_{n,s}(t_0 - \alpha_n)\left(B_{n+1,n} + B_{n+1,n}^*\right).$$ (105)

For this case, the Euler-Lagrange equations are written as

$$\frac{\partial\langle L\rangle}{\partial y_n} = 0$$ (106)

which leads to

$$y_n = -\frac{\alpha}{K}\left(B_{n,n+1} + B_{n,n+1}^*\right).$$ (107)

However, due to the fact that the system has periodic boundary conditions, the following condition should be satisfied

$$\Sigma_n y_n = 0$$ (108)

Therefore, it is necessary to add one term in Eq. 107 as follows

$$y_n = -\frac{\alpha}{K}\left(B_{n,n+1} + B_{n,n+1}^*\right) + \frac{\alpha}{NK}\left[\Sigma_n\left(B_{n,n+1} + B_{n,n+1}^*\right)\right].$$ (109)

Tight-Binding Models for the Charge Transport in Organic Semiconductors 129

In summary, in order to solve these equations numerically, first a stationary state that is self-consistent with all degrees of freedom of the system (lattice and electrons) needs to be obtained. The initial bond configuration and the electronic structure of a polymer chain containing a quasi-particle can be achieved by solving the following self-consistent equations of the bond configuration $\{u_n\}$ and the electronic wavefunctions $\{\phi_n\}$ [51]:

$$u_{n-1} - u_n = -\frac{2\alpha}{K} \sum_\mu \phi_\mu(n) \ \phi_\mu(n+1) + \frac{2\alpha}{NK} \sum_{\mu,n} \phi_\mu(n) \ \phi_\mu(n+1), \qquad (110)$$

$$\varepsilon_\mu \phi_\mu(n) = -[t_0 - \alpha(u_{n-1}u_n)]\phi_\mu(n+1) - [t_0 - \alpha(u_n - u_{n-1})]\phi_\mu(n-1) \qquad (111)$$

where ε_μ is the eigenvalue of the μ-th energy level. We begin by constructing the Hamiltonian from a $\{y_n\}$ set of conveniently closer positions. By solving the time-independent Scrhodinger equation

$$H_{ele}|\psi_k\rangle = E_{ele} \ |\psi_k\rangle, \qquad (112)$$

a new set of eingenfunctions $\{\psi_k\}$ is obtained, and from that a new set of coordinates $\{y_n\}$ using Eq. 110. Iterative repetitions of this procedure yields a self-consistent initial state when $\{y_n\}$ is close enough to the previous step. After that, from the self-consistent initial state, we can evolve the system determining the eigenstates and eigenvalues of the electronic hamiltonian at each step. The equations 97, 98 and 103 govern the system dynamics.

3.6. Temperature Effects and Occupation Number

In order to take into account the temperature effects, the SSH model should be further modified. The temperature is included in the equations that describe the lattice backbone dynamics (classical part of the Hamiltonian). Hence, the temperature influence on the electronic part of the system is only considered indirectly by means of coupling terms, through the term expressed in Eq. 91. It should be mentioned, however, that the contribution of the electronic part to thermal properties is usually considered much smaller than that of the cores.

The temperature effects on the lattice are simulated using the Langevin Equation, as mentioned in the previous section. Here, it is used the stochastic signal $\zeta_n(t)$, generally known as "white noise." Such nomenclature is due to the fact that the spectral intensity of a signal is defined as a Fourier transform of the eigen-correlation function. The noise $\zeta_n(t)$ adopted in this work has as the correlation function a Dirac delta function. As the Fourier transform of a delta function is constant, it is understood that all the frequencies are present with similar intensity, thus the analogy with the white light.

Here, the temperature effects are simulated by adding thermal gaussian random forces with zero mean value $\langle\zeta_n(t)\rangle \equiv 0$ and variance $\langle\zeta_n(t)\zeta_n(t')\rangle = 2k_BT\gamma M\delta(t-t')$ [28, 52]. We adopted a white stochastic signal $\zeta_n(t)$ as the fluctuation term. Also, in order to keep the temperature constant at its initial value after a transient period (named thermalization), it is necessary to introduce a damping factor, γ. Therefore, Eq. (94) is modified to

$$M\ddot{u} = -\gamma\dot{u} + \zeta_n(t) + F_n(t) \tag{113}$$

The modified equations no longer defines a set of ordinary differential equations (ODEs); rather, in this formalism we deal with a set of stochastical differential equations (SDEs). Therefore, it is important to find a proper integrator for solving SDEs. Since our model assumes a classical treatment for the lattice, it is possible to use the regular Langevin-type approach to take thermal effects into account. Several discretizations technics for the modified equation have been suggested by the literature. We have used the velocity-verlet (Eqs. 98 and 103) that is very similar to the popular BBK integrator [52]. Furthermore, we have introduced both the dissipative force and the gaussian random force in such a way to possess the power spectral density given by the fluctuation-dissipation theorem. In this way, the fluctuations can be obtained by using

$$\zeta_n(t) = \sqrt{(2k_B T\gamma M)/\Delta t} \times Z^n \tag{114}$$

where Z^n is a random number [52]. The damping constant can be determined by low temperature lattice thermal conductivity measurements. The γ value used here has the same order of magnitude as expected from experimental data of Raman spectral line width in polydiacetylene ($\gamma = 0.01\omega_Q$) [28]. It should be emphasized that this procedure of including temperature effects by means of a Langevin formalism has been extensively used in the literature and is known to yield excellent qualitative results.

As discussed above, the electronic wave functions and the lattice displacements at the $(j+1)$th time step are obtained from the jth time step. At time t_j the wavefunctions $\{\psi_{k,s}(i,t_j)\}$ are expressed as an expansion of the eigenfunctions $\{\phi_{l,s}\}$ of the electronic Hamiltonian at that moment:

$$\psi_{k,s}(i,t_j) = \sum_{l=1}^{N} C_{l,k}^{s} \phi_{l,s}(i), \tag{115}$$

where $C_{l,k}^{s}$ are the expansion coefficients. The occupation number for each eigenstate $\phi_{l,s}$ is

$$\eta_{l,s}(t_j) = \sum_k{}' \left| C_{l,k}^{s}(t_j) \right|^2 \tag{116}$$

$\eta_{l,s}(t_j)$ contains information concerning the redistribution of electrons among the energy levels, thus being of fundamental importance to analyze the products formed after the recombination mechanism as well as their yields.

3.7. Charge Carriers and Excited States

Due to the diversity of experimental results regarding the conductivity in conjugated polymers, a need to propose different conduction mechanisms from those used to describe the charge transport in inorganic materials arises naturally. As mentioned in the first section, the first conduction model in organic semiconductors was proposed by Su, Schrieffer and Heeger (the SSH model). In this model, the organic semiconductor is formed by long chains, for which the occurrence of topological defects in the polymerization process becomes natural. These structural defects could also be created upon doping or photoexcitation processes forming radicals. These defects may change the dimerization shape generating specific forms to the CH groups configuration. The high conductivity of the conjugated polymers is due to the presence of these defects in the lattice. When an external electric field is applied, if the defect has charge, it can move through the lattice. Thus, these defects behave as charge carriers. This collective behavior of the lattice and electronic system is characterized as quasi-particles.

The conventional charge carrier in organics semiconductors *par excellence* is the polaron [53-60]. This quasi-particle can be understood as a rearrangement of the π-electrons, which polarizes the lattice locally, resulting in a short range modification to the spacial configuration of the carbon atoms. In organic semiconductors, the polaron could also be understood as a bond state of solitons (the original solution for the PA chain, previously mentioned) pair. In other words, in organic conductors, the polaron is a quasi-particle arising from the electron-phonon interactions manifested in the form of a lattice distortion.

Regarding its electronic structure, the topological defects represents a symmetry breaking in the lattice. As a consequence, one electronic state arises in the band gap. The electronic spectrum of the polaron presents two energy levels inside the band gap: one is closer to the conduction band whereas the other one closer to the valence band, as shown in Figure 4. A polaron has spin $\pm 1/2$ and a charge $\pm e$. Thus, polarons can respond, simultaneously, to the action of electrical and magnetic fields.

Another type of topological defects can be generated due to the strong electron-phonon interaction presented by conjugated polymers. Bipolarons, for instance, which are spinless charge carriers possessing charge $\pm e$, may be created in organic semiconductors due to a large concentration of polarons [53, 57-59, 61, 62]. For example, two acoustic polarons with the same charge and antiparallel spins can combine with each other to form an acoustic bipolaron. In this way, bipolarons are similar to polarons by presenting similar dimerization pattern and also two localized electronic states inside the band gap. However, regarding the energy levels profile, a bipolaron can be identified by a two states deeper inside the gap when compared to those of a polaron, as shown in Figure 5.

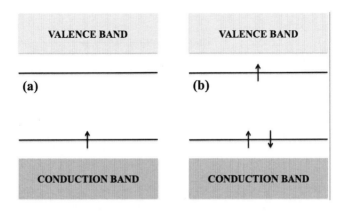

Figure 4. Schematic representation of the electronic spectrum for a polyacetylene lattice containing (a) a positive polaron (hole-polaron) and (b) a negative polaron (electron-polaron).

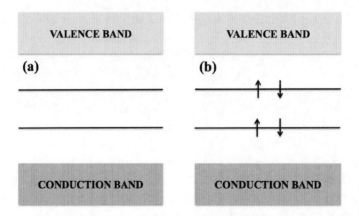

Figure 5. Schematic representation of the electronic spectrum for a polyacetylene lattice containing (a) a positive bipolaron (hole-bipolaron) and (b) a negative bipolaron (electron-bipolaron).

In contrast to the conventional inorganic conductors, the fact that conjugated polymer are quasi-one-dimensional materials leads to the property of its lattice structure being easily distorted to form self-trapped elementary excitations. Another type of self-localized electronic state are the excitons [63]. In conjugated polymers, an exciton is a bonded state of an electron-hole pair formed due to the strong electron-lattice interactions. As the charge carriers in organic conductors are mainly quasiparticles, the most typical exciton in these material is the one composed of the bound state between a hole (positively charged) polaron and an electron (negatively charged) polaron. Moreover, in these materials the excitons are generally considered to be more strongly localized than excitons in three-dimensional semiconductors, especially because in the former the exciton is substantially confined to a single polymer chain [63].

The spin wavefunction of the exciton, formed from the two spin-1/2 electronic charges, can be either singlet (S = 0) or triplet (S = 1). The radiative emission (fluorescence) is from the singlet only. The triplet exciton, in its turn, do not produce light emission other than by indirect processes such as phosphorescence or by triplet-triplet annihilation [13]. The energy levels configuration for system with an exciton present two localized electronic states inside the band gap, similarly to an organic semiconductor lattice which contains a bipolaron structure, as shown in Figure 6.

A key aspect in the physics of organic compared to inorganic semiconductors is the difference on the nature of the optically excited states. Whereas in inorganic materials, the production of free charge is carried out directly, the absorption of a photon in organic materials causes a delocalization of these states, which leads to the formation of an exciton due to the strong electron-lattice interactions, as mentioned before [46, 63]. The organic exciton binding energy is naturally large, on the order of or larger than 500 meV. This binding energy represent twenty times or more the thermal energy at room temperature, $k_B T(300K) = 26 meV$, compared with a few meV in the case of inorganic semiconductors [63].

Tight-Binding Models for the Charge Transport in Organic Semiconductors 133

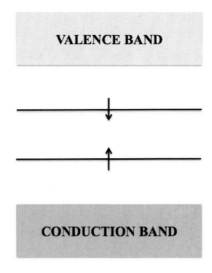

Figure 6. Schematic represantation of the electronic spectrum for a polyacetylene lattice containing an exciton.

Here, is worth to mention some important aspects regarding the most common product found in our studies (a charged excited state), which is generated from the recombination mechanism between charge carriers. An OLED normally consists of a luminescent conjugate polymer layer, introduced between two metal electrodes. Electrons and holes are injected from the electrodes into the polymer layer and, as a result, this process induce self-localized electron states (polarons). It is known that the injected electrons and holes forms electron-polarons and hole-polarons due to the strong electron-lattice interactions in these materials. Bipolarons, in their turn, can be created in OLEDs when the charge injection results in a large concentration of polarons. When an electron-polaron (-bipolaron) meets a hole-polaron (-bipolaron), they may collide and recombine to form a mixed state composed by polarons (bipolarons) and excitons (biexcitons), in which the electron and the hole are bonded in a self-trapped lattice deformation known as exciton (biexciton or polaron-exciton (bipolaron-exciton), in analogy to conventional excitons in inorganic semiconductors. The photon emission results from the radiative decay of the charged and neutral excited states. Thus, the yield of these excitations determines the electroluminescence efficiency in conjugated polymers.

4. SOME ILLUSTRATIONS

In this section we illustrate the power of the methodology discussed in this chapter by presenting some of the results published in the last half-decade by our group of research using the here discussed approach. The list is merely exemplary and more results can be vastly found in the literature, including the references listed at the end of the chapter. We chose to present five sets of results, each one concerning a different relevant property considered in the kind of system discussed in the chapter.

4.1. Temperature

Temperature effects are important for a number of reasons, the most prominent being thermoelectric phenomena present in organic polymers and the necessity excitons have of diffusing through the chain in order to give rise to free charge carriers. In the present, we chose to present the second of the cases, as discussed in a recent publication of the group [64].

The work discussed simulations of low energy excitons diffusing through a polyacetylene chain. It is found that a normal diffusion is achieved according to Figure 7. Also, when investigating the dependence of diffusivity on temperature, a Marcus-like behavior is achieved, as can be seen from Figure 9.

The adopted approach also allowed us to obtain diffusivity values much greater than those of tetracene and P3HT. This fact is attributed to the highly ordered character of the treated system.

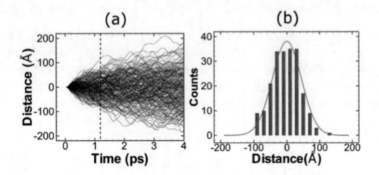

Figure 7. (a) Position and (b) Distribution of exciton for 300K.

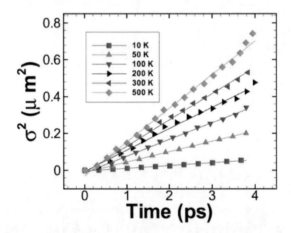

Figure 8. Variance of exciton distribution.

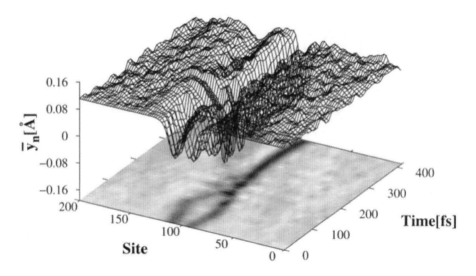

Figure 9. Bond-length order parameter time evolution of the biexciton formation process.

4.2. Reactive Scattering

Another issue of the most importance on organic semiconductors is the collisions between quasi-particles. Depending on the external conditions implemented on the system, several kinds of reactive scattering processes might take place thus leading to products with very different properties.

In a recent work [65] we observed that collisions between singlet excitons can give rise to different quasiparticles, depending on the implemented conditions: (1) a negative polaron and a positive bipolaron was found in the absence of temperature and the presence of coulomb interactions and electric field; (2) two free and oppositely charged polarons when thermal effects are considered together with coulomb interactions and moderate electric field and (3) a biexciton arises in the absence of electric field.

4.3. Concentration Effects

Another property of major importance is the influence of the charge carrier concentration on the system's dynamics [66]. By means of our analysis we were able to determine different regimes in which polaron-excitons or neutral excitations are formed from collisional processes related to different number of polarons. We observe that higher concentration of charge carriers favors the creation of polaron-excitons when compared to excitons.

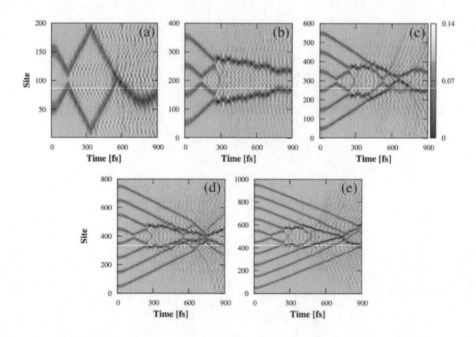

Figure 10. Reactive scattering of different initial states as a function of the concentration.

4.4. 2D-Nanoribbons

The Tight-binding approach used can be further modified to comprise the investigation of charge carrier transport in two-dimensional systems such as Graphene nanoribbons [67]. By means of this approach we were able to obtain experimentally expected charge distribution (such as the one depicted in Figure 11 as well as a polaron-like transport feature, as shwon in Figure 12.

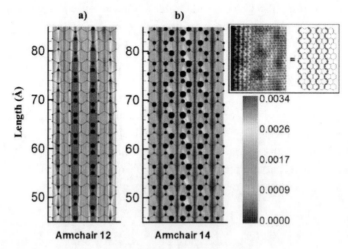

Figure 11. Charge distribution in Armchair Graphene Nanoribbons.

Tight-Binding Models for the Charge Transport in Organic Semiconductors 137

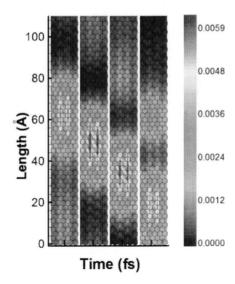

Figure 12. Quasi-particle mediated transport.

4.5. Defects

As a final note, it is well known that disorder effects plays an important role on defining the performance of the system as far as transport properties are concerned. This issue has been discussed in back to back years [68, 69] for graphene nanoribbons. In the first work we show that the presence of an impurity significantly changes the charge distribution associated to the polaron after the colision, as can be seen from Figure 13. In the second work, instead of considering a single site impurity, we study how a vacancy acts as defect in the lattice. Figure 14 presents snapshots before, during and after the colision. The evidence of charge redistribution is, again, quite clear and the arising structure is expected to possess substantially different properties. This kind of behavior are of fundamental importance because, in practice, pristine systems are obtained only under very specific conditions.

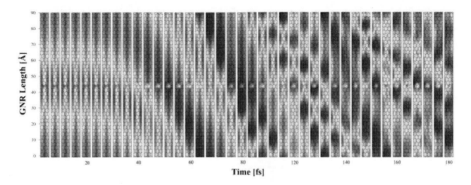

Figure 13. Charge carrier transport under the presence of an impurity in graphene nanoribbon.

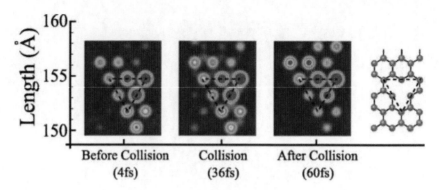

Figure 14. Charge carrier transport under the presence of an impurity.

REFERENCES

[1] N. Hall. Shirakawa. "Twenty–five years of conducting polymers." *Chemical Communications*, 1–4. Accessed January 20, 2017. doi: 10.1039/B210718J.

[2] C. P. de Melo. "Polímeros condutores." *Ciência Hoje* 36, 38–42. 1987.

[3] J. A. Epstein and J. S. Miller. 1979. "Linear–chain conductors." *Scientific American* 241, 52–61. Accessed January 20, 2017.

[4] H. Shirakawa and S. Ikeda. 1979. "Preparation and morphology of as-prepared and highly stretch-aligned polyacetylene." *Synthetic Metals* 80, 175–184. Accessed January 20, 2017. doi:org/10.1016/0379-6779(80)90008-9.

[5] H. Shirakawa. 2001. "Nobel lecture: The discovery of polyacetylene film: the dawning of an era of conducting polymers." *Reviews of Modern Physics* 73, 713-718. Accessed January 20, 2017. doi:org/10.1103/RevModPhys.73.713.

[6] H. Shirakawa and S. Ikeda. 1971. "Infrared spectra of poly(acetylene)." *Polymer Journal* 2, 231–244. Accessed November 10, 2016. doi:10.1295/polymj.2.231.

[7] H. Shirakawa, T. Ito, and S. Ikea. 1978. "Electrical–properties of polyacetylene with various cis–trans compositions." *Macromolecular Chemistry and Physics* 179, 1565–1573. Accessed November 15, 2016. doi: 10.1002/macp.1978.021790615.

[8] T. Yamabe, K. Akagi, H. Shirakawa, K. Ohzeki, and K. Fukui. 1981. "Electronic–structure of doped polyacetylene – mechanism of isomerization from cis to trans form." *Chemica Scripta* 17, 157–158. Accessed November 19, 2016. Doi.

[9] C. K. Chiang, Y. W. Park, A. Heeger, H. Shirakawa, E. J. Louis, and A. G. MacDiarmid. 1978. "Conducting polymers: Halogen doped polyacetylene." *Journal of Chemical Physics* 69, 5098. Accessed November 5, 2016. doi:10.1063/1.436503.

[10] C. Chiang, C. Fincher, Y. W. Park, A. J. Heeger, H. Shirakawa, E. J. Louis, S. C. Gau, and A. G. MacDiarmid. 1977. "Electrical conductivity in doped polyacetylene." *Physical Review Letters* 39, 1098. Accessed November 15, 2016. doi: 10.1103/PhysRevLett.39.1098.

[11] G. MacDiarmid. 2001. "'Synthetic metals': A novel role for organic polymers," *Angewandte Chemie International* Edition 40, 2581–2590. Accessed November 15, 2016. doi:10.1002/15213773(20010716)40:14.

[12] J. H. Burroughes, D. D. C. Bradley, A. R. Brown, R. N. Marks, K. Mackay, R. H. Friend, P. L. Burns, and A. B. Holmes. 1990. "Light-emitting diodes based on

conjugated polymers." *Nature* 347, 539–541. Accessed November 15, 2016. doi:10.1038/347539a0.

[13] R. H. Friend, R. W. Gymer, A. B. Holmes, J. H. Burroughes, R. N. Marks, C. Taliani, D. D. C. Bradley, D. A. D. Santos, J. L. Bre'das, M. Logdlund, and W. R. Salaneck. 1999. "Electroluminescence in conjugated polymers." *Nature* 397, 121–128. Accessed November 15, 2016. doi:10.1038/16393.

[14] N. C. Greenham, S. C. Moratti, D. D. C. Bradley, R. H. Friend, and A. B. Holmes. 1993. "Efficient light-emitting diodes based on polymers with high electron affinities." *Nature* 365, 628–630. Accessed November 15, 2016. doi:10.1038/365628a0.

[15] G. Li, V. Shrotriya, J. Huang, Y. Yao, T. Moriarty, K. Emery, and Y. Yang. 2005. "High-efficiency solution processable polymer photovoltaic cells by self–organization of polymer blends." *Nature Materials* 4, 864–868. Accessed December 5, 2016. doi:10.1038/nmat1500.

[16] W. Ma, C. Yang, X. Gong, K. Lee, and A. J. Heeger. 2005. "Thermally stable, efficient polymer solar cells with nanoscale control of the interpenetrating network morphology." *Advanced Functional Meterials* 15, 1617–1622. Accessed December 15, 2016. doi:10.1002/adfm.200500211.

[17] H. Sirringhaus, P. J. Brown, R. H. Friend, M. M. Nielsen, K. Bechgaard, B. M. W. Langeveld-Voss, A. J. H. Spiering, R. A. J. Janssen, E. W. Meijer, P. Herwig, and D. M. de Leeuw. 1999. "Two-dimensional charge transport in self-organized, high-mobility conjugated polymers." *Nature* 401, 685–688. Accessed December 15, 2016. doi:10.1038/44359.

[18] H. Sirringhaus, N. Tessler, and R. H. Friend. 1998. "Integrated optoelectronic devices based on con-jugated polymers." *Science* 280, 1741–1744. Accessed December 15, 2016. doi:10.1126/science.280.5370.1741.

[19] H. Sirringhaus, N. Tessler, and R. H. Friend. 2004. "The path to ubiquitous and low-cost organic electronic appliances on plastic." *Nature* 428, 911–918. Accessed December 15, 2016. doi:10.1038/nature02498.

[20] R. E. Peierls, *Quantum Theory of Solids* (Oxford, 1955).

[21] D. I. Bower, *An Introduction to Polymer Physics* (Cambridge, 2002).

[22] G. Strobl, *The Physics of Polymers* (Springer, 1997).

[23] Szabo and N. S. Ostlund, *Modern Quantum Chemistry* (Dover Publications, 1996).

[24] J. J. Sakurai, *Modern Quantum Mechanics* (Pearson, 2014).

[25] J. M. Ziman, *Principles of the Theory of Solids* (Cambridge, 1972).

[26] Einstein, *Investigations on the Theory of Brownian Movement* (Dover Publications, 1956).

[27] D. S. Lemos, *An Introductions to Stochastic Processes in Physics* (The Johns Hopkins University Press, 2002).

[28] G. W. Ford and M. Kac, "On the quantum lavegevin equation," *Journal of Statistical Physics* 46 (1986).

[29] J. Luczka, "Non–markovian stochastic processes: Colored noise," *Chaos* 15 (2005).

[30] T. Tomé and M. J. Oliveira, *Dinâmica Estocástica e Irreversibilidade* [*Stochastic Dynamics and Irreversibility*] (Editora da Universidade de São Paulo, 2001).

[31] W. P. Su, J. R. Schrieffer, and A. J. Heeger. 1979. "Solitons in polyacetylene." *Physical Review Letters* 42, 1698–1701. Accessed October 10, 2016. doi:org/10.1103/PhysRevLett.42.1698.

[32] W. P. Su, J. R. Schrieffer, and A. J. Heeger. 1980. "Soliton excitations in polyacetylene." *Physical Review B* 22, 2099–2111. Accessed October 7, 2016. doi.org/10.1103/PhysRevB.22.2099.

[33] J. Heeger, S. Kivelson, J. R. Schrieffer, and W. P. Su. 1988. "Soliton in conducting polymers." *Reviews of Modern Physics* 60, 781. Accessed October 7, 2016. doi:org/10.1103/RevModPhys.60.781.

[34] J. C. Chiang and A. G. MacDiarmid. 1986. "Polyaniline – protonic acid doping of the emeraldine form to the metallic regime." *Synthetic Metals* 13, 193–205. Accessed October 17, 2016. doi:org/10.1016/0379-6779(86)90070-6.

[35] H. Shirakawa, E. J. Louis, A. G. MacDiarmid, C. K. Chiang, and A. J. Heeger. 1977. "Synthesis of electrically conducting organic polymers: halogen derivatives of polyacetylene chx." *Journal of the Chemical Society – Chemical Communications* 16, 578–580. Accessed October 12, 2016. doi: 10.1039/C39770000578.

[36] Y. Lu, *Solitons & Polarons in Conducting Polymers* (World Scientifc, 1965).

[37] J. Heeger. 2001. "Semiconducting and metallic polymers: The fourth generation of polymeric materials (nobel lecture)." *Angewandte Chemie* 40, 2591–2611. Accessed October 10, 2016. doi:10.1002/1521-3773(20010716)40:14.

[38] J. Heeger. 2010. "Semiconducting polymers: the third generation." *Chemical Society Reviews* 39, 2354. Accessed October 17, 2016. doi:10.1039/b914956m.

[39] S. Stafstrom. 2010. "Electron localization and the transition from adiabatic to nonadiabatic charge transport in organic conductors." *Chem. Soc. Rev.* 39, 2484–2499. Accessed June 14, 2016. doi:10.1039/B909058B.

[40] S. Pinheiro and G. M. e Silva. 2002. "Use of polarons and bipolarons in logical switches based on conjugated polymers." *Phys. Rev. B.* 65, 094304–5. Accessed June 14, 2016. doi:10.1103/PhysRevB.65.094304.

[41] Johansson and S. Stafstrom. 2004. "Nonadiabatic simulations of polaron dynamics." *Phys. Rev. B.* 69, 235205–7. Accessed July 14, 2016. doi:org/10.1103/*PhysRevB*. 69.235205.

[42] Johansson and S. Stafstrom. 2001. "Polaron dynamics in a system of coupled conjugated polymer chains." *Phys. Rev. Lett.* 86, 3602–3605. Accessed July 14, 2016. doi:10.1103/PhysRevLett.86.3602.

[43] G. M. e Silva and A. Terai. 1993. "Dynamics of solitons in polyacetylene with interchain coupling." *Phys. Rev. B.* 47, 12568–12577. Accessed July 14, 2016. doi:org/10.1103/PhysRevB.47.12568.

[44] G. M. e Silva. 2000. "Electric-field effects on the competition between polarons and bipolarons in conjugated polymers." *Phys. Rev. B.* 61, 10777–10781. Accessed July 14, 2016. doi:10.1103/PhysRevB.61.10777.

[45] M. P. Lima and G. M. e Silva. 2006. "Dynamical evolution of polaron to bipolaron in conjugated polymers." *Phys. Rev. B.* 74, 224303–6. Accessed November 24, 2016. doi:10.1103/PhysRevB.74.224304.

[46] S. Brazovskii and N. Kirova. 2010. "Physical theory of excitons in conducting polymers." *Chem. Soc. Rev.* 39, 2453–2465. Accessed November 24, 2016. doi:10.1039/B917724H.

[47] Z. An, C. Q. Wu, and X. Sun. 2004. "Dynamics of photogenerated polarons in conjugated polymers." *Phys. Rev. Lett.* 93, 216407–4. Accessed November 24, 2016. doi:10.1103/PhysRevLett.93.216407.

[48] Z. G. Yu, M. W. Wu, X. S. Rao, and A. R. Bishop. 1996. "Excitons in two coupled conjugated polymer chains." *J. Phys. Condes. Matter* 8, 8847–8857. Accessed November 14, 2016. doi:org/10.1088/0953-8984/8/45/018.

[49] W. Barford. 2013. "Excitons in conjugated polymers: A tale of two particles." *J. Phys. Chem. A.* 117, 2665–2671. Accessed November 14, 2016. doi: 10.1021/jp310110r.

[50] W. Barford. 2004. "Theory of singlet exciton yield in light-emitting polymers." *Phys. Rev. B.* 70, 205204–205212. Accessed October 14, 2016. doi:org/10.1103/PhysRevB.70.205204.

[51] Z. An, B. Di, and C. Q. Wu. 2008. "Inelastic scattering of oppositely charged polarons in conjugated polymers." *European Physical Journal B* 63, 71–77. Accessed October 18, 2016. doi:org/10.1140/epjb/e2008-00216-8.

[52] J. A. Izaguaire, D. P. Catarello, J. M. Wozniak, and R. D. Skell. 2001. "Langevin stabilization of molecular dynamics." *J. Chem. Phys.* 114, 2090–2098. Accessed October 18, 2016.

[53] S. A. Brazoviskii and N. Kirova. 1981. "Excitons, polarons and bipolarons in conducting polymers," *Soviet Physics: Journal of Experimental and Theoretical Physics (JETP)* 33, 4–8. Accessed October 18, 2016.

[54] S. Brodeaux, R. R. Chance, J. L. Bre'das, and R. Silbey. 1983 "Solitons and polarons in polyacety- lene: Self–consistent–filed calculations of the effect neutral and charged defects on molecular geometry." *Physical Review B* 28, 6927–6936. Accessed October 18, 2016. doi:org/10.1103/PhysRevB.28.6927.

[55] Y. R. Liu and K. Maki. 1980. "Two–soliton interaction energy and the soliton lattice in polyacetylene." *Physical Review B* 22, 5754–5758. Accessed November 09, 2016. doi:10.1103/PhysRevB.22.5754.

[56] Y. Onodera and S. Okuno. 1983. "Two–polaron solution and its stability in the continuum model of polyacetylne." *Journal of Physics Society of Japan* 52, 2478–2484. Accessed November 09, 2016. doi.org/10.1143/JPSJ.52.2478.

[57] S. Stafstrom and K. A. Chao. 1984. "Polaron–bipolaron — soliton doping in polyacetylene." *Physical Review B* 30, 2098–2103. Accessed November 09, 2016. doi:10.1103/PhysRevB.30.2098.

[58] Y. Onodera. 1984. "Polarons, bipolarons, and their interactions in cis-polyacetylene," *Physical Review B* 30, 775–785. Accessed November 09, 2016. doi:org/10.1103/PhysRevB.30.775.

[59] J. L. Brédas and G. B. Street. 1985. "Polarons, bipolarons, and solitons in conducting polymers." *Accounts of Chemical Research* 18, 309–315. Accessed November 09, 2016. doi: 10.1021/ar00118a005.

[60] Khun. 1989. "*Solitons, polarons, and excitons in polyacetylene: Step-potential model for electron-phonon coupling in π-electron systems.*" *Physical Review B* 40, 7776. Accessed November 12, 2016. doi:org/10.1103/PhysRevB.40.7776.

[61] K. Campbell and A. R. Bishop. 1981. "Solitons in polyacetylene and the relativistic field theory model." *Physical Review B* 24, 4859–4862. Accessed November 12, 2016. doi.org/10.1103/PhysRevB.24.4859.

[62] W. Lang, Z. B. Su, and F. Martino. 1986. "Bipolaron dynamics in nearly degenerate quasi–one– dimensional polymers." *Physical Review B* 33, 1512–1515. Accessed November 12, 2016. doi.org/10.1103/PhysRevB.33.1512.

[63] Kipplen and J. L. Brédas. 2009. "Organic photovoltaics." *Energy & Environmental Science* 2, 251-261. Accessed November 12, 2016. doi:10.1039/B812502N.

[64] P. H. de Oliveira Neto, D. A. da Silva Filho, W. F. da Cunha, P. H. Acioli and G. M. e Silva, "Limit of exciton diffusion in highly ordered π-conjugated systems," *Jour. Phys. Chem. C* 119(34), 19654 – 19659. Accessed September 09, 2015. doi: 10.1021/acs.jpcc.5b05508.

[65] L. A. Ribeiro Junior, W. F. da Cunha and G. M. e Silva, "Singlet-singlet exciton recombination: Theoretical insight into the influence of high density regime of excitons in conjugated polymers," *Jour. Phys. Chem. B* 118, 5250 – 5257. Accessed March 16, 2014. doi: 10.1021/jp4107926.

[66] L. A. Ribeiro Junior, W. F. da Cunha, A. L. A. Fonseca and G. M. e Silva, "Concentration effects on intrachain polaron recombination in conjugated polymers," *Phys. Chem. Chem. Phys.* 17, 1299-1308. Accessed November 20, 2015. doi:10.1039/c4cp04514a.

[67] P. H. de Oliveira Neto, J. F. Teixeira, W. F. da Cunha, R. Gargano and G. M. e Silva, "Electron-lattice coupling in armchair graphene nanoribbons," *Jour. Phys. Chem. Lett.* 3(20), 3039 – 3042. Accessed March 12, 2017. doi: 10.1021/jz301247u.

[68] W. F. da Cunha, L. A. Ribeiro Junior, A. L. A. Fonseca, R. Gargano and G. M. e Silva, "Impurity effects on polaron dynamics in graphene nanoribbons," *Carbon* 91, 171-177. Accessed October 23, 2015. doi:10.1016/j.carbon.2015.04.065.

[69] W. F. da Cunha, P. H. de Oliveira Neto A. Terai and G. M. e Silva, "Dynamics of charge carriers on hexagon nanoribbons with vacancy defects," *Phys. Rev. B* 94, 014301-1 – 014301-10. Accessed December 13, 2016. doi: 10.1103/PhysRevB. 94.014301.

In: Polarons: Recent Progress and Perspectives
Editor: Amel Laref

ISBN: 978-1-53613-935-8
© 2018 Nova Science Publishers, Inc.

Chapter 3

THE POLARON EFFECT ON CHARGE TRANSPORT PROPERTY FOR ORGANIC SEMICONDUCTORS

Nianduan Lu, Ling Li and Ming Liu*

Key Laboratory of Microelectronic Devices and Integrated Technology,
Institute of Microelectronics of Chinese Academy of Sciences,
Beijing, China

ABSTRACT

Organic semiconductors have held several unique advantages for low cost, large area, low-end, lightweight, and ultra-flexible electronics applications. After Bässler et al. suggested that the activation energy of the charge transport in organic semiconductors were contributed to disorder and polaron, interests in the polaron effect have been remarkably increased. In this chapter, we mainly concentrated on the polaron effect on the charge transport of organic semiconductors. Several kinds of theoretical models of the polaron effect have been described. Otherwise, based on these current theoretical models, the polaron effect on the charge transport property in organic semiconductors has been discussed in details. Finally, a future outlook of orgainc semiconductors is briefly discussed.

Keywords: organic semiconductor, charge transport, polaron effect, theoretical model

INTRODUCTION

History

Organic Semiconductor

Organic semiconductors are a class of materials that possess the electronic features of semiconducting materials and the chemical and mechanical advances of organic compounds

---

* Corresponding Author Email: lunianduan@ime.ac.cn.

[1, 2]. Interest in organic semiconductors was stimulated by a suggestion in 1946, at which some processes in biological systems might be accounted for by the transfer of π-electrons over large distance along molecular stepping stones [3]. The earliest studies on the electrical behavior of organic materials may go back to the 1960s [4]. Early studies on organic system concentrated on the effort to produce pure material. In 1968, Berets and Smith studied the effect of oxygen content on polycrystalline powder and found that oxygen played an important role in the conductivity of polyacetylene [5]. Then, the appearance of conductive polymers in the late 1970s, and of conjugated semiconductors and photoemission polymers in the 1980s, greatly motivated the development in the field of organic electronics [6, 7]. Polyacetylene was one of the first polymers reported to be capable of conducting electricity [8], and it was found that the oxidative dopant with iodine could largely enhance the conductivity by 12 orders of magnitude [9]. The fundamental discovery of the highly-conductive organic polymers was honored by the Nobel Prize in 2000. Three outstanding scientists (Alan J. Heeger, Alan G. MacDiarmid and Hideki Shirakawa) were jointly awarded the Nobel Prize in Chemistry in 2000 for their discovery in 1977 and development of oxidized, iodine-doped polyacetylene. After that, plenty of new organic semiconductor materials were developed. The development of this new class of organic semiconductor materials continues to offer the promise of a wide range of novel applications including molecular electronics, displays, transistors and photovoltaics.

Polaron

A polaron is a quasi-particle composed of the charge and its surrounding polarization cloud. The general polaron concept was introduced by Landau in 1933. Subsequently, in 1951 Landau and Pekar investigated the self-energy and the effective mass of the polaron. In 1954, Fröhlich developed a method to study polaron in the adiabatic or strong-coupling regime. In 1959, Holstein described the polaron motion by using one-dimensional molecular crystal model [10]. The polaron formation in organic materials is usually treated in terms of the small-polaron model suggested by Holstein and Friedman and further developed for the nonadiabatic polaron transfer between sites with different energies by Emin. In 1990, L. B. Schein et al. proved the mechanism of the small-polaron hopping debased on their observation in molecularly doped polymers [11]. Then, in 1994 Bässler et al. firstly suggested that the activation energy of the charge transport could be split into two parts of contribution: disorder and polaron [12]. In 2001, Parris et al. demonstrated the compatibility of the polaron effect with experimental observation by using the Marcus rate model in disordered organic materials (molecularly doped polymers) [13]. Since then the polaron effect has aroused extensive research interests.

Structure of Organic Semiconductors

Electronic Structure

Organic semiconductors are carbon-based materials and consist of the hydrocarbon molecules with a backbone of carbon atoms. For the carbon atoms in organic semiconductors, s-orbital and p-orbital in the outer shell (ground state: $1s^22s^22p^2$) hybridize to three sp^2 orbitals. The sp^2 hybridized orbitals induce strongly bound σ-bonds and molecular backbone.

The forth orbital is a p_z orbital perpendicular to the carbon plane. The p_z orbitals overlap to a lesser degree, which leads to less binding or antibinding of π and π^* orbitals, thus forming the frontier orbitals of the molecule. Generally, in organic semiconductor π and π^* orbitals can provide the localized electronic states which can be occupied by electrons, or by holes in the case of an underoccupation. These localized states are often taken as transport sites. Their electronic wave function is localized in space within a certain volume and can therefore be assigned spatial coordinates. In the ground state of the molecule, all bonding orbitals up to the highest occupied molecular orbital (HOMO) are filled with two electrons of antiparallel spin while the antibonding orbitals, from the lowest unoccupied molecular orbital (LUMO) onwards, are empty. One can call the HOMO and LUMO energies electron and hole site energies, respectively. Different sites display different site energies for the electron and hole which depend on the inter- and intra-molecular interactions. This interaction dependence implies an energetic disorder.

Crystal Structure

Based on the degree of the crystallinity (static disorder), two major classes of organic semiconductor material have provided: crystalline or amorphous materials [14, 15]. Of course, more detailed types of organic semiconductor material include the following, such as amorphous molecular films, molecular crystals and polymer films. In general, organic semiconductors are characterized by weak van der Waals bonding, which gives them weak intermolecular interactions. This weak coupling of molecules results in weak interaction energy to give narrow electronic bandwidths. Existing narrow electronic bands will be eliminated by the statistical variation of width in the energy level distribution of molecules, which hence creates Anderson charge localization [14]. For the crystalline organic semiconductor dominated, a number and variety of defects are the largest contributors to the existence of disorder within a lattice. In the soluble small-molecule and polymeric crystals, the steric hindrance of side-groups in the combination with local chemical defects can cause slight perturbations in the local spacing and tilt of the neighbouring molecules. This type of accumulation of the slight local perturbations and distinct lattice imperfections lead to a static disorder (even in the ordered regions of a film), which may ultimately limit the motion of charges. For the disordered organic semiconductors, because a large concentration of crystal imperfections, such as impurities, grain boundaries, and dangling bonds, breaks the periodicity of a crystal, the localization of charge carriers is attributed to spatial and energetic disorder due to weak intermolecular interactions [16].

Charge Transport Mechanism

Hopping transport

Due to the spatial and energetic disorder in organic semiconductors, charge carriers are always located on localized sites. Therefore, the charge transport always happens to jump from one localized site to another. This type of transport mechanism is called hopping transport. The transition of hopping between two sites depends on the overlap of the electronic wave functions of these two sites, which allows tunnelling from one localized site to another. Whenever a charge carrier hops to a site with a higher (lower) site energy than the site that it came from, the difference in energy is accommodated for by the absorption (emission) of a phonon. Based on the hopping distance and site energy, the transport of charge carriers can be divided into nearest-neighbour hopping (NNH) and variable-range hopping (VRH). The nearest neighbour hopping describes a hopping regime, in which the tunnelling part of the hopping rates is slower than the energy contribution, that only the nearest neighbours are addressed in hops. This is the case for very high temperatures or very small localization radii. However, it is not suitable for the charge carrier under low enough temperature, where the thermally-activated hopping rates become much smaller than the spatial tunnelling rates. For the low enough temperature, the charge carriers are more possible to hop to the sites with more favourable energies closer to the Fermi level. This transport mechanism is called variable-range hopping. The variable-range hopping was firstly proposed by N. Mott, et al. in 1971 [17] and hence was also called the Mott VRH, describing the low temperature conduction in strongly disordered systems with localized states. N. Mott pointed out that if the activation energy to a nearest-neighbour site was large, a more favourable hop might be to a site farther away with a lower activation energy. This tradeoff of energy and distance for the optimal jump depends on the respective transition rates (energy dependent hops and spatial-dependent tunnelling rates). Since the energy-dependent transition is thermally-activated, the optimal hopping distance will depend on temperature.

Multiple Trapping and Release Theory

For the crystalline organic materials, such as small-molecule organic semiconductors, which have a strong tendency to form polycrystalline film. These semiconductors also show a result of the regular molecular arrangement, and the delocalized orbitals of neighbouring molecules partially overlap, thereby facilitating more efficient intermolecular charge-carrier transfer and carrier mobility that is much larger than in amorphous films. In this situation, the transport mechanism is not easily explained with the hopping theory. In contrast to the hopping theory, the multiple trapping and release (MTR) theory adapted for crystal organic materials. The MTR theory assumes that the charge carrier transport occurs in the extended states, and that most of the charge carriers are trapped in the localized states. The energy of the localized state is separated from the mobility edge energy. When the localized state energy is slightly lower than the mobility edge, then the localized state acts as shallow trap, from which the charge carrier can be released (emitted) by thermal excitations. But, if that energy is far below the mobility edge energy, then the charge carrier cannot be thermally excited (emitted). The number of carriers available for the transport then depends on the difference in the energy between the trap level and the extended-state band, as well as on the temperature.

MATERIALS

Organic semiconductor materials have received much attention due to various morphologies and a number of conjugated molecules, which result in the wide application in electronic devices. The building blocks in organic semiconductors are π bonded molecules or polymers, which are formed by the carbon and hydrogen atoms, and sometimes are heteroatoms such as oxygen, nitrogen and sulfur. They essentially come in three flavours based on their microstructure: amorphous molecular films, molecular crystals, and polymer films, as mentioned above. They are generally electrical insulators but can transform semiconducting property when the charges are either injected from the appropriate electrodes, upon doping or by photoexcitation.

Amorphous Molecular Films

By this term, organic molecules deposited as an amorphous film through evaporation or spin-coating. Amorphous molecule films are employed for device applications such as organic light emitting diodes (OLEDs), organic field effect transistors (OFETs), and organic solar cells (OSCs). Illustrative materials are tris(8-hydroxyquinolinato)aluminium, C60, phenyl-C61-butyric acid methyl ester (PCBM), pentacene, carbazoles, and phthalocyanine. Figure 1 shows some examples of amorphous molecular films [18].

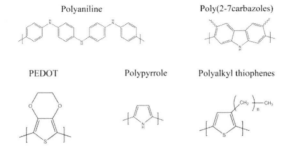

Figure 1. Examples of amorphous molecular films.

Molecular Crystals

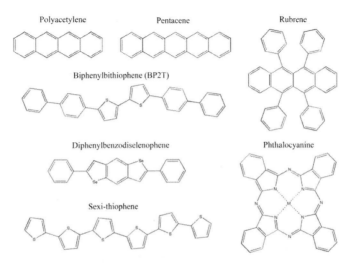

Figure 2. Examples of molecule crystals.

By definition, a crystal consists of a lattice and a basis. In the same way how atoms like silicon can form a crystal by covalent bonding, or sodium and chloride atoms by ionic bonding, molecules such as naphthalene or anthracene can form the basis of a crystal that is held together by van-der-Waals interactions. The advantage of employing molecular crystals instead of amorphous film is that their charge carrier mobility is much larger. Figure 2 shows some examples of molecule crystals [2].

Polymer Films

Polymer films may be considered as a chain of covalently coupled molecular repeat units. Usually, they are processed from solution, which allows for a range of deposition techniques including the simple spin-coating, ink-jet deposition, or industrial reel-to-reel coating. They are also more suitable to blending than molecules since the polymer blends are thermodynamically more stable and less susceptible to the crystallization. The semiconducting properties in all of these types of organic semiconductors have a similar origin, though their excited states and associated photophysical properties vary slightly depending on the order and coupling in the solid. Figure 3 shows some examples of polymer films [2].

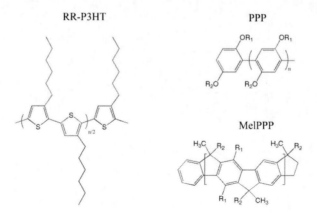

Figure 3. Examples of polymer films.

CHARGE TRANSPROT THEORY OF ORGANIC SEMICONDUCTORS

1D Holstein Molecular Model

A significant insight into the polaron transport was obtained from the analytical results derived by Holstein in his seminal work [10]. For the standard Holstein-type polaron model, the physical meaning of the local coupling constants can be readily understood by considering the limiting case of the weak electronic coupling. In this case, the Hamiltonian for a single charge carrier in the lattice can be diagonalized exactly with the resulting energy given by [19]

The Polaron Effect on Charge Transport Property for Organic Semiconductors   149

$$E_m = \epsilon_m^{(0)} - \frac{1}{N}\sum_{qj}\hbar\omega_{qj}|g_m(q,j)|^2 + \sum_{qj}\hbar\omega_{qj}\left(n_{qj} + \frac{1}{2}\right), \tag{1}$$

where $\epsilon_m^{(0)}$ is the electron site energy, N denotes the total number of unit cells, $g_m(q,j)$ dnotes the corresponding local electron-phonon coupling constants, n_{qj} denotes the annihilation operator for a phonon of branch j with energy $\hbar\omega_{qj}$ and wavevector q.

The electron (hole) is localized on a single lattice site with a stabilization energy referred to as the polaron binding energy, E_p,

$$E_p = \frac{1}{N}\sum_{qj}\hbar\omega_{qj}|g_m(q,j)|^2. \tag{2}$$

The polaron binding energy results from the deformations in molecular and lattice geometries that occur as the carrier localizes on a given site. This quantity is thus closely related to the reorganization energy in electron-transfer theories. The contribution to the polaron binding energy is arising from the internal degrees of freedom. In the framework of small polaron theory, in the case of the electron-phonon interactions with an optical phonon of energy $\hbar\omega_0$ and characterized by a coupling constant g, the hopping rate is given by [10]

$$k_{ET} = \frac{t^2}{\hbar^2\omega_0}\left[\frac{\pi}{g^2 csch\left(\frac{\hbar\omega_0}{2k_BT}\right)}\right]^{1/2} exp\left[-2g^2\tanh\left(\frac{\hbar\omega_0}{4k_BT}\right)\right], \tag{3}$$

where k_B is the Boltzmann constant, t is the average transient time. For $\hbar\omega_0 \ll k_BT$, k_{ET} obeys a standard Arrhenius-type law [19],

$$k_{ET} = \frac{t^2}{\hbar}\left[\frac{\pi}{2E_p k_BT}\right]^{1/2} exp\left(-\frac{E_p}{2k_BT}\right). \tag{4}$$

Under the situation of $E_p = \lambda_{reorg}/2$, Eq. (4) will be transferred to the classical Marcus equation for the electron-transfer rate. By using the Einstein relation and $D = a^2 k_{ET}$ in a 1D system (here a is the spacing between molecules), the classical charge mobility is written as [19]

$$\mu_{hop} = \frac{ea^2 t^2}{k_BT\hbar}\left[\frac{\pi}{2E_p k_BT}\right]^{1/2} exp\left(-\frac{E_p}{2k_BT}\right), \tag{5}$$

here e is the elemental charge. At very high temperature, $E_p \ll 2k_BT$ and Eq. (5) yields a $T^{-3/2}$ dependence of the mobility.

Marcus Rate Equation

Charge transport in many doped organic materials shows a strong response to the external electric field and temperature. To understand the electric-field dependence of the charge mobility in molecularly doped organic materials, K. Seki et al. considered the charge transport across a sample of width $L = Nl$, where l is the mean interdopant spacing and N+1 is the number of sites [20]. The charge transport from the site denoted by i to its adjacent site i+1 is described by the Marcus theory (see Figure 4). According to the Marcus rate, the transition probability is given by [21]

$$W_{i \to i \pm 1} = \frac{2\pi}{\hbar} J^2 \frac{1}{\sqrt{4\pi\lambda k_B T}} exp\left[-\frac{[V(i\pm 1)-V(i)\mp eFl+\lambda]^2}{4\lambda k_B T}\right], \tag{6}$$

where J is the transfer integral, $V(i)$ is the site energy in the absence of an external electric field F, and λ is the reorganization energy. The reorganization energy is twice of the polaron binding energy. $\lambda \sim 0.3\ eV$ is a rough estimate of the intramolecular relaxation in doped organic molecules. Since the site energy appears in the Marcus theory as an energy difference $V(i+1) - V(i)$, only the fluctuating component contributes to the transition rate when $\langle V(i+1) \rangle = \langle V(i) \rangle$. Thus, without loss of generality, $V(i)$ denotes the fluctuating component of a site energy whose sample average is zero: $\langle V(i) \rangle = 0$.

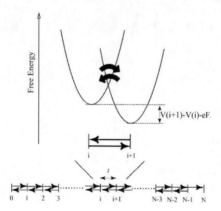

Figure 4. Schematic representation of theoretical model of one-dimensional charge transport among N+1 lattice sites [20]. A perfectly reflecting boundary condition is imposed on site 0, and a perfectly absorbing boundary condition is imposed on site N. Each transition rate is described by the Marcus rate, for which a transition takes place at the intersection of the two parabolic free energy curves.

One of the sources of the local energy difference, typically of the order 0.1 eV, derived from the electronic polarization. The localized charge induces a fast electronic polarization of surrounding molecules. The induced dipoles in the vicinity of the charged molecule create a local variation of the potential. The local energy disorder of the Gaussian energy distribution can be characterized by

$$\langle V(i)V(j) \rangle = \sigma_L^2 \delta_{ij}, \tag{7}$$

where $\sigma_L^2 \sim 0.1\ eV$ and δ_{ij} is Kronecker's delta. In addition to the above local energy disorder, it is evidenced that the energy disorder depends on the permanent dipole moment of the dopant molecule as well as that of the host organic material. The distribution of permanent dipoles generates fluctuations in the electrostatic potential due to the charge-dipole interactions

$$V(i) = e \int_{EV} dr P(r) \cdot \frac{r_i - r}{|r_i - r|^3},\tag{8}$$

where $P(r)$ is the dipole moment at the position r, and r_i is the position of the charge. *EV* denotes that the region occupied by the charged dopant molecule which is assumed to be spherical with radius a_0 is excluded when calculating the spatial integration. Due to the long-range nature of the charge-dipole interactions, the site energy has spatial correlations. The spatial correlations of site energies due to the charge-dipole interactions can be calculated in the same way as in the calculation of Marcus reorganization energy due to the charge-dipole interactions. If ϵ denotes the dielectric constant of the doped organic materials, the fluctuation-dissipation theory in the mean-field approximation leads to

$$\langle P_{\alpha 1}(r) P_{\beta 1}(r') \rangle = \frac{k_B T}{4\pi} \left(1 - \frac{1}{\epsilon}\right) \delta_{\alpha 1 \beta 1} \delta(r - r'),\tag{9}$$

where $\alpha 1$ and $\beta 1$ is introduced to represent the Cartesian components of P, and $\delta_{\alpha 1 \beta 1}$ denotes Kronecker's delta, $\delta_{\alpha 1 \beta 1} = 1$ if $\alpha 1 = \beta 1$ and $\delta_{\alpha 1 \beta 1} = 0$ if $\alpha 1 \neq \beta 1$. Using the definition of $V(i)$, together with Eq. (9), for the isotropic materials one can obtain

$$\langle V(i) V(j) \rangle = e^2 k_B T \left(1 - \frac{1}{\epsilon}\right) \left(\frac{1}{a_0} \delta_{ij} + \frac{1}{|r_i - r_j|} (1 - \delta_{ij})\right),\tag{10}$$

where $(1 - \delta_{ij})$ should be regarded as zero if i=j, despite the multiplication of the divergent function. When the sample dielectric constant comes mainly from the permanent dipoles of the doped molecules, Debye's formula can be applied, being written as

$$1 - \frac{1}{\epsilon} = \frac{4\pi}{3} c \frac{P^2}{k_B T},\tag{11}$$

where c is the molecular concentration of the doped molecules with the permanent dipole moment P. Then, the site energy correlation can be written as

$$\langle V(i) V(j) \rangle = \frac{4\pi}{3} c e^2 P^2 \left(\frac{1}{a_0} \delta_{ij} + \frac{1}{|r_i - r_j|} (1 - \delta_{ij})\right).\tag{12}$$

In general, both types of energetic disorder, the local disorder induced by the electronic polarization and the long-range disorder due to the permanent dipoles, are present. Hence the correlation of the site energies can be written as

$$\langle V(i) V(j) \rangle = \sigma^2 \left(\delta_{ij} + \frac{a}{|i - j| l} (1 - \delta_{ij})\right),\tag{13}$$

where $a = a_0 \sigma_d^2 / \sigma^2$, $\sigma^2 = \sigma_L^2 + \sigma_d^2$, and $\sigma_d^2 = e^2 k_B T (1 - 1/\epsilon)/a_0$ in terms of the dielectric constant of doped materials, or $\sigma_d^2 = 4\pi c e^2 p^2 /(3a_0)$ in terms of the dipole moment of the doped molecule.

Charge Hopping Transport Based on Marcus and VRH Theory

To build the theoretical model of the polaron effect for the hopping transport of organic semiconductors, a key step is to find the transition rate between the localized states or sites. So far, two kinds of transition rates (Miller-Abrahams (MA) and Marcus transition rate) have been usually recognized. If the polaron effect can be neglected, the transition rate for a carrier moving from site i to site j is based on the MA transition rate follows as [22]

$$v = v_0 \, exp(-R) = v_0 \begin{cases} exp\left(-2\alpha R_{ij} - \frac{E_j - E_i}{k_B T}\right), E_j - E_i > 0 \\ exp\left(-2\alpha R_{ij}\right), E_j - E_i < 0 \end{cases}$$

(14)

where v_0 is the attempt-to-jump frequency, R is the hopping range, α is the inverse localized length, R_{ij} is the hopping distance, E_i and E_j are the energies at sites i and j, respectively. When an electric field F is applied to the disorder semiconductor, it will weaken the Coulomb barrier and hence reduce the thermal activation energies. Therefore, the hopping range with the normalized energy ($\epsilon = \frac{E}{k_B T}$ and $r_{ij} = 2\alpha R_{ij}$) under the electric field can be rewritten as

$$R = \begin{cases} (1 + \beta \cos \theta) r_{ij} + \epsilon_j - \epsilon_i, \epsilon_j - \epsilon_i > -\beta \, r_{ij} \cos \theta \\ r_{ij}, \epsilon_j - \epsilon_i < -\beta \, r_{ij} \cos \theta \end{cases}$$

(15)

where $\beta = 0.5 Fe/\alpha k_B T$ and θ is the angle between R_{ij} and the electric field ranging from 0 to π.

For the polaron model, the transition rate will obey the Marcus jump rate. In this case, Eq. (14) should be replaced by the Marcus rate as [21]

$$v = \frac{|J_{ij}|^2}{\hbar} \sqrt{\frac{\pi}{E_a k_B T}} \, exp\left(-\frac{(E_j - E_i + E_a)^2}{4 E_a k_B T}\right),$$

(16)

where $J_{ij} \propto exp(-2\alpha R_{ij})$ is the transfer integral, i.e., the wave function overlap between sites i and j, E_a is the polaron activation energy. According to the hopping range derived from the MA jump rate, the hopping range derived from the Marcus rate with the normalized energy can be represented as [23]

$$R = 2 r_{ij} + \frac{(\epsilon_j - \epsilon_i + \epsilon_a)^2}{4 \epsilon_a}.$$

(17)

Since the applied electric field can decrease the thermal activation energies, hence the Marcus rate under the electric field can be expressed as [23]

$$v = \frac{|J_{ij}|^2}{\hbar k_B T} \sqrt{\frac{\pi}{\epsilon_a}} exp\left(-\frac{(\epsilon_j - \epsilon_i + \beta r_{ij} \cos \theta + \epsilon_a)^2}{4 \epsilon_a}\right).$$

(18)

Similarly, the hopping range is written as

$$R = 2r_{ij} + \frac{(\epsilon_j - \epsilon_i + \beta r_{ij}\cos\theta + \epsilon_a)^2}{4\epsilon_a}. \tag{19}$$

Then, the average hopping range of the charge carrier will be calculated in a hopping space. For a site with the energy ϵ_i, the most probable hop for a carrier on this site is to move an empty site at a range R with the minimum energy. The first step is to obtain the empty number within a range R of a particular site with the energy ϵ, as a function of the temperature and electric field. The four-dimensional hopping space can be represented for a particular angle θ by a two-dimensional diagram, as shown in Figure 5.

The average hopping range R_n can be obtained by solving the following equation:

$$N(\epsilon_i, T, \beta, R_n) = B_c, \tag{20}$$

where parameter B_c=2.8 is determined according to the percolation criteria. $N(\epsilon_i, T, \beta, R_n)$ is the empty states enclosed by the contour R and can be expressed as

$$N(\epsilon_i, T, \beta, R_n) = \frac{1}{8a^3}\int_0^\pi d\theta \sin\theta \int_0^{R_n/2} dr 2\pi r^2 \int_{\epsilon_i-\epsilon_{\beta\downarrow}}^{\epsilon_i-\epsilon_{\beta\uparrow}} d\epsilon \rho(\epsilon), \tag{21}$$

where $\epsilon_{\beta\uparrow} = \epsilon_a + \beta r \cos\theta - \sqrt{4\epsilon_a(R_n - 2r)}$ and $\epsilon_{\beta\downarrow} = \epsilon_a + \beta r \cos\theta + \sqrt{4\epsilon_a(R_n - 2r)}$ include both the polaron effect and electric filed effect. $\rho(\epsilon) = g(\epsilon)(1 - f(\epsilon))$, $g(\epsilon)$ is DOS with a Gaussian function $g(\epsilon) = \frac{N_t}{\sqrt{2\pi}\sigma_0}\exp(-\frac{\epsilon^2}{2\sigma_0^2})$, N_t is the number of states per unit volume and $\sigma_0 = \sigma/k_B T$ indicates the width of the DOS. $f(\epsilon) = 1/(1 + \exp(\epsilon - \epsilon_f))$ is the Fermi-Dirac distribution and $1 - f(\epsilon)$ is the probability that the final site is empty, ϵ_f ($\epsilon_f = E_f/k_B T$) is the Fermi level.

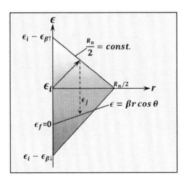

Figure 5. The contour of a constant $R_n/2$ for an initial site at the energy ϵ_i. The solid arrowed line from initial site with the energy ϵ_i represents a hop of range $R_n/2$ to a final site at the unperturbed energy ϵ_j. The line $\epsilon = \beta r \cos\theta$ represents the shift of the Fermi level with the distance under the electric field. All final sites within the contour are included in the evaluation of $N(\epsilon_i, T, \beta, R_n)$, therefore the range of integration of ϵ_j is $\epsilon_i - \epsilon_{\beta\downarrow} \rightarrow \epsilon_i - \epsilon_{\beta\uparrow}$ with $\epsilon_{\beta\downarrow} = \epsilon_a + \beta r \cos\theta + \sqrt{4\epsilon_a(R_n - 2r)}$ and $\epsilon_{\beta\uparrow} = \epsilon_a + \beta r \cos\theta - \sqrt{4\epsilon_a(R_n - 2r)}$ [23].

For a carrier with the initial energy ϵ_i at a given site, it may hop to the site with an average distance R_n in the hopping space. However, the direction of the carrier movement is random, and hence in the real space the final net displacement of the carrier will be zero. As the electric field is applied to the system, a spatial displacement along the electric field will be larger than that against the electric field. Summing all the possible hopping sites at the initial energy ϵ_i under the electric field, an average real forward distance of a carrier hopped, $\bar{x}_f$, can be calculated as follows:

$$\overline{x_f}(\epsilon_i, T, \beta) = \frac{\overline{x_f}(\epsilon_i, T, \beta)}{2\alpha}, \tag{22}$$

here $\overline{x_f}(\epsilon_i, T, \beta)$ is the normalized average forward hopping distance of a carrier.

In order to calculate the average forward distance of a carrier hopped, one should firstly receive the range of integration of ϵ_j by using the average hopping range R_n. By analyzing the integration limit of Eq. (22), it is easy to obtain the upper limit of integral of $\epsilon_i - \epsilon_a + \sqrt{4\epsilon_a R_n}$. While the lower limit of integral should equal to the value of $\epsilon_i - \epsilon_a - max\left(\sqrt{4\epsilon_a R_n}, \left|\frac{1}{2}R_n \beta \cos\theta\right|\right)$. Since the function of $f(r) = \left|\beta r \cos\theta - \sqrt{4\epsilon_a (R_n - 2r)}\right|$ is non-monotonic, it has the maximum value as $r = 0$ or $r = \frac{1}{2}R_n$, as shown in Figure 6(a).

Then, the normalized value of the average forward hopping distance of a carrier can be calculated by averaging $r \cos\theta$ over the contour in Figure 6(b), to give

$$\overline{x_f}(\epsilon_i, T, \beta) = \frac{I_1}{I_3}, \tag{23}$$

where

$$I_1 = \int_0^\pi d\theta \sin\theta \int_{\epsilon_i - \epsilon_a - max\left(\sqrt{4\epsilon_a R_n}, \left|\frac{1}{2}\beta R_n \cos\theta\right|\right)}^{\epsilon_i - \epsilon_a + \sqrt{4\epsilon_a R_n}} d\epsilon \rho(\epsilon) \left[\frac{-B \pm \sqrt{B^2 - 4AC}}{2A}\right]^3 \cos\theta,$$

$$I_3 = \int_0^\pi d\theta \sin\theta \int_{\epsilon_i - \epsilon_a - max\left(\sqrt{4\epsilon_a R_n}, \left|\frac{1}{2}\beta R_n \cos\theta\right|\right)}^{\epsilon_i - \epsilon_a + \sqrt{4\epsilon_a R_n}} d\epsilon \rho(\epsilon) \left[\frac{-B \pm \sqrt{B^2 - 4AC}}{2A}\right]^2,$$

with $A = (\beta \cos\theta)^2$, $B = 2\beta \cos\theta (\epsilon - \epsilon_i + \epsilon_a) + 4\epsilon_a$, $C = (\epsilon - \epsilon_i + \epsilon_a)^2 - 4\epsilon_a R_n$.

The probability of all these hops is equal to $exp(-R_n)$ in the hopping space. If one can know the average hopping distance, the average rate of the carrier transport is easily obtained, i.e., $v_0 \bar{x}_f exp(-R_n)$. Finally, the mobility of the carrier at the energy ϵ_i is expressed as

$$\mu(\epsilon_i, T, \beta) = \frac{v_0}{F} \bar{x}_f exp(-R_n). \tag{24}$$

After averaging over the normalized site energies, the mobility of the charge carrier based on the Marcus and VRH (M-VRH) theory is

$$\mu = \frac{\int_{-\infty}^\infty \frac{v_0}{F} \bar{x}_f exp(-R_n) g(\epsilon) f(\epsilon) d\epsilon}{\int_{-\infty}^\infty g(\epsilon) f(\epsilon) d\epsilon}. \tag{25}$$

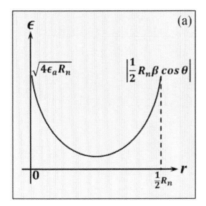 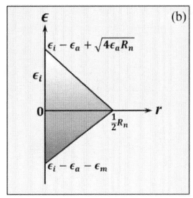

Figure 6. (a) Schematic diagram of the function of $f(r) = |\beta r \cos\theta - \sqrt{4\epsilon_a(R_n - 2r)}|$ as a function of the hopping distance r, (b) the integration contour of Eq. (22) at R_n = const., for a particular angle θ and the evaluation $\overline{x_f}(\epsilon_i, T, \beta)$. Here, $\epsilon_m = max(\sqrt{4\epsilon_a R_n}, |0.5R_n\beta \cos\theta|)$ [23].

Effective Medium Approximation Based on Marcus Rate

Based on the Marcus rate, I. I. Fishchuk et al. developed effective medium approximation (EMA) approach for the disordered organic medium [24-26]. In the EMA approach, the localized states for the charge carriers is replaced by an effective ordered cubic 3D lattice with spacing $a = N^{1/3}$ equal to the average distance between the localized states, where N is the density of the localized states. The energy of the localized states is described by Gaussian DOS. It describes the polaron hopping with a rate given [21]

$$\begin{cases} W_{ij} = W_2\left[-\frac{\epsilon_j-\epsilon_i}{2k_BT} - \frac{(\epsilon_j-\epsilon_i)^2}{16E_ak_BT}\right], \\ W_2 = W_0 exp\left(-\frac{E_a}{k_BT}\right) \end{cases} \quad (26)$$

where $W_0 = (J_0^2/\hbar)\sqrt{\frac{\pi}{4E_ak_BT}}exp(-2\alpha R_{ij})$.

To calculate the conductivity using the EMA method, it is simply assumed to be composed by a set of equal resistors (with conductance G_e) connecting the nearest neighbor nodes on the cubic lattice. G_e is determined by the following condition-the extra voltages (the local fields), being induced upon replacing G_e in this medium with random individual conductance G_{12}, should average to zero. Sequential theoretical treatment based on the above condition has resulted in the following relation for the effective conductivity $\sigma_e = G_e/a$ characterizing the whole disordered system,

$$\langle\frac{\sigma_{12}-\sigma_e}{\sigma_{12}(d-1)\sigma_e}\rangle = 0, \quad (27)$$

where $\sigma_{12} = G_{12}/R_{ij}$ is the conductivity in two-site cluster approximation, d is the dimensionality of the hopping transport system, G_{12} is the two-site conductance, and the angular brackets denote the configuration averaging. In general, the configurational averaging of some value Q is performed by solving a double integral $\langle Q \rangle = \int_{-\infty}^{\infty} d\epsilon_1 \int_{-\infty}^{\infty} d\epsilon_2 P(\epsilon_1)P(\epsilon_2)Q$, where $P(\epsilon_1)$ and $P(\epsilon_2)$ denote the certain distribution functions for ϵ_1 and ϵ_2, respectively. The conductance G_{12} can be determined for the MA rate and for the Marcus rate, respectively,

$$
\begin{cases}
G_{12} = G_1 \dfrac{exp\left(-\frac{|\epsilon_1 - \epsilon_2|}{2k_B T}\right)}{4\cosh\frac{(\epsilon_1 - \epsilon_2)^2}{2k_B T}\cosh\left(\frac{\epsilon_2 - \epsilon_F}{2k_B T}\right)}, G_2 = \dfrac{e^2 W_1}{k_B T}. \ MA \ rate \\[2em]
G_{12} = G_1 \dfrac{exp\left(-\frac{|\epsilon_1 - \epsilon_2|}{16k_B T}\right)}{4\cosh\frac{(\epsilon_1 - \epsilon_2)^2}{2k_B T}\cosh\left(\frac{\epsilon_2 - \epsilon_F}{2k_B T}\right)}, G_2 = \dfrac{e^2 W_2}{k_B T}. \ Marcus \ rate
\end{cases}
\tag{28}
$$

To calculate the effective conductivity σ_e, performing a configuration averaging in Eq. (28) is essential. An elementary method of configurational averaging would be to separately average over the starting site ϵ_1 and target site ϵ_2 energies using the product of Gaussian functions $P(\epsilon_1)P(\epsilon_2) = g(\epsilon_1)g(\epsilon_2)$. This averaging method is abbreviated as "averaging B". Otherwise, based on the occupied DOS (ODOS) distribution for the starting sites energies and unoccupied DOS (UDOS) distributions for the target site energies, the authors created another method, that is "averaging A". In the "averaging A", the localized states occupied by the carriers are described by the ODOS distribution $P(\epsilon_1)$ normalized to unity:

$$
P(\epsilon_1) = \frac{g(\epsilon_1)f(\epsilon_1, \epsilon_F)}{\int_{-\infty}^{\infty} g(\epsilon)f(\epsilon, \epsilon_F)d\epsilon}.
\tag{29}
$$

The empty localized states are described by the UDOS distribution $P(\epsilon_2)$ which in normalized to unity form is given as

$$
P(\epsilon_2) = \frac{g(\epsilon_2)f(\epsilon_2, \epsilon_F)}{\int_{-\infty}^{\infty} g(\epsilon)f(\epsilon, \epsilon_F)d\epsilon}.
\tag{30}
$$

In the limiting case of vanishing carrier concentration the above relations reduce to $P(\epsilon_1) = (1/\sigma\sqrt{2\pi})exp[-(1/2\sigma^2)(\epsilon_1 - \epsilon_0)^2]$ and $P(\epsilon_2) = (1/\sigma\sqrt{2\pi})exp[-(1/2\sigma^2)(\epsilon_2)^2]$, where $\epsilon_0 = -\sigma^2/k_B T$. By substituting Eqs. (29) and (30) into Eq. (28), the transcendental equation for σ_e in the case of the MA hopping conductance,

$$
\int_{-\infty}^{\infty}\int_{-\infty}^{\infty} dt_1 dt_2 \frac{exp\{-0.5(t_1^2 + t_2^2)\}}{\{1+exp[x(t_1 - x_F)]\}\{1+exp[-x(t_2 - x_F)]\}} \frac{\frac{x exp\left(-\frac{x}{2}|t_1 - t_2|\right)}{4\varphi(t_1, t_2, x_F)} - X_e}{\frac{x exp\left(-\frac{x}{2}|t_1 - t_2|\right)}{4\varphi(t_1, t_2, x_F)} + 2X_e} = 0,
\tag{31}
$$

Using the Marcus rate for the hopping conductance is

$$
\int_{-\infty}^{\infty}\int_{-\infty}^{\infty} dt_1 dt_2 \frac{exp\{-0.5(t_1^2 + t_2^2)\}}{\{1+exp[x(t_1 - x_F)]\}\{1+exp[-x(t_2 - x_F)]\}} \frac{\frac{x exp\left(-\frac{x(t_1 - t_2)^2}{16x_a}\right)}{4\varphi(t_1, t_2, x_F)} - Y_e}{\frac{x exp\left(-\frac{x(t_1 - t_2)^2}{16x_a}\right)}{4\varphi(t_1, t_2, x_F)} + 2Y_e} = 0,
\tag{32}
$$

here $X_e = \frac{\sigma_e}{\sigma_1}$, $\sigma_1 = e^2 W_1/\alpha\sigma$, $Y_e = \frac{\sigma_e}{\sigma_2}$, $\sigma_2 = e^2 W_2/\alpha\sigma$, $x = \sigma/k_B T$, $x_a = E_a/\sigma$, $x_F = \epsilon_F/\sigma$, and $\varphi(t_1, t_2, x_F) = cosh\left[(\frac{x}{2})(t_1 - x_F)\right] cosh[(\frac{x}{2})(t_2 - x_F)]$. Based on the Einstein relation, the effective mobility μ_e and effective diffusivity D_e can be obtained as

$$\mu_e = \frac{\sigma_e}{en}, D_e = \frac{k_B T}{e}\mu_e. \tag{33}$$

Monte Carlo Simulation

Monte Carlo method under the polaron aspects has been developed by M. Jakobsson and S. stafström to study the hopping charge transport in disordered conjugated polymers [27]. This implementation mainly includes three parts: constructing geometrical structure of sites, creating couplings between them, and finally running the simulation by observing a random walk of a charge carrier moving between these sites over the couplings. To describe the transition rate of an electron moving from i to j under the polaron aspects, the semi-classical Marcus equation was employed. In the Marcus theory, $\lambda > 0$ is the reorganization energy, and a transition downward in the energy will increase the transition rate as long as the difference is not greater than the reorganization energy, i.e., as long as $\epsilon_j - \epsilon_i > -\lambda$. The electronic transfer integral, J_{ij}, depends on the distance. For the extended sites used in the simulations, this distance is defined as the length of the shortest possible vector connecting the transport sites. The electronic transfer integral, J_{ij}, is assumed to be

$$J_{ij} = \frac{J_0}{\sqrt{n_i n_j}} exp\left[-\frac{\beta_1}{2}(R_{ij} - R_0)\right], \tag{34}$$

here, J_0 is a constant that coincide with the electronic transfer integral between two monomers not subject to form a delocalized state and separated by a distance R_0. n_i and n_j is the number of monomers in the donor and acceptor, respectively, and the fall-off parameter β_1 is related to the electronic localization length, α, according to $\beta_1 = 2/\alpha$. For the special case of nearest neighbor transitions within the same polymer, the transfer integral is simply given by $\frac{J_0}{\sqrt{n_i n_j}}$.

The energy of a state k of a chromophore in the absence of an external electric field is then given by

$$\epsilon_i^{(0)} = \epsilon_i k = \epsilon_0 + 2J_0 cos\frac{k\pi}{n_i+1}, \tag{35}$$

where ϵ_0 is the energy associated with an isolated monomer of the chromophore. In the simulations, ϵ_0 is either drawn from a Gaussian distribution to introduce a greater energetic disorder in the system or simply set to zero. For the acceptor, any orbital $k \in [1, n_i]$ is a valid state, while for the donor it is assumed that the charge carrier has enough time until the next transition to relax into the LUMO state of $k = n_i$.

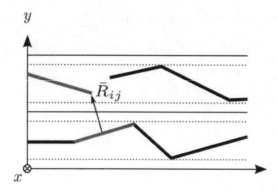

Figure 7. Two chromophores in different polymers are marked in red and the shortest possible vector connecting them is drawn in the figure [27]. The length of this vector, R_{ij}, is used in Eq. (34).

After the polymer structure is generated and the couplings between the chromophores are established, the Monte Carlo simulation is carried out by inserting a charge carrier at a random site and allowing it to move through the system. Otherwise, the Monte Carlo simulation of the charge transport can be characterized as VRH. The transition rates are used to find the probabilities for the different transitions at each site and to draw the dwell time, i.e., the time the charge will spend on its current donor until the drawn transition is carried out. After a predetermined time, τ_{tot}, the distance the charge has traveled, d, is measured and used together with the electric field strength, F, to find the drift mobility according to

$$\mu = \frac{d}{\tau_{tot} F}. \tag{36}$$

POLARON EFFECT ON CHARGE TRANSPORT PROPERTY

Polaron Activation Energy Dependence of Charge Mobility

Indeed, after Bässler et al. suggested that the effective zero-electric-field Arrhenius activation energy of the mobility can be approximated by a sum of the contribution of disorder and polaron [12], the polaron effect has to be taken into account in describing the transport of organic materials. The current theoretical models for discussing the polaron effect in organic semiconductors are almost based on the Marcus rate or can be transferred to the classical Marcus equation. For the Marcus rate, one key parameter controlling the polaron effect is the polaron activation energy E_a(or called reorganization energy λ in the small-polaron model) which connects the polaron effect with the charge transport performance. In the early days, researchers usually discussed the charge transport under a constant of the polaron activation energy. At which the polaron activation energy (reorganization energy) only played a role in fitting the experimental data, and is assumed to be a half of the polaron binding energy [28, 29]. The variable value of the polaron activation energy has been proposed from the EAM theory [25]. Figure 8(a) shows Arrhenius plots of the temperature dependences of the polaron mobility for several values of the polaron activation energies E_a. The results show a nonlinear type for the small polaron activation energies, which suggests a

The Polaron Effect on Charge Transport Property for Organic Semiconductors   159

dominant role of the disorder effects in such a situation. As long as the polaron activation energy is larger, the Arrhenius temperature dependence of the mobility tends to be closer to a straight line while the deviation from the linear dependence in the sub-Arrhenius coordinates becomes more pronounced. Thus, an apparently linear Arrhenius plot is an unambiguous evidence for both the presence of the polaron effects and a relatively large polaron binding energy. Figure 8(b) shows the dependence of the polaron mobility in a disordered organic system on the polaron activation energy for several values of disorder. It bears out the expected linear decrease in $\ln \mu$ with increasing the polaron activation energy.

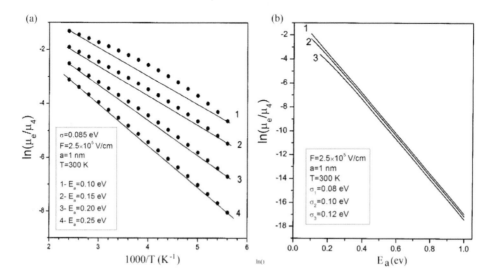

Figure 8. (a) The Arrhenius plot of the temperature dependences of the polaron mobility for several values of the polaron activation energies E_a. Solid straight lines are given for linearity comparison. (b) Dependence of the polaron mobility in a disordered organic system on the polaron activation energy for different values of disorder [25].

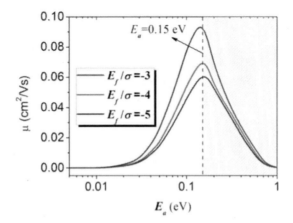

Figure 9. The polaron activation energy dependence of the charge carrier mobility in the normalized Fermi level E_f/σ (carrier density) [23]. The input parameters are $N_t=1\times10^{20}$ cm^3, $\alpha^{-1}=0.15$ nm, $F=1\times10^6$ V/m, $T=300$ K, and $v_0=10^{12}$ s^{-1}.

In terms of the Marcus and VRH theory, Lu et al. also discussed the polaron activation energy dependence of the charge carrier mobility [23], as shown in Figure 9. The authors suggested that the charge carrier mobility represents different polaron effect dependence. For example, when the polaron activation energy (E_a) is lower, e.g., $E_a < 0.15\ eV$, the charge carrier mobility remarkably increases with the increase of E_a, and then decreases with E_a. The variable polaron effect dependence is attributed to the different charge transport mechanism. When the polaron activation energy is lower, the carrier jump upward due to the thermally activated energy will dominate the carrier transport. With the increase of E_a, the polaron effect would remarkably suppress the carrier of thermal excitation to shallower energy states and hence dominate the carrier transport. Therefore, with the increase of the polaron activation energy, the polaron effect on the charge carrier transport tends to be saturated and even hinder the charge carrier transport, and thus decreases the mobility.

Disorder Effect

The most significant feature of organic semiconductors is that the charge carriers are always located on the localized sites and the transport of charge carriers happens to jump from one localized site to another. Therefore, the disorder always is considered to dominate the charge transport. In the charge transport theory, the disorder degree is generally evaluated by the value of the width of the DOS. A typical result of the disorder effect is shown in Figure 10 [30]. In Figure 10, the carrier mobility increases with increasing the disorder degree at a small disorder, and then decreases at a large disorder.

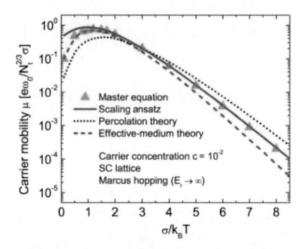

Figure 10. Dependence of the mobility μ on σ for the Marcus hopping, and a comparison of different theory models [30].

A complete understanding for the carrier transport characteristics by combining the disorder and polaron effect has been given [23]. Figure 11(a) firstly shows a comparison for different theory models with the same carrier density $n/N_t=10^{-2}$. One can see that, the difference is always inevitable for different theory models. The M-VRH model and transport energy model derived from the MA transition rate, for the small disorder the simulated results are basically similar as the Ref. 28, but are lower than that in the transport energy model for the large disorder. Otherwise, the reduced trend of the carrier mobility based on the MA transition rate with the disorder is obviously faster than that in the scaling theory [30] and EMA [25]. Figure 11(b) shows the disorder dependence of the charge carrier mobility for the different polaron activation energy. The disorder dependence of the mobility will be stronger with increasing the polaron activation energy.

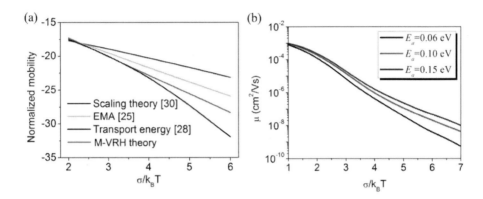

Figure 11. Disorder dependence of the normalized charge carrier mobility: (a) for different theory model with the same carrier density $n/N_t=10^{-2}$, and (b) for different polaron activation energy [23].

Temperature Dependence of Charge Mobility

In the case of the Marcus hopping, the temperature plays an important role in the charge transport performance. In order to reveal the temperature dependence of the charge transport performance, I. I. Fishchuk et al. investigated the temperature dependence of the diffusion coefficient calculated for the Marcus rate in a 3D system by using two kinds of averaging method (averaging A and averaging B) in the EMA model, as shown in Figure 12 [26]. Obviously, the charge mobility decreases with the increase of the temperature. Otherwise, the authors compared two kinds of averaging method, and found that the averaging A method results in a stronger temperature dependence that does depend on E_a/σ and becomes progressively weaker with increasing E_a/σ.

The stronger temperature dependence that does depend on E_a/σ has been determined by Lu et al. [23]. Figure 13 shows the temperature dependences of the charge carrier mobility plotted semi-logarithmically versus $1/T^2$ parameter for different polaron activation energy and different E_a/σ ratios. Otherwise, Lu et al. displayed much stronger temperature dependence with the polaron activation energy. The stronger temperature dependence implies that the polaron effect definitely affects the transport characteristics of the charge carrier.

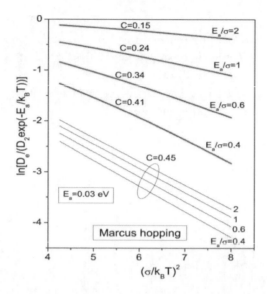

Figure 12. The diffusion coefficient calculated for the Marcus hopping in a 3D system at the vanishing carrier concentration using averaging A (red solid curves) and using averaging B (blue thin curves) for different E_a/σ ratios at $E_a = 0.03\ eV$. Factor C denotes the slope of the temperature dependencies [26].

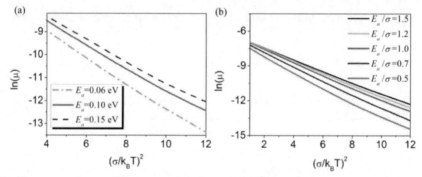

Figure 13. Charge carrier mobility as a function of $1/T^2$ temperature: (a) for different polaron activation energy, (b) for different E_a/σ ratios [23]. The input parameters are $N_t=3\times10^{21}\ cm^3$, $\alpha^{-1}=0.1$ nm, $F=1\times10^7$ V/m, $T=300$ K, $E_f/\sigma=-4.5$, and $v_0=10^{12}\ s^{-1}$.

Carrier Concentration Dependence of Charge Mobility

In general, the carrier concentration in organic semiconductors is expressed as $n = \int_{-\infty}^{\infty} g(\epsilon)f(\epsilon,\epsilon_F)d\epsilon$. Using the averaging A and averaging B methods, I. I. Fishchuk et al. discussed the hopping transport at an arbitrary carrier density for the Marcus rate [26]. Figure 14(a) presents the carrier concentration dependencies of the Marcus hopping mobility. The calculations clearly show that the effective charge-carrier mobility depends very weakly on the carrier concentration even at low temperatures in the case when the polaron effects dominate over the energy disorder effects ($E_a > \sigma$), while this dependence appears to be strong, especially at low temperatures, when the polaron activation energy is relatively small ($E_a < \sigma$).

The Polaron Effect on Charge Transport Property for Organic Semiconductors 163

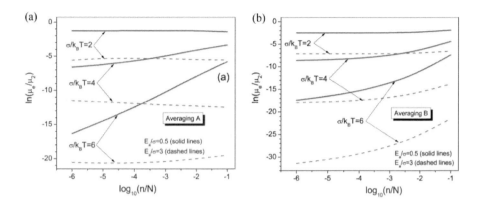

Figure 14. The charge-carrier mobility as a function of the carrier-concentration for the Marcus rate by using averaging A (a) and by using averaging B (b) for different temperatures at $E_a/\sigma = 0.5$ (solid curves) and $E_a/\sigma = 3$ (dashed curves) [26].

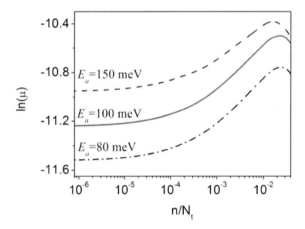

Figure 15. The carrier density dependence of mobility for different polaron activation energy at a constant energetic disorder ($\sigma/k_BT = 3$) [23]. The input parameters are $N_t=1\times10^{21}$ cm^3, $\alpha^{-1}=0.1$ nm, $F=10^7$ V/m, $T=300$ K, and $v_0=10^{12}$ s^{-1}.

Similar phenomenon of the carrier concentration dependence has also been found in Lu et al., as shown in Figure 15 [23]. The calculated results show that the charge carrier mobility increases with the increase of the carrier density. Beyond that, in the lower carrier density the charge carrier mobility depends very weakly on the carrier density, while the dependence will be stronger for the higher carrier density.

Electric Field Dependence of Charge Mobility

Generally, in the real semiconductor devices, the electric field could be as high as 10^8 V/m. The high field remarkably affects the transport property of organic semiconductors. In order to understand the electric-field dependence of the charge mobility in molecularly doped organic materials, K. Seki and M. Tachiya studied some aspects of the analytical expressions of the mobility based on the transition rate of carriers from the Marcus rate equation [20].

Figure 16(a) shows a comparison of the analytical expression and the numerical evaluation. One can see that the charge transport under the weak fields is more influenced by the disorder than that under the higher fields. It is also found that the field dependence of the mobility is very weak in the absence of the disorder, as compared with in the presence of the disorder. Otherwise, the plot of $\log \bar{\mu}$ against $\sqrt{\bar{F}}$ exhibits a linear dependence over a wide range in the intermediate region of the field strength. In Figure 16(b), the linear region then shifts toward a lower field strength by increasing $\bar{a}$, which is the characteristic length of the long-range spatial correlations. The straight line of $\log \bar{\mu}$ against $\sqrt{\bar{F}}$ is apparently the signature of the influence of the disorder on the hopping motion described by the Marcus rate equation. The further introduction of the spatial correlations moves the region described by the straight line.

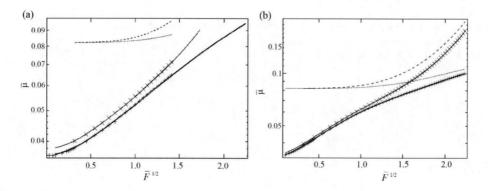

Figure 16. Semi-logarithmic plot of the dimensionless mobility against the square root of the dimensionless electric-field strength with a comparison between the analytical expression with the numerical evaluation [20]. (a) $E_a = 10$, $\bar{\sigma} = 1$, $\bar{a} = 0.1$, and (b) $E_a = 10$, $\bar{\sigma} = 1$, $\bar{a} = 0.5$. The dimensionless field strength is equal to 1 ($\bar{F} = 1$) with the normalized field strength of $F \sim 3 \times 10^5$. The + symbols on the line are the numerical evaluation, the upper thin solid line is the analytical result. The × symbols on the line are the corresponding numerical results. The short dashed line is the mobility from the Marcus rate equation without the disorder. The long dashed line is the Bagley's mobility.

It is different from the discussion from K. Seki and M. Tachiya, which kept the constant of the polaron activation energy, I. I. Fishchuk et al. investigated the field dependences of the polaron mobilities in a disordered material for different temperatures and polaron activation energies [24], as shown in Figure 17. As expected, the polaron mobility strongly decreases as the polaron activation energy increases in the sequence of 0.1, 0.2, and 0.4 eV. However, the authors found that the slopes of the field dependences of the polaron mobilities become were relatively less sensitive to the temperature in the case of the large polaron activation energies.

Based on the Marcus and VRH theory, Lu et al. also discussed the effect of the electric field on the charge carrier transport with different polaron activation energy and carrier concentration [23]. Figure 18 shows the corresponding electric field dependence of the charge carrier mobility for different polaron activation energy and carrier density, respectively. In Figure 18, the mobility shows a weaker dependence of the electric field at both low and high field. While the mobility displays the stronger electric field dependence at field-saturated region, i.e., the mobility decreases linearly with $1/F$. The various dependence of the electric field is attributed to the different carrier transport under the electric field. Generally, the carrier transport under the effect of electric field includes three processes, such as diffusive process at low fields, field-assisted transport at high fields, and field-saturated transport at

The Polaron Effect on Charge Transport Property for Organic Semiconductors 165

which the carrier mobility decrease as $1/F$. The Marcus theory has predicted that the mobility will decrease with increasing the electric field strength for large fields. The results are consistent with this prediction. Otherwise, since the mobility as a function of the field associates with the mean velocity of the charge carriers, an implication of this relation is that the mobility is defined as $\mu = \langle v \rangle/F$. Here, the mean velocity is defined as $\langle v \rangle = v_0 \bar{x}_f exp(-R_n)$. When $\langle v \rangle$ is independent of the temperature and field, the mobility will decrease linearly with $1/F$. On the other hand, one can see that with the increase of the polaron activation energy, the decreasing trends of the mobility as a function of the electric field will be weakened, as well as the carrier density dependence (see in Figure 18(b)). The results show that the polaron effect can change the field dependence of the mobility.

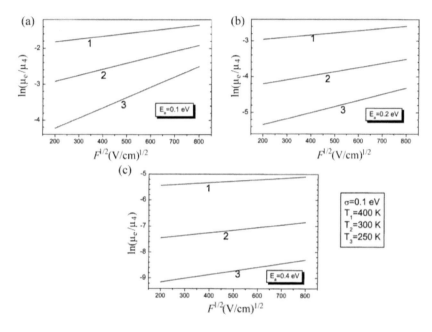

Figure 17. Field dependences of the polaron mobilities in a disordered material for different temperatures and polaron activation energies of (a) 0.1 eV, (b) 0.2 eV, and (c) 0.4 eV [24].

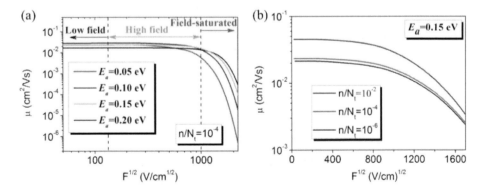

Figure 18. Electric field dependence of the charge carrier mobility for different polaron activation energy (a) and for different carrier density at E_a=0.15 eV (b) [23]. The input parameters are N_t=1×10^{20} cm^3, α^{-1}=0.4 nm, T=300 K, $\sigma/k_B T = 3$, and v_0=10^{12} s^{-1}.

The analytical models of the charge transport have been discussed in details above. As compared with the analytical model, including plenty of free parameters during simulation and calculation, numerical model (such as, Monte Carlo simulation) can eliminate these hindrances. Figure 19 shows the mobility as a function of the electric field using the Monte Carlo simulation by M. Jakobsson and S. Stafströn [27]. In Figures 19(a)-(d), the qualitative behavior of the electric field dependence, characterized by an increasing mobility with the electric field followed by a maximum and then a decreasing mobility. The authors thought that the observed maximum arise from a crossover into the Marcus inverted region. It is also made apparent from the shift of the maximum to a higher applied electric field when the reorganization energy is increased (compare Figures 19(a)-(b) with Figures 19(c)-(d)). In Figure 19(e), it also shows the electric field dependence of the mobility when the field is applied orthogonal to the polymers at 300 K. In general, the electric field dependence is weaker in the orthogonal case compared to the parallel, especially for the systems with no extrinsic diagonal disorder. Furthermore, only the system with a reorganization energy of 0.1 eV and no extrinsic diagonal disorder exhibit a maximum (barely visible on the scale in Figure 19) in the mobility in the range of field strengths under study. Finally, the authors emphasized that the electric field dependence is characterized by a crossover into the Marcus inverted region, not present in the Gaussian disorder model. Available analytical approximations to describe the electric field dependence of the mobility in the Marcus theory fail to fit the simulation data and hence cannot be used to directly draw conclusions about the importance of polaron effects for charge transport in conjugated polymers.

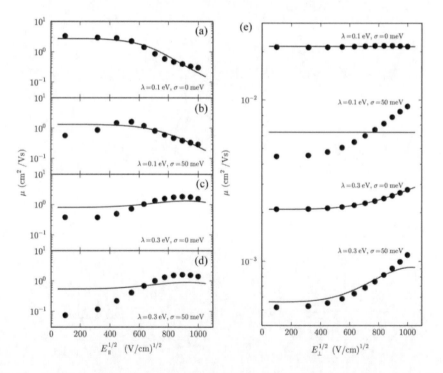

Figure 19. The mobility as a function of an electric field applied parallel to the general direction of the polymers at a temperature of 300 K [27]. The lines show the Auweraer expression fitted to the data points.

CONCLUSIONS AND OUTLOOK

Organic semiconductors possess the electronic features of semiconducting materials and the chemical and mechanical advances of organic compounds. With the development of the organic electronics, several effects on the charge transport performance have been found. Since Bässler et al. firstly suggested that the activation energy of the charge transport could be split into two parts of contribution: disorder and polaron, researchers have given growingly interests. In this chapter, the history, structure and charge transport mechanism of organic semiconductors, have been introduced. Then, several kinds of theoretical models of the polaron effect, such as, 1D Holstein molecular model, Marcus rate equation, charge hopping transport based on Marcus and VRH theory, effective medium approximation based on the Marcus rate, Monte Carlo simulation, have been described. In terms of these current theoretical models, the polaron effect on the charge transport property in organic semiconductors has been discussed in details. The authors hope that these contexts can be helpful to improve the polaron effect in organic materials and provide the motivation for the growth of organic semiconductor devices. The authors believe that the organic electronic will stimulate the intensive research and development works in the field of novel organic materials and devices that will result in serious improvement in organic materials and enable the development of novel organic devices with high performance.

ACKNOWLEDGMENTS

This work was supported in part by the Opening Project of Key Laboratory of Microelectronic Devices and Integrated Technology, Institute of MicroElectronics Chinese Academy of Sciences, by National key research and development program (Grant Nos. 2017YFB0701703, 2016YFA0201802), by the National Natural Science Foundation of China (Grant Nos. 61574166, 61725404, 61306117, 61221004, and 61376112), by the Beijing Training Project for the Leading Talents in S&T under Grant No. Z151100000315008, by the Strategic Priority Research Program through the Chinese Academy of Sciences under Grant XDB12030400, and by the National 973 Program (Grant Nos. 2013CBA01604, 2013CB933504, and 2013CB933504).

REFERENCES

[1] Lüssem, B., Riede, M. and Leo, K. (2013). "Doping of organic semiconductors." *Phys. Status Solidi A.* 210 (1):9-43.

[2] Grozema, F. C., and Siebbeles, L. D. A. (2008). "Mechanism of charge transport in self-organizing oranic materials." *International Reviews in Physical Chemistry* 27(1):87-138.

[3] Erdman, J. G., and Corwin, A. H. (1946). "The nature of the N-H bond in the Porphyrins." *J. Am. Chem. Soc.* 68:1885.

[4] Friedman, L. (1964). "Transport properties of organic semiconductors." *Phys. Rev.* 113(6A): A1688-A1679.

[5] Berets, D. J., and Smith, D. S. (1968). "Electrical properties of linear polyacetylene." *Trans. Faraday Soc.* 64:823.

[6] Law, K. K. (1993). "Organic photoconductive materials: recent trends and developments." *Chem. Rev.* 93: 449-486.

[7] Li, F. M., Nathan F., Wu Y. L., and Ong, B.S. (2011). *Organic thin film transistor integration: A hybrid approach*, WILEY-VCH 2011 pp. 270.

[8] Ito, T. Shirakawa, H., and Ikeda, S. (1974). "Simultaneous polymerization and formation of polyacetylene film on the surface of concentrated soluble Ziegler-type catalyst solution." *Journal of Polymer Science: Polymer.* Chemistry Edition. pp. 11-20.

[9] Chiang, C. K., Fincher, C. R., Park, Jr. Y. W., Heeger, A. J., Shirakawa, H., Louis, E. J., Gau, S. C., and MacDiarmid, Alan G. (1977). "Electrical conductivity in doped polyacetylen." *Phys. Rev. Lett.* 39(17):1098-1101.

[10] Holstein, T. (1959). "Studies of polaron motion Part I. The molecular-crystal model." *Ann. Phys.* 8:343-389.

[11] Schein, L. B., Glatz, D., and Scott, J. C. (1990). "Observation of the transition from adiabatic to nonadiabatic small polaron hopping in a molecularly doped polymer." *Phys. Rev. Lett.* 65(4):472-475.

[12] Bässler, H., Borsenberger, P. M., and Perry, R. J. (1994). "Charge transport in Poly(methylphenylsilane): The case for superimposed disorder and polaron effects." *Journal of Polymer Science: Part B: Polymer Physics* 32:1677-1685.

[13] Parris, P. E., Kenkre, V. M., and Dunlap, D. H. (2001). "Nature of charge carriers in disordered molecular solids: Are polarons compatible with observation?" *Phys. Rev. Lett.* 87(12):126601.

[14] Lu, N. D., Li, L., Liu, M. (2016). "A review of carrier thermoelectric-transport in organic semiconductors." *Phys. Chem. Chem. Phys.* 18: 19503.

[15] Minder, N. A., Ono, S., Chen, Z. H., Facchetti, A., and Morpurgo, A. F. (2012). "Band-like electron transport in organic transistors and implication of the molecular structure for performance optimization." *Adv. Mater.* 24: 503-508.

[16] Tessler, N., Preezant, Y., Rappaport, N., and Roichman, Y. (2009). "Charge transport in disordered organic materials and its relevance to thin-film devices: A tutorial review." *Adv. Mater.* 21: 2741-2761.

[17] Mott, N. F., and Davis, E. A. (1971). *Electronic Processes in Non-Crystalline Materials,* 2nd ed. Oxford University Press, 1971' pp. 32.

[18] Culebras, M., Gómez, C. M., Cantarero, A. (2014). "Review on polymers for thermoelectric applications" *Materials,* 7: 6701-6732.

[19] Coropceanu, V., Cornil, J., Filho, Demetrio A. da S., Olivier, Y., Silbey, R., Brédas, J. L. (2007). "Charge transport in organic semiconductors." *Chem. Rev.* 107: 926-952.

[20] Seki, K., and Tachiya, M. (2001). "Electric field dependence of charge mobility in energetically disordered materials: Polaron aspects." *Phys. Rev. B* 65:014305.

[21] Marcus, Rudolph A. (1993). "Electron transfer reactions in chemistry. Theory and experiment." *Reviews of Modern Physics* 65(3):599-610.

[22] Miller, A., and Abrahams, E. (1960). "Impurity conduction at low concentrations." *Phys. Rev.* 120:745.

[23] Lu, N. D., Li, L., Banerjee, W., Sun, P. X., Gao, N., and Liu, M. (2015). "Charge carrier hopping transport based on Marcus and variable-range hopping theory in organic semiconductors." *J. Appl. Phys.* 118: 045701.

[24] Fishchuk, I. I., Kadashchuk, A., Bässler, H., and Nešpůrek, S. (2003). "Nondispersive polaron transport in disordered organic solids." *Phys. Rev. B* 67: 224303.

[25] Fishchuk, I. I., Arkhipov, V. I., Kadashchuk, Heremans P., and Bässler, H. (2007). "Analytic model of hopping mobility at large carrier concentration in disordered organic semiconductors: Polarons versus bare charge carriers." *Phys. Rev. B* 76: 045210.

[26] Fishchuk, I. I., Kadashchuk, A., Hoffmann, S. T., Athanasopoulos, S., Genoe, J., Bässler, H., and Köhler, A. (2013). "Unified description for hopping transport in organic semiconductors including both energetic disorder and polaronic contributions." *Phys. Rev. B* 88: 125202.

[27] Jakobsson, M., and Stafström, S. (2011). "Polaron effects and electric field dependence of the charge carrier mobility in conjugated polymers." *The Journal of Chemical Physical* 135: 134902.

[28] Arkhipov, V. I., Emelianova, E. V., Kadashchuk, A., Blonsky, I., Nešpůrek, S., and Weiss, D. S., Bässler, H. (2002). "Polaron effects on thermally stimulated photoluminescence in disordered organic systems." *Phys. Rev. B* 65: 165218.

[29] Arkhipov, V. I., Heremans, P., Emelianova, E. V., Adriaenssens, G. J., and Bässler, H. (2003). "Equilibrium trap-controlled and hopping transport of polarons in disordered materials." *Chem. Phys.* 288: 51.

[30] Cottaar, J., Koster, L. J. A., Coehoom, R., and Bobbert, P. A. (2011). "Scaling theory for percolative charge transport in disordered molecular semiconductors." *Phys. Rev. Lett.* 107: 136602.

In: Polarons: Recent Progress and Perspectives
Editor: Amel Laref

ISBN: 978-1-53613-935-8
© 2018 Nova Science Publishers, Inc.

Chapter 4

POLARONS IN ELECTROCHEMICALLY DOPED NON-DEGENERATE Π- CONJUGATED POLYMERS

S. S. Kalagi[1] and P. S. Patil[2]
[1]Department of Physics, G.S.S. College, Belgaum, Karnataka, India
[2]Department of Physics, Shivaji University, Kolhapur, Maharashtra, India

ABSTRACT

Conjugate polymers comprise a class of polymers with alternate single and double bonds. In π-conjugated polymer systems with non-degenerate ground states, interchanging single and double bonds leads to a change in structure from high energy quinonoid states to low energy aromatic states. On electrochemical doping, charges are added or removed from the polymeric backbone producing defect states associated with quasiparticles such as polarons and bipolarons in the $\pi - \pi^*$ bandgap. Delocalization of the π - electrons gives rise to semiconductor-like energy bands and increased electrical conductivity on electronic excitation, oxidation or reduction. The relatively smaller energy gap between HOMO and LUMO π -orbitals as compared to σ-orbitals, leads to intraband transition and appearance of absorption bands in UV-Visible range. By controlling the amount of charges inserted/removed, conductivity of the polymer can be made to vary from an insulating (non-doped) state to a highly conducting (fully doped) state. Coulomb interactions of the polaron with the surrounding electrons lead to a new band states because of the accessible midgap defect levels. The strong electron-phonon coupling, along with conformational and site disorder give rise to the polaronic species while its energy is determined by Coulomb interactions and optical transitions. An analysis of UV-Vis spectra of neutral and oxidized or reduced species as a function of doping levels shows that in most non-degenerate polymers, polarons are the major species generated by electrochemical doping which are responsible for decreased bandgap and increased conductivity.

Keywords: conjugated polymers, polarons, charge transport, transition energy, PEDOT:PSS

ABBREVIATIONS

Conductivity	σ
Room temperature	RT
Charge	q
Spin	s
Angular frequency	ω
Effective mass	m
Dielectric constant	ε
Frohlich coupling constant	α
Planck's constant	h
Energy gap	Eg
Soliton	S
Polaron	P
Fermi level energy	E_F
Bipolaron	BP
Interband transition energy	$h\nu_{IB}$
Saturated calomel electrode	SCE
Cyclic voltammetry	CV
Electron spin resonance	ESR
Electron energy loss spectroscopy	EELS

1.1. CONDUCTING POLYMERS

1.1.1. Introduction

"Can metal/semiconductor like conductivity be achieved in organic polymers?" This question had posed an interesting challenge to researchers for a long time. Traditionally insulators, the concept of "A polymer which can conduct electricity" was initially met with a lot of skepticism. The discovery of polymers like polysulphurnitride [(-S = N-)x] in 1973 which exhibited metal like conductivity triggered an unprecedented interest in the scientific arena to search for new chemical compounds showing similar characteristics [1]. In 1977, MacDiarmid, Shirakawa and Heeger found that in order to control the electronic conductivity of molecules, the introduction of molecules behaving as donors or acceptors during the doping/dedoping process played a crucial role [2]. This process of doping/dedoping could tune the electronic properties of such conducting polymers from insulating to conducting and vice versa. This era saw the coming together of scientists from different fields like solid state Physics, field theory, Physical and molecular chemistry working on a relatively new area which could be instrumental in bonding condensed matter physics with organic chemistry. Theoretical scientists analysed the experimental data in the early 1980s in search of tests for their calculations in non-linear dynamics resulting in identification of polarons (single electronic charges) and bipolarons (doubly charged), successfully integrating the processing and mechanical properties of organic polymers with electrical and optical properties of metals [3]. A milestone was reached when it was recognized that conducting polymers could be synthesised by electrochemical methods and their electrical and optical properties could be manipulated by doping and subsequent dedoping. Year 2000 was a landmark year in which the three scientists Alan J. Heeger, Alan MacDiarmid, Hideki Shirakawa were awarded the Nobel Prize in Chemistry for the discovery and development of conducting polymers. The discovery of these exotic materials with unusual properties has resulted in intense research and a spurt in number of publications bringing up a number of unforeseen and exciting applications. Some of the applications envisaged are in the field of organic photovoltaics, light emitting diodes, field effect transistors etc. In addition to the advantages of light weight, flexibility, and transparency, processing organic semiconductors is less energy intensive. Conducting polymers have also gained popularity due to their ease of processibility, rapid response times, high optical contrasts and the ability to modify their structure with a change in the external environment. The structure of a few conducting polymers are shown in Figure 1. Of the number of conjugated polymers, the most widely studied ones are derivatives of poly(thiophene)(PTh), poly(pyrrole)(PPy), and poly(aniline)(PANI). Conjugated polymer systems, while not as developed as their inorganic counterparts, promise high contrast ratios, rapid response times, and long life times for use in EC display technology. Besides these, conjugated polymers have the ability to physically structure polymer-based electrochromic devices (ECDs), exert control over their EC responses, very flat panel displays, and seed layers for electroless plating on printed circuit boards [4]. There are also other examples like fiber-optic switches and routers based on non-linear optics, batteries, photo- responsive polymers in optical computing and holographic data storage, diode rectifiers formed by donor and acceptor etc.

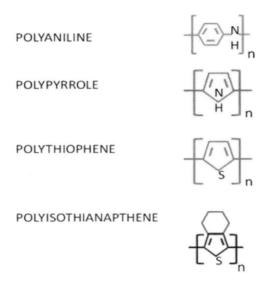

Figure 1 Structure of some conjugated polymers.

Figure 2 shows the conductivity range of conjugated polymers with respect to those of metals, semiconductors and insulators. The comparison is a bit illusive as none of the polymers are good electrical conductors in chemically pristine state. They are mostly insulators or to some extent, semiconductors. Their transition from insulators to conducting state takes place only after treatment with oxidizing or reducing agents. This procedure by which a chemically pure insulating polymer is transformed into a semiconductor or conductor by the addition or subtraction of charges is termed as "doping." In semiconductor physics, while undoped semiconductors are intrinsically conducting, doped semiconductors are termed as extrinsically conducting. Contrary to this, doped polymers are often referred to as intrinsically conducting polymers. This is to distinguish them from polymers which acquire conductivity by loading with conducting particles, such as carbon black, carbon nanotubes, metal flakes, nanoparticles, or fibers of stainless steel.

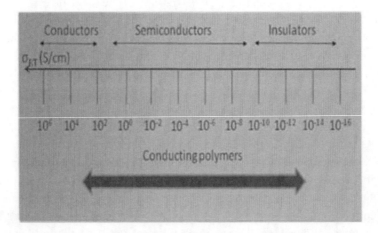

Figure 2. Comparison of conductivities of conjugated polymers with other materials.

1.1.2. Structure of Conjugated Polymers

In saturated polymers like polyethylene, all the valence electrons are used up to form σ-bonds. This results in a wide energy gap between the conduction band and the valence band. The absence of free electrons for conduction makes the material behave as an insulator. On the other hand, conducting polymers are linear carbon chains made up of alternate single and double bonds. In this structure, carbon atoms form three σ-bonds with the neighboring atoms and the remaining p-orbital referred to as the p_z orbital is engaged in the formation of π systems along the polymer backbone.

Conjugated polymers can be generally divided into two groups based on the energies of their resonance structures: degenerate and non-degenerate ground state systems [5-11]. Whenever the position of single and double bond can be changed without affecting the energy of the ground state, the conducting polymer is referred to as a degenerate polymer. The electrons in the π- bonds therefore exist not only between any two carbon atoms but can also move along the polymer chain. The length of the polymer molecule over which the conjugation is spread governs the width of the energy gap. Longer is the molecule, smaller is the energy gap created between energy levels. For infinitely long polymers, these discrete levels combine together to give rise to energy bands.

If the interchange of single and double bonds results in two distinct energy levels, one of them with higher energy and lesser stability, then such polymers are called as non-degenerate polymers. Most of the conjugated polymers belong to the latter group i.e., they are non-degenerate ground systems. Just like a metal, small conjugated systems can be described by discrete energy levels corresponding to σ, σ*, π and π* for each bond. If the polymer chain extends over a large number of molecules, merging of discrete energy levels takes place to form continuous energy bands, corresponding to conduction and valence bands in a metal, separated by an energy bandgap E_g [12]. This energy gap is defined as the energy necessary to elevate an electron from the highest occupied molecular orbital (HOMO) to lowest unoccupied molecular orbital (LUMO). The energy gap existing between π-bands in conjugated polymers is very small due to the delocalization of electrons as compared to the energy gap between σ-bonds in non-conjugated polymers. Thus an application of small energy equivalent to the energy gap results in the transfer of an electron from HOMO to LUMO, making an electron available for conduction. In some systems like polyaniline, atoms of Nitrogen are also involved in the process of conduction by being a part of the conjugation system. The conjugated π-π^* bands lowers the ionization potential to less than 6eV while at the same time it increases the electron affinity (nearly 2eV). This makes the polymer susceptible to easy oxidization by electron accepting molecules like I_2, AsF_5, SbF_5 or reduction by electron donors (Li, Na, K, etc.) [9].

1.2. CHARGE CARRIERS: SOLITON, POLARON AND BIPOLARON

In their neutral states, conjugated polymers show insulating/semiconducting behavior with an energy gap (Eg) between the valence band (HOMO) and the conduction band (LUMO). To transform them to conducting state, charges need to be introduced into the system. The introduction of charges into the polymer skeleton or to use a more appropriate term "doping," can be done by different methods (section 1.4). Charges that are introduced into the polymers by these diverse methods lead to formation of self localized excitations namely solitons, polarons and bipolarons [13]. While these quasiparticles were predicted theoretically, their introduction has enabled satisfactory explanation of increased conductivity and variation in many other optical and magnetic properties of materials, both inorganic and organic. These newly created energy states comprise of charge(s) accompanied by a lattice distortion. In degenerate polymers such as transacetylene, formation of solitons only is permissible as they are the states with lowest energy configuration. Here, the solitons formed lie at the centre of the polymer molecule flanked by single and double bond on either side. These excitations are characterized by their charge and spin. A soliton can have a spin of either zero or 1/2 while having a charge of 0 or $\pm q$.

A polaron is formed when a polymer is oxidized (reduced) on addition (removal) of some charges in a small proportion from the conjugated system. This addition/abstraction of electron leads to chain deformation. The strong vibrational and torsional coupling modes typically exhibited by π-conjugated polymers, along with this conformational and site disorder, ends up with the creation of polarons. The ionic radical along with the distortion in the local geometry creates new energy levels lying inside the bandgap. Similar to solitons, a half filled level is created with a spin 1/2 and a charge of $\pm q$. The energy difference between the band edge and newly created states are found to be dependent on the bandgap, doping proportion and chain length of the polymer. The polaron can thus be treated as a nontopological excitation which does not require the ground-state degeneracy for its existence. If charges are injected into a polymer chain in moderate proportion, electron transfer due to polarons causes transport of charge along the length of the polymer.

However,when the addition or removal of some charges takes place at a large scale, i.e., under heavily doped conditions, the number of electrons added is in much greater proportion than before, the polarons come together to form doubly charged, spinless bipolarons. This process is in line with Brazovskii–Kirova (BK) model which predicts that energy wise formation of a bipolaron is more favorable over two separate polarons while neglecting the e-e interactions [14]. Thus if two polarons with the same charge exist on a polymer chain, they will merge together in order to acquire a lower energy state [15]. Two new energy levels are created with the formation of bipolarons on either side of the Fermi level. For a positive bipolaron, both the levels are unfilled while for a negative bipolaron, both the levels are completely filled. These bipolarons are di-ions associated with strong lattice relaxation.

1.3. CLASSIFICATION OF POLARONS

Interactions between a conduction electron (or hole) and atoms in a solid material has fascinated the scientific community for long. These interactions between an electron and atoms can be best understood in terms of quasiparticles such as *polarons.* In 1933, Lev Landau proposed the concept of polarons for the first time, describing the movement of electron in a dielectric crystal [16]. He suggested that whenever an electron (hole) moves through a crystal or a semiconductor, it creates a deformation in the lattice structure. This lattice distortion, along with the electron bringing about the deformation, move together inside the crystal from one place to another as a single entity. Concomittantly, atoms of the crystal, while vibrating about their equilibrium position, effectively screen the moving electron by long wavelength phonon clouds. This screening of the electron results in induced self trapping, thereby lowering electron mobility and increasing its effective mass. Existence of such polaronic quasiparticles has been envisaged both in ordered and disordered solids, amorphous semiconductors viz. crystal lattices, semiconductors, conducting polymers, superconductors etc. Even though electrons, holes, polarons are all supposedly charge carriers engaged in charge transport, the physical properties of polarons are vastly different from the rest. They can be differentiated on the basis of their characteristic binding energy, effective mass and their response to polarizing electrical and magnetic fields (e.g., mobility and impedance) [17]. Additionally, the polaron concept has been extended to several systems where one or many fermions interact with a bath of bosons. A large volume of work on polarons explaining the interactions between a charge carrier (electron, hole) and the long-wavelength optical phonons is credited to Frohlich because of his vast contribution in this field [18]. However, to date, a full analytical or numerical solution at different couplings of the Frohlich polaron has evaded the theoretical scientists since it was first taken up for studies in 1950. An active field of research, persistent efforts are going on to find exact numerical solutions to the case involving one or two polarons in a large crystal lattice as well as multi polaron problems.

Since polarons can be produced by different methods, they are classified on the basis of their sizes, mode of creation, fields where they are perceived etc. The next section deals with a brief discussion on the classification of polarons and their characteristic properties.

1.3.1. Large Polaron

A *large polaron or Frohlich polaron* is characterised by its radius R_P. The radius of a *Frohlich polaron* is much larger than the lattice constant of the material where the localised electron exists but less when compared to the interatomic distances [19]. The parameter describing the properties of a large polaron represented by "α" is defined by equation (1)

$$\alpha = \frac{e^2}{\hbar c} \sqrt{\frac{m_b c^2}{2\hbar\omega_{LO}}} \left(\frac{1}{\varepsilon_\infty} - \frac{1}{\varepsilon_0}\right) \tag{1}$$

where "α" is called as *Frohlich coupling constant*, ε_0 is the static dielectric constant, ε_∞ is the high-frequency dielectric constant, m_b is the effective mass given by the band structure and ω_{LO} is the Longitudinal Optical (LO)-phonon angular frequency. The *Frohlich coupling constant* "α" is approximately two times the number of phonons in the phonon cloud of a given charge carrier (electron or hole) [20], and therefore polaronic effects become much more noticeable when "α" becomes of the order of unity or larger. As the electron moves together along with the lattice distortion as a single entity, the mobility of a *Frohlich* polaron is often hindered by scattering of the optical phonons due to the strong electron-optical-phonon coupling. Since both the electron and the distortion travel together, the effective mass of such a polaron turns out to be greater than the mass of the electron which was responsible for its creation. Large polarons have typical optical signatures in the THz frequency range, where the frequency of photon is comparable to ω_{LO}, the angular frequency corresponding to the Longitudinal Optical (LO) - phonon. For example, a polaron, trapped in a potential well can be excited by a photon to a higher-energy state, still within the same boundary conditions, and this giving rise to an absorption maximum [20]. A continnum approach is utilised in the theoretical treatment which assumes that the lattice parameters are negligible in comparison with the size of the polaron while the spin and relativistic effects do not alter the system appreciably. Charge transport due to such large polarons takes place due to interband transitions rather than charge hopping along the length of the polymer chain.

1.3.2. Small Polarons

In some materials, the shape and size of the potential well where the charge carrier can be localized due to ionic or atomic displacements is approximately equal to the volume of a unit cell. Under such conditions, an electron gets trapped by the field induced as a result of its own atomic displacement. This entrapment of electron in a region of linear dimension, which is in the same regime as the lattice constant, is designated as *small polaron*. Here, the induced lattice polarization is effectively confined in a volume of the order of a unit cell. Hence, the electron is localized on an individual site for a time larger than the relaxation time of the lattice [21]. Unlike large polarons, short range forces the appearance of small polarons. These small polarons travel from one site to other by means of thermally-activated hopping. Two types of hopping have been envisaged in the theory of small-polaron transport: *adiabatic* hopping, in which the electron sits in the potential well of its lattice distortion, and *anti-adiabatic* hopping, in which the electron jumps out of the potential well after which an equilibrium is achieved with the position of the electron at the new site. A distinguishing feature of the small polaron state is its ability to stay stable in two states, either as an unbound electron in a undistorted lattice or as a trapped electron in a localized potential well. The former is analogous to the band electron state in a rigid lattice; the latter models a small-polaron state. According to Alexandrov and Mott, the small polaron energy comprises of three components: the kinetic energy of the charge carrier due to tunneling, energy due to the atomic (ionic) displacements field and finally potential energy of the electron or hole trapped in the potential well due to the displacements [22]. The interaction of the localized electron (hole) with the phonons then induces them to jump from one site to another. This process is called hopping. If the time taken by the carrier "t_o" to jump over is much smaller than the time period ω_{LO}^{-1} of the phonon, i.e.:

$$t_o \ll \omega_{LO}^{-1} \tag{2}$$

then, the electron "jumps out of" the old potential well due to its self-induced lattice deformation, thus initiating a multiphonon process of the lattice relaxation leading to the disappearance of the small-polaron states.

1.3.3. Spin or Magnetic Polaron

Polarons and bipolarons are presumed to play a vital role in high Tc superconductors. This belief has triggered enormous curiosity in the physical properties of the multi-polaron set up, especially so in the context of their optical and magnetic properties. Theoretical treatments have been extended from one-polaron to many-polaron systems. The name "De Gennes" crops up whenever there is a mention of the initial studies related to spin polarons [23]. It has been perceived that exchange interaction causes the charge carriers to interact with the magnon field in a manner similar to that between electrons or holes with the atomic lattice. This gives rise to the formation of the spin polaron or magnetic polaron. The spin polaron consists of both the carrier with its spin and the lattice magnetization created by the carrier spin. Here the physical properties of the carrier are influenced by the self induced magnetization cloud.

In an anti ferromagnet–lattice coupling, an electron interacts with the surrounding magnon and effectively polarises their spins giving rise to a bound polaron [24]. For materials with preference for a specific magnetic/antiferromagnetic alignment, the kinetic energy of the polaron can be tuned by manipulating the spins of the surrounding ions. This allows a greater control over the mobility of the carrier.

1.3.4. Polarons in Conducting Polymer Systems

Akin to inorganic semiconductors, such as silicon and germanium, conjugated polymers exhibit poor conductivity in their pristine state. This state can be reversed by injecting charges into the polymer system. The number of charges injected into a conducting polymer system plays a vital role in the alteration of a number of physical and chemical properties. Similar to semiconductors, doping in conjugated polymers can be carried out by various methods viz chemically, electrochemically or by photoexcitation [25]. The first method to introduce charge into the polymer chain is by doping the material chemically. Here, doping refers to a chemical reaction—oxidation or reduction. In either of the cases, new electronic states are created and a previously semiconducting/insulating material shows enhanced conductivity. The reducing agents (n-type) end up donating the electrons to the conjugated polymers while the oxidants behave as electron acceptors (p-type) dopants. Another method of introducing the dopants is by introducing the charges electrochemically. Here, the necessary counterion is drawn from the surrounding solution or the electrolyte. In the case of photo-doping the conjugated polymer is locally oxidized and reduced by forming an electron–hole pair due to light (photon) absorption followed by charge diffusion.

Apart from these major classes, the polaron concept has been extended to include many other categories like acoustic polaron, electronic polaron, bound polaron, trapped polaron, Jahn Teller polaron, bipolaron and many polaron systems. These conceptual extensions are useful in explaining conducting properties in conjugated polymers, colossal magnetoresistance, perovskites, high T_c superconductors, layered MgB_2 superconductors, fullerenes, quasi-1D conductors, semiconductor nanostructures [19].

1.4. DOPING: METHODS FOR CARRIER GENERATION

In their natural state, conjugated polymers behave as wide band-gap semiconductors and show low conductivitiy just like insulators. As discussed before, doping of these polymers is an indispensable process to make them electrically conductive. A radical change in the electronic property of the material can be brought about by doping of the materials. The first results displaying change in properties with doping of such organic materials were obtained in 1974, when conductivity of polyacetylene films showed an increase by several orders of magnitude after reacting with Lewis acids (such as arsenic penta fluoride AsF5) or bases [25]. The section below deals with a comprehensive discussion about different doping methods and subsequent carrier generation.

1.4.1. Electrochemical Doping

In conjugated polymers, electrochemical doping refers to a chemical reaction-oxidation or reduction. Following the reaction, new energy levels are created within the bandgap and the semiconducting properties of the material changes to conducting state. Analogous to solid state Physics, different types of dopants like electron donating (n-type) and electron accepting (p-type) have been used to introduce charges into a conjugated polymer to make it conductive. Such type of doping is observed in polyacetylene, polypyrrole, derivatives of polythiophene etc. In general, p-doping is carried out by addition of a reactant that oxidizes the material. The extraction of electrons gives rise to a p-type conductor, with a positively charged unit in the conjugated system [25]. The reaction follows according to the equation:

$$P + xA^+ \leftrightarrow Px^+ (A^-)_x + xe^- \tag{3}$$

where "P" refers to the polymer chain and A⁻ refers to the counter ion, "x" denotes the number of counter ions necessary for the oxidation to take place. An exactly reverse process takes place during n-doping of a polymer. Injecting of electrons into the polymer backbone reduces the polymer to form a negatively charged unit in the conjugated system as shown by the equation below:

$$P + ye^- + yB^+ \leftrightarrow P^{y-} (B^+)_y \tag{4}$$

While both p- and n-doping can be carried out easily, p-doped polymers find more applications in devices and experiments. This is because of the fact that the atmospheric oxygen reduces the n-doped polymers making them unstable in ambient atmosphere.

1.4.2. Photo Doping

When a conjugated polymer is subjected to radiations with photonic energy equal to the bandgap energy of the polymer, electrons absorb energy and make a transition to the higher energy state. This process is termed as Photo-doping [26]. This method of doping is dependent on the time for which the polymer is exposed to radiations. Once the radiation falling on the material ceases, recombination of electrons and holes ensues. To keep the process of Photo-doping uninterrupted, along with the radiation from the source, an electric potential needs to be applied so as to prevent the recombination of electrons and holes. Under such conditions, continuous photo conductivity is observed.

1.4.3. Charge Injection Doping

Consider a configuration of metal/dielectric/polymer stacked in the form of a sandwich. When a high potential is applied to such a system, charges accumulate on the surface of the polymer. These charges play a crucial role in the conductivity enhancement of the polymer. Such charge injection induced conductivity on the application of a high potential has been observed in derivatives of polythiophene.

1.4.4. Non-Redox Doping

This method of doping is different from the other techniques. In this process, the number of electrons in the polymer backbone remains the same. However, the energy levels get rearranged on doping. Such a doping method is observed in Polyaniline when treated with protonic acids. Similar observations have been noticed in Poly (heteroaromatic vinlyenes) as well [27]. This particular method of doping is found to increase the conductivity of the polymers by an order of ten or more.

1.5. Polarons in Π Conjugated Polymers

Polymers are composed of non-metallic elements, mostly found at the upper right corner of the periodic table. Of these, Carbon is most prevalent element found in polymers. The chemical bonds in polymers are basically covalent bonds, unlike those found in metals and ceramics. In typical inorganic semiconductors, the electrical conductivity and semiconductive properties are determined by the crystalline and defect structures. Charge carrier concentration is controlled by injected impurity atoms, while the mobility of charges is determined by the nature of the material and the degree of crystallinity. Organic compounds or polymers, on the other hand, show little or no conductivity as all the valence electrons are used up in forming covalent bonds. This unavailability of free electrons for conduction makes the polymers insulators in their pure state. A major feature of conductive polymers is the presence of conjugated double bonds along the backbone of the polymer. However if the polymers are made of conjugate bonds i.e., bonds wherein there is strict alternation of single and double bonds, then the scenario becomes entirely different. Such polymers show a metal to insulator transition and vice versa on doping. The notion of doping is the exceptional, fundamental, essential, and the unifying theme which distinguishes conducting polymers from all other types of polymers [7]. To exhibit electrical conductivity, the polymers must have ordered conjugation with extended π electrons and sufficiently large carrier charge concentration. The restricted addition of known, usually small (~10%) non stoichiometric measure of chemical species results in striking changes in the physical, electronic and structural properties of the polymer. Doping in such systems is a reversible process wherein the opposite action produces back the original product with little or no change. Both the processes, doping and undoping, involving dopant counterions for stability, may be carried out chemically or electrochemically. During the doping process, an organic polymer, either a semiconductor or an insulator, having little or no conductivity, typically in the range 10^{-5} to 10^{-10} S cm^{-1}, is converted into a polymer whose conductivity lies in the "metallic" regime (~1 to 10^4 Scm^{-1}). Transitory doping methods of conducting polymers had previously been accomplished by redox doping. This involves the partial addition (reduction) or removal (oxidation) of electrons to or from the p system of the polymer backbone.

Traditional insulating polymers have saturated sp^3 hybridised carbon atoms making up the polymer backbone. On the other hand, conducting polymers are conjugated systems where delocalized π -electrons overlap along the polymer chain. Consequently, such materials are ordered and strongly anisotropic. A characteristic feature of such polymers is the quasi one-dimensional (1-D) property of the conjugated electronic structure. This peculiar molecular character of polymers makes electronic motion along the individual macromolecules one-dimensional. In such polymers electron delocalization can be evidenced in two ways: either along the chain direction due to π-π overlap between neighboring monomers in the 1-Dal chain or because of π-π overlap between successive monomers along the stacking direction. If the overlap of π-orbitals is along the entire backbone, delocalization of π-electrons ensues. The π-bonds then get dispersed and the polymer becomes conductive. By contrast, the σ-delocalization is caused by overlapping of sp^3 silicon orbitals. This delocalization of electrons is strongly depending on the chain length and its conformational property [9]. These delocalized electrons may move around the whole system and become the carriers for charge transport along the length of the polymer,Interestingly, it has been found that conjugated polymers without any dopants are only slightly conductive, so conjugation may not be the only cause which brings about an insulator to metal transition. To achieve good conductivity, dopant ions have to be added or removed in order to bring about electronic excitations like solitons, polarons, bipolarons etc. The weak interchain binding among the molecules aids easy diffusion of dopant ions into and out of the structure (between chains), while the comparatively stronger intrachain carbon-carbon bonds maintain the structure stability of the polymer. Once the charge carriers are produced by either by electronic excitation, photoexcitation, oxidation or reduction in the polymer, delocalization of the π-electrons gives rise to semiconductor-like energy bands and a subsequent increase in electrical conductivity. The smaller energy gap between HOMO and LUMO π-orbitals compared with the σ-orbitals, leads to absorption and emission of wavelengths in the visible range (400-700 nm).

1.5.1. Conjugational Defects

In semiconductors, the strong covalent bonds between the atoms lead to fairly rigid structures. These structures are unaffected by external dopant ions and so the charge transport in such systems can be explained in terms of electrons and holes. Such doping develops defect states between the conduction and valence band. These defect bands can get easily populated as the charge carriers acquire thermal energy at room temperature to bridge the gap. On the other hand, in conjugated polymers, the high anisotropy and non-rigid structures make interaction between the chains comparatively much weaker than those between the atoms. This weak interchain interaction renders the polymer susceptible to Peierl's distortions, a regular feature observed in quasi 1Dimensional materials [28]. Unlike inorganic semiconductors, doping level for conjugated polymers needs to be fairly high, of the order of few percent in order to see a noticeable change in conductivity. Any change in concentration of charges on doping/dedoping leads to easy structural and conformational distortions. As a result, these distortions become the simplest method for describing the charge transport along the polymer skeleton. The conductivity then depends on the length of the polymer molecules over which the distortion is spread out. Interaction of these polymers with photons can be obtained from spectroscopic results which help in classifying them on the basis of their degeneracy. Thus the strict consecutive bond alteration nature in such polymers, when disturbed by addition of charges, results in defect structures. Similar to p-type or n-type semiconductors, these defects require the existence of a dangling bond to make up the four bonds of a Carbon atom. In polyacetylene like polymers with repeating olefin group, both quininoid and benzenoid structures have equal energy states. In such systems, solitons are considered to be the theoretically predicted charge carriers. They result from charge injection in the chain, and represent a kink in the system, a domain wall connecting the two possible bond alternation phases separated by a domain wall. This kink or defect can travel along the chain on account of its small mass and extends over several C-C bonds. Individual solitons are generated during the synthesis of the polymer while further charge injection leads to formation of soliton/antisoliton pairs, similar to particle/antiparticle pair of elementary particles. Solitary or neutral soliton generation takes place during the production of the polymer itself. A soliton is free to move, because its position does not alter the total energy of the system when the polymer chain extends over several molecules while in the reverse case, the soliton is relegated to the middle because of the end effects. It migrates from one molecule to another by pairing with the neighbouring electron and leaving its earlier partner unattached. During this migration, the soliton has to overcome a potential barrier when moving from one spot to the next which sometimes leads to its self-trapping on the chain. Figure 3 (a) shows a neutral soliton. External charges added to or removed from the system generate either positive soliton (b) or negative soliton (c). The spin-charge relations of a soliton are quite strange. In a neutral chain, the soliton is associated with an unpaired electron, and has spin S = 1/2. This unpaired electron can be removed, or another can be added, giving solitons with spin S = 0 and charge ±e. The energy level diagram associated with generation of different types of solitons is shown in Figure 3. There exists only one defect state in degenerate polymers which occurs midway between the conduction band and the valence band. The allowed transition then can be from Valence band to the midgap defect state or from the midgap defect state to the Valence Band.

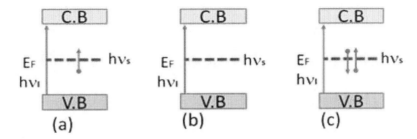

Figure 3. Energy level diagram of a soliton (a) Neutral q = 0, s = 1/2 (b) Positive q = +e, s = 0 (c) negative solito q = -e, s = 0.

The energy required for this transition is given by

$$h\vartheta_s = \frac{1}{2}h\vartheta_I \qquad (5)$$

where $h\vartheta_s$ is the interband transition energy while $h\vartheta_I$ is the bandgap energy of the polymer.

Non Degenerate polymers: Polymers like polyparaphenylene (PPP), PolyPyrrole (PPy), Polythiophene (PTh) belong to the category of non-degenerate polymers. In the case of polymer with such non degenerate ground state, it has been shown that, the doping process creates defect states within the gap [30]. A number of unusual structural, physical and transport properties have been observed in these materials [31–38]. It is very difficult to interpret these properties within the conventional semiconductor model. However, the concept of nonlinear excitations such as polarons, and bipolarons taking part in the charge-transport mechanism seem to be very successful in describing these prominent phenomena. These suggestions are corroborated by the fact that in such polymers with benzoic rings or polyheterocycles, both theoretical calculations and experimental results indicate that the charge carriers are polarons and bipolarons [6], although some experimental results suggest that a truly metallic state with free carriers may be reached at very high carrier concentration. The ground state in such a system is non-degenerate with the structural geometry corresponding to the aromatic form. The resonance structure in the quininoid form has a higher total energy, of the order of 0.4eV per ring. Such polymers having resonant structures with non-equivalent energy lead to the lifting of degeneracy. The carrier transport in these systems seems to take place through charged defects originating as a result of doping.

Brazowskii and Kirova have shown explicitly that stable soliton configuration cannot exist in such polymers with non degenerate ground state. In conjugated polymers, the polaron is one of the predominant excitations and plays an important role in the properties and applications of optoelectronic devices. Many theoretical studies concerning external-electric-field effects on the dynamics of polarons in conjugated polymers have been carried out [1–8]. From these studies, it is possible to conclude that polaron can travel as an entity along the polymer chain in low electric fields but dissociate in high electric fields. These charged defects are influenced by the polaronic interaction with the surrounding atoms. At low temperatures the charge transport is governed by the motion of the defects i.e., interband transition occurs; at higher doping levels hopping conductivity dominates the transport properties. The role of electron-electron (e-e) interaction on polaron dynamics has been investigated by a number of research groups as well [39–41]. Their results indicate that the e-e interaction is an important factor which influences the stability of a polaron. In addition, properties of charge carriers are greatly influenced by dopants, defects, and inter carrier collisions [42–46].

Consider a local geometric distortion in the ground state of such a polymer whose band structure is comparable to that of a standard semiconductor. This localised distortion, at the expense of energy E_{dis} gives rise to localised electronic states, equidistant on both sides of the Fermi level within the bandgap region. This defect is spread out over the shortest possible chain segment over which both the minimum energy and Coulomb repulsion of localized charges are in equilibrium. Charge carriers like polaron and bipolaron are generated upon doping/ dedoping and the extent to which this process takes place.

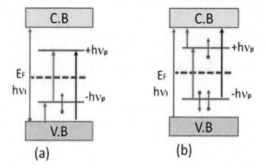

Figure 4. Energy level states and allowed transitions in Polaron (a) Positive polaron q = +e, s = 1/2. (b) Negative Polaron q = -e, s = 1/2.

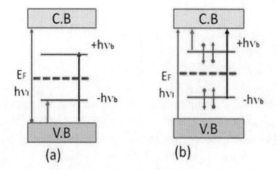

Figure 5. Energy level states and allowed transitions in Bipolarons (a) Positive Bipolaron q = +2e, s = 0 (b) Negative Bipolaron q = -2e, s = 0.

The favorable bond alternation order in non-degenerate conjugated polymer systems can be disrupted by optical excitation or charge injection in moderation. The system attains a stable configuration by confining the defect on the smallest possible chain. Such charged defects on the chain are called polarons. A polaron has a spin $=1/2$ and an electric charge of $\pm e$. These polarons can be either positive (P^+) or negative (P^-) [47]. Figure.4 shows the newly formed defect states on either side of the Fermi level, charge and spin of the respective polarons along with the possible transitions. For a negative polaron, P^-, the state just above the valence band is filled and the state just below the conduction band is half-filled. For a positive polaron, P^+, the state just above the valence band is half-filled and the state just below conduction band is empty. In both cases, there is the density of states at the Fermi level in the electronic structure and these defects correspond to two new localized states in the band gap.

When two polarons interact, they form a double charged state to produce a bipolaron, a spinless excitation i.e., $s = 0$ with charge of $\pm 2e$. These bipolaronic states can again be either positive (BP^{++}) or negative (BP^{--}) [6]. Even though bipolarons are energetically more favorable than polarons, initially polaronic defects are produced on the polymer chain on charge injection. The appearance of bipolarons is possible only after all the polaronic states are filled and the doping proportion is further increased. Figure 5 shows the generation of bipolaronic defects on either side of the Fermi level and the possible energy transitions. Two new localized states are formed in the band gap, with both states being doubly occupied for negative bipolarons, BP^{--}, and both states being empty for positive bipolarons, BP^{++}. For the bipolarons, there is no density of states at Fermi level, since it is situated halfway between the conduction and the top bipolaron state for negative bipolaron, and half-way between the valence band and the lowest lying bipolaron state for positive bipolaron [17]. However, because of structural disorder, the mobility of the polaronic states is found to be much lower in the polymer samples than in organic crystals [48]. In disordered polymer systems, on the application of an electric field, the probability of an overlap between the localized states is quite less because of different chain segments length and variation in the interatomic distance between chain segments [49]. In such systems, the transport properties are dominated by thermally assisted hopping process.

2.1. POLY (3, 4-ETHYLENEDIOXYTHIOPHENE): POLY (STYRENE SULFONATE) (PEDOT:PSS)

Conjugated Polymers like polyaniline, polypyrrole, polythiophene have been attracting widespread attention for a number of novel applications because of their innate properties [50, 51]. Polyaniline and polypyrrole are conducting polymers with high conductivity but lack properties of solubility and processibility [52, 53]. However, by controlling the pendant chain chemistry, Polythiophene, can be used to prepare new polymers with tunable properties. Polythiophene and its derivatives were found to show stable characteristics both in doped and pure states. They also exhibit high stability of the optical spectrum and high conductivity retention at elevated temperatures for long intervals [54]. Poly (3, 4-ethylenedioxythiophene): poly (styrenesulfonate) (PEDOT:PSS), a polythiophene derivative is one such conjugated polymers with exceptional optical, conductive and solubility properties. It forms an dispersion of oxidatively doped cationic derivative PEDOT electrostatically attached to polyanion Poly styrene sulphonate (PSS) in water [55]. PSS also acts as a charge compensating counter-ion and stabilizes the p-doped PEDOT. The solution-processability of this aqueous colloidal dispersion can be used to achieve a thin, optically transparent film for use in a variety of bendable electronics applications, as well as for use in anti-static coatings, supercapacitors, sensors, OLED applications, solar cells, organic transistors etc. [56-62]. Doped PEDOT: PSS, has been used widely for depositing thin, transparent films for use as an electrode in electronics applications [53]. Thin films of PEDOT:PSS show high conductivity of the order of several hundred S/cm. It is also a promising candidate for electrochromic devices (ECDs) like windows/displays due to its small bandgap, fast response time, low redox potentials, and simplistic fabrication [54, 55]. PEDOT:PSS thin films can be easily deposited using different techniques like template synthesis, CVD, spin coating, inkjet printing, electro -polymerization and spraying [56-58].

PEDOT:PSS is a low bandgap p-type semiconducting material with its energy gap lying in the VIS-NIR range (1.6eV for $\pi \rightarrow \pi^*$ transition) and shows an absorption maximum in the middle of the visible spectrum at 600nm. Doped PEDOT: PSS is almost transparent in the visible region (with a light blue tint) and while neutral polymer possesses a blue-black color. In its natural state, a film of PEDOT-PSS consists of undoped and doped PEDOT units mixed in the ratio 1:4, corresponding to an intermediate state between the oxidized and reduced state. Electrochemical doping of PEDOT:PSS results in addition of mid-gap energy levels, producing absorptive transitions in the visible region invoking an observable color change. The high-contrast electrochromic phenomenon associated with electrochemical doping and the corresponding changes in absorption spectrum are of particular interest to understand the charge storage mechanism during doping/deoping process. Whenever electrochemical doping takes place, ions are incorporated into or extracted from the PEDOT:PSS film. This incorporation of ions manifests itself in the form of increase in electrical conductivity. The conductivity increase in these materials arises from the creation of polaronic and/or bipolaronic quasiparticles on the polymer backbone, which is suitably charge-compensated by a counter ion [59]. The presence of these charged molecular defects sets up symmetric mid band gap energy states in the energy gap [60]. With increased doping levels, broad bipolaron bands develop, effectively shrinking the bandgap energy and transporting the material into a quasi-metallic regime [61]. This electrochromic behavior brought about by an electron transfer reaction that takes place during the electro-chemical oxidation and reduction of the polymer makes PEDOT:PSS suitable as a cathodically coloring polymer in construction of ECDs [62-64]. Such ECD has been studied in detail by several groups like Reynolds et al. [44], Delongchamp et al. [65], Cho et al. [66] etc. Kalagi et al. have reported the shift in energy level for interband transition with the change in doping concentration and the corresponding emergence of polaron/bipolaron bands, intensity variations from the experimental data [67].

To prepare the samples of PEDOT:PSS thin films, aqueous dispersion of PEDOT:PSS was coated on to precleaned substrates by spin coating technique using a programmable spin coater at a rotational speed of 2000 rpm for 30s. After that, the films were kept in vacuum oven for 15min to allow the samples to dry completely. Films were optimized for highest transparency and good electrochromic contrast.

2.2. UV-VIS Spectroscopic Analysis

Optical properties of conjugated polymers can be utilized to provide fundamental understanding of the basic electronic structure of the material. The π-π^* conjugation in the polymers is entailed by their color and their absorption spectra; thus UV-VIS spectroscopy can be utilized to probe the electronic processes that occur in the polymer in the process of doping/dedoping. The high-contrast electrochromic phenomenon associated with electrochemical doping and the corresponding spectral changes are of particular interest. Dietrich et al. [66] have carried out a detailed study by measuring the UV-VIS spectra of PEDOT at different charging levels. The effect of such spectral changes depends initially on the magnitude of the energy gap (E_g). The undoped insulating polymer is transparent (or lightly colored) as energy gap E_g exceeds 3 eV, whereas after doping, the conjugated polymer becomes highly absorbing in the visible range. Generally in p-type conjugated polymers, as the applied positive potential increases, it leads to increasing doping levels, and an increased absorption in the Visible-NIR region [45].

Optical characterizations of the optimized films were carried out using in the wavelength range of $300 < \lambda < 1100$ nm to study spectral transmittance modulation in the range with respect to the changes in the applied polarizing voltage. Figure 6 shows a series of spectra taken at different oxidizing potentials, with energy (eV) on the abscissa and absorbance on the ordinate axis as the doping proceeded via the reaction shown in equation:

$$x(PEDOT) + x(ClO_4^-) \Leftrightarrow (PEDOT^{x+})(ClO_4^-)_x + xe^- \qquad (6)$$

Deep Blue (undoped)   Light Blue(doped)

PEDOT:PSS in its neutral form (doped state), exhibits a strong absorption in the red part of the visible spectrum centered at 600 nm. This characteristic absorption due to π-π^* interband transition accounts for its dark blue color. On electrochemical doping, ClO_4^- ions enter the polymer backbone from the electrolyte, effectively lowering the gap existing between the newly created energy bands by addition of charge carriers. Consequently, the absorption maxima shift towards shorter wavelengths with concomitant increase in transparency of the film. At an applied potential of -0.8 V when the doping level is low and PEDOT:PSS is in its most reduced state, two strongly absorbing bands at 1.4 eV and 2.14 eV and a weak band at 3.30 eV because of π-π^* transition can be easily identified. The overlap between bipolaron states lead to the formation of two bipolaron states in the gap 0.74eV wide. With the oxidising potential changing from -0.8 to -0.6V, there is a gradual decrease in π-π^* transition evidenced by the decreased intensity of interband transition accompanied by a lateral shift in the 2.14 eV band to 2.11 eV. Simultaneously, the binding energy decreases to 0.64 eV indicating the formation of bipolaron with more favorable lower energy states [80]. A further change in potential from -0.6 to -0.8V causes overlapping of the two bipolaron bands and we see the two bands coalesce into a single peak at 1.39 eV. Interestingly, as one moves towards increased oxidation potentials, it is observed that while the IR peak with lower energy remains fairly steady at nearly 1.4 eV, a red shift takes place in the case of higher energy peak from 2.14 eV to 2.04 eV band, a clear evidence of the presence of polarons and bipolarons in the polymer backbone (Bredas). Another important feature of the absorption spectra is the appearance of polaron bands (1.9 eV) at intermediate doping levels (-0.6 to-0.4

V) and their disappearance at higher doping levels. The disappearance of this band may be attributed to increased dopants giving rise to a larger number of polarons which recombine together to produce more favourable low energy bipolarons. At high doping levels (+1V), the bipolaron bands integrate into one broad band stretching from far visible to near IR (0.6-2.5μm). Simultaneously, a new high intensity absorption peak is seen at 2.8 eV. At this highest applied potential, the bipolaron bands merge together and the spectra resembles that of a metallic state with the entire film almost fully bleached.

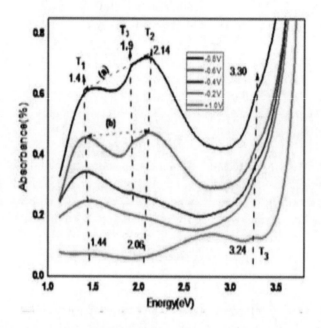

Figure 6. Optical absorbtion spectra of PEDOT:PSS film at different electrochemical doping levels from -0.8 to +1V vs SCE. T_1, T_2 and T_3 show evolution of polaron and bipolaron bands. (a) and (b) show the systematic decrease in inter band peak energy with increased doping content.

2.3. INTERBAND TRANSITION ENERGY

Figure 6 shows the energy level diagram deduced from optical absorption data indicating an energy level structure at different doping levels in PEDOT:PSS films. The three possible optical transitions in the decreasing order of energy are: from valence band (V.B) to conduction band (C.B), V.B to lower polaron level (T_1) and V.B to upper polaron level (T_2). Another transition level from lower polaron level to C.B (T_3) is also possible for intermediate optical excitations. If $h\nu_1$, $h\nu_2$, $h\nu_3$, represent the energies required for the transitions T_1, T_2 and T_3, then energy for interband transition $h\nu_{IB}$ can be written as :

$$h\nu_1 + h\nu_2 = h\nu_{IB} \approx E_g \tag{7}$$

where E_g is the bandgap energy.

2.4. POLARON/BIPOLARON FORMATION: INTERPRETATION FROM UV-VIS SPECTRA

Electrochemical doping/dedoping in conjugated polymers brings about a profound change in the optical spectra of thin polymer films. Studies of these changes in absorption spectra as a function of doping levels have played an important role in elucidating the doping mechanism and nature of charge storage species in the polymer backbone. Interaction between the polymer unit segment and it neighbors lead to formation of electronic bands. The highest occupied orbital termed as HOMO is the analog for valence band (V.B) in metals and semiconductors while lowest unoccupied molecular orbital (LUMO) correspond to the conduction band (C.B) separated by a forbidden band called energy gap E_g. Electrochemical doping leads to creation of new electronic states within the bandgap leading the material from insulating to quasimetallic regime. The energy difference between the band edge and these newly introduced states depend upon the band gap and the chain length of the polymer. Charges that are introduced in polymers on doping are stored in the form of quasiparticles called polarons, bipolarons, or solitons (Section 1.5). These states include a charge and a lattice distortion. A polaron is formed when an electron is added to or removed from the conjugated chain. Bipolaron formation takes place by the combination of two polarons having the same charge. Solitons result from the odd number of carbon atoms in the polymer chain where the possibility of having an unpaired electron exists. These entities-polaron, bipolarons or solitons, comprising of a charge and structural distortion can propagate along the conjugated polymer chain. This mobility of these quasiparticles is exploited by effectively utilizing them as charge carriers as a function of doping levels delocalized over a number of monomer units, but *not* the entire conjugated polymer lattice. It has been proved theoretically that lowest energy configuration for inserted charges is possessed by charged solitons, followed by bipolarons and lastly by polarons. The lack of degeneracy in the polythiophene structure implies that it cannot support stable soliton states thus establishing bipolarons as the lowest energy charge transfer configuration.

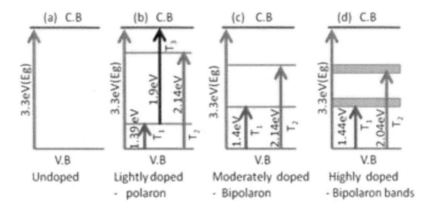

Figure 7. Energy level diagram of PEDOT:PSS showing possible transitions on optical absorption. T_1, T_2, and T_3 correspond to transition between valence band, conduction band and newly generated localized bands and necessary energy required for the transition.

Comparing Figure 7 with optical spectra of Figure 6, we find that this depicts a tractable and ordered system wherein the evolution of bipolaron bands from polarons is clearly visible. T_1 and T_2 represent the transitions between a localised band state and valence band state. Figure 7a shows the band edge transition from V.B to C.B. Figure 7b shows the energy band evolution due to formation polarons at low doping levels, wherein three absorption bands evolve below the gap at 1.4, 1.9 and 2.14 eV. At intermediate doping levels, the energy level structure (Figure 7c) clearly shows the disappearance of polaron absorption at 1.9 eV with increased doping. At high doping levels, broader absorption curve due to bipolaron formation is seen. This disappearnce of polaron band is in agreement with theoretical calculations by Bredas and co-workers, showing that because of lower energy, bipolaron formation favored as compared to existence of two polarons. Here it is imperative to note that in an actual conjugated polymer structure, polaron population must first saturate the entire conjugated polymer chain before bipolaron formation gets initiated. At high doping levels, the individual bipolaron states combine to form broad bands as shown in Figure 7d effectively lessening the interband gap and pushing the material into semimetallic states. This process signals the importance of interactions beween bipolarons which finally leads to the metallic state at sufficiently high concentrations. The broadening of these bands are entirely dependent on the amorphous nature of the film, the length distribution of PEDOT segments and interaction of charge carriers at high doping levels [43].

2.5. POLARON/BIPOLARON FORMATION: INTERPRETATION FROM ELECTROCHEMICAL ANALYSIS

The most basic information which can be had about a film comes from its electrochemical characterization. The CV peaks correspond to the change in conductivity and color of the film with the formation of polarons and bipolarons corresponding to the peaks in the CV curve.

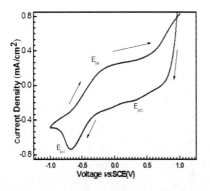
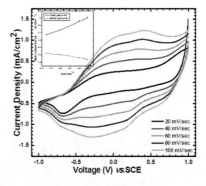

Figure 8. (a) CV of PEDOT:PPS film at scan rate of 40 mVs^{-1}.SCE b) CV of film at different scan rates from 20-100 mV/s. Inset shows cathodic and anodic peak current dependence on scan rate.

Figure 8a represents the CV results for PEDOT:PSS film cycled between ±1V in 0.5M LiClO$_4$ + PC+ACN electrolyte at a scan rate of 40mVs^{-1} vs. SCE. The curve displays broad peaks with large capacitance-like currents, a common feature often found in conducting polymers [75]. Semiconductor like behavior of any conjugated polymer can be discerned from a CV from the presence of two distinct redox peaks, with quite a noticeable separation. In the oxidation cycle, the first peak is attributed to removal of one electron from the polymer, and further to the generation of polaron-dominated mid-gap states [76]. The second peak corresponds to the transference of the second electron with the generation of bipolaron-dominated states. Thus, the total electrochemical process involved in doping /dedoping of this conjugated polymer can be said to encompass a two-step process:- formation of polarons and bipolarons. Sometimes, these two processes may be so close to each other in energy that electrochemically, a single, broad oxidation peak is seen. This broadening of peak can be attributed to doping induced variations of the polaron population [11]. Apart from the broadening of the oxidation peak, the appearance of a plateau at potentials higher than that of the oxidation peak is a general feature in CV of a conjugated polymer film. This peak broadening, a feature common to most polythiophenes is clearly visible in the anodic part of the CV where a broad oxidation peak is seen at -0.015 V, followed by a current plateau [77]. In fact the slow evolutions of the polaron/bipolaron bands with increased doping appear to confirm this. As additional anionic dopants enter CP from the solution in the oxidation cycle, absorption goes on reducing and at the highest positive potential of +1V, the film is fully bleached and appears transparent with a faint blue tinge. In the cathodic branch of the return sweep, with the lowering of applied bias, continuous decrease in optical transparency of the film is seen. A steep rise in the cathodic current alongwith the appearance of two cathodic peaks at 0.3V and -0.66 V is observed. These peaks can be directly assigned to the expulsion of anionic dopants from the polymer matrix into the electrolyte. CVs of the PEDOT: PSS film run at different scan rates, with an increment of 20 mVs^{-1} is shown in Figure 8(b) with the scan rate changing from 20 to 100 mVs^{-1}.

The anodic peak corresponding to the dedoping of PEDOT or extraction of perchlorate ions from the electrolyte shows an increase in amplitude along with a shift towards more positive potentials, i.e., from -0.11V to +0.4 V with the increase in the sweep rate. In the cathodic curve, a distinct peak is observed at -0.66 V, displaying a small shift in position to -0.68 V as the sweep voltage varies, ongoing from 20 to 60 mVs^{-1}. Beyond 60 mVs^{-1} upto 100 mVs^{-1}, the peak height decreases and vanishes into a broad featureless curve. An analysis of this behavior implies that at lower scan rates, the flux to the electrode surface being considerably smaller, more time is available for cation insertion/extraction and hence redox reactions take place at lower potentials. With increased scan rate, available time for cations to enter/leave the polymer film decreases and equilibria is not established as rapidly as before forcing the redox reactions to take place at higher potentials. This makes anodic peak shift towards more positive potentials and cathodic peak towards more negative potentials. Inset of Figure 8b shows the dependence of peak current (i_p) on the scan rate (v). The linear relationship between peak current and root of scan rate $v^{1/2}$ demonstrates that current is diffusion controlled and is limited only by the number of carriers on the film surface [78]. The difference in the slope of anodic and cathodic peak currents *vs.* the root of scan rate implies that the oxidation process in PEDOT:PSS takes place much more easily than reduction. Similar results have been reported by Reynolds group for electrochemically polymerised PEDOT [79].

From the above discussion, it is clear that the generation of quasiparticles like polarons and bipolarons can be easily detected using experimental techniques. The effect of secondary doping on the generation and subsequent disappearance of these quasiparticles because of conjugational defects due to relaxation and conformational disorder within the polymer can be understood by UV-VIS spectra and electrochemical analysis. Many other experimental techniques like ESR, EELS etc. may also be used for further studies. Such electronic band structure analysis from the polaron and bipolaron bands helps in understanding the conductivity behavior of materials which aid in designing devices with new and improved properties. The dependence of polaron pair formation on vibrionic coupling, charge separation dynamics and its sensitivity to disorder give an insight for tailoring light-to-current conversion processes in organic materials and many such other next generation devices of organic electronics.

REFERENCES

[1] V. V. Walatka, M. M. Labes and J. H. Perlsten, 1973, "Polysulfur Nitride—a One-Dimensional Chain with a Metallic Ground State," *Phy. Rev. Lett.* 31,1139-1142.

[2] H. Shirakawa, E. J. Louis, A. G. MacDiarmid, C. K. Chiang, A. J. Heeger, 1977, "Synthesis of Electrically Conducting Organic Polymers: Halogen Derivatives of Polyacetylene, (CH)x.," *J. Chem. Soc. Chem. Commun.* 16, 578-580.

[3] T. A. Skotheim *Handbook of Conducting Polymers*, 1986, Ed (Dekker, New York) vol 1& 2.

[4] J. Campbell Scott 1997, www.sciencemag.org. *Science* 278, 19 December.

[5] A. J. Heeger, S. Kivelson, J. R. Schrieffer, and W. P. Su, 1988, "Solitons in conducting polymers" *Rev. Mod. Phys. 60*, 781 -850.

[6] K. Fesser, A. R. Bishop, and D.K. Campbell, "Optical absorption from polarons in a model of polyacetylene," 1982, *Phys. Rev. B* 27,4804 -4825.

[7] J. L. Bredas, R. R. Chance, and R. Silbey, 1982, "Comparative Theoretical Study of the Doping of Conjugated Polyacetylene and Polyparaphenylene," *Phys. Rev.* B 26, 5843-5844.

[8] J. L. Bredas, B. Themans, J. G. Fripiat, J. M. Andre and R. R. Chance,1984, *Phys. Rev. B* 29, 6761.

[9] Y. Furukawa, 1996, Electronic Absorption and Vibrational Spectroscopies of Conjugated Conducting Polymers. *J. Phys. Chem.* 100, 15644-15653

[10] W. R. Salaneck, R. H. Friend, and J. L. Bre´das,1999, Electronic structure of conjugated polymers: consequences of electron–lattice coupling, *Phys. Rep.* 319, 231-251.

[11] P. Chandrasekhar, 1999, *Conducting polymers: Fundamentals and Applications A practical approach* by Springer Science+Business Media New York.

[12] A. J. Heeger, 2000, *Semiconducting and metallic polymers: the fourth generation of polymeric materials,* Nobel Lecture.

[13] W. Barford, M. Marcus, and O. R. Tozer, 2016 "Polarons in π-Conjugated Polymers: Anderson or Landau?" *J. Phys. Chem.* A 2016, 120, 615−620

[14] S. A. Brazovskii, N. N. Kirova, 1981, Excitons, polarons, and bipolarons in conducting polymers. *JETP Lett* 33, 6-10.

[15] C. S. Pinheiro, G. M. Silva, 2003, "Dynamics of Polarons and Bipolarons with Interchain Coupling in Conjugated PolymersInternational." *J of Quantum Chemistry*, 95, 001–006.

[16] L. D. Landau, 1933, Electron Motion in Crystal Lattices, *Phys. Z. Sovjet.* 3, 664-665 [English translation:(1965), Collected Papers, New York: Gordon and Breach, 67 - 68].

[17] S. I. Pekar, 1951, "Research in Electron Theory of Crystals, Moscow: Gostekhizdat" [German translation: (1954), Untersuchungen 39¨uber die Elektronentheorie der Kristalle, Berlin: Akademie Verlag; English translation: (1963), *Research in Electron Theory of Crystals,* US AEC Report AEC-tr-5575].

[18] H. Fr¨ohlich, (1954), Electrons in lattice fields, *Advances in Physics* 3, 325 – 361.

[19] J. T. Devreese, "Polarons," in *Digital Encyclopedia of Applied Physics*, edited by G. L. Trigg (Wiley, online, 2008).

[20] C. Kittel, 1987, *Quantum Theory of Solids*, Wiley publications.

[21] J. T. Devreese, F. Peeters, (Eds.) 1984, *"Polarons and Excitons in Polar Semiconductors and Ionic Crystals,"* Plenum Publishing Corp., New York.

[22] A. S. Alexandrov, N. F. Mott (1994), "Bipolarons," *Reports on Progress in Physics* 57, 1197 – 1288.

[23] P. G. de Gennes, 1960, Effects of Double Exchange in Magnetic Crystals, *Phys. Rev.* 118, 141 - 154.

[24] B. Ram shastry and D. C. Mattis, "Theory of bound polaron," *Phy. Rev.* B, 1981, 24, 5340-5349.

[25] J. L Bredas and G. B. Street, 1985 "Polarons, Bipolarons and solitons in conducting Polymers," *Acc. Chem. Res.,* 18, 309-315.

[26] A. G. Mcdermid, R. G. Mannone, R. B. Kanor and S. G. Porter, 1985, "The concept of doping in conducting polymers: The role of reduction Potentials, *Phil. Trans. R. Soc. A, 314,3-15.*

[27] A. G. MacDiarmid, 2001, "Synthetic Metals: A Novel Role for Organic Polymers (Nobel Lecture)" *Angew. Chem. Int. Ed.* 2001, 40, 2581 – 2590.

[28] J. Heeger, 2000, *"Semiconducting and metallic polymers: The fourth generation of polymeric Materials,"* Nobel Lecture, December 8.

[29] A. O. Patil, A. J. Heeger and F. Wudl 1988, Optical Properties of Conducting Polymers. *Che. Rev.,* 88, 183-200.

[30] J. L. Bredas, B. Themans and J. M. Andre, 1984, "The role of mobile organic radicals(Solitons,polarons and Bipolarons) in the transport properties of conducting polymers" *Syn. Met.* 9. 265-274.

[31] S. Miyauchi, H. Aiko, Y. Sorimashi, I. Tsubata, 1989, Preparation of barium titanate–polypyrrole compositions and their electrical properties, *J. Appl. Polym. Sci.* 37 289-293.

[32] K. A. Gschneidner, V. K. Pecharsky, 1999, Magnetic refrigeration materials, *J. Appl. Phys.* 85, 5365-5368.

[33] X. Peng, Y. Zhang, J. Yang, B. Zou, P. L. Xia, T. J. Li, 1992, Formation of nanoparticulate iron(III) oxide-stearate multilayer through Langmuir-Blodgett method, *J. Phys. Chem.* 96, 3412-3415.

[34] Y. Y. Wang, X. L. Jing, 2005, Intrinsically conducting polymers for electromagnetic interference shielding, *Polym. Adv. Technol.* 16, 344-351.

[35] P. K. Shen, H. T. Huang, A. C. Tseung, 1992, A study of tungsten oxide and polyaniline composite films, I. Electrochemical and electrochromic behavior. *J. Electrochem. Soc.* 139, 1884-1845.

[36] H. Hu, M. Trejo, M. E. Nicho, J. M. Saniger, A. Garcia-Valenzuela, 2002, Ädsorption kinetics of optochemical NH_3 gas sensing with semiconductor polyaniline film," *Sens. Actuators B* 82, 124.

[37] S. Chandra, S. Annapoorni, F. Singh, R. G. Sonkawade, J. M. S. Rana and R. C. Ramola, 2010, "Low temperature resistivity study of nanostructured polypyrrole films under"electronic excitations," *Nucl. Instr. Meth. Phys. Res*, 268, 62–66 65.

[38] W. P. Su, J. R. Schrieffer, 1980, Soliton dynamics in polyacetylene. *Proc. Natl. Acad. Sci. USA*, 77, 5626-5629.

[39] B.S. Ong, Y. Wu and P. Liu. 2005, "Design of High-Performance Regioregular Polythiophenes for Organic Thin-Film Transistors," *Proc IEEE* 93, 1412-1419.

[40] W. F. Pasveer, J. Cottaar, C. Tanases, R. Coeho"orn, P. A. Bobbert, P. W. M. Blom, D. M. De Leeuw and M. A. J. Michels, 2005, "Unified Description of Charge-Carrier Mobilities in Disordered Semiconducting Polymers," *Phys Rev Lett* 94, 206601-206605

[41] T. A. Skotheim, R. L. Elsenbaumer, J. R. Reynolds, 1998, *"Handbook of Conducting Polymers,"* CRC Press, NewYork, 1998.

[42] A. J. Heeger, 1993, Polyaniline with surfactant counterions, Conducting polymer materials which are processible in conducting form, *Synth. Met.* 55-57, 3471-3482.

[43] S. Kirchmeyer, W. Lövenich, U. Merker, K. Reute and A. Elschner, 2011, *"Principles and Applications of an Intrinsically Conductive Polymer, PEDOT,"* Taylor and Francis.

[44] L. B. Groenendaal, F. Jonas, D. Freitag, H. Pielartzi and J. R. Reynolds, 2000, "Poly(3,4-ethylenedioxythiophene) and Its Derivatives: Past, Present, and Future," *Adv. Mater.* 12, 481.

[45] J. C. Gustafsson, B. Liedberg and O. Inganäs, 1994, *"In situ* spectroscopic investigations of electrochromism and ion transport in a poly (3,4-ethylenedioxythiophene) electrode in a solid state electrochemical cell," *Solid State Ionics*, 69, 145-152.

[46] H. W. Heuer, R. Wehrmann and S. Kirchmeyer, 2002" Electrochromic Window Based on Conducting Poly(3,4-ethylenedioxythiophene)–Poly(styrene sulfonate)," *Adv. Funct. Mater.* 12, 89- 94.

[47] M. Gratzel, 2001, "Molecular photovoltaics that mimic photosynthesis," *Pure Appl. Chem.* 73, 459-467.

[48] J. S. Yang and T. M. Swager, 1998, Fluorescent Porous Polymer Films as TNT Chemosensors: Electronic and Structural Effects, *J. Am. Chem. Soc.* 120, 11864-11873.

[49] Y. Kaminovz, E. Smela, T. Johansson, L. Brehnme, 2000, Optical and electrical properties of substituted 2,5-diphenyl-1,3,4-oxadiazoles, *Synthetic Metals*, 111,75-78.

[50] B. C. Thompson, P. Schottland, K. Zong, J. R. Reynolds,2000, *In Situ* Colorimetric Analysis of Electrochromic Polymers and Devices, *Chem. Mater.* 12,1563-1571.

[51] B. Sankaran, J. R. Reynolds, 1997, High-Contrast Electrochromic Polymers from Alkyl-Derivatized Poly(3,4-ethylenedioxythiophenes), *Macromol.* 30, 2582-2588.

[52] J. P. Lock, J. L. Lutkenhaus, N. S. Zacharia, S. G. Im, P. T. Hammond, K. K. Gleason,2007, Electrochemical investigation of PEDOT films deposited via CVD for electrochromic applications, *Synth. Met.* 157, 894-898.

[53] J. Ouyang, C. W. Chu, F. C. Chen, Q. Xu, Y. Yang,2005, On the mechanism of conductivity enhancement in poly(3,4-ethylenedioxythiophene):poly(styrene sulfonate) film through solvent treatment. *Adv. Funct. Mater.* 15 203.

[54] J. M. Leger, 2008, Organic electronics: the ions have it, *Adv. Mater.* 20, 837-841.

[55] A. A. Argun, A. Cirpan, J. R. Reynolds, 2003, The First Truly All-Polymer Electrochromic Devices. *Adv. Mater.* 15, 1338-1341.

[56] S. S. Kalagi, D. S. Dalavi, S. S. Mali, A. I. Inamdar, R. S. Patil, P. S. Patil,2012, "Study of Novel WO3-PEDOT:PSS Bilayered Thin Film for Electrochromic Applications" *Nanosci. Nanotech. Let 4.*, 1146-1154.

[57] G. Zotti, S. G. Schiavon Zecchin, L. B. Groenendaal, 2002 Electrochemical and Chemical Synthesis and Characterization of Sulfonated Poly(3,4-ethylenedioxythiophene): A Novel Water-Soluble and Highly Conductive Conjugated Oligomer Macromol. *Chem. Phys.* 203, 958.

[58] T. Mustonen, K. Kordás, S. G. Saukko Tóth, S. Jari, P. P. Helistö, H. Seppä, H. Jantunen, 2007, Inkjet printing of transparent and conductive patterns of single-walled carbon nanotubes and PEDOT-PSS composites, *Phys. stat. sol.* (b) 244, 4336-4340.

[59] J. L. Bredas, R. R. Chance, R. Silbey,1982, *Phys. Rev. B* 26, 5843.

[60] J. S. Kim, W. S. Chung, K. Kim, D. Y. Kim, K. J. Paeng, S. M. Jo, S. Y. Jang, 2010, Performance optimization of polymer solar cells using electrostatically sprayed photoactive layers adv. *Funct. Mater.* 20, 3538-3546.

[61] S. S. Kalagi, S. S. Mali, D. S. Dalavi, A. I. Inamdar, H. Im, P. S. Patil, 2012, "Transmission attenuation and Chromic contrast characterization of R. F. sputtered WO3 thin films for Electrochromic device applications," *Electrochim. Acta* 85, 501-508.

[62] M. Deepa A. K. Srivastava, K. N. Sood, A. V. Murugan, J*ournal of The Electrochemical Society,* 2008, "Nanostructured Tungsten Oxide-Poly(3,4-ethylenedioxythiophene): Poly(styrenesulfonate) Hybrid Films: Synthesis, Electrochromic Response, and Durability Characteristics"155 (11) D703-D710

[63] L. Groendal, F. Jonas, D. Freitag, H. Pielartzik, and J. R. Reynolds, 2000, "Poly(3,4-ethylenedioxythiophene) and Its Derivatives: Past, Present, and Future." *Adv. Mater.* 2000, 12, 481-494.

[64] D. Delongchamp, P. T. Hammond, 2001, "Layer-by-Layer Assembly of PEDOT/Polyaniline Electrochromic Devices," *Adv. Mat.*, 13, 1455-1459.

[65] S. I. Cho, W. J. Kwon, S. J. Choi, P. Kim, S. A. Park, J. Kim, S. J. Son, R. Xiao, S. H. Kim, S. B. Lee, 2005, Nanotube-Based Ultrafast Electrochromic Display, *Adv. Mater.* (Weinheim, Ger) 17, 171-175.

[66] S. S. Kalagi, P. S. Patil, 2016, Secondary electrochemical doping level effects on polaron and bipolaron bands evolution and interband transition energy from absorbance spectra of PEDOT: PSS thin films. *Synth. Met.* 220, 661–666.

[67] M. Dietrich, J. Heinze, G. Heywang, F. Jonas, 1994, Electrochemical and spectroscopic characterization of polyalkylenedioxythiophenes, *J. Electroanal. Chem.* 369, 87-92.

[68] A. J. Heeger, 1985, Charge storage in conducting polymers: Solitons, polarons and Bipolarons," *Polymer J.* 17, 201-208.

[69] M. J. Rice, 1983, Physical dynamics of solitons, *Phys. Rev. B* 28 152.

[70] Y. Lu, *Solitons and Polarons in Conducting Polymers*, 1988, Book (a review plus collection of articles), World Sci Publ. Co., Singapore.

[71] J. Cornil, J. L. Bredas, 1995, Nature of the optical transitions in charged oligothiophenes, *Adv. Mater.* 7, 295-297.

[72] S. Brazovskii, N. Kirova, 1981, Excitons, polarons, and bipolarons in conducting polymers, *JETP Lett.* 33, 4.

[73] C. Ziegler, in: H. S. Nalva (Ed.), 1997, *Handbook of Organic Conductive Molecules and Polymers*, 3, J. Wiley & Sons, New York, 678.

[74] Y. J. Lin, F. M. Yang, C. Y. Huang, W. Y. Chou, J. Chang, Y. C. Lien, 2007,Increasing the work function of poly(3,4-ethylenedioxythiophene) doped with poly(4-styrenesulfonate) by ultraviolet irradiation, *App. Phys. Lett.* 91, 092127.

[75] P. G. Pickup, 1999, B. E. Conway, J. OÏM. Bockris, R. E. White (Eds.), *Modern Aspects of Electrochemistry,* 33, Plenum, New York, 549.

[76] G. Garcia-Belmonte, E. V. Vakarin, J. Bisquert, J. P. Badiali, 2010, Doping-induced broadening of the hole density-of-states in conducting polymers 2010, *Electrochim. Acta* 55, 6123-6127.

[77] M. Deepa, S. Bhandari, M. Arora, R. Kant, 2008, Electrochromic Response of Nanostructured Poly(3,4-ethylenedioxythiophene) Films Grown in an Aqueous Micellar Solution Macromol. *Chem. Phys.* 209, 137-149.

[78] F. Miomandre, P. Audebert, K. Zong, J. R. Reynolds, 2003, Adsorption-Assisted Electro-oxidation of Dioxypyrrole and Dioxythiophene Monomers Probed by Fast Cyclic Voltammetry, *Langmuir* 19, 8894-8898.

[79] A. A. Argun, P. H. Aubert, B. C. Thompson, I. Schwendeman, C. L. Gaupp, J. Hwang, N. J. Pinto, D. B. Tanner, A. G. MacDiarmid and J. R. Reynolds, 2004," Multicolored Electrochromism in Polymers:Structures and Devices," *Chem. Mater.* 16, 4401-4412.

In: Polarons: Recent Progress and Perspectives
Editor: Amel Laref

ISBN: 978-1-53613-935-8
© 2018 Nova Science Publishers, Inc.

Chapter 5

ANHARMONICITY, SOLITON-ASSISTED TRANSPORT AND ELECTRON SURFING AS A GENERALIZATION OF POLARON TRANSPORT WITH PRACTICAL CONSEQUENCES

Manuel G. Velarde[1,] and E. Guy Wilson[2]*
[1]Instituto Pluridisciplinar, UCM, Madrid, Spain
[2]School of Physics and Astronomy, Queen Mary University of London, London, UK

ABSTRACT

We discuss a generalization of the polaron concept to include soliton assisted transport. Electron surfing with a lattice soliton is described. The generalization also offers a clear case of mechanical control of electrons at the nanolevel. Building upon such a generalization, the concept of a solectron was introduced and a new solectron field effect transistor (SFET) has been designed, which relies on the ballistic flight of the solectron at a velocity independent of the field, and even in a zero field. This opens up the possibility of low energy computation.

Keywords: anharmonicity, solitons, solectrons, electron surfing, hyperconduction, SFET

1. INTRODUCTION

The polaron concept originated in Landau's idea [1] of self-trapping for an electron in the polar lattice distortion it creates when added as an excess charge to an (originally) ionic crystal, due to the Coulomb interaction. He considered that such a compound (electron and induced polarization field) was the actual carrier in charge of transport, hence of electricity in the presence of an external electric field. The concept also applies to the self-trapping of a (positive) hole. Later on, the concept was refined and redefined, first by Pekar [2, 3], Landau

---

* Corresponding Author address. Email: mgvelarde@pluri.ucm.es.

and Pekar [4] and, subsequently, by Pekar [5] and collaborators [6, 7]. Eventually it was made complete by Fröhlich [8, 9], Holstein [10], Feynmann [11] and others [7, 12-17] thus leading to the quantum transport theory in solid state physics [18-24].

The building bricks of the polaron as we consider it today are, on the one hand, the crystal Hamiltonian with harmonic interactions (hence the linear mechanical elasticity) underlying the phonons as quantum (stationary) wave excitations in the lattice extending from one end to the other of the crystal. The other item is the nonlinear electron-phonon interaction. Thus a polaron, also denoted as a quasi-particle, is a dressed electron or an electron surrounded by a cloud of (a few) phonons. Indeed from a field theoretic viewpoint, the induced polarization appears as a cloud of virtual phonons around the electron. Needless to say, the polaron is a genuine electronic excitation, as the electron causes the polaron to exist.

From 1960 to 1980, several lines of polaron extension occurred. On the one hand, there was the discovery of (dopable) conducting polymers that eventually led to the Nobel Prize in 2000 to Heeger, MacDiarmid and Shirakawa and, more relevant to us, the introduction of the so-called SSH Hamiltonian by Schrieffer and colleagues [25, 26] to account for such conduction of otherwise originally (when undoped) insulator-like synthetic materials, which were later on denoted as "synthetic metals". The second line was the introduction of the electro-soliton concept by Davydov to account for transport and related phenomena in biomolecules [27-29]. The third line was the numerical "discovery" and spectacular development of the soliton concept, coinage by Zabusky and Kruskal [30], with work – analytical and numerical – by many others, among which included Toda [31, 32]. The latter offered a Hamiltonian model with anharmonic interactions between the units that he had completely been solved analytically. Solitons are mechanical excitations in a crystal lattice that tend to be of significance past the Dulong-Petit plateau in the specific heat against temperature, thus indicating their more classical than quantum character [33].

The SSH line of thought [25, 26] builds upon the harmonic interactions between units, the degeneracy of the ground state of polymers like trans-polyacetylene (tPA) described by a quartic contribution added to the originally harmonic Hamiltonian, and the tight binding approximation to account for band flow as means of electron transport. tPA (like many other polymers in the same class) was a poorly crystalline material that upon doping changed from an insulator-like behavior to copper-like metallic conductor. When electrically conducting, it was first suggested that electron transport was based on the existence of solitons/kinks, otherwise denoted as topological solitons, due, indeed, to the degeneracy of the ground state. Later on, the transport was rather associated to polaron or bipolaron transport. Still today no one mechanism seems to apply to tPA over the whole range of doping.

Davydov [27-29] was able to identify bio-related lattices with characteristics capable of leading to soliton-bearing Hamiltonians – at variance with the SSH theory for tPA not based upon degeneracy of the ground state- whose solution he offered. Unfortunately his predictions – at that level – were not supported by experiment nor did they survive going to temperatures beyond a few K, though this result was not without criticism [34-38]. Then in the 1980s and 1990s, Davydov [29, Ch. 6, 37] and colleagues [39-41] extended the original theory by using Hamiltonians with cubic and quartic anharmonic interactions thus allowing for supersonic soliton-assisted charge transport with, once more, all results studied at zero K.

In an independent approach, already during the 1970s, Zmuidzinas [42, 43] also explored the idea of going beyond polaron-assisted transport. In particular, building upon the Fröhlich approach (distortion dependent on-site energy) he analyzed a Hamiltonian with quartic interactions and also advanced the concept of soliton trapping electrons thus leading to electron surfing transport. Zmuidzinas estimated the activation energy of a lattice soliton in the range of 1eV (quite a high temperature 1.16×10^4K; earlier we mentioned that solitons, e.g., in a Toda lattice, tend to appear past the Dulong-Petit plateau in the specific heat [33]) though he suggested the possible observation of this form of soliton-assisted hyperconduction in the range of tens of degrees K.

During the 1990s, following the path set by Davydov, Kadantsev and colleagues [44, 45] considered a Hamiltonian with quintic interactions in a lattice in interaction with a thermal bath (non-zero T). The quintic interaction and the heating permitted to sustain solitons capable of trapping electrons and hence soliton-assisted transport at non zero K.

In the 1980s and 1990s, Wilson and Donovan and colleagues [46-49] explored the experimental characteristics of a non dopable, perfectly crystalline polymer, poly-diacetylene (PDA) and sulphonate derivatives (PDATS). Donovan and Wilson [46] found, by experiment in PDA, a photocarrier to travel at a velocity close to the sound velocity, when pulled by an electric field. However, the velocity was independent of field, even over four decades of field research. Moreover, at the lowest field, the velocity was greater than what would be found in any semiconductor at the same field. (In conventional semiconductor language, the low field mobility was ultra-high, greater than 20 m^2 s^{-1} V^{-1}). The carrier had thus a low dissipation of energy to the lattice. The carrier was ballistic. The backbone of PDA crystal is made of π conjugated carbon chains. The chains are long, straight and parallel for macroscopic distances. The lowest electronic excitations belong to the backbone chains. The chains are separated by a (electronically) large distance (0.7 nm). Thus they form ideal one dimensional semiconductor. Optical absorption above the band gap (2.4 eV) created electron and hole photocarriers. The theory developed by Wilson [50] built upon the SSH approach –with, however, no degeneracy of the ground state in PDA- and later found having similarity to Davydov's approach (distortion dependent electron transfer). He advanced the concept of solitary wave acoustic polaron (SWAP) as charge carrier clearly extending the original polaron realm into the soliton-like assisted transport, and quite a different mechanism for transport relative to the dopable tPA family. Yet in all cases so far mentioned, though the lattice dynamics enters well beyond the simplest induced lattice polarization distortion, we still have genuine electronic excitations, as without the added excess electron there is no evidence of the crystal excitation.

Starting in 2005, in an independent approach, Velarde and colleagues [51-56] considered from the very beginning a Hamiltonian with Morse anharmonic interactions and developed a theory that, besides providing genuinely new predictions, embraced many of the results obtained by earlier mentioned authors, emphasizing the significance of supersonic solitonic motions. In Emin's book [17] there is discussion of the role of anharmonicity in polaronic problems but we suggest it does not apply here, for it belongs to a different line of thought than the one presented here. The dynamic bound state of an added excess electron to a [Morse (Toda)] lattice soliton was denoted solectron [51, 55], as the natural extension of the Landau-Pekar polaron concept where the (nonlinear) mechanical lattice excitation dominates and permits control of an added excess electron even in the absence of an external electric field [56]. In [57], in a certain sense interpolating between the theories of Zmuidzinas [42, 43] and Wilson [50] (albeit retaining the discrete character of the lattice throughout; both Zmuidzinas and Wilson resort to continuum approximations), there is a discussion of the role of the increasing strength of electron-lattice (phonon) interaction leading from polaron to solectron. Their results also included predictions about the stability of solitons and solectrons up to physiological temperatures, e.g., bio-related lattices (ca. 300K) [58-60]. They also considered cubic and quartic interactions, and the onset of pairing of two solectrons (bisolectron concept as extension of the bisoliton concept of Davydov and Brizhik [61-65]).

The outline of the chapter is as follows. In Section 2 we discuss the implementation of the solectron concept, in the form of solitary wave acoustic polaron, as done by Wilson [50]. Consequences of such implementation are sketched and some limitations of the theory indicated. Then Section 3 deals with the implementation of the solectron concept based upon the anharmonicity of the lattice interactions thus extending Wilson's study. The concept of electron surfing on discrete lattice solitons is recalled and some of its significant consequences are discussed. In Section 4 a novel filed effect transistor based upon the solectron concept (SFET) is succinctly described.

2. SOLITON WAVE ACOUSTIC POLARON THEORY

We discuss now a solectron-like theory produced by Wilson [50] (then called a solitary wave acoustic polaron, or (SWAP) in an attempt to explain extraordinary experimental results found by Donovan and Wilson [46-49] on charge motion in a highly electronically one dimensional system PDATS.

PDATS is a perfect crystal comprised of parallel conjugated polymer chains. The chain backbones, which host any extra electron added, are a large distance apart from each other; thus, such extra electron resides in a perfect electronically one dimensional world. They found an extra electron, photo injected, to travel at a constant velocity, no matter what the value of the applied field. This was so for a variation of field by four orders of magnitude. Moreover the distance of travel in such fashion could approach a millimeter.

The theory was later learned to be similar to soliton-like theories of Davydov. But the theory attempted to use experimental values of all parameters to give results which had values to compare to experiment, and to predict experimental behavior. Attention was also paid to the stability of the solectron, and to the dissipation to an external phonon bath.

The main conclusion was that, at room temperature, a solectron of thermal energy kT, travelled as a ballistic particle, with a velocity just below the sound velocity. An applied field made little change to this velocity. Dissipation by acoustic phonon emission is forbidden. Rather, ambient acoustic phonons can bounce off the solectron and extract energy. But the probability of this process was shown to be very weak. Thus, the theory seemed to give a good account of the experimental results.

Both the experimental results, and the theory, are utterly different than the situation in three dimensional materials such as silicon.

The theory begins with the Su, Schrieffer, Heeger (SSH) Hamiltonian [25, 26], but used here to describe one extra electron on a polymer chain,

$$H = \left(\frac{1}{2M_a} \right) \sum_n \left\{ p_{a,n}^2 + \left(\frac{M_a S^2}{a^2} \right) \left(u_{a,n+1} - u_{a,n} \right)^2 \right\} - \sum_n t_{n+1,n} \left(c_{n+1}^* c_n + c_n^* c_{n+1} \right)$$

$$(1)$$

where

$$t_{n+1,n} = t_0 - \alpha\left(u_{a,n+1} - u_{a,n}\right)$$

(2)

The lattice part of H describes the momentum, $p_{a,n}$ and displacement, $u_{a,n}$ of the n'th unit call, of mass M_a. a is the unit cell size, and S is the sound velocity. While the unit cell of PDA has 4 carbon atoms along the backbone, there is considerable greater mass residing in the TS side groups of the polymer. The side group is $-CH_2 -O-SO_2 -C_6 H_4 -CH_3$. There are 46 atoms in the unit cell of a single chain, of which only 4 are in the carbon backbone. A long wavelength acoustic wave in the chain direction will give compressions and rarefactions of the whole mass of the cell.

Values of these parameters M_a and a are given in Table 1. The value of unit cell length a is from the X ray structure of Kobelt and Paulus [66]. Donovan and Wilson [46] estimated the sound velocity $S \approx 5 \times 10^3$ m/s from values of bulk modulus, $c \approx 5 \times 10^{10}$ N/m^3 [67] and density $\rho = 1.48 \times 10^3$ kg/m^3 [65]. The density is from X ray structure determination, and so accurate. Derivation of S from $S = (c/\rho)^{1/2}$ is very fraught in an anisotropic material, so S is not accurate. Lochner et al. [67] measured the changes of lattice parameters with hydrostatic pressure, observing anisotropy. The elastic modulus C3 = 1.88 $\times$ 10^{10}N/m^2. This gives S = 3.6 $\times$ 10^3 m/s in the chain direction, which is probably a better estimate of S, and used in Table (1).

Now consider the electron part of H. When undeformed, the electron resides in a Bloch band, the lowest unoccupied band of the PDA. A band structure calculation of Wilson [69] gave equal electron and hole band masses of (1/4) the free electron mass. Consider a chain of N unit cells. All the N states of a band, e.g., using the whole Brillouin zone, can be used to form N Wannier states. Such states are localized around the 4 carbon atoms of the backbone of the unit cell. They are identical within each cell, and form a Complete Normal Orthogonal Set. These states are used as the basis states of the electronic part of the Hamiltonian. That is

$$\phi_n = c_n^* \,|\, 0 >$$

(3)

is a Wannier state ϕ_n at cell n produced by the creation operator c_n^*. These basis states are molecular like, but come from the whole band. They are not some fictitious sub molecule formed by arbitrarily chopping up the chain.

Consider now the term, $t_{n+1,n}$, in the Hamiltonian, which transfers the electron between adjacent Wannier states. In the undeformed chain then H creates a band, and the bottom of the band has an effective mass given by

$$m = \frac{\hbar^2}{2a^2 t_0}$$

(4)

Table 1. Numerical values for Polydiacetylene (PDA)

Quantity	Symbol	Value	Reference/Comment
unit cell length	a	0.49 nm	
unit cell mass	M_a	7.0×10^{-25} kg	includes side chain mass
sound velocity	S	3.6 km/s	
electron band mass	m	2.3×10^{-31} kg	1/4 of free electron mass
deformation potential	D_a	3.7 eV	
solectron width parameter	γ_{0a}	9.59×10^{-2}	Eq. (9)
solectron energy parameter	U_a	5.80 meV	Eq. (12)

Thus, the value of t_0 can be found from the band mass.

The deformation potential D_a is a measure of the shift of the band edge with dilation, and is used in semiconductor physics to measure the strength of the electron acoustic phonon interaction, and to model the electron scattering probability, by acoustic phonons and hence to model the mobility. The SSH Hamiltonian measure of the effect of dilation is due to α, where α multiplied by the bond stretch is the change of the transfer integral. Then

$$D_a = 2\alpha a. \tag{5}$$

Thus, the value of α can be found if the deformation potential is experimentally known. Dilation can be carried out by mechanical stretching, and band edge shifts resulting detected optically. However, in PDA, excitons dominate the optical absorption, while the electron hole interband absorption is a weak feature. Dilation of a sample in the chain direction by mechanical stretching gives an observed shift in the exciton energy. This shift is 3.7 eV/m [69]. If the associated electron band edge is shifted the same amount, then $D_a = 3.7$ eV.

This estimate of D_a and hence α is an approximation. For the exciton binding energy may shift also. And the shift may be due to hole band edge shift as well as electron band edge shift. This concludes discussion of the experimental numerical parameters for the Hamiltonian.

The solectron electron state ψ is of the form

$$\psi = \sum_n a_n \phi_n \tag{6}$$

An analytic form for the amplitudes a_n, and then all properties of the solectron, can be found, in the continuum limit, and assuming the adiabatic approximation. For the electron, then the localized electron amplitudes are

$$a_n = (\gamma_a/2)^{1/2} \sec h[\gamma_a(n - n_0)] \tag{7}$$

where

$$\gamma_a = \gamma_{0a}/(1 - \beta^2) \tag{8}$$

$$\gamma_{0a} = \left(D_a^2 ma^2\right)/\left(2M_a S^2 \hbar^2\right) \tag{9}$$

and

$$\beta = v/S . \tag{10}$$

Here v is the solectron velocity. The energy of the electron in this state is

$$\varepsilon_e = -U_a /\left(1-\beta^2\right)^2 \tag{11}$$

where

$$U_a = \left(m/8\hbar^2\right)\left(D_a^4 a^2 / M_a^2 S^4\right) \tag{12}$$

Corresponding to this electron state there is a localized displacement of the unit cell positions given by

$$u_{a,n} = -\left[D_a a/2M_a S^2\left(1-\beta^2\right)\right]\tanh\left[\gamma_a\left(n-n_0\right)\right] \tag{13}$$

The lattice energy due to this distortion is

$$\varepsilon_L = \left(2U_a/3\right)\left(1+\beta^2\right)/\left(1-\beta^2\right)^3 \tag{14}$$

The resultant and total solectron energy is

$$\varepsilon = -\left(U_a/3\right)\left(1-5\beta^2\right)/\left(1-\beta^2\right)^3 \tag{15}$$

Thus, all properties of the solectron are numerically determined by these equation and the experimental parameters of Table (1). The values of the resulting solectron width parameter, γ_{0a}, and energy parameter, U_a, are also given in Table 1.

As the velocity v increases from zero, the total energy increases. The solectron narrows, the electron becomes more tightly bound within the distortion of the lattice, and the lattice distortion energy increases. As v approaches the sound velocity S then the energy rises asymptotically to infinity. If a force F is applied to a solectron at rest, by means of an applied electric field, the energy increases, and the velocity increases towards the sound velocity.

It is desirable to know the precise effect of the field on the solectron dynamics and internal structure. It is interesting that the solectron is described by the Lagrangian L_s where

$$L_s = \left(U_{0a}/3\right)/\left(1-\beta^2\right)^2 ; \beta = V/S ; V = dX_s/dt \tag{16}$$

The solectron momentum P_s conjugate to the solectron position X_s is

$$P_s = \partial L_s / \partial V = (4U_{0a}/3S)\beta/(1-\beta^2)^3 \qquad (17)$$

The solectron Hamiltonian H_s is then

$$H_s(P_s, X_s) = VP_s - L_s \qquad (18)$$

It is the case that H_s is identical to the solectron total energy, confirming that the Lagrangian is correct.

A force F adds an extra term $-FX$ to the Hamiltonian. The Hamilton equation for dP/dt then gives

$$dP_s / dt = -\partial H_s / \partial X_s = F \qquad (19)$$

Thus, an applied force changes the value of solectron momentum P_s in a manner linear in time. Thus, the value of all properties of the solectron in response to an applied field can be determined. There is no explicit expression for the energy against momentum P_s.

Figure 1 shows the energies making up the solectron as a function of its velocity. At room temperature T then the thermal energy kT is 25 meV. Thus, a solectron in thermal equilibrium is already moving at around 70% of the sound velocity. The electron is more tightly bound than in the solectron at rest.

Figure 2 shows the total energy and velocity as a function of the momentum. One can now follow the response to an electric field applying the force F to the solectron. Suppose the solectron is initially at rest, having zero momentum, and then a positive force is applied. The momentum increases at a uniform rate in time. The velocity increases, linearly in time initially, but asymptotically approach the sound velocity. If the field is now reversed, the momentum decreases uniformly in time, and eventually goes negative. The velocity accordingly diminishes, goes through zero and then reverses.

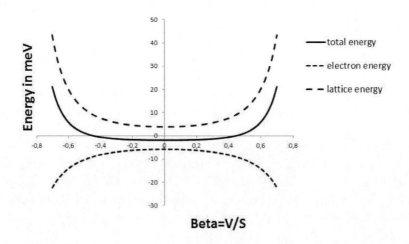

Figure 1. Electron energy, lattice energy and total energy, in meV against $V/S.kT$ is 25 meV.

In anticipation of Section 4 it is considered an ensemble of solectrons under a conducting source or drain. It is imagined that solectrons of a range of energies around kT will leave these regions. Their subsequent motion in response to fields can be determined from the above considerations. In particular, a field opposing the motion can stop and then return a solectron to the drain.

The plotted values of momentum are helpfully related to the Brillouin zone boundary wave vector K_{BZ} and momentum P_{BZ}, which are given by

$$P_{BZ} = \hbar K_{BZ} = \hbar(\pi/a) = 0.00420 (meV)(s/m) \tag{20}$$

In Figure 2 it is seen that a Solectron of thermal energy kT has a momentum of about 0.012 (meV)(s/m) which is about 2.9 P_{BZ}. Thus the solectron K value lies well beyond the first Brillouin zone boundary. This in turn gives a de Broglie wavelength $\lambda = 0.69a$. The wavelength is less than the unit cell size.

Consider now the width of the solectron, which can be taken as $(2/\gamma_a)$ unit cells. For a solectron at rest this gives a width of 20.9 unit cells. For a solectron of thermal energy and velocity 0.69S the width shrinks to 10.7 unit cells. The large size of the solectron compared to the unit cell size is a justification for using the continuum limit approximation in deriving all the results. The size of the solectron is much greater than its de Broglie wavelength. Thus it behaves as a massive ballistic large particle and most unlikely to scatter at Brillouin zone boundaries.

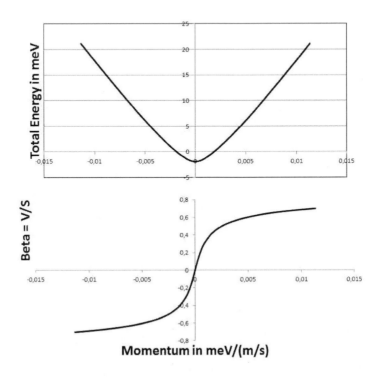

Figure 2. Solectron total energy in meV, and velocity V/S, against solectron momentum, in (meV)(s/m).

3. NONLINEAR ELASTICITY, LATTICE SOLITONS AND ELECTRON SURFING

3.1. Potential Interactions. Going beyond the Harmonic Potential

In the preceding section, the interaction potential as lattice intersite interaction used to construct the theory is the harmonic one, as expressed in the second term of the left hand side of (1). We have shown how far we can go in the consideration of a generalization of the polaron concept by introducing the concept of soliton acoustic wave polaron (SWAP). The latter came as consequence of the continuum approximation to the originally discrete lattice problem after the electron-lattice interaction was introduced. Now we consider a further albeit different extension –still in 1d- obtained by replacing the harmonic with the Morse potential which is anharmonic. However, at variance with the above described SWAP approach we shall strictly be retaining now the discrete character of the lattice throughout and we shall not consider the addition of the electron and hence no electron-lattice interaction in a first step. The Morse lattice intersite potential is

$$V(r) = (a/2b)(e^{-2br} - 2e^{-br}) = D(e^{-2br} - 2e^{-br}) \qquad (21)$$

where, for later convenience, here r denote interatomic distance, D is the potential depth (or dissociation energy; $D = a/2b$, with a and b here denoting, respectively, the linear Hooke elastic constant (corresponding to the factor MS^2/a^2 in Eq. 1) and the (anharmonic) stiffness of the intersite "springs" (Figure 3). The lowest order hence harmonic vibration is $\omega_M = (ab/M)^{1/2} = (2Db^2/M)^{1/2}$. Referring to relative intersite motions we set $q_n = b(u_n - \sigma)$, with σ here indicating the equilibrium value. To be recalled is that in 1d the kinetic energy is twice the potential energy.

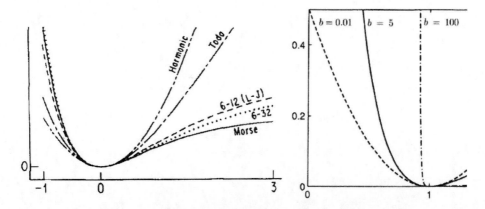

Figure 3. Interatomic potentials. Left panel: Harmonic, Toda [32], (6-12) and (6-32) Lennard-Jones and Morse interactions. Right panel: stiffness of the Morse potential from harmonic (b going to cero; $b = 0.01$) to "hard sphere" (b going to infinity; $b = 100$) interactions. Noteworthy is the lack of validity of both the harmonic and Toda interactions when dealing with certain material characteristics besides having an unphysical attractive component as the equilibrium lattice interatomic distance grows.

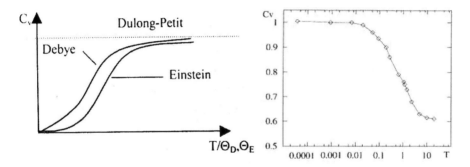

Figure 4. Specific heat C_v, at constant volume/length against absolute temperature T, in dimensionless units (rescaled with the Morse well depth). Left panel: sketch of the predictions of Einstein and Debye theories, at low temperatures, ending in the classical Dulong-Petit plateau at high enough temperatures (curves rescaled with their corresponding Debye and Einstein temperatures). Right panel: computer results for a Morse lattice of the Dulong-Petit plateau and its continuation as the temperature increases on the way to melting. It is along the slopped region (0.1-1) where solitons in such crystal lattice are of significance as also indicated by the dynamical structure factor depicted in Figure 5.

Noteworthy is that if rather than the Morse potential we use the Toda potential (Figure 3, left panel), the dynamical lattice system is integrable, that is, we explicitly know its exact analytical solution [32]. For the non integrable case with Morse interaction the solution has been obtained numerically and the results are within ten percent error relative to the exact Toda solutions [33]. The lattice solitons, for both the Toda and the Morse lattice intersite potentials, are predicted to be of significance in the slopped region of the specific heat (at constant length), against temperature, past the Dulong-Petit plateau (Figure 4, right panel). This result is confirmed by the computation of the dynamic structure factor (space-time correlation of density fluctuations in thecrystal) (Figure 6), a typical quantity for experiments with thermal neutron scattering.

As noted above, the interaction potential (21) refers to change in relative intersite distances and hence longitudinal motions. Noteworthy is that the Morse potential is a "soft" potential in the sense that when widening the intersite/atomic separation it is of weaker strength than the linear Hooke/harmonic spring, and hence the frequency of small amplitude oscillations around the minimum decreases with increasing amplitude. The converse case is considered as a "hard" potential. It turns out that discreteness in a lattice provides bounds and gaps to the spectrum of linear oscillations whereas nonlinearity makes the amplitude of oscillation frequency-dependent. The combination of discreteness and nonlinearity has led to the finding of a form of local excitation denoted discrete breather (DB), otherwise said intrinsic localized mode (ILM), a surprising finding as the lattice might be quite perfect defectless [71]. DB could be mobile, subsonic or otherwise, pinned or even impossible depending on parameter values in the potentials. Inside the phonon band they are not permitted since any resonance of their harmonics with the extended phonons will radiate them away. They can only appear below or above the phonon band. Accordingly, they cannot decay by emitting linear radiation/phonons. This is at variance with lattice solitons, corresponding to atomic lattice longitudinal elongations, which are generally supersonically moving localized excitations like those exhibited by the Toda or Morse lattices as we show below. If to longitudinal elongations we add onsite vibrations and, accordingly we add an onsite potential at each lattice site then the picture becomes rich of possibilities. For instance localized modes could be pinned or mobile depending not only on parameter values in the absolute sense but also on the ratio of the stiffness and depth of the intersite and onsite potentials involved in the dynamics. DB/ILM is to be expected when strong onsite potentials compete with weak intersite interactions. In this context it shoud be mentioned the polaron-breather concept as another extension of the original polaron idea. Two particular cases exist: a standard static polaron and a localized periodic lattice oscillation associated with a localized quasi-periodic electronic oscillation [72].

Figure 6 illustrates the supersonic traveling soliton found when the onsite potentials are switched off, whereas figure 7 illustrates the extreme opposite case when the intersite potentials are practically switched off and a pinned DB/ILM appears. The space-time evolution of localized excitations between these two extreme cases has been provided in ref. [73].

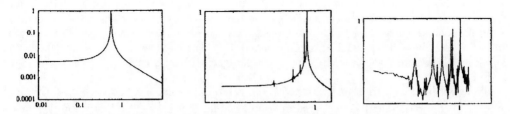

Figure 5. Morse lattice. Computed dynamic structure factor (suitably rescaled for illustrative purposes) against frequency (both in dimensionless units) with increasing temperature, respectively, below (left panel) at (center panel) and above (right panel) the expected transition, along the slopped region (0.1-1) in Figure 4, We can observe the transition from phonon to a messy background where the soliton emerges as the highest peak at extreme right. The value "1" along the abscissa indicates the dimensionless sound velocity and hence the soliton appears with supersonic speed.

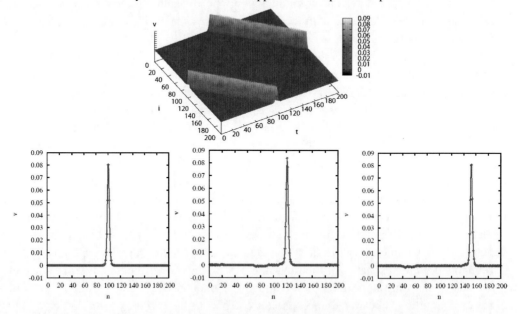

Figure 6. Morse lattice. Typical soliton in a 1d lattice with only intersite Morse interactions. The upper panel shows the actual motion along a lattice of two hundred units during two hundred time units (made dimensionless using as scale the inverse of the harmonic vibrations around the minimum of the Morse potential well). The lower panels illustrate, from left to right, the velocity v, as a defining quantity at time instants t = 10, 20 and 50, in appropriate dimensionless units. Motion is supersonic with value 1.05 the sound velocity. (Adapted from [73] with the permission of EDP Sciences).

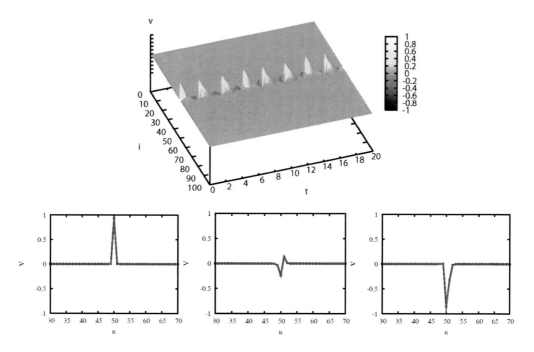

Figure 7. Morse lattice. Typical DB/ILM in a 1d lattice with strongly dominating onsite Morse interactions. The upper panel shows the actual motionless character of the local excitation. The lower panels show how there is local periodic alternance between a maximum and a minimum as time proceeds, for time instants t = 0.1, 1.2 and 2 in appropriate dimensionless units. (Adapted from [73] with the permission of EDP Sciences).

3.2. Electron Surfing on Nanoscale Lattice Solitons

Now we consider the role of the Morse interactions, and hence the role of lattice solitons, on the transport of electrons. Thus, we introduce an additional excess electron into the lattice crystal system. Figure 8 qualitatively shows the electric polarization field consequence of a local compression along the crystal lattice as horizontal axis, denoted y. In the reference frame of the atoms, this lattice deformation or distortion as it moves appears as a soliton correlated to a polarization wave (polarization and lattice distortion can be used to account for the same conceptual consequence). There is a "single" deepest well/trough whose depth depends on the actual local lattice compression (stronger in the left panel relative to the right panel). Otherwise all the troughs would have the same depth and would be placed one at each atom site. If atoms were considered moving (c_{ion} to the right) then the solitonic wave moves (c_{wave}) in the opposite direction to that of the overall atomic motions. Such a behavior is quite the same as that of the solitonic bore wave (hydraulic jump) in a river which travels upstream while the water flows downstream towards, say, the sea. Here the added, excess electron would fall in the deepest trough and, only in it, would be moving bound to it as a surfing motion.

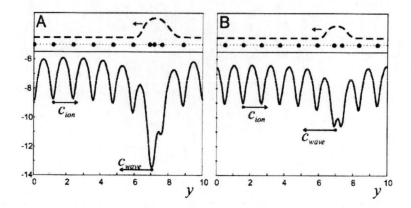

Figure 8. Morse lattice. The upper portion in the left panel corresponds to three atoms compressed to being quite near to each other whereas in the right panel the corresponding compression is weaker and affects only a couple of atoms. The middle horizontal rows show the atoms along the lattice assumed to move left to right with velocity c_{ion}. The lower portions of the panels depict the corresponding traveling polarization field in the form of solitonic wave, moving right to left, with velocity c_{wave}. These two velocities are here inserted to clearly illustrate the fact the solitonic wave travels opposite to the given motion of the atoms in the crystal lattice.

To substantiate the above sketched heuristic argument, on the one hand we replace in Eq. (1) the lattice intersite potential and on the other hand to recall the role of onsite energies/potentials here we replace the third term of (1) by the more complete electron contribution

$$H_e = \Sigma_n \varepsilon_n(q_n) c_n c^*_n - \Sigma_n t_{n+1,n}(q_n)(c^*_{n+1} c_n + c^*_n c_{n+1}) \tag{22}$$

with $\varepsilon_n = \varepsilon^o_n + \chi_0 q_n + \chi_1 (q_{n+1} - q_{n-1}) + \ldots$ and here

$$t_{n,n-1} = t_o exp[-\alpha(q_n - q_{n-1})] \tag{23}$$

where, following Slater [23, 24], the matrix elements (overlapping or hopping integrals between nearest-neighbors) are taken in exponential form. Clearly, Taylor expanding (23) and (5) in the first approximation we get the corresponding component of the SSH Hamiltonian (1)-(2). The parameter α regulates how much the $t_{n,m}$ are influenced by the corresponding intersite separations, $(q_n - q_{n-1})$. In the simplest case it is enough to account for the compound product αt_0 as in (1). Note the different units of α in Eqs. (2) and (23).

To make the description of universal value we rescale all quantities using new units. Thus we set $t = t_o/2D$, $\tau = t_o/\hbar\omega_M \sim 10\text{-}20$ and $\varepsilon_o = \varepsilon^o_n/\hbar\omega_M$. Accordingly, in terms of energy we have:

$$\varepsilon_e = [\varepsilon^o_n + \chi_0 q_n + \chi_1 (q_{n+1} - q_{n-1})]|c_n|^2 - 2\tau \Sigma_n Re(c_{n+1} c^*_n) exp\alpha(q_n - q_{n+1}) \tag{24}$$

Then from the Hamiltonian now incorporating Morse lattice intersite interactions and the electron-lattice interactions the following equations of motion are obtained:

Anharmonicity, Soliton-Assisted Transport and Electron Surfing ...   211

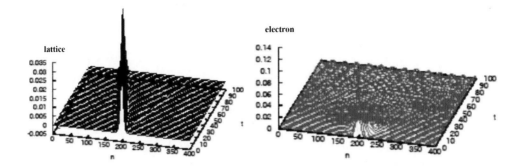

Figure 9. Morse lattice. Left panel: Lattice soliton evolution according to Eq. (25) with $\alpha = t_0 = 0$. Right panel: Evolution of an electron as provided by Eqs. (26) when also $\alpha = t_0 = 0$. Initially both are localized at site 200. As time proceeds, the soliton evolves almost unaltered whereas the electron peaked at the initial time "uniformly" spreads all over the lattice in accordance with the lattice discretized Schrödinger equations (26). (Adapted from [54] with the permission of AIP Publishing).

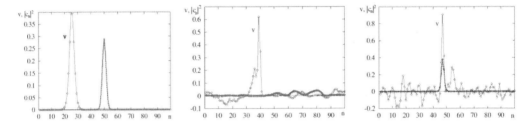

Figure 10. Morse lattice. Extreme left and center panels: Initial and final outcome of the evolution of the lattice soliton and the "excess albeit free" electron as depicted in the preceding Figure 9, when $\alpha = t_0 = 0$. Extreme right panel: Once the final states are those of the center panel, the electron-lattice interaction is switched on and hence when α and t_0 are nonvanishing, the soliton via Eqs. (25) and (26) dynamically gathers the tiny (practically) uniform electron probability density (satisfying the lattice discretized Schrödinger equations) piling it up to a peak around itself (kind of vacuum cleaner process) and then travels as a solectron bound state. (Adapted from [54] with the permission of AIP Publishing).

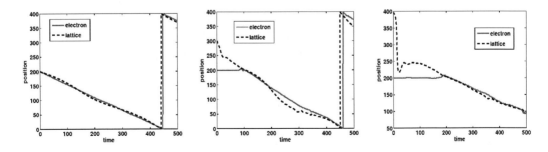

Figure 11. Morse lattice. Sketch of the space-time trajectories of the soliton and the electron. When α and t_0 are nonvanishing, when the former is placed at sites 200, 300 and 400 whereas the latter is always localized at site 200 at the initial time. Following a short transient the soliton manages to catch the electron (in probability density) thus forming the solectron bound state. Needless to say there is also a polaron process which we do not consider here for simplicity. (Adapted from [54] with the permission of AIP Publishing).

$$d^2q_n/dt^2 = [1-exp(q_n-q_{n+1})]exp(q_n-q_{n+1})-[1-exp(q_{n-1}-q_n)]exp(q_{n-1}-q_n)+ 2\alpha t_0[Re(c_{n+1}*c_n)$$
$$exp\alpha(q_n-q_{n+1})-Re(c_nc*_{n-1})exp\alpha(q_{n-1}-q_n)]+(t/\tau)[-\chi_o|c_n|^2+\chi_1(|c_{n+1}|^2-|c_{n-1}|^2)] \tag{25}$$

$$i(dc_n/dt) = [\mathcal{E}^o_n+\chi_oq_n+\chi_1(q_{n+1}-q_{n-1})]c_n-\tau[c_{n+1}exp\alpha(q_n-q_{n+1})+c_{n1}exp\alpha(q_{n-1}-q_n)]-(n-n_{el})Ec_n \tag{26}$$

The latter is the lattice discretized Schrödinger equation. For completeness we have added an external electric field, with E and $\hat{E}$ such that $E = (h\omega_M/\sigma e)\hat{E}(V/m)$. It seems clear that external electric field is capable of altering the lattice dynamics and hence affecting acoustic excitations (waves, phonons, solitons) in the system.

The equations (25) and (26) are a mixed classical-quantum system that is to be integrated using appropriate periodic boundary conditions, $q_{N+1} = q_1$ and $dq_{N+1}/dt = dq_1/dt$, and initial conditions, recalling the conservative character of the dynamics and the probability density constraint for the electron $\Sigma_n |c_n|^2 = 1$. To a first approximation we can restrict to $\chi_o = \chi_1 = 0$ and $\mathcal{E}^o_n$ can be eliminated by rescaling the reference energy level. Then following a short transient quickly the electron of (26) (in probability density) is trapped by the soliton wave of (25) thus forming a bound state which is the solectron. As for the SWAP case discussed in the preceding section, this solectron is indeed a further generalization of the polaron concept (and charge carrier). Here, however, we have the electron surfing on the mechanical, acoustic lattice soliton wave that exists alien to the presence of the added excess electron. Clearly, this is at variance with both the polaron and the SWAP concepts where it is the electron that selftraps, thus creating the new charge carrier; both are electronically induced crystal lattice excitations. In the present case one expects indeed the action of the two influences, polaronic and solitonic but we shall not dwell on this matter here, referring the reader to [57].

Figure 9 illustrates how a soliton and an electron evolve separately in a Morse crystal lattice. Figures 10 and 11 offer alternative views. Noteworthy is that if the Schrödinger equation is integrated separately alone $(\alpha = t_0 = 0)$, as time proceeds the electron "diffusively" becomes delocalized all over the lattice ring (Figure 10, center panel). Then when, after the delocalization is established, α and t_0 are made non zero hence switching on the electron lattice/soliton/phonon interaction, soon the electron becomes trapped by the lattice soliton as the latter forces reconstruction of a peaked electron probability density around itself in a kind of "vacuum cleaner" process (Figure 10 extreme right panel). We have a clear case of mechanical control of electros at the nanolevel [75].

3.3. Consequence of Electron Surfing

Figure 12 gives the results of a computer simulation to explain the earlier mentioned experimental findings of Donovan and Wilson [46-49]. It considers a 400 unit one-dimensional lattice and applies Eqs. (25) and (26). As an initial condition at $t = 0$ is taken the known analytical form of the soliton of the Toda lattice $exp[3(q_n-q_{n+1})] = 1+\beta^2/\cosh^2(\kappa n-\beta t)$, with the corresponding formula for its velocity; $\beta = \sinh\kappa/\kappa$ and κ is its mid-height inverse width (in units $1/\sigma$). Further, at $t = 0$ the electron probability density (normalized to unity, $\Sigma_n|c_n|^2=1$ is taken as a Gaussian distribution, centered at site n_{el}, with a width σ_{el}, which is approximately that of the chosen soliton initial condition; in practical terms $\sigma_{el} = 3$ and $\tau = 10$ in all the computations. For simplicity we set $\chi_0 = \chi_1 = \varepsilon^0{}_n = 0$. Other values used are $M \sim 10^{-22}$g, $D \sim 0.03$-0.3eV, $b \sim 2$-5Å^{-1}, $t_0 < 1$eV, $\omega_M \sim 5.10^{12}$ s^{-1} and $\omega_e \sim 10^{14}$s^{-1}. Also for simplicity, the soliton and the electron are placed at lattice site 100 to rule out boundary effects. This is not a limitation since, as earlier illustrated, wherever the electron is placed along the lattice, the soliton placed in the same or different site is always able to trap it.

In accordance with the theory presented above, after a short transient, there is the formation of a bound state of the soliton with the electron which is the solectron. The solectron velocity (in units of sound velocity) is estimated as the slope of its trajectory in the space-time (n, t) plot. Figure 12 (left panel) depicts the solectron velocity v_s against the compound parameter αt_0 (in Eq. 25) accounting for the electron-lattice interaction. As expected, the latter affects the soliton with, however, supersonic values in a significant range $(0 < \alpha t_0 < 0.3)$. In Figure 12 (right panel) there is a domain up to, say, $E = 0.1$ where the drift velocity is field independent for quite a wide range of values of the field strength. On the other hand, in the subsonic case the drift velocity grows with the increasing field value reaching saturation at the sound velocity. That panel also shows two significant consequences. First, as the field strength goes down, an ultrahigh mobility is expected, in agreement with the earlier mentioned experimental results found by Donovan and Wilson [46-49]. Second, the electric field appears as a (left-right) symmetry-breaking agent. Indeed, in general, as motion is always expected even in the absence of an external electric field, once the soliton is excited, as solitons, and hence the solectrons, can move to the left or to the right, on the average no net conduction would be the outcome. Switching on the field breaks the symmetry thus allowing one or the other to be realized.

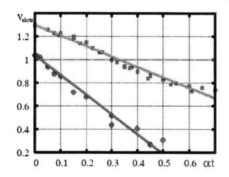 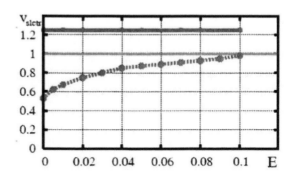

Figure 12. PDA. Left panel: solectron velocity, v_{slc} against the electron-lattice interaction parameter αt, for $\kappa = 0.5$ with initial $v_{soliton} = 1.04$ (pink/bottom line) and for $\kappa = 1.3.5$ with initial $v_{soliton} = 1.3$ (green/upper line). Dots are computer results and straight lines approximate fits for illustration, $v_{slc} = 1.3 - 0.9\alpha t$ (upper line) and $v_{slc} = 1.04 - 1.7\alpha t$ (lower lines). Right panel: dimensionless subsonic (blue/lower curve) and supersonic (red/upper line) solectron velocity against dimensionless field strength. The green/horizontal line at the value unity corresponds to the (linear) sound velocity. For illustration $E = 0.1$ corresponds to $E = 10^6$V/m. (Adapted from [74] with the permission of EDP Sciences).

Finally, note that the solectron transport theory outlined in subsections 3.2 and 3.3 offer universality. Indeed, besides the application to PDA, it can be used for other materials whose crystal vibrations permit strong enough compressions capable of exhibiting soliton excitations. Let us insist that it is a clear case of mechanical control of electrons at the nano-scale [75].

4. A NOVEL FIELD EFFECT TRANSISTOR AS A PRACTICAL APPLICATION OF ELECTRON SURFING

Worth noting is that if both the expression of the Morse potential (21) and that of the Slater's overlapping integral (23) are taken in appropriate approximations, then we go to the Davydov's approach or to the SSH description (save the degeneracy of the ground state) thus leading to Wilson's SWAP theory. One way or the other we have a soliton assisted (solectron) charge transport. Accordingly, it is of interest to consider the possibility of using the propagation properties of the solectron in novel new devices, which may complement devices using the propagation of electrons. We now describe the design of a new field effect transistor using the solectron properties [SFET-Solectron Field Effect Transistor and Inverter, GB2533105 A- 15.06.2016 UK patent pending; MG Velarde and EG Wilson, co-inventors], and compare its properties to the existing silicon field effect transistors (SiFET).

The proposed SFET uses an insulator, which if injected with an electron, forms a solectron, which propagates at around the sound velocity, in a preferred direction, set by the insulator properties. The source to drain channel lies along this preferred direction. The distance of ballistic flight of the solectron is greater than the channel length. Thus, there is severe constraint in material suitable for the SFET (as there is, of course, for the SiFET). As expected from the theory and arguments presented in the preceding sections possible candidates fulfilling these restrictions are the polydiacetylene (PDA) family of polymer perfect single crystals.

In Figure 13 metallic source and drain electrodes are evaporated onto the surface of the solectron supporting crystal, using appropriate masks to create the desired metal shapes. The masks are orientation relative to the crystal such that the solectron preferred propagation direction is oriented along the channel from source to drain. An insulator layer is then evaporated or spin coated. Finally the gate metal is evaporated. The source, drain and gate metals are identical and chosen to have a lower work function than the crystal electron affinity. Thus, the source and drain metals, in thermal equilibrium, inject electrons into the conduction band of the crystal, which then form solectrons.

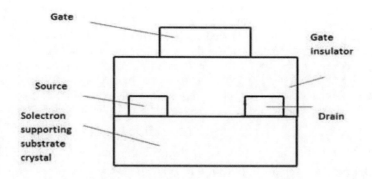

Figure 13. The source, drain, and gate are the conductors of the three terminal transistor.

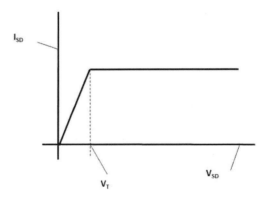

Figure 14. The source to drain current I_{SD}, against the source to drain voltage V_{SD}, at constant gate to source voltage.

Figure 14 shows a graph of the source–drain current I_{SD}, against the source – drain voltage V_{SD}. The voltage V_T is $(kT/e) = 25$ mV. For $V_{SD} > V_T$ the current is carried by solectrons moving from source to drain at constant velocity. The voltage along the channel does not change. The solectron density along the channel does not change. For $V_{SD} < V_T$ the current falls, as solectrons of thermal energy under the drain can enter the channel and reach the source. Continuing increase of $V_{SD} > V_T$ does not change the channel voltage; or current; rather, a voltage fall, in a small length adjacent to the drain is increased in value.

Figure 15 shows a graph of the source – drain current, I_{SD} against the source–gate voltage, V_{SG}. Solectrons under the source of thermal energies kT require the gate voltage to be $-V_T$ to stop them entering the channel. However, due to the distribution of energies of solectrons, about the average of kT, the current will not be abruptly zero as V_{SG} is reduced. It is important to stress that the source to drain, D, current, I_{SD}, is carried by injected charge from the source. This injected charge is not in thermal equilibrium with the insulator electron states or with insulator dopant electron states. This is in contrast to the SiFET where the channel charge is in local thermal equilibrium, such charge residing in dopant electron states or the conduction band.

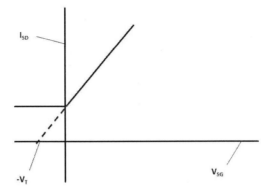

Figure 15. The source-drain current I_{SD} against the source-gate voltage V_{SD} at constant source drain voltage.

The limit to the response time of the SFET is the transit time, T, from source to drain. For example, for a solectron velocity of $v = 2 \times 10^3$ m/s, and a source–drain distance of $L = 0.05$ μm, then

$$T = L/v = 0.025 \ ns \qquad (27)$$

The capacity between gate and channel, C, depends on the gate area, which for a SFET of width W cannot be smaller than WL. C also depends on the dielectric constant of the layer between gate and channel, ε, and the thickness, d, of the insulator layer according to

$$C = \varepsilon\varepsilon_0 \ WL/d \qquad (28)$$

The distance d is constrained only by being of order L or less. This is in contrast with the SiFET, which requires field from the gate to enter the semiconductor in order to change the local thermal equilibrium, and hence carrier densities. It the gate insulator is too thick, insufficient field enters the semiconductor to effect such change. Thus, the gate to channel insulator capacity, C, can be much smaller in the SFET than the SiFET.

The transfer conductance G, is defined here as the change of source - drain current in response to change of the gate – source voltage, e.g.,

$$G = (dI_{SD}/dV_{SG}) = C/T = \varepsilon\varepsilon_0 \ v \ (W/d) \qquad (29)$$

Thus, G is dependent on only W and d and independent of L.

Table 2. Comparison of a silicon FET (SiFET) with a solectron FET (SFET). (a) Typical parameters for state of the art Intel Pentium chip (b) Choose same width and length as SiFET (c) Choose insulator thickness to equal source-drain length (d) The saturated velocity in n silicon (e) The solectron velocity (f) Choose same clock frequency and period as SiFET (g) The minimum usable clock period is limited by factors external to the SiFET

	SiFET (a)	SFET
source-drain length L/μm	0.05	0.05 (b)
width w/μm	0.2	0.2 (b)
insulator thickness/nm	2	50 (c)
Insulator dielectric constant ε	3.9 (SiO_2)	2
gate-channel capacity C/F	1.73×10^{-16}	3.54×10^{-18}
supply voltage VDD/volts	1	0.1
charge in channel Q/C	1.73×10^{-16}	3.54×10^{-19}
carrier velocity/ms^{-1}	6×10^4(d)	2×10^3 (e)
transit time T/ns	0.00083	0.025
channel current I/μA	208	0.0142
switch energy CV^2/Joules	1.73×10^{-16}	3.54×10^{-20}
switch energy CV^2/eV	1080	0.22
Clock frequency/GHz	3	3 (f)
Clock period/ns	0.33(g)	0.33 (f)

The advantages of the SFET are twofold.

First, the source drain voltage, V, can be small, of order *(kT/e),* and the SFET still functions. This is because of the ballistic flight of the solectron; a field is not required along the channel to keep it moving. This is in contrast to the SiFET.

Second, the gate insulator thickness can be large, and yet small gate voltages can be used. This allows of a smaller gate to channel capacity, C.

These two factors are critical in the device energy consumption. The energy cost of a switch from ON to OFF in a FET as a component in an inverter, or in a Boolean logic gate, is CV^2. Thus, there are enormous gains in energy cost of computing to be gained due to reducing these two factors.

Table 2 displays the effect of these two factors when a numerical comparison is made between the silicon and the solectron FET. To affect the comparison a typical Intel Pentium chip SiFET is chosen, and compared to the SFET of the same width and channel length. It is important to note the larger gate insulator thickness and the smaller source drain voltage for the SFET, and the consequent reduction in the energy of switching.

CONCLUSION

In the preceding sections, we have recalled theoretical results concerning the evolution of a one-dimensional crystal lattice with an added excess electron when there is significant lattice deformation affecting the electronic structure of the crystal, which in turn drastically affects the electron states as a consequence of the electron-lattice/soliton/phonon interaction. This occurs either when the added electron forces the lattice deformation as in the original idea of the polaron concept, defined by Landau and Pekar and subsequent developers of the theory, or when due to a preexisting lattice deformation the electron is trapped in the already deformed electric polarization field associated to such lattice distortion. The standard polaron case occurs for crystal lattices with harmonic intersite interaction potentials. The case of intersite anharmonic interactions, like the Morse potential, offers the novelty of permitting the preexistence of a significant lattice deformation in the form of, e.g., lattice solitons that may appear as supersonic excitations although no added excess electron has been introduced. When in this other case the electron is added, the electron-lattice interaction leads to the generalization of polaron concept in the form of solectron, the latter due to the solitonic character of the crystal excitation. In such a dynamic situation there appears the possibility of field-free electron transport in certain range of parameter values. As a practical consequence of such form of transport, electron surfing with the soliton, there has been the invention of a solectron field effect transistor (SFET).

The claims of the invention are few and straightforward, with universality and not just limited to the above presented particular illustration using PDA and derivatives: 1) A field effect transistor in which the current carrier in the source-drain channel is an electron solectron which is a coherent particle formed from an electron and accompanying crystal lattice distortion. 2) A field effect transistor in which the current carrier in the source-drain channel is a hole solectron which is a coherent particle formed from a hole and accompanying crystal lattice distortion. And 3) an inverter constructed from the field effect transistors just mentioned.

The inverter is predicted to operate at very low supply voltages. So also can circuits that rely on the inverter such as the Boolean logic gates used in computers. The gate–channel insulator thickness can be large, in comparison to the silicon field effect transistor (SiFET). For in the latter the insulator thickness is constrained to be increasingly small as both the supply voltage and hence the threshold voltage, are reduced. A larger insulator thickness is desirable as it leads to a smaller gate input capacity, and so less energy consumption on switching. It can operate at quite small currents, leading to rather low quiescent energy consumption. As the inverter is constructed from two SFET the energy cost of an inverter switch scales proportionally to the square of the supply voltage, and inversely proportional to the gate insulator thickness. Thus, in view of the above much lower energy consumption is predicted using SFET compared to current silicon technology.

Computation currently consumes many orders of magnitude more of energy in digital switching events than is theoretically necessary according to statistical physics [76, 77]; but the technology to reduce the consumption does not exist. The SFET offers a radically new idea to produce computer elements that enable switching with three orders a magnitude reduction in energy consumption. Its production will lead the way to substantial energy reductions by IT industries in use of digital computers, server farms, and smart phones.

REFERENCES

[1]   Landau, L. D., 1933. *Phys. Z. Sowjetunion* 3, 664.

[2]   Pekar, S. I. 1946., *J. Phys. USSR* 10, 341.

[3]   Pekar, S. I. 1946., *Zh. Eksp. Teor. Fiz./Soviet Phys. JETP* 16, 341; 1948. *Ibidem* 18, 105.

[4]   Landau, L. D., Pekar, S. I., 1948. *Zh. Eksp. Teor. Fiz./Soviet Phys. JETP* 18, 419.

[5]   Pekar, S. I., 1954. *Untersuchungen über die Elektronentheorie* [*Investigations on the theory of electrons*], Berlin, Akademie Verlag (and references therein) [German translation from Russian original edition dated 1951].

[6]   Rashba, E. I., in [7] (and references therein).

[7]   Alexandrov, A. S., ed.2007. *Polarons in Advanced Materials*, Dordrecht, Springer (and references therein).

[8]   Fröhlich, H. Pelzer, H., Zienau, S., 1950. *Phil. Mag.* 41, 221.

[9]   Fröhlich, H., 1954. *Adv. Phys.* 3, 325 (and references therein).

[10]   Holstein, T., 1959. *Ann. Phys.* 8, 325, 343 [reprinted 2000. *Ibidem* 281, 706, 725].

[11]   Feynmann, R. P. 1955. *Phys. Rev.* 97, 660.

[12]   Firsov, Yu. A., 1975. *Polarons*, Moscow, Nauka (in Russian).

[13]   Kaganov, M. I., Lifshitz, I. M., 1979. *Quasiparticles. Ideas and principles of Solid State Quantum Physics*, Moscow, Mir.

[14]   Devreese, J. T. L., ed. 1972. *Polarons in Ionic Crystal and Polar Semiconductors*, Amsterdam, North-Holland (and references therein).

[15]   Alexandrov, A. S., Mott, N. F., 1996. *Polarons and Bipolarons*, Singapore, World Scientific.

[16] Yu. L., ed. 1992. *Solitons and Polarons in Conducting Polymers*, Singapore, World Scientific.

[17] Emin, D., 2013. *Polarons,* Cambridge, Cambridge Univ. Press.

[18] Wannier, G. H., 1959. *Elements of Solid State Theory*, Cambridge, Cambridge Univ. Press.

[19] Peierls, R. E., 1965. *Quantum Theory of Solids,* Oxford, Clarendon Press.

[20] Harrison, W. A., 1970. *Solid State Theory*, New York, Mc-Graw-Hill.

[21] Ascroft, N. W., N. Mermin, N. D., 1976. *Solid State Physics*, New York, Thomson Learning.

[22] Kittel, C., 2005. *Introduction to Solid State Physics*, 8th Ed., New York, Wiley.

[23] Slater, J. C., 1974. *Quantum Theory of Molecules and Solids*, vol. 4, New York, McGraw-Hill.

[24] Launay, J. P., Verdaguer, M., 2014. *Electrons in Molecules. From Basic Principles to Molecular Electronics*, Oxford, Oxford Univ. Press.

[25] Su, W. P., Schrieffer, J. R., Heeger, A. J., 1979. *Phys. Rev. Lett.*42, 1698.

[26] Heeger, A. J., Kivelson, S., Schrieffer, J. R., Su, W. P., 1988. *Rev. Mod. Phys.*60, 781 (and references therein).

[27] Davydov, A. S., 1973. *J. Theoret. Biol.* 38, 559.

[28] Davydov, A. S., 1982. *Usp. Fiz. Nauk*138, 603 [1982. *Sov. Phys. Usp.* 25, 898] (and references therein).

[29] Davydov, A. S., 1991. *Solitons in Molecular Systems*, 2nd Ed., Dordrecht, Kluwer (and references therein).

[30] Zabusky, N. J., Kruskal, M. D., 1965. *Phys. Rev. Lett.*15, 57.

[31] Toda, M., 1967. *J. Phys. Soc. Japan* 22, 432; 23, 501.

[32] Toda, M., 1989. *Theory of Nonlinear Lattices*, 2nd Ed., Berlin, Springer.

[33] Chetverikov, A. P., Ebeling, W., Velarde, M. G., 2006. *Int. J. Bifurc. Chaos* 16,1613.

[34] Lomdahl, P. S., Kerr, W. C., 1985. *Phys. Rev. Lett.*55, 1238.

[35] Christiansen, P. L., Scott, A. C. eds., 1990. *Davydov's Soliton Revisited. Self-trapping of Vibrational Energy in Protein*, New York, Plenum.

[36] Scott, A. C., 1992. *Phys. Rep.* 217, 1.

[37] Cruzeiro-Hansson, L., Takeno, S., 1997. *Phys. Rev. E***56**, 894.

[38] Davydov, A. S., 1979. *Teor. Mat. Fiz.*40, 408 (in Russian).

[39] Davydov, A. S., Zolotaryuk, A. V., 1984. *Phys. Scr.* 30, 426.

[40] Zolotaryuk, A. V., Spatschek, K. H., Savin, A. V., 1995. *Europhys. Lett.*31, 531.

[41] Zolotaryuk, A. V., Spatschek, K. H., Savin, A. V., 1996. *Phys. Rev. B*54,266 (and references therein).

[42] Zmuidzinas, J. S., 1978. *Phys. Rev. B*17, 3919.

[43] Kosevich, Yu. A., 2017. *J. Phys. C: Conf. Series*833,012021.

[44] Kadantsev, V. N., Lupichev, L. N., 1990. *Phys. status solidi B* 161, 769.

[45] Lupichev, L. N., Savin, A. V., Kadantsev, V. N., 2015. *Synergetics of Molecular Systems,* Heidelberg, Springer.

[46] Donovan, K. J., Wilson, E. G., 1981. *Phil. Mag. B*44, 9; 31..

[47] Donovan, K. J., Freeman, P. D., Wilson, E. G., 1985. *J. Phys. C: Solid State Phys.* 18, L275.

[48] Donovan, K. J., Wilson, E. G., 1990, *J. Phys.: Condens. Matter* 2, 1659.

[49] Donovan, K. J., Elkins, J. W. P., Wilson, E. G., 1991. *J. Phys.: Condens. Matter* 3, 2075.

[50] Wilson, E. G., 1983. *J. Physics C: Solid State Phys.* 16, 6739.

[51] Velarde, M. G., Ebeling, W., Chetverikov, A. P., 2005. *Int. J. Bifurc. Chaos* 15, 245.

[52] Velarde, M. G., Ebeling, W., Hennig, D., Neissner, C., 2006. *Int. J. Bifurc. Chaos*, 16, 1035.

[53] Hennig, D., Neissner, C., Velarde, M. G., Ebeling, W., 2006. *Phys. Rev. B* 73, 024306.

[54] Hennig, D., Chetverikov, A. P., Velarde, M. G., Ebeling, W., 2007. *Phys. Rev. E* 76, 046602.

[55] Velarde, M. G., 2010. *J. Comp. Appl. Maths.* 233, 1432.

[56] Chetverikov, A. P., Ebeling, W., Velarde, M. G., 2016. *Eur. Phys. J. B* 89, 196.

[57] Cantu-Ross, O. G., Cruzeiro, L., Velarde, M. G., W. Ebeling, W., 2011. *Eur. Phys. J. B* 80, 545.

[58] Velarde, M. G., Ebeling, W., Chetverikov, A. P., 2008. *Int. J. Bifurc. Chaos* 18, 3815.

[59] Chetverikov, A. P., Ebeling, W., Velarde, M. G., 2009. *Eur. Phys. J. B* 70, 217.

[60] Chetverikov, A. P., Ebeling, W., Velarde, M. G., 2010. *Int. J. Quantum Chem.* 110, 46.

[61] Brizhik, L. S., Davydov, A. S., 1984. *J. Low Temp. Phys.* 10, 748.

[62] Brizhik, L. S., 1986. *J. Low Temp. Phys.* 12, 437.

[63] Brizhik, L. S., Eremko, A. A., 1995. *Physica D* 81, 295.

[64] Velarde, M. G., Brizhik, L. S., Chetverikov, A. P., Cruzeiro,L., Ebeling, W., Röpke, G., 2011. *Int. J. Quantum Chem.* 112, 551, 2591.

[65] Brizhik, L. S. Chetverikov, A. P. Ebeling, W., Röpke, G. Velarde, M. G., 2012. *Phys. Rev. B* 85, 245105.

[66] Kobelt, D., Paulus, E. F., 1974. *Acta Crystallogr. B* 30(1974) 232.

[67] Batchelder, D. N., Bloor, D., 1979. *J. Polymer Sci.* 17, 568.

[68] Lochner, K., Bassler, A., Sowa, H., Ahsbahs, H., 1980. *Chem. Phys.* 52, 179.

[69] Wilson, E. G., 1975. *J. Phys. C: Solid State Phys.* 8, 727.

[70] Batchelder, D. N., Bloor, D., 1978. *J. Phys. C: Solid State Phys.* 11, L629.

[71] Dmitriev, S. V., Korznikova, E. A., Baimova, Yu. A., Velarde, M. G., 2016. *Phys. Uspekhi* 59, 446.

[72] Aubry, S., 1997. *Physica D* 103, 201

[73] Velarde, M. G. Chetverikov, A. P., Ebeling, W., Dmitriev, S. V., Lakhno, V. D. 2016. *Eur. Phys. J. B* 89, 233.

[74] Velarde M. G., Chetverikov, A. P., Ebeling, W., Wilson, E. G., Donovan, K. J., 2014. *Europhys. Lett.-EPL* 106, 27004.

[75] Velarde, M. G., 2016. *Eur. Phys. J. ST* 225, 921.

[76] Landauer R., 1961. *IBM J. Res. Dev.* **5**, 183 [Reprinted 2000. *Ibidem* 44, 261].
[77] Vacarro, J., Barnett, S., 2011. *Proc. Roy. Soc. A* 467, 1770.

In: Polarons: Recent Progress and Perspectives
Editor: Amel Laref

ISBN: 978-1-53613-935-8
© 2018 Nova Science Publishers, Inc.

Chapter 6

POLARONS IN THE FUNCTIONALIZED NANOWIRES

Victor A. Lykah[1,*] *and Eugene S. Syrkin*[1,2,†]
[1]Institute of Physical Engineering,
National Technical University "Kharkiv Polytechnic Institute",
Kharkiv, Ukraine
[2]Theoretical Department,
B. I. Verkin Institute for Low Temperature Physics
and Engineering of National Academy of Sciences,
Kharkiv, Ukraine

Abstract

The self-consistent theoretical approach has been developed to describe the electronic spectra in the functionalized semiconductor nanowires (NW). Inside a perfect semiconductor quantum wire, the charge carrier is uncompensated, and the nanowire is covered with a soft molecular layer formed by the functionalizing organic molecules. In a system with the molecular functionalization, the soft degrees of freedom have been considered in the following sequence: (i) the radial deformation; (ii) the conformation; (iii) the misfit dislocations. For each molecular soft degree of freedom, the self-consistent model of the polaron has been developed. The system of equations describes longitudinal quantization of the charge carriers in NW (Schrodinger equation), the interaction of the carrier in NW and the molecular electric dipoles, the molecular soft degree of freedom and their change. It has been shown, that the functionalization modifies the electronic spectra of NW and creates the various conditions for localization or tunneling for holes and electrons. We have introduced the polarons with tunable parameters: (i) *the elastic polaron* with the soft radial degrees of freedom, (ii) *the conformational polaron*, and (iii) *the discrete polaron in incommensurate structure*. In fact, all these soft degrees of freedom coexist. The polarons can be responsible for the experimentally observed decrease in the conductivity of nanowires.

PACS: 68.65.La, 67.70.+n, 03.65.-w, 11.10.Lm, 72.20.Ht

Keywords: nanowire, functionalization, polaron, quantum mechanics, nonlinear effects

---

*E-mail address: lykahva@yahoo.com.
†E-mail address: syrkin@ilt.kharkov.ua.

1. Introduction

The first fundamental description of the polaron was given by Pekar [1]. In further research [2], the Pekar's polaron was obtained from a microscopic model of a quantum crystal. The quantum description of the polaron and the transition to the classical continual description of the medium as a soliton in molecular chain was demonstrated by Davydov [3].

One type of the polaron with a continual description of the medium arises in the theory of the phason and the fluctuon [4]. The phason is a charge carrier surrounded by a phase region with other structure that causes a higher polarizability. The fluctuon is formed by an electron localized in the region of fluctuations; as a result, in the disordered state, the concentration of one component increases.

Another type of the polaron with a continual description of the medium arises in the theory of Nagaev's ferron [5], [6]. In the theory, in an antiferromagnetic semiconductor matrix, a ferromagnetic domain is formed by the exchange interaction with an electron or a hole. In the paper [7], fluctuons were introduced in ferroelectric-ferromagnetic systems. Here, both the structure and the ferromagnetic domains play an important role simultaneously.

Almost all of the above-mentioned polaron types are described as three-dimensional polarons of large radius. Their properties are described in detail in the book [8]. In this connection it is interesting to note the theory of the two-dimensional (micro-) nanodomain polaron of intermediate radius in the ferroelectric model of high-temperature superconductors [9].

The creation of new types of nano- and mesoscopic materials and the prospects for their application in nanoelectronics generate the interest in the study of the fundamental properties of such objects [10]. Thus, the optical properties and conductivity of a nano-object are determined by the set of discrete energy levels for charge carriers [11]. The effect of the energy level structure on the conductivity was observed for metallic nanowires [12], [13] and nanotubes [14]. The influence of structure of the quantum levels on various physical properties was found in quantum nanowires consisting of intrinsic semiconductor or carbon nanotubes (CNTs). Carbon nanotubes are nanomaterials with a small diameter of about 1 nm, a length of about 1μm and the mean free path of charge carriers, which exceeds 10μm [15]. The large mean free path is an important factor for quantization along the nanowire or CNT axis [16, 17].

The carrier spectrum is extremely sensitive to the state of the contacting molecular subsystem. For example, nanowire conductance is extremely sensitive to adsorbed layers of NH_3 [18] and more complex molecules. CNT conductance is extremely sensitive to the presence of adsorbed molecules forming the Langmuir-Blodgett films [21].

Functionalization (the act of modifying the surface) is the powerful method for tuning the quantum energy levels and the physical properties of the nanowires and CNTs [19], [20]. The novel high-sensitive biosensors, electronic and optoelectronic devices are created due to functionalization. CNTs and DNA [22] or surface self-organizing layered organic structures [23] have the potential for creating computer chips.

Contact of nanotubes with a complex organic medium result to qualitatively new effects connected with other different physical properties. Thus, a small additions of nanotubes components to nematic liquid crystals leads to giant electromechanical effect [24]. A mixture of nanotubes and DNA results in formation of a structurally ordered phase with each

Polarons in the Functionalized Nanowires 225

nanotube helically "wrapped" by DNA [25]. CNTs can form a liquid crystal phase themself [26].

The nanotubes and nanowires serve as the basis for novel high-sensitive biosensors, electronic and optoelectronic devices [27]. The state of molecules plays an important role. Strong conductance variation was found in conformationally constrained molecular tunnel junctions [28]. Optically switchable device due to conformational transition in functionalized single walled carbon nanotubes was created [29]. Ferroelectric or antiferroelectric liquid crystals [30, 31] and other organic molecules containing polar groups are particularly interesting. Examples are colloids, smectics [32], [33], [34], and DNA.

The molecular interaction with carbon nanotubes and nanowires can be divided into two large groups. They are chemisorption and physisorption . The chemisorption involves a chemical reaction between the surface and the molecules. The physisorption does not change chemical and band structure of NW or CNT, interaction with adjacent medium (molecules) has physical nature (van der Waals forces). The physisorption functionalization will be considered in this chapter.

Theory of the energy spectra tuning in the semiconductor nanowire as the result of functionalization by molecular films was developed by the authors [35], [36], [37], [38]. it was possible to identify those properties of the functionalizing molecules that allow to obtain analytical description of the charge carrier in a functionalized nanowire. In [35] and [36] we considered the soft radial degree of freedom for functionalizing molecules. Theory of energy spectra tuning in the semiconductor nanowires as the result of functionalization by sufficiently thick molecular films was developed.

In [37] the conformational degree of freedom for functionalizing molecules and its action on a charge carrier were researched. For the system with misfit dislocations in functionalizing molecular layer, the properties were considered in [38]. The spectrum is extremely sensitive to the state of the molecular subsystem. We considered the effect of the interaction of the uncompensated charge carried by an electron or hole in a quantum nanowire with the neighboring medium that has low mechanical rigidity and consists of molecules having an intrinsic electric-dipole moment. The nonlinear nonlocal equations describing the system were derived. Longitudinal quantization was reduced to the spectral problem for a nonlinear Schrodinger equation [36], to a conformational domain quantization in [37] and to quantization in an functionalizing incommensurate molecular system [38]. For the charge carrier energy levels, the shift of the energy was calculated. Possibility of the carrier localization and the polaron formation were shown. Of course, in real systems, all the indicated degrees of freedom can manifest simultaneously. However, the presence of the exact solutions can create opportunity for easy analysis of real complex systems.

In the present chapter, the mentioned above models are applied to research the possibility of the polaron formation in the functionalized 1D nanosystems.

2. The Model Description

We can see real STM image of a wadding structure which is formed by the functionalized carbon nanotubes [21]. The system's cylindrical geometry is shown in Figure 1. Scales of the nanowire sizes are $\sim L \gg r_0$, R_0 and $r_0 \sim R_0$. Here $2L$ is the nanowire length, r_0 and R_0 are the nanowire and the functionalizing layer external radii.

The nanowire cylindrical geometry causes the following energy scale [16, 17, 39]: the transverse and longitudinal quantizations have characteristic energies ~1eV and ~0.001eV correspondingly. Further estimations for the longitudinal localization with the polaron formation causes energy ~(0.1÷1)eV which is comparable with the transverse quantization energies.

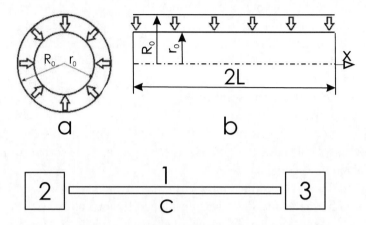

Figure 1. Cylindrical quantum nanowire with the adsorbed functionalizing molecules. Empty arrows indicate direction of the molecular electric dipole vectors. (a) View along the nanowire axis $0x$. (b) Cross-section along the nanowire axis. (c) Electrical connection of the nanowire 1 as a quantum dot, leads 2 and 3 do not have direct contact with the nanowire.

2.1. General Description of the Model

2.1.1. Schrodinger Equation

The time-independent 3D Schrodinger equation for an extra charge inside the intrinsic semiconductor quantum wire can be usually written as [11, 40]:

$$-\frac{\hbar^2}{2m_{ef}}\Delta\psi + U(\boldsymbol{r})\psi = W\psi. \qquad (1)$$

Here, $\psi \equiv \psi(\boldsymbol{r})$, m_{ef}, and $\boldsymbol{r}(x,y,z)$ are the wave function, an effective mass, and the radius vector of the carrier, $\hbar$ is Plank constant, Δ is the Laplace operator, W and $U(\boldsymbol{r})$ are the total and potential energies. In semiconductors, a deep potential well can be approximated by the infinite depth well: $U(\boldsymbol{r}) \equiv U_0(\boldsymbol{r})$, where $U_0(\boldsymbol{r}) = 0$ inside and $U_0(\boldsymbol{r}) = \infty$ outside the semiconductor nanowire.

2.1.2. Molecules-Electron Interaction

The potential of the interaction with the surrounding medium $U_{int}(\boldsymbol{r})$ is given by relation:

$$U(\boldsymbol{r}) = U_0(\boldsymbol{r}) + U_{int}(\boldsymbol{r}). \qquad (2)$$

Polarons in the Functionalized Nanowires 227

This interaction is relatively strong due to electrostatic contributions, if the functionalizing molecules have an intrinsic dipole moment d. In the organic molecules, the electric-dipole moment exists due to the presence of atomic groups that break the charge symmetry [23, 30]. For the polaron formation, the general difference of the functionalizing molecular medium under consideration from a usual ionic one is the following. The functionalizing molecules have the rigid dipole groups and only the molecule as a whole has a soft degree of freedom . Meanwhile in the usual ionic medium each ion has soft degrees of freedom.

Sum of the potential contributions coming from all the molecular dipoles ϕ_i^{dip} gives the interaction potential energy at a point r_0 within the nanowire:

$$U_{int}^e(\boldsymbol{r}) = e \sum_i \phi_i^{dip} \to e \int n(\boldsymbol{r}')\phi^{dip}(\boldsymbol{r} - \boldsymbol{r}')d\boldsymbol{r}',$$

where $n(\boldsymbol{r}')$ is the volume number density of the molecules depending on space coordinates, e is the carrier's charge.

Electrical potential created by one dipole d can be written [41] as $\phi_i^{dip} = d_i \boldsymbol{R}/\varepsilon R^3$, where $\boldsymbol{R} = \boldsymbol{r} - \boldsymbol{r}'$ is radius-vector from the dipole $\boldsymbol{r}'$ to the point $\boldsymbol{r}$ where we are finding the potential, ε is relative dielectric permeability of the medium. Sum over all sites (the molecular dipoles in the layer) transforms the interaction potential to the following form:

$$U_{int}^e(\boldsymbol{r}_0) = e \int d\boldsymbol{r}' n(\boldsymbol{r}')\boldsymbol{d}(\boldsymbol{r}')\frac{\boldsymbol{r}_0 - \boldsymbol{r}'}{\varepsilon|\boldsymbol{r}_0 - \boldsymbol{r}'|^3}. \tag{3}$$

Here, $\boldsymbol{d}(\boldsymbol{r}')$ is the mean value of the dipole moment at the point $\boldsymbol{r}'$. The interaction potential (3) can also be written in terms of the electric-polarization vector $\boldsymbol{P}$ [41] after substitution $n(\boldsymbol{r}')\boldsymbol{d}(\boldsymbol{r}') \to \boldsymbol{P}(\boldsymbol{r}')$ in (3).

In turn, the electric field intensity $\boldsymbol{E}(\boldsymbol{r}')$ created by the carrier determines the potential energy of an individual molecular dipole:

$$U_{int}^d(\boldsymbol{r}') = -\boldsymbol{d}(\boldsymbol{r}')\boldsymbol{E}(\boldsymbol{r}').$$

In detailed form, the potential energy of an individual molecular dipole is:

$$U_{int}^d(\boldsymbol{r}') = -\boldsymbol{d}(\boldsymbol{r}') \int d\boldsymbol{r}_0 \frac{e|\psi(\boldsymbol{r}_0)|^2(\boldsymbol{r}' - \boldsymbol{r}_0)}{\varepsilon|\boldsymbol{r}' - \boldsymbol{r}_0|^3}. \tag{4}$$

Therefore, on the one hand, for the charge carrier inside the nanowire, we have found the potential energy due to all the dipoles (functionalizing molecules). On the other hand, for an individual dipole of a functionalizing molecule, we have found potential energy due to the charge carrier (inside the nanowire). It will help to find concrete forms of their mutual action.

2.1.3. The Material Equations

The charge carrier field provide the maximum impact on the functionalizing molecular layer when (i) the molecules have an intrinsic electric dipole moment; (ii) the molecular system is soft, in practice only a part of the degrees of freedom must be soft. The material equations take into account these factors:

$$n(\boldsymbol{r}') = n(\boldsymbol{E}(\boldsymbol{r}')); \quad \boldsymbol{d}(\boldsymbol{r}') = \boldsymbol{d}(\boldsymbol{E}(\boldsymbol{r}')). \tag{5}$$

228   *Victor A. Lykah and Eugene S. Syrkin*

Generally, the equation set (1 - 4) has to be completed with the material equations (5). Below, the concrete form of the material equations will be written.

2.2. The Simplified Description of the Model

2.2.1. The Schrodinger Equation Splitting

Let us consider a system with cylindrical geometry (Figure 1). Under the assumption made, the separation of coordinate variables is possible: $\psi(r) = \psi(x)\psi_\perp(y, z)$ and $W = W_x + W_\perp$, here, x is the coordinate along the wire axis. The detailed consideration of the carrier inside the nanowire may result to a modification of $\psi_\perp(y, z)$ and $W_\perp$. We also assume that any variations in $\psi(x)$ occur on a length scale on the order of L. With accounting of the cylindrical 3D geometry, the equation (1) can be split and rewritten to describe the 1D and 2D motions of the charge carriers in the nanowire as follows:

$$-\frac{\hbar^2}{2m_{ef}}\frac{\partial^2\psi(x)}{\partial x^2} + U(x)\psi(x) = W_x\psi(x).$$
$$-\frac{\hbar^2}{2m_{ef}}\nabla_\perp\psi_\perp(y, z) + U_\perp\psi_\perp(y, z) = W_\perp\psi_\perp(y, z). \tag{6}$$

Here the first equation describes particle in the one dimensional quantum rectangular infinite potential well:

$$U(x) = \begin{cases} \infty; & |x| > L; \\ 0; & |x| \le L. \end{cases} \tag{7}$$

Here $2L$ is the nanowire length. The 1D wave function satisfies the normalization condition :

$$\int_{-L}^{+L} |\psi(x)|^2 dx = 1. \tag{8}$$

In the localization case, the integration limits change practically in such manner $\pm L \to \pm\infty$. The second equation in (6) describes 2D transverse quantization in the nanowires and was researched in detail for CNT, see for example [39].

2.2.2. Simplification of the Carrier Potential

In a long nanowire approximation, a nonlocal contribution reduces into a local one. For this purpose, in the integral (3), we separate variables in cylindrical coordinates (Figure 1). The interactions (3) and (4) are nonlocal and largely depend on the configuration of the system. The Schrodinger equation (1) with potentials (2-4) can be transformed into a nonlinear integro-differential equation which can be solved only by using methods of approximation [40]. It is reasonable to assume that radii of the nanowire r and the functionalizing layer R are much smaller than the nanowire length $2L$, i.e., $r < R \ll 2L$; this condition is satisfied in experiments (see the STM image in [21]). The predicted effects are the most pronounced when the configuration can be described as follows. The effect of the charge-carrier field is strongest if the molecules possess the intrinsic electric-dipole moment d and the molecular system is soft. However, calculations can be carried out more readily for a molecular system where only some of the degrees of freedom are soft, for example, in the case of layered smectic A liquid crystals [31]. We assume that the molecular system is rigid in the direction over the nanowire surface (a close-packed layer of long linear molecules) and soft in the radial direction (elastic molecules or elastic coupling between the layers) [31] Chapter 10, so that the molecular axes are oriented normally to the layer [31] Chapter 5; [23]). The electric-dipole moment in these molecules exists due to the presence of atomic groups that break the charge symmetry ([30] Chapter 10; [31] Chapter 10; [23]). For simplicity, we assume that the dipole moment of a molecule is oriented along its axis. Ferroelectric and antiferroelectric ordering and disordered phases are possible [30], and a nonzero polarization can appear by the flexoelectric effect [31] Chapter 6. We assume that the molecular system is rigid in the direction over the nanowire surface (in the most part of the cases). The special case is the incommensurate structures, see below. Ferroelectric ordering is possible [30], especially, if it is oriented by the nanowire as a substrate.

Approximation of a long nanowire, relatively thin layer and cylindrical coordinates allow reducing the integral contributions to a local description [35, 36]. Let us express integral (3) in cylindrical coordinates. The integration limits r_0 and R_0 coincide with the outer radii of the nanowire and the molecular layer (see Figure 1). In the absence of charge carriers or in the case of a rigid molecular system, the density and dipole moment along a radius in (3-5) are constant: $n(r') = n_0$ and $d = d_0$. So they can be kept out of the integration sign. Integration with respect to x (the coordinate along the nanowire axis) is reduced to an integration within infinite limits, even for $x_0 - x' \geq 3R_0$.

At the same distances, the edge effect is not felt (at the end of the nanowire, the interaction is halved). Those, for $3R_0 \ll L$ estimates, the local condition can be considered fair. We use in integration [42, 43], so the potential energy of the interaction of the charge carrier with the dipole subsystem of the functionalizing molecules is

$$U_{int}^{e0} = -\frac{4\pi}{\varepsilon} n_0 de(R - r) \rightarrow -\frac{4\pi}{\varepsilon} n_{S0} de. \tag{9}$$

Here n_{S0} is the surface molecular density in the limit case of thin functionalizing layer.

The liquid crystals of layered smectic A [31] satisfy to these requests. The system of long linear molecules is rigid in a direction along a layer (surface of a nanowire) and is soft degree of freedom in a direction perpendicularly to layer. Suppose a dipole to be directed along the axes of the molecules and perpendicular to the layer [31]. See geometry of the system in Figure 1.

2.2.3. Simplification of the Potential for the Molecular Dipoles

Let us obtain an approximate analytical expression for the potential energy of the interaction between the charge carrier and an individual molecular dipole. Instead of carrying out a spatial integration in (4), the electric field intensity flux can be calculated. Thus, it makes unnecessary to specify the shape of the radial distribution. In other words, the local value of the radial component of the field intensity we replace with a value calculated for an infinitely long wire taking the local value of the wave function:

$$U_{int}^d(x, r') = -\boldsymbol{E}\boldsymbol{d} \rightarrow -\frac{2\tau(x)d}{\varepsilon r'}; \tag{10}$$

$$\tau(x) = e|\psi(x)|^2 F_\perp; \quad F_\perp = \int |\psi(y,z)|^2 dy dz. \tag{11}$$

Here $\tau(x)$ is local linear charge density and ε is the relative permittivity of the medium. The fact that the positive direction of a dipole moment d coincides with the direction towards the center of the nanowire, as indicated by (3), is taken into account. The error resulting from an approximation of a long nanowire can be estimated using the divergence theorem, also known as Gauss' theorem or Ostrogradsky's theorem. Let us determine the leakage of the electric-field flux through the bases of a cylinder coaxial with the nanowire [36]. Then this error can be estimated as a product of the base area $S \sim R^2$ and the axial component of the electric-field intensity $E_x \sim L^{-2}$. thus, the disregarded contribution to the flux is $\Phi_x \sim (R/L)^2 \ll 1$, Meanwhile, the flux through the side surface of the cylinder is $\Phi_\perp \sim 1$. Relative error of the approximation is $\Phi_x/\Phi_\perp \sim (R/L)^2 \ll 1$. The radial field intensity is replaced with a value for an infinitely long wire taking the local value of the wave function.

Simplification of the material equations has to be done for each concrete situation in the further sections.

3. The Elastic Polaron for the Functionalizing Molecules with the Soft Radial Degrees of Freedom

In this section, we show that the interaction of a nanowire with a soft medium composed of organic molecules leads to a modification of the charge-carrier energy spectrum, which is a fundamental characteristic of a nanoconductor. It appears that the spectrum is extremely sensitive to the state of the molecular subsystem. We consider the effect of the interaction of the uncompensated charge carried by an electron or hole in a quantum nanowire with the neighboring medium, which has low mechanical rigidity and consists of molecules having an intrinsic electric-dipole moment, and derive nonlinear nonlocal equations describing such a system. We consider a system in which a perfect quantum wire is uniformly covered by molecules with an intrinsic electric dipole moment. Taking into account the deformation of the molecular layer, we calculate the linear and nonlinear contributions to a shift of the charge-carrier energy levels, which depend on the molecules' polarization. For the case of a long nanowire surrounded by a thin molecular layer, the problem of longitudinal quantization is reduced to solving the spectral problem for a nonlinear Schrodinger equation. In the solution obtained, the normalization of the wave function manifests itself tangibly, which is a radical difference from the known solution for nonlinear classical oscillations in a finite chain [44]. Calculation of the nonlinearity parameter and the energy of a given quantum level of a charge carrier, expressed via the parameter of the nonlinear interaction of the carrier with the molecules, is reduced to solving a set of transcendental equations. The analysis indicates that localization of the charge-carrier motion along the nanowire is possible. Physically, the situation considered represents one of the manifestations of the polaron effect; however, there are specific features related to the confinement of the charge carriers and the one-dimensional character of their motion. Continual transition from longitudinal quantization in NW to formation of longitudinal polaron is researched.

Polarons in the Functionalized Nanowires

3.1. Description of the Model with the Soft Radial Degrees of Freedom

3.1.1. The Material Equations and the Potential Energy of the Carrier

The material equation (5) can be obtained in the case of a thin molecular layer. Let's combine two conditions. The first condition is the equilibrium elastic displacement δ and r' (account the differential relation $d\delta = dr'$) of an individual dipole which is pulled into (or pushed out of) the region of a stronger field by a force F_i:

$$\nabla \frac{k\delta^2}{2} = -\nabla \frac{\tau d}{\varepsilon r'}; \quad k\delta = \frac{\tau d}{\varepsilon r'^2} = \nabla U_{int}^d = -F_i. \tag{12}$$

Here k is coefficient of rigidity which can arise from as rigidity of an individual molecule as a molecular interaction. The first condition means the local energy (elastic plus electrostatic) optimization for the molecular dipoles. The second condition is conservation of the number of the molecules

$$\pi(R_0^2 - r_0^2)n_0 = \pi(R_1^2 - r_0^2)n_1. \tag{13}$$

Here, subscript 0 (or 1) corresponds to the case without (or with) a charge carrier, $R_1 = R_0 + \delta$. Combining relations (13) and (9) we obtain approximation $(R_1 - r_0)n_1 \simeq n_0(R_0 - r_0)$. Expanding it with respect to $\delta/R_0 \ll 1$, we get the potential energy of a charge carrier in the self-consistent field of the elastically displaced dipoles:

$$U_{int}^e = U_{int}^{e0} + U_{int}^{e1}; \tag{14}$$

$$U_{int}^{e0} = -\frac{4\pi}{\varepsilon} dn_0 e(R_0 - r_0); \quad U_{int}^{e1} = -G|\psi(x)|^2; \tag{15}$$

$$G = \frac{8\pi n_0 d^2 e^2 F_\perp}{k\varepsilon R_0^2}\left(1 - \frac{r_0}{R_0}\right). \tag{16}$$

In the case of the film consisting of several layers arranged along the cylinder surface of the nanowire, the volume free-energy density can be written in the continuum approximation [45]. In this case in equation (16) the transformation
$kR_0^2 \to (B/n_0)f(r_0/R_0)$ is necessary, as it was shown in [35, 36]. Here the dimensionless function is $f(r_0/R_0) \sim 1$ at considered deformation and B is the elastic moduli of the liquid crystal with respect to variation in the interlayer spacing with free energy density $w = B(u_{r'})^2/2$, another deformations are negligible.

3.1.2. The Nonlinear Schrödinger Equation for the Carrier

Accounting the self-consistent potential energy (14, 15, 16), Eq. (1) can be rewritten to describe the one dimensional motion of the charge carrier inside the nanowire as follows:

$$-\frac{\hbar^2}{2m_{ef}} \frac{\partial^2 \psi(x)}{\partial x^2} + [U_{int}^{e0} - G|\psi(x)|^2]\psi(x) = W_x\psi(x). \tag{17}$$

Here, $-L < x < +L$. The interactions U_{int}^{e0}, given by (9), and G, given by (16), determine the parameters of the linear and nonlinear interaction of the charge carriers with the elastic molecular subsystem, correspondingly. The sign of the linear interaction parameter depends on the sign of the charge and the orientation of the dipoles. Thus, for an attractive interaction we have $ed > 0$, the carrier energy U_{int}^{e0} inside the quantum conductor decreases. The nonlinear interaction always leads to a decrease in the energy of the system, due to the elastic response of the molecular layer; thus, the parameter G contains even powers of charge and dipole.

3.2. Results for the Soft Radial Degrees of Freedom

3.2.1. Analysis of the Solutions

Using the method standard for quantum mechanics [40], [46] let us introduce new variables:

$$k_p^2 = (W_x - U_{int}^{e0})\frac{2m_{ef}}{\hbar^2}; \quad 2g = G\frac{2m_{ef}}{\hbar^2}; \tag{18}$$

where k_p is a component of the wave vector of the particle along the quantum wire, g is the renormalized interaction parameter. Then eq. (17) transforms into

$$\psi''(x) + k_p^2\psi(x) + 2g|\psi(x)|^2\psi(x) = 0. \tag{19}$$

Without nonlinear interaction ($g = 0$) the Schrodinger equation (17) has well known solutions in the box $-L < x < +L$ [40], [46] with following energy eigenvalue for n-th level:

$$W_n = \frac{\hbar^2\pi^2}{8m_{ef}L^2}n^2; \tag{20}$$

Multiplying the equation (19) on $d\psi(x)$ and integrating result, one obtain the first-order integral that makes it possible to divide the variables. Further integration, carried out using the properties of elliptic functions [42], [47], yields even- and odd-numbered solutions [46] of the Schrodinger equation:

$$\psi_{even} = b\,\mathrm{cn}(\kappa x); \quad \psi_{odd} = k'_m b\,\mathrm{sd}(\kappa x). \tag{21}$$

Here, k_m and $k'_m = \sqrt{1 - k_m^2}$ are elliptic moduli, and $\mathrm{cn}(y)$, $\mathrm{sd}(y) = \mathrm{sn}(y)/\mathrm{dn}(y)$, and $\mathrm{sn}(y)$ are elliptic functions. These solutions for $n=1,2$ are shown in Figure 2.

The integration constant κ is related to the amplitude b and the modulus k_m of the elliptic functions by the equations

$$b^2 = \frac{1}{2g}(\kappa^2 - k_p^2); \quad k_m = \frac{b\sqrt{g}}{\kappa}. \tag{22}$$

Another two equations which sets the relation between k_m, κ, b are given by the boundary conditions and the normalization condition. The boundary conditions $\psi(\pm L) = 0$ result in the following equation for k_m and κ:

$$MK(k_m) = \kappa L. \tag{23}$$

Polarons in the Functionalized Nanowires

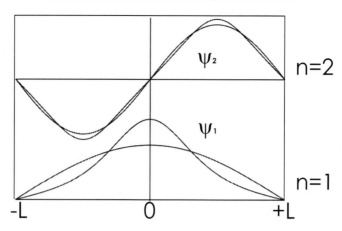

Figure 2. The shape of the charge-carrier wave functions for the first and second levels according to (21) at $gL \simeq 10$. The sinusoids correspond to the elliptic modulus $k_m = 0$ ($g = 0$), and the more sharp bell-shaped curves correspond to $k_m = 0.993$ for n=1 (nonlinear interaction with elastic molecules is present) and $k_m = 0.83$ for n=2. For comparison see these values in Figure 3.

Here, $M = 2m + 1$ ($m = 0, 1, 2, ...$) for even-numbered solutions and $M = 2m$ ($m = 1, 2, ...$) for odd-numbered solutions. The normalization condition (8) results in the following equation for k_m and b (we use [42, 48] and equation (23)):

$$\frac{2Lb^2}{k_m^2 K(k_m)}(E(k_m) - k_m'^2 K(k_m)) = 1. \tag{24}$$

Here, $E(k_m)$ is the complete elliptic integral of the second kind. These four equations (22, 22, 23, 24) form the self-consistent closed equation set with respect to the parameters b, κ, k_m and k_p.

$$\begin{cases} b^2 = \dfrac{1}{2g}(\kappa^2 - k_p^2); \\ k_m = \dfrac{b\sqrt{g}}{\kappa}; \\ nK(k_m) = \kappa L; \\ \dfrac{2Lb^2}{k_m^2 K(k_m)}(E(k_m) - k_m'^2 K(k_m)) = 1. \end{cases} \tag{25}$$

After excluding b and κ, it is possible to find a following solution for k_p:

$$k_p^2 = (1 - 2k_m^2)\frac{K^2(k_m)}{L^2}n^2. \tag{26}$$

Here, n is the number of the level and k_m is a root of the equation

$$\frac{2}{gL}K(k_m)[E(k_m) - k_m'^2 K(k_m)] = \frac{1}{n^2}. \tag{27}$$

The parameter k_p^2 (the square of the quasi-momentum) determines the charge-carrier energy spectrum (18), which can be studied experimentally.

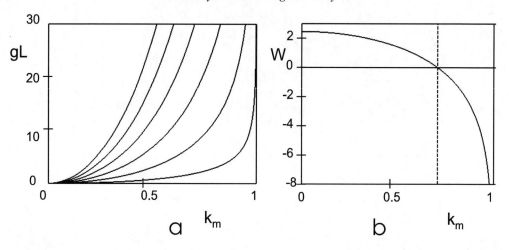

Figure 3. (a) The graphical representation of Eq. (27). The dimensionless interaction parameter gL dependence on the elliptic modulus $k_m^{(n)}$. The plotted curves correspond to arising number $n = 1 \div 6$ from bottom to top. (b) Renormalized energy level $W = k_p^2 L^2/n^2$ as a function of the elliptic modulus k_m (all levels in one curve).

3.2.2. The Carrier Spectrum

In order to analyze the dependence of k_p^2 on the parameters of the system, we make use of a graphical representation. Equation (27) determines the relationship between the parameter k_m, which describes the nonlinearity, and the parameter of nonlinear interaction gL (this relationship is shown in Figure 3). The shape of this dependence is represented in Figure 3a and the wave function shape in Figure 2). It allows making some conclusions: (i) an increase in gL results in a larger nonlinearity parameter $k_m^{(n)}$; (ii) the greatest nonlinearity is observed for the lowest levels.

Let us consider the manifestation of these features in the behavior of energy spectrum (18). Figure 4a shows the dependence of the renormalized energy level $W = k_p^2 L^2/n^2$ on the modulus of the elliptic module k_m. For $k_m = 1/\sqrt{2}$, the energy levels cross the zero value. In the Figure 3 b and Figure 4b, the level energies are plotted as functions of the interaction parameter gL. The nonlinear character is clearly observed for the first level.

3.3. The Elastic Polaron or Soliton-Like State

In the limit of an extremely weak interaction ($gL \to 0$ and $k_m \to 0$), the nonlinearity vanishes and $K(k_m) \to \pi/2$. Then, eq. (26) is reduced to the well-known solution for a rectangular infinite depth potential well [40], [46]: and the wave vector $k_p^2 = \pi^2 n^2/4L^2$.

In the limit of a long nanowire, the case of extremely strong interaction can be realized even for not great value of the parameter g: the sufficient parameters are ($Lg \to \infty$ and $k_m \to 1$). Using the known asymptotic behavior of elliptic functions, [47], [49]: ($K(k_m) \to \infty$, $E(k_m) \to 1$ and $k_m'^2 K(k_m) \to 0$), we find that, in this case, the set of Eqs.

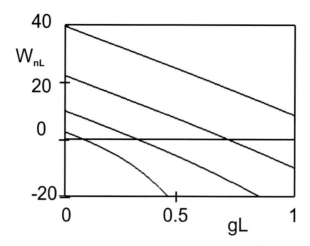

Figure 4. Energy levels $W_{nL} = k_p^2 L^2$ as a function of the interaction parameter gL according to (26), the level $n=1$ is the lowest curve. Compare with Figure 3a,b.

(26) and (27) transforms into

$$\frac{K(k_m)n}{L} = \frac{g}{2n}; \quad k_p^2 = -\frac{g^2}{(2n)^2}; \quad (Lg \to +\infty). \tag{28}$$

For the carrier, the tendency is to localize into a soliton-like state, which energy ($k_p^2 < 0$) does not depend on L. Taking into account equations (22), (22), and (28), the wave function (21) of localized state takes the soliton-like shape for $n=1$:

$$\psi(x) = \frac{\sqrt{g}}{2} \cdot \frac{1}{\cosh(x/l_{sol})}; \quad l_{sol} = \frac{2}{g}. \tag{29}$$

Here l_{sol} is the characteristic length of the soliton (polaron) solution.

The solution (29) can be obtained directly from (19). When $1 \ll Lg$, the effect of the walls of the potential well vanishes, the energy becomes independent on L, and the ψ shape is the same for any choice of the origin $x \to (x - x_0)$. It is solution the elastic polaron. In this way, for the state of carrier, we have obtained a description of the continuous transition from the localization in the quantum well to the polaron self-localization.

The potential well shape U_{int}^{el} and the ground energy level W_{x1} can be found as the potential energy of a charge carrier in a self-consistent field of elastically displaced dipoles (15, 16) and (18, 28):

$$U_{int}^{el} = -\frac{G^2}{8} \cdot \frac{2m_{ef}}{\hbar^2} \cdot \frac{1}{\cosh^2(x/l_{sol})}; \tag{30}$$

$$W_{x1} \simeq -\frac{G^2}{16} \cdot \frac{2m_{ef}}{\hbar^2}; \quad n = 1. \tag{31}$$

These results are shown in Left panel of Figure 5. Such relation between the well depth and the first level exists in the modified Poschl-Teller potential

$$U \sim -\frac{\lambda(\lambda-1)}{\cosh^2(x)}$$

(at parameter $\lambda = 2$, see problem 39 in [46]).

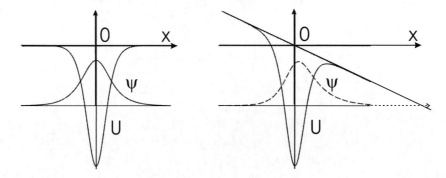

Figure 5. The polaron for the functionalizing molecules with the soft radial degrees of freedom. The self-localized soliton-like polaronic state: ψ shows the wave function of the ground state and its energy level, $U \equiv U_{int}^{el}$ is the self-consistent inhomogeneous part of the potential energy of a charge carrier. Left panel. The homogeneous state. Right panel. The polaron state in electric field, dot arrow shows possible tunneling into a delocalized state.

The appearance of such localized states could be responsible for decreasing of conductivity in chemisorption sensors based on nanowires [18] and in nanowires covered by Langmuir-Blodgett thin film [21]. The carrier's wave function (21), and the molecular layer deformation are shown in Figure 6.

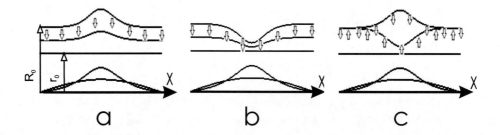

Figure 6. The shape of the polaron coat (along $0x$) for the ground level according to equations (21). Elliptic modulus is $k_m = 0$ for $g = 0$ (for comparison) and $k_m = 0.993$. Deformation of the molecular layer for the following cases: (a) the molecular polarization is directed to the nanowire axis and the hole carrier; (b) the molecular polarization is directed to the nanowire axis and the electron carrier; (c) antiferroelectric dipole ordering and the hole carrier.

Polarons in the Functionalized Nanowires 237

For the purposes of estimation of the nonlinear interaction parameter and the elastic polaron energy , let us use the following values for the parameters [45]: $B \sim (10^6 \div 10^8)$ J/m^3, $d = el$ and $l = 2 \cdot 10^{-10}$m, the volume occupied by a molecule of liquid crystal $1/n_0 = 5 \cdot 5 \cdot 20 \cdot 10^{-30}$ m^3, $F_\perp \sim 1$, $\varepsilon \simeq 2$ and $L \sim (1 \div 10) \cdot 10^{-6}$m. Then the effective radius of the deformation of the molecular layer $r_d \sim 10^{-8} \div 10^{-10}$m, and, according to (18, 16), the dimensionless interaction parameter $gL \sim \mu(1 \div 10^3)$; here, $\mu = m_{ef}/m_e \sim 10^{-2} \div 10^{+2}$ is the ratio of the charge carrier effective mass to the free-electron mass. Thus, the nonlinear interaction energy may vary in a wide range, from $\sim 10^{-4}$eV for $gL = 10$ to ~ 1eV for $gL \sim 10^3$. The nonlinear interaction is stronger and the extent of charge-carrier localization is greater for softer coatings, longer nanowires, and heavier carriers.

3.3.1. The Elastic Polaron Mass

Let us find the elastic polaron mass. The first step is to determine the velocity v of each dipole in the deformed molecular layer:

$$v = \frac{\partial \delta}{\partial t} = \frac{\partial \delta}{\partial x} V. \tag{32}$$

Here V is velocity of the polaron as a whole along $0x$ axis. The equilibrium elastic displacement δ is defined by the equation (12) and is uniquely related to the local linear charge density $\tau \sim |\psi|^2$ (11):

$$\delta = \Delta_{sol} \frac{1}{\cosh^2(x/l_{sol})}; \quad \Delta_{sol} \simeq \frac{deg F_\perp}{4k\varepsilon R_0^2}. \tag{33}$$

Here for the ground state, ψ is defined in eq. (29). Then a dipole velocity (32) and its kinetic energy $W_{kmol} = m_{mol} v^2/2$ can be found. Here m_{mol} is mass of a functionalizing molecule. If the molecule does not move as a whole object but is deformed (one end is motionless) then it is necessary take $m_{mol}/2$.

To find the general kinetic energy of the polaron, it is necessary to integrate over the entire nanowire surface.

$$W_{ksol} = \frac{m_{mol} V^2}{2} \cdot \frac{(2\Delta_{sol})^2}{l_{sol}^2} n_S 2\pi r_0 2 \int\limits_{0}^{+\infty} \frac{\sinh^2(x/l_{sol})}{\cosh^6(x/l_{sol})} dx. \tag{34}$$

Using [42, 43] we obtain the polaron kinetic energy

$$W_{kpol} = \frac{m_{mol} V^2}{2} \cdot \frac{(2\Delta_{sol})^2}{l_{sol}^2} n_S r_0 l_{sol} \frac{32}{15}\pi. \tag{35}$$

Here the last several factors $n_S r_0 l_{sol}(32/15)\pi$ give approximately the quantity of the molecules at the cylindrical nanowire surface on the characteristic length of the polaronic soliton: $\simeq n_S 2\pi r_0 l_{sol}$.

Then the effective polaron mass is sum of the effective carrier mass without functionalization (practically negligible) plus mass of the polaron coat:

$$m_{pol} = m_{eff} + m_{mol} \frac{32}{15}\pi \frac{(2\Delta_{sol})^2}{l_{sol}^2} n_S r_0 l_{sol}. \tag{36}$$

It has been shown that the problem of the longitudinal-quantization in a long functionalized nanowire is reduced to nonlinear Schrodinger equation with boundary conditions. For the interaction parameters, which vary in a wide range, the charge-carrier wave functions, and energies and quantum well shape are determined self-consistently. The parameters variation allow to consider all range between limit cases (i) the rectangular box (rigid molecular layer), (ii) the completely localized charge carrier with the soliton-like wave function and the polaron formation ("soft" molecular layer and a heavy carrier). The polaron formation induced by the nanowire functionalization with complex organic molecules may be responsible for the experimentally observed reduction in conductance [21]. In dependence on the carrier charge sign and on the molecule polarization orientation, the linear interaction and the corresponding shift of the levels can change sign.

The charge-carrier spectrum and the polaron formation depend most strongly on the rigidity of the functionalizing molecules. Crystallization of the liquid-crystal film leads to a sharp increase in the rigidity, corresponding drop in the nonlinear interaction parameter; and the localization of the carrier which causes a jump in the conductance temperature dependence. Thus, the nanowires can be used as a sensor for the state of the functionalizing molecular system.

4. The Conformational Polaron by the Conformational Degrees of Freedom in the Functionalizing Molecules

The state of the molecules plays an important role. Strong conductance variation was found in conformationally constrained molecular tunnel junctions [28]. Optically switchable device by the conformational transition in functionalized single walled CNTs is created [29].

In this section, taking into account the conformational transition in the molecular layer, we calculate the shift of the charge-carrier energy levels that depends on the polarization of the molecules. In the case of a long nanowire covered by a thin molecular layer, the problem of longitudinal quantization is reduced to solving the spectral problem for the Schrodinger equation . The self-consisted system of four equations will be obtained. Calculation of the quantum well parameter and the energy of the quantum level of a charge carrier will be reduced to solving of the set of two transcendental equations. The analysis indicates that the localization of the charge-carrier motion along the nanowire and the conformational polaron formation are possible when the interaction parameters change continually. In the semiconductor nanowire, the hole and electron spectra are symmetric. As it is shown, the functionalizing molecules with conformational transition break this symmetry because the molecular dipoles create the localization conditions for only one type of carrier. It depends on charge sign and dipoles orientation. Therefore, the functionalized nanowires can be used as a semiconducting rectifier.

4.1. Model of the System with the Conformational Transition

4.1.1. Conformational Transition (CT)

The conformational configurations mean that a molecule has different space configurations of atomic groups. The conformation transition (CT) changes the space configuration of the molecule. The conformational configurations may have the same values of energy as in the molecule of NH_3 and threemetilensulfide or different ones as in cis- and trans-states [50]. In the general case an asymmetric molecule has different energies of the conformational configurations. A number of examples of the conformational transition can be found in book [51]. The conformational configurations in vitamins B_6 (pyridoxine) and D_3 are shown in Figure 7.

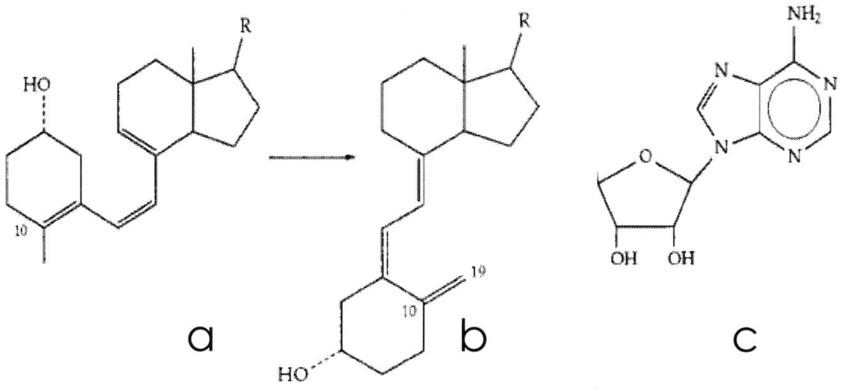

Figure 7. The conformational transition examples. a,b) precalciferol $\rightarrow$ calciferol (vitamin D_3) transition; c) in vitamins B_6 (pyridoxine) with the rotating atomic ring.

Let us denote the following variables: the energy difference between the conformational configuration of an asymmetric molecule W_c, the molecular electric dipole moments of a functionalizing molecule in the initial more stable conformation d_1 and in the unstable conformation d_2. Then in CT, the electric dipole moment changing is

$$\Delta d = d_2 - d_1. \tag{37}$$

CT is possible if the carrier electric field intensity, which is external for the molecules, exceeds the critical value E_c. Its substitution into (10) leads to the critical values of the linear charge density τ_c and to the wave function ψ_c in the nanowire:

$$E_c = \frac{W_c}{\Delta d_r}; \quad \tau_c = \frac{W_c \varepsilon r'}{2 \Delta d_r}; \quad |\psi_c|^2 = \frac{\tau_c}{eF_\perp}. \tag{38}$$

Here d_r is the radial projection of the electric dipole moment. The conformational transition is possible only for one mutual orientation of the vectors E and d as it is shown in Figure 8. The change of the mutual orientation of the vectors increases stability of the conformation. A carrier with opposite sign of charge does not create the conformational transition.

The initial molecular conformation changes the carrier potential according to relation (9) in comparison with pure nanowire. After conformational transition in functionalizing molecules, the additional energy decreasing is possible for the carrier:

$$\Delta U \equiv \Delta U^e_{int} = -\frac{4\pi}{\varepsilon} n_s e \Delta d_r. \tag{39}$$

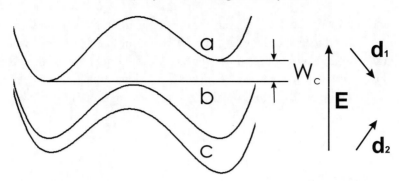

Figure 8. Left panel. Dependence of the potential energy of a functional group on orientation of an electric dipole moment is shown. a) The initial potential: W_c is the energy difference between the conformational configuration of the molecule. b) The conformational potential in the critical external electric field that gives the same energies to the configurations according to (38). c) The conformational potential in an external electric field which changes the conformational configuration. Right panel. The arrows show the electric field E, the initial d_1 (potential a) and the final d_2 (potential b) dipole moments.

The CT goes as order-disorder phase transition in the external electric field of the carrier. In different molecules, the conformational transitions go independently. More probable scenario: a carrier tunnels into the nanowire (CNT) on a high level and jumps to the ground state, then the conformational transition goes (the phonon relaxation is faster than conformation one). The ground state wave function must exceed the critical value (38) $|\psi_1(x)| \geq |\psi_c|$.

4.1.2. Schrodinger Equation for the Conformational Transition

As it has been shown above, in a cylindrical nanowire the 3D Schrodinger equation (1) splits into 2D and 1D equations. Then the 1D Schrodinger equation (6a) describes quantization along the nanowire. Without the conformational transition along the nanowire axis, it is the one dimensional quantum rectangular infinite potential well (7) of length $2L$. In this potential, the problem is reduced to the well known quantization in a box [40, 46].

In the ground state, the wave function has only one symmetric maximum, as the result, the conformational domain takes symmetric position. The infinite potential well of length $2L$ arises, it includes symmetrically a finite potential well of depth ΔU (39) and length $2a$ ($a < L$). So, the carrier potential inside the nanowire is:

$$U(x) = \begin{cases} \infty; & |x| > L; \\ 0; & a \leq |x| \leq L; \\ -\Delta U; & |x| \leq a. \end{cases} \quad (40)$$

where the zero potential energy reference point $U_0 + U_{int}^{e0}$ is chosen as in eq. (7), a is a half of the conformational domain length. For the final configuration of the conformational domain, we consider the symmetric and sufficiently deep quantum well (see discussion below).

Polarons in the Functionalized Nanowires

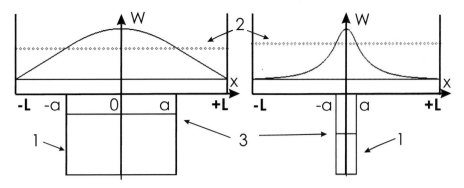

Figure 9. The conformational transition in the functionalizing molecular layer effects on the potential well 1, the carrier's wave function $\psi_1(x)$ 2, and position of its ground energy level 3 of the charge carrier in the nanowire. Diamonds show level of the critical value ψ_c for the conformational transition. Left and right wells differ by the interaction parameter. The height of the wells and the wave functions are normalized.

Following to [46] let us introduce the wave numbers: k_0, k, and κ, that describe the conformational domain well depth, the ground energy level height, and the ground energy level depth correspondingly:

$$\Delta U = \frac{\hbar^2}{2m_{ef}} k_0^2; \quad W_x = \frac{\hbar^2}{2m_{ef}} k^2; \quad \kappa^2 = k_0^2 - k^2. \tag{41}$$

Here we consider a localized state with $W_x < \Delta U$ or $k^2 < k_0^2$, see (40) and Figure 9. The wave function of the ground state is symmetric (even) and exponential outside the conformational domain well. So, according to [46], it can be written in such form:

$$\psi(x) = \begin{cases} A \cos kx; & 0 \leq |x| \leq a; \\ A \cos ka \dfrac{\sinh \kappa(L-x)}{\sinh \kappa(L-a)}; & a \leq |x| \leq L. \end{cases} \tag{42}$$

Here the continuity condition in the point $x = a$ and boundary condition $\psi(l) = 0$ have been used for the wave function. For the first derivation in the point $x = a$, the continuity condition gives such equation:

$$\tan ka = \frac{\kappa}{k} \frac{1}{\tanh \kappa(L-a)}. \tag{43}$$

The wave function amplitude A can be found from normalization condition (8):

$$\frac{1}{A^2} = a + \frac{1}{k} \sin ka \cos ka + \frac{\cos^2 ka}{\kappa}\left[\frac{1}{\tanh \kappa(L-a)} - \frac{\kappa(L-a)}{\sinh \kappa(L-a)}\right]. \tag{44}$$

4.2. The Self-Consistent Equation Set for the Conformational Transition

Let us write the self-consistent system of equations:

$$\begin{cases} \kappa^2 = k_0^2 - k^2; \\ \tan ka = \dfrac{\kappa}{k} \dfrac{1}{\tanh \kappa(L-a)}; \\ \dfrac{1}{A^2} = a + \dfrac{1}{\kappa}\tanh \kappa(L-a) \\ \quad + \dfrac{\cos^2 ka}{\kappa}[\dfrac{2}{\sinh 2\kappa(L-a)} - \dfrac{\kappa(L-a)}{\sinh^2 \kappa(L-a)}]; \\ A\cos ka = \psi_c. \end{cases} \tag{45}$$

Here the symmetry of the quantum well (Figure 9) is accounted: it is enough to consider the region $x \geq 0$. The equation set defines following unknown variables: the conformational domain half-length a, the ground quantum level energy k^2, the energy depth κ^2 of the ground quantum level and amplitude A of the carrier wave function. The material equations give two parameters only: the depth of the conformational domain well k_0^2 and the critical value (amplitude) of the carrier wave function.

The first equation of the set (45) defines relation between the wave numbers (41). The second equation of the set is the continuity condition in the point $x = a$ for the first derivation (43). The third equation of the set is the normalization condition (44), which is rewritten in more convenient form for investigation of limit cases. Thus, in the limit case $L \to \infty$, the normalization condition takes form $1/A^2 = a + 1/\kappa$ as for a quantum well of finite depth [46]. In the limit case $L \to a$ the normalization condition takes form $1/A^2 = a$ as for a quantum infinite potential well. The fourth equation reflects the following fact. At the boundary of the conformational domain, the wave function takes the critical value $\psi(a) = \psi_c$. The last equation defines the self-consistency of all parameters of the problem.

In the set (45), the first and the last equations allow easy exclude the variables A and κ. The rest of the set, the second and the third equations, can be simplified and transformed to the following self-consistent form:

$$\begin{cases} D\tan D = \sqrt{C^2 - D^2}\ \dfrac{1}{\tanh[\sqrt{C^2 - D^2}(\frac{\Lambda}{C} - 1)]}; \\ 1 + \dfrac{1}{D\tan D} + \\ \dfrac{1}{1+\tan^2 D}\{\dfrac{D\tan D}{C^2 - D^2} - \dfrac{1}{D\tan D} - (\dfrac{\Lambda}{C} - 1)(\dfrac{D^2\tan^2 D}{C^2 - D^2} - 1) - \dfrac{1}{\Phi C}\} = 0. \end{cases} \tag{46}$$

Here the dimensionless variables C, D and parameters Λ, Φ are introduced:

$$C = k_0 a; \quad D = ka; \quad \Lambda = k_0 L; \quad \Phi = \dfrac{|\psi_c|^2}{k_0}. \tag{47}$$

The physical sense of the variables and parameters can be cleared by rewriting them

$$C = k_0 a = 2\pi\dfrac{a}{\lambda_0}; \quad D = 2\pi\dfrac{a}{\lambda}. \tag{48}$$

The variable C is the half-length a of the conformational domain normalized on the wavelength of the carrier with energy equal the depth of the conformational domain well ΔU.

Polarons in the Functionalized Nanowires 243

The variable D is the half-length of the conformational domain normalized on the wavelength of a carrier with the ground level energy inside the conformational domain well W. The parameter Λ is the half-length L of the nanowire normalized on the wavelength of the carrier with energy equal the depth of the conformational domain well. Therefore, the variable C directly gives the half-length a of the conformational domain. The half-length a allows to find k and W from another variable D. The sense of the parameter Φ is the critical density of probability $|\psi_c|^2$ which is multiplied on the wavelength $\lambda_0/2\pi$, i.e. it is the critical probability on the interval $\lambda_0/2\pi$.

Accounting of inequalities $a \leq L$ and $k \leq k_0$ and physical sense of Φ, we obtain inequalities between the variables and parameters of the problem:

$$D \leq C \leq \Lambda; \quad 0 < \Phi < 1. \tag{49}$$

In this investigation, we do not take limit case when inequality becomes equality.

4.3. Solutions: The Conformational Domain Size and the Carrier Energy

We derived the nonlinear transcendental equation set (46) with two unknown variables. The first equation is the generalized equation for one unknown variable in the case of the finite depth quantum well which was researched in detail [46]. In the equations for the finite depth quantum well, the parameter C is fixed, ka is unknown, and $\tanh(x)$ is absent. In book [46], the equation was solved by the graphic method on a plane after dividing on left and right parts which are convenient for further qualitative analyze. Numerical methods are very unstable with infinite and discontinuous functions which contain singularities $\tan(x)$ or $1/(x-x_0)$. As minima, smooth interval for the functions has to be pointed out previously.

Our graphic must operate with 3D space for two unknown variables. We find the minimum energy solution or the ground state. For the ground quantum state, the system's solutions can be ambiguous. The basic idea is to avoid periodic singularities by multiplication on $\sin D, \cos D$. Let's introduce smooth functions which depend on two variables that correspond to equations (46):

$$\begin{cases} M(C, D) = D \sin D \tanh[\sqrt{C^2 - D^2}(\dfrac{\Lambda}{C} - 1)] - \sqrt{C^2 - D^2} \cos D; \\ N(C, D) = \sin D[1 + \dfrac{\sin D \cos D}{D} + (\dfrac{\Lambda}{C} - 1 - \dfrac{1}{\Phi C}) \cos^2 D](C^2 - D^2) + \\ \quad + D \sin^2 D[\cos D - (\dfrac{\Lambda}{C} - 1)D \sin D]. \end{cases} \tag{50}$$

Solution of the system (46) corresponds to the point of triple intersection of the graphics $0, M(C, D), N(C, D)$ or intersection of curves (a pair crossing) on C, D plane:

$$\begin{cases} M(C, D) = 0; \\ N(C, D) = 0. \end{cases} \tag{51}$$

We use standard graphic program and calculate for the following dimensionless parameters: $\Lambda = 2; \; \Phi = 0.2$. Direct construction of these three surfaces is not resulting. The reason is very different vertical scales. Some simplification can be due to a guide idea [46]: for the ground quantum level the variable D changes in the range $[0; \pi/2]$. The ranges

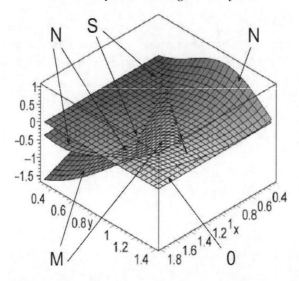

Figure 10. The graphical representation of the solution of the equation set (51) for the conformational polaron. The dimensionless parameters' values are $\Lambda = 2$; $\Phi = 0.2$. The surfaces $z = N(C, D)$, $z = M(C, D)$ and $z = 0$ are marked as $N, M, 0$. Axes are $C \equiv x$, $D \equiv y$. Letter S marks the solution points.

for $M(C, D)$ and $N(C, D)$ functions differ in one order as minimum, so we multiple less value, this procedure does not change solution points in (51) (see Figure 10).

Next qualitative result can be obtained from the graphic analyze of $M(C, D)$ and $N(C, D)$ functions on (C, D) plane. In the considered region with high precision on the surface $z = 0$ at Figure 10, the straight line which contains the joint point can be described as:

$$D(C) = D_0 + k(C - C_0); \quad k := \frac{D_1 - D_0}{C_1 - C_0}; \tag{52}$$

where the points coordinates $C_0 \simeq 0.74$; $D_0 \simeq 0.7$; $C_1 \simeq 1.04$; $D_1 \simeq 0.91$; can be found from the large scale graphic. Further substitution of the dependence (52) into equations (51) and (46) reduces the problem to one unknown variable:

$$\begin{cases} M(C) = 0; \\ N(C) = 0. \end{cases} \tag{53}$$

Graphically, it means intersection of the surfaces by the vertical plane (52). At Figure 11 a, this intersection is shown in wide range. Standard 2D graphic programs give more precise presentation in comparison with 3D packages. In the large scale the curves crossing is shown in Figure 11 b. The fitting of the intersecting vertical plane (52) can be executed by the probing point D_1. The solutions C_c and D_c are obtained, they are shown in the Table 1. Checking up by substitution of C_c and D_c into (50) gives $M = 1.379 \cdot 10^{-8}$ and $N = 9.74 \cdot 10^{-9}$.

We know C and D values and can find the relative values of the physical parameters:

$$\frac{a}{L} = \frac{C_c}{\Lambda}; \quad \frac{k}{k_0} = \frac{D_c}{C_c}; \quad \frac{W}{W_0} = \frac{k^2}{k_0^2}. \tag{54}$$

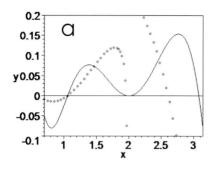 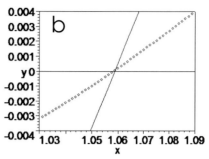

Figure 11. Graphic solution of the equation set (53) for the conformational polaron. Axis $C \equiv x$. Horizontal straight line means $y = 0$, line curve is $y = N(C)$, and box curve is $y = M(C)$. (a) wide ranges of x, y; (b) the large scale view on the curve crossing.

Here a/L, k/k_0 and W/W_0 are the relative conformational domain half-length, momentum and the ground energy level, correspondingly. The results are shown in Table 1.

Table 1. The initial parameters and graphic solutions for the conformational polaron

Λ	Φ	C_c	D_c	a/L	k/k_0	W/W_0	κ
2	0.2	1.0589	0.9193	0.5294	0.8682	0.7538	0.4962
10	0.2	1.1005	0.7813	0.1100	0.7100	0.5041	0.7042
10	0.2	0.2001	0.1966	0.0200	0.9826	0.9655	0.1857

For another dimensionless parameters values $\Lambda = 10$; $\Phi = 0.2$, fitting of the intersecting plane gives three solutions. Two of them are shown in the second and the third strings of Table 1 correspondingly. Checking up by direct matching of the functions (50) gives $M = 2.658 \cdot 10^{-7}$, $N = 2.635 \cdot 10^{-8}$ for the first and $M = -5.38 \cdot 10^{-9}$, $N = 1.807 \cdot 10^{-5}$ for the second solutions.

It is obviously, that the second solution (the third strings of Table 1) is unstable in comparison with the first one because of $W_2 > W_1$. The third formal solution coincides with special point of the functions $M(C, D), N(C, D)$ and is very unstable. For the dimensionless parameters values $\Lambda = 10$; $\Phi = 0.2$ and the stable solution, the relative length and depth of the arising quantum well are presented in Figure 9 (right well) and the second string in Table 1.

At small Λ (sufficiently shallow well), the equation set has only one solution. At big Λ (sufficiently deep well) for the base quantum state, the system's solution becomes ambiguous. Therefore, on the line $\Lambda = 2 \div 10$; $\Phi = 0.2$ in the parameter's plane a bifurcation of the system (50) arises.

Estimation. The molecular electric dipole and its change are $\Delta d \sim d \sim |er_0| \sim 10^{-28} Cl \cdot m$. After the conformational transition, the carrier energy (39) decreases by an amount $\Delta U \sim (10^{+1} - 10^{-1})$eV; in experiments [29] $\Delta U \simeq 1.5$eV. The linear charge density (38) and the energy difference between the conformations are $\tau_c \sim e/2l \sim$

$10^{-13} Cl/m$; $W_c \sim 10^{-3} eV$. The shorter is nanowire the more W_c can be overcome by a carrier.

4.4. The Conformational Polaron

The conformational polaron formation means that walls of the quantum dot (nanowire) are negligible, i.e. $L \gg a$. Thus, the self-consistent system (45) is simplified due to transitions $\tanh \kappa L \to 1$ and $\sinh 2\kappa L \to \infty$. So for the localized polaron, the self-consistent equation set takes form

$$\begin{cases} \kappa^2 = k_0^2 - k^2; \\ \tan ka = \dfrac{\kappa}{k}; \\ \dfrac{1}{A^2} = a + \dfrac{1}{\kappa}; \\ A\cos ka = \psi_c. \end{cases} \tag{55}$$

Then the rest of the system (46), can be simplified and transformed to the following self-consistent form for the conformational polaron:

$$\begin{cases} D \tan D = \sqrt{C^2 - D^2}; \\ \dfrac{1}{C\Phi} \cos^2 D = 1 + \dfrac{1}{\sqrt{C^2 - D^2}}. \end{cases} \tag{56}$$

Here we have the previous variables (47) but only one parameter Φ which gives critical value of the linear charge density. Equation (56) can't be found by cut off of some terms in (46). For finding the ground quantum state, we still use the basic idea: to avoid periodic singularities by multiplication on $\sin D, \cos D$. Let's introduce smooth functions $M_p(C, D)$, $N_p(C, D)$ depending on two variables. They correspond to equations (56) and are the simple version of (50):

$$\begin{cases} M_p(C, D) = D \sin D - \sqrt{C^2 - D^2} \cos D; \\ N_p(C, D) = (1 - \dfrac{1}{C\Phi} \cos^2 D)\sqrt{C^2 - D^2} + 1. \end{cases} \tag{57}$$

Solution of the system (56) still corresponds to the point of triple intersection of graphics 0, $M_p(C, D), N_p(C, D)$ or a pair crossing of intersection curves on C, D plane and is analogue of Figure 10.

4.4.1. The Conformational Polaron Mass

The conformational polaron mass arises only at the polaron movement. Under localization, the polaron takes more or less symmetric configuration and is motionless. There are reasonable causes to the polaron movement; usually it is an external electric field. The conformational polaron in the external electric field along $0x$ is shown in Figure 12. In the external electric field, the symmetry of the quantum well and the wave function is broken. It causes to disbalance between the front and back walls of the conformational domain which begin movement.

For the conformational polaron mass evaluation, let us introduce new space scale l_c, the characteristic length of the conformational transition region, i.e. the conformational domain

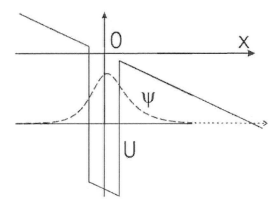

Figure 12. The conformational polaron in the external electric field along $0x$ is shown. In strong electric field, the carrier tunneling from quantum well (dotted arrow) is possible.

boundary. $l_c \ll a$ is the characteristic boundary length. It can be considered as analogy of the domain boundary in ferroelectrics [52]. Now we can write the conformational domain boundaries equation:

$$Q = \pm Q_0 \tanh \frac{x \mp a}{l_c}. \tag{58}$$

Here Q, Q_0 are the generalized coordinate which describes the conformational transition and its amplitude in Figure 8. The generalized coordinate can be as an angle as linear shift of a dipole group, both these values are uniquely related. In eq. (58) and in Figures 9 and 12, the first sign $+$ $(-)$ is to be applied to the right (left) side of the conformational domain.

If the conformational domain moves with velocity V, then the domain boundaries (front and back) move with coordinates: $\pm a \to \pm a_0 + Vt$. In (58), let us deal with a linear shift Q of the dipole groups as a smooth variable. Then in the conformational domain boundary, the velocity v of each dipole group is defined by equation (32) with substitution $z \to Q$. The equilibrium displacement Q is defined by (58). It is unically related to local linear charge density $|\psi_c|^2$. Then one dipole group velocity and its kinetic energy can be found

$$W_{kdg} = \frac{m_{dg}V^2}{2} \cdot \frac{Q_0^2}{l_c^2} \cdot \frac{1}{\cosh^4 \frac{x-a}{l_c}}. \tag{59}$$

Here m_{dg} is mass of the dipole group in a functionalizing molecule.

To find the kinetic energy of the conformational polaron as a whole, it is necessary to integrate over the entire nanowire surface.

$$W_{kcp} = 2 \int_{-\infty}^{+\infty} W_{kdg} n_S 2\pi r_0 dx. \tag{60}$$

Finally, the conformational boundaries' energy is given by relation

$$W_{kcp} = \frac{m_{dg}V^2}{2} \cdot \frac{Q_0^2}{l_c^2} n_S r_0 l_c \frac{16}{3}\pi. \tag{61}$$

Then the effective polaron mass is sum of the effective carrier mass without functionalization (practically negligible) and mass of the polaron coat:

$$m_{pol} = m_{eff} + m_{dg}\frac{16}{3}\pi\frac{Q_0^2}{l_c^2}n_S r_0 l_c. \tag{62}$$

To clear the physical meaning of this result, let us pick up several factors in relations (61) and (62): $16\pi n_S r_0 l_c/3$. This combination gives approximately the number ($\simeq 2n_S 2\pi r_0 2l_c$) of the molecules at the cylindrical nanowire surface on two characteristic length of two (forward and back) conformational boundaries. Namely, these molecules move only when the polaron shifts.

It means that there are two molecular belts where the molecules change their conformational configuration when the polaron moves. They are the forward and back boundaries of the conformational domain. The belts width is one of the smallest parameters in the length scale: $L >> a >> l_c \geq 3R_0$. However the smallest length parameter is Q_0, it is an amplitude of the dipole shift under the conformational transition in the functionalizing molecule: $l_c \geq 3R_0 >> Q_0$. Therefore, in relations (61) and (62), we have competition of the great (n_S) and small (Q_0^2/l_c^2) factors. So, we wait variation of the polaron effective mass in wide ranges.

The following physical results have been obtained in this section.

The quantization, the localization and the polaron formation by the uncompensated charge carrier in the functionalized nanowire are investigated. In the functionalizing molecular system, the interaction of the charge carrier with the conformational degrees of freedom is considered. The physical mechanism responsible for this interaction is pointed out: the molecular dipole moment interaction with the charge carrier electric field.

The self-consistent equation set has been derived for the well width, the spectrum parameters and the critical charge density. The system of four nonlinear transcendental equations has been reduced to two equations and solved by the graphical method. With this purpose 3D graphics have been analyzed qualitatively, the intersections have been found for reducing to 2D problem.

Thus, it has been shown that if the conformational domain well is sufficiently deep, then the self-consistent system of the equations has ambiguous solution. Nevertheless, the physical criterion of selection of the stable solution has been pointed out, namely, minimum of the system energy.

The charge-carrier energy spectrum depends on the rigidity of the functionalizing molecules. At the conformational transition, the result depends on mutual orientation of the electric field intensity of the uncompensated charge carrier and the molecular electric dipole moments. The conformational transition in the electric field and energy spectrum modification are possible for one sign of the carrier charge. The carrier with another sign of charge feels a homogeneous potential along nanowire. Thus, the functionalized nanowire can be used as a semiconductor rectifier.

5. The Discrete Polarons by Incommensurate Structures of the Functionalizing Layer

In this section the functionalization of the semiconductor nanowire by a thin molecular layer with misfit dislocations, and caused by them spectra and localization are researched. The functionalizing molecules have degree of freedom with incommensurate modulation . The self-consistent system of equations has been derived. The system describes the charge carrier quantization in the semiconductor nanowire, the incommensurate molecular structure, the interaction of the carrier and the molecular electric dipoles. It is shown, that the misfit dislocations create periodic relatively narrow quantum barriers for electrons and quantum wells for holes simultaneously. Change of the dislocation type or the dipoles orientation leads to exchange of the hole and electron spectra. In turns, the carrier leads to the dislocation rearrangement. It is shown that the misfit layer reconstructs the nanowire electronic superlattice spectra and creates different conditions of localization or tunneling for holes and electrons in dependence on orientation of the molecules. The final result of reconstruction of the misfit dislocation structure is the carrier localization and formation of the polaron.

5.1. The Model for the Misfit Dislocations System

5.1.1. The Material Equation: Carrier-Dipole-Substrate Interactions

The equation set (1)-(3) is to be completed with the modified material equations (5):

$$n(\boldsymbol{r}') = n(\boldsymbol{E}(\boldsymbol{r}')) \;\; or \;\; n(\boldsymbol{r}') = n(U_s(\boldsymbol{r}')). \tag{63}$$

Here U_s is the substrate potential which effects on the molecules together with the electric field.

5.1.2. Frenkel-Kontorova Type Equation for the Misfit Dislocation

Hamiltonian of a molecular chain on a substrate is [3, 53]:

$$H_{FK} = \frac{1}{2} \sum_n [m\dot{u}_n^2 + \kappa(u_n - u_{n-1})^2 + U(u_n)]. \tag{64}$$

Here u_n is a displacement of a molecule in the n-th site from an equilibrium position in the chain and κ is an elastic constant for a relative molecular displacement. $U(u_n)$ is the potential, that acts on the n-th molecule from the substrate. In continual approximation, the contribution into the Hamiltonian of the molecular chain on the substrate is [3]:

$$U_s(u_n) = U_0[1 - \cos(\frac{2\pi u_n}{a_s})]. \tag{65}$$

Here a_s is the substrate lattice constant.

The Hamiltonian (64) leads to the discrete equation of motion

$$m\ddot{u}_n + \kappa(2u_n - u_{n-1} - u_{n+1}) + \frac{\pi U_0}{a_s} \sin(\frac{2\pi u_n}{a_s}) = 0. \tag{66}$$

Continual approximation can be introduced with following transformations:

$$u_n(t) = u(x,t); \quad \phi(x,t) = \frac{2\pi u(x,t)}{a_s};$$
$$U_s(\phi(x,t)) = U_0[1 - \cos(\phi(x,t))]; \quad (67)$$

where ϕ is an effective phase. Then the continual equation of motion is

$$\phi'' - \frac{1}{c_0^2}\ddot{\phi} = \frac{1}{\lambda_0^2}\sin\phi. \quad (68)$$

Here c_0 is a characteristic longitudinal sound velocity in the molecular chain (layer). However we study the static solutions that correspond to the fixed dislocations, in this case the continual equation of motion has the integral E:

$$(\phi')^2 = \frac{2}{\lambda_0^2}(E - \cos\phi); \quad \lambda_0 = \frac{a_m^2}{2\pi}\sqrt{\frac{2\kappa}{U_0}}. \quad (69)$$

Here λ_0 is the characteristic length, a_m is the molecular lattice constant, and κ is elastic constant in the molecular chain.

At $E = 1$ the continual equation has the stationary solution as the solitary domain wall:

$$\phi(x) = 4\arctan[\sigma\exp\frac{x - x_0}{\lambda_0}]. \quad (70)$$

In dependence on topological charge $\sigma = \pm 1$, this solution gives kink ($\sigma = +1$) or antikink ($\sigma = -1$) that are presented in Figure 13a.

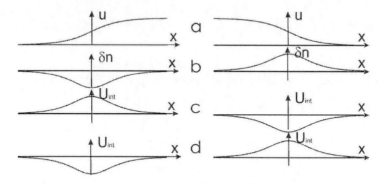

Figure 13. Characterization of the misfit dislocations in the functionalizing molecular layer is shown. Kink or stretched area (left panel, $\sigma = +1$) and antikink or compressed area (right panel, $\sigma = -1$). a) Coordinates $u(x)$ of the functionalizing molecules. b) Corresponding change of the molecular surface density (73a). c,d) the modulated band edge in the nanowire according (74) at $d > 0$ (c) and $d < 0$ (d).

At $E > 1$, the continual equation of motion also has the periodic stationary solution:

$$\cos(\frac{\phi(x)}{2}) = -\sigma\,\text{sn}(\xi, k); \quad \xi = \frac{x - x_0}{k\lambda_0}. \quad (71)$$

Here $\mathrm{sn}(\xi, k)$ is the Jacobi elliptic function, k is elliptic module defined by relation [3]: $k^2 = 2/(E+1); \ 0 \leq k \leq 1$. The period of $\mathrm{sn}(\xi, k)$, $\mathrm{cn}(\xi, k)$ in dependence on ξ is $4K(k)$. Here $K(k)$ is the full elliptic integral of the first kind. Following to [3] in (71, the space period in dependence on the variable x are

$$l_x = 4kK(k)\lambda_0 ;$$
$$l_x \simeq 2\frac{a_s^2}{|a_s - a_m|}. \tag{72}$$

Here $|a_m - a_s| \ll a_m \simeq a_s$. Using (72), one can evaluate the elliptic module k and then the integral E. At $k \to 1$, limit of the integral is $K(k) \to \infty$. In the limit case $k \to 0$ or $E \to +\infty$, the periodic solutions (71) can be presented as $\mathrm{sn}(\xi, k) \to \sin(k\xi)$ that means a homogeneous changing of displacement u_n in the molecular chain. In the limit case $k \to 1$ or $E \to 1$ ($E > 1$), the periodic solution (71) can be presented as a set of the solitary domain walls (70) with the period (72). One of these dislocations is shown in Figure 13.

If the space periods of the molecular layer and NW coincide then $n_S(x) = n_{S0}$. However, the solitary (70) or periodic dislocations are possible. Regions of local extension (kink, $\sigma = +1$) or constriction (antikink, $\sigma = -1$) exist where deviation of concentration is:

$$\delta n_S(x) = -n_{S0}\frac{\partial u}{\partial x} = -\frac{n_{S0}a_s}{\pi\lambda_0}\begin{cases} \sigma \ \mathrm{sech}(\frac{x}{\lambda_0}); & E = 1; \quad \text{(a)} \\ \dfrac{\sigma}{k} \ \mathrm{dn}(\frac{x}{k\lambda_0}, k); & E > 1. \quad \text{(b)} \end{cases} \tag{73}$$

They are analogies of the bright or dark soliton. The potential (9) can be transformed into

$$U_{int}^e(x) = U_{int}^{e0} + U_{int}^{e\delta}(x); \quad U_{int}^{e\delta}(x) = -\frac{4\pi}{\varepsilon}de\delta n_S(x). \tag{74}$$

Integration along the nanowire axis on the misfit dislocations period yields $\Delta n_S = \sigma a_m n_{S0}$, which gives an extra ($a_s < a_m$, $\sigma = -1$) or lack ($a_s > a_m$, $\sigma = +1$) row of the functionalizing molecules. Accounting relatively a large size and a complicated construction of organic molecules, the creation of partial dislocations is very probable.

5.1.3. Renormalization of the Molecules-Substrate Interaction

The material equations lead to a renormalization of the interaction between the molecular layer and the substrate (NW). A simultaneous effect of the substrate U_s and the carrier charge field U_{int}^d (10) has to be accounted:

$$U_s^d(z, x) = U_s(z, x) + U_{int}^d(z, x). \tag{75}$$

Here x and z are coordinates along the nanowire and its radius correspondingly. Radial equilibrium position of a molecule can be found from $\partial U_s(z, x)/\partial z = 0$ for neutral nanowire. Then the substrate potential (67) can be written as

$$U_s(z, x) = \begin{cases} U_{sc}(z_{0c}) = 0; & \text{commensurate;} \quad \phi(x) = 0; \\ U_{si}(z_{0i}) = 2U_0; & \text{incommensurate;} \quad \phi(x) = \pi. \end{cases} \tag{76}$$

In relation (76) the zero level of the potential is chosen for the commensurate regions with a deeper (stronger) interaction. The incommensurate regions have a relatively weak interaction with the substrate. For the commensurate regions, the depth and rigidity of the interaction are higher, correspondingly. Then expansion of the molecule-substrate potential around the equilibrium position z_0 depends on the coefficient of rigidity k_s:

$$U_{s(c,i)}(z_{0(c,i)} + \delta_s) = U_{0s(c,i)}(z_{0(c,i)}) + U_{0s(c,i)}(z_{0(c,i)}, \delta_s);$$
$$U_{0s(c,i)}(z_{0(c,i)}, \delta_s) = \frac{k_{s(c,i)}}{2}\delta_s^2; \quad k_{s(c,i)} = \frac{\partial^2}{\partial z^2}U_{0s(c,i)}(z)|_{z_{0(c,i)}};$$
$$k_{sc} > k_{si}; \quad z_{0c} < z_{0i}. \tag{77}$$

Here δ_s is the deviation from z_0. The commensurate regions are more rigid.

In the case of the nanowire with the carrier, the condition $\partial U_s^d(z, x)/\partial z = 0$ determines the molecule equilibrium position. Accounting relations (77) for $U_s(z_0 + \delta_s, x)$ and (10) for $U_{int}^d(z, x)$ in harmonic approximation, the equilibrium elastic displacement of an individual dipole pulled into (or pushed out of) the region of a stronger field can be found:

$$\delta_s = -\frac{2\tau d}{k_s \varepsilon r'^2}. \tag{78}$$

The relation is analogue of equation (12). However, here the system has much less deformation of the molecule-substrate bonding, not a molecule deformation. With accounting (77) and (10), substitution of δ_s into Eq. (75) gives increasing of the potential energy of elastically displaced dipoles in the field of the charge carrier:

$$U_{0s(c,i)}(z_{0(c,i)}, \delta_s) = G_{4s}|\psi(x)|^4; \tag{79}$$

$$G_{4s} = \frac{2}{k_s}[\frac{deF_\perp}{\varepsilon r_0'^2}]^2. \tag{80}$$

The inequality $G_{4sc} < G_{4si}$ is obtained with accounting of relation between rigidities (77). Decreasing of the molecule interaction with the charge carrier can be expanded as:

$$U_{int}^d = U_{int0}^d + U_{int1}^d;$$
$$U_{int0}^d(x) = -G_2|\psi(x)|^2;$$
$$U_{int1}^d(z_0, x) = -2G_{4s}|\psi(x)|^4; \tag{81}$$

where

$$G_2 = \frac{2deF_\perp}{\varepsilon r_0'}. \tag{82}$$

And so, finally, the simultaneous effect of the substrate and the carrier charge field yields decreasing of the molecule energy:

$$U_s^d(z, x) = U_{s0}(z_0, x) - G_2|\psi(x)|^2 - G_{4s}|\psi(x)|^4. \tag{83}$$

For one molecule, this relation is found to be similar to (10). So, for describing molecular chain, we obtain the potential along the nanowire. The molecular system as a whole has energy which can be found as sum or integral over all molecules.

Then with accounting of (80), the amplitude of the substrate potential can be found through the interaction potentials in the incommensurate $U_{si}(z_0, x)$ and commensurate $U_{sc}(z_0, x)$ regions in (83):

$$U_{0E}(z_0, x) = U_0 - \Delta G_{4s}|\psi(x)|^4 < U_0;$$
$$\Delta G_{4s} = [\frac{1}{k_{si}(z_0)} - \frac{1}{k_{sc}(z_0)}] \times [\frac{deF_\perp}{\varepsilon r_0'^2}]^2. \tag{84}$$

Polarons in the Functionalized Nanowires 253

After substitution of (84) into (67), we obtain the Frenkel-Kontorova potential which is renormalized by the carrier electric field. The long wave modulation along $0x$ axis depends on $\sim |\psi(x)|^4$. According to equations (72), (69), and (84), we obtain

$$l_{xE} \sim \lambda_{0E} \sim \frac{1}{\sqrt{U_{0E}}} \text{ and } U_{0E} < U_0, \ l_{xE} > l_x. \tag{85}$$

Qualitatively the obtained result does not depend on a concrete form of the interaction.

5.2. Results: Superlattice by the Misfit Dislocations

Let us consider an action of the functionalizing layer on the carrier inside the nanowire. With this purpose, we calculate the electric potential created by the functionalizing molecules. In this case in order to describe the one dimensional motion of the charge carrier in the nanowire, the Schrodinger equation (1) can be rewritten as follows:

$$-\frac{\hbar^2}{2m_{ef}}\frac{\partial^2\psi(x)}{\partial x^2} + [U_0(x) + \\ +U_{int}^{e0}(x) + U_{int}^{e\delta}(x)]\psi(x) = W_x\psi(x). \tag{86}$$

The initial potential $U_0(x)$ is described by eq. (7). The potential U_{int}^{e0} is given by relation (9) and is responsible for the homogeneous shifting of the bottom of the quantum well. The potential $U_{int}^{e\delta}(x)$ is given by relations (73b, 74) and is responsible for creation of the superlattice structure in the case of the *periodic dislocation arrangement* at $E > 1$. The inequalities $\lambda_0 << l_x << L$ have to be satisfied. The band structure modulation of the semiconductor nanowire is shown in Figure 14 at $E > 0$ and $\sigma d > 0$. It is a set of the dislocation potentials shown in the left column of Figure 13 (c) or the right column of Figure 13 (d) with narrow barriers for electrons in the conduction band edge. The valence band top edge has narrow peaks.

The periodic quantum barriers for electrons are relatively narrow and can form the superlattice carrier spectrum. Let us consider here *the strong coupling approximation*. The lowest quantum levels for electrons (between relatively narrow barriers) are shown qualitatively in the conduction band in Figure 14. The lowest quantum levels for holes (between relatively wide barriers) are shown qualitatively in the valence band in Figure 14. The quantum levels of hole have considerably higher intervals than the electron ones as the sequence of the narrower quantum well.

At $E > 0$ and $\sigma d < 0$, the band structure modulation can be obtained from Figure 14 by mirror reflection around the axis $0x$. The potential is from the right column in Figure 13c or the left column in Figure 13d. This potential has narrow wells for electrons in the conduction band and narrow barriers for holes in the valence band.

5.3. Results: Self-Consistence, the Misfit Dislocations Rearrangement

Let us consider the carrier in turn action on the functionalizing molecular layer. The amplitude of the substrate potential has been found in (84). After its substitution into (67), we obtain the Frenkel-Kontorova potential with amplitude renormalized by the carrier electric field. Along $0x$ axis, the long wave modulation occurs according to $|\psi(x)|^4$ modulation.

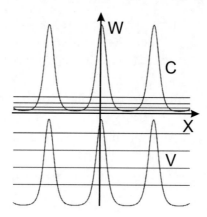

Figure 14. The band structure of the semiconductor nanowire modulated according to (74) at $E > 0$ and $\sigma d > 0$. Quantum levels of electrons (qualitatively close to a rectangular well) in conduction band (C) and holes (qualitatively close to an oscillator) in a valent band (V) are shown in the strong coupling approximation. The functionalizing molecular layer creates opposite conditions for the carrier localization or tunneling in the superlattice created by the misfit dislocations.

The carrier attracts (repulses) the molecular layer that leads to the coherent regions widening and the dislocation rearrangement. With accounting of (84) and (85), the characteristic length (69) in the molecular layer takes such form:

$$\lambda_{0E}(x) = \frac{a^2}{2\pi}\sqrt{\frac{2\kappa}{U_{0E}(x)}} \simeq \lambda_0[1 + \frac{\Delta G_{4s}|\psi(x)|^4}{2U_0}]. \tag{87}$$

In the relations (72), the space period of the misfit dislocation turns to depend on the carrier electric field $l_{xE}(x) \sim \lambda_{0E}(x)$.

Then the higher carrier density inside the nanowire creates the regions with a larger space period of the misfit dislocation. Direct inserting of (87) with $\psi \sim \cos(\pi x/2L)$ into (71) and (73) yields the picture of the dislocation rearrangement and the band structure modulation that are presented in Figure 15 (b). This physical situation is realized only if the extra electron in the nanowire exists quite long time enough for the misfit dislocation redistribution. It looks like the dislocation rearrangement is not realized completely. The reason is this: the superlattice tunneling is destroyed and the electron turns to be locked in one period of the deformed dislocation structure at the first changing of the dislocation periodicity.

In the case of a fast tunneling of the carrier, the dislocations have no time to move, so the undisturbed picture in Figure 15 (a) and the superlattice spectrum are kept.

For the hole in the nanowire, the periodic quantum wells are relatively narrow. They are divided by the wide commensurate regions. It rather can lead to the carrier localization at one dislocation shown in Figure 15 (c). The dislocation width and the potential depth change to provide the minimal energy of the hole-dislocation system.

The case $E > 1$ and $\sigma d > 0$ is shown in Figures 14, 15. In the case $\sigma d < 0$, the hole and the electron have to be exchanged in our consideration. It means that Figures 14, 15 must be reflected about the horizontal axis as a mirror.

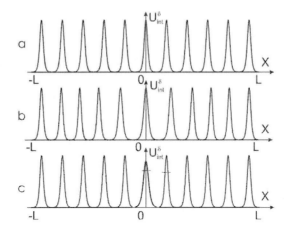

Figure 15. Form of the conduction band bottom edge or valence band top edge. a) The homogeneous misfit dislocation distribution along the nanowire as in Figure 14. b) The first stage of rearrangement of the misfit dislocation in the electron electric field. The next stage is localization at one dislocation period. c) Localization of the hole near a peak of the valence band top edge at one dislocation. In the cases b and c for convenient illustration, the dislocation system is overstated.

To estimate the change of the electron (hole) potential by the molecules, let us use the parameters following to [3, 30, 36]. Thus, from (9), one can evaluate the change of the bottom quantum level in the nanowire with the functionalizing molecular layer: $|U_{int}^{e0}| \sim (1 \div 10)$eV. The barrier height can be evaluated as $|U_{int}^{e\Delta}(x)| \sim |U_{int}^{e0}|a/\lambda_0 \sim (0.1 \div 1)$eV.

5.4. Results: The Localization, the Discrete Polaron Formation

As it is shown above, in the case of the misfit dislocation formation and rearrangement, the carrier (electron or hole) can be localized at a misfit dislocation or at an interval between the dislocations. By other words, the initial one-dimension quantum dot (nanowire) is divided into a set of identical smaller quantum dots by the misfit dislocation. Then one of the smaller dots is transformed for localization by the carrier electric field. The localized carrier has all attributes of the polaron, however this formation has discrete structure. In an external electric field along NW axes, the carrier moves by the discrete steps (resonant tunneling, phonon assisted tunneling or hopping), see Figure 16. The polaron mass can be evaluated through the tunneling probability, see [3].

In dependence on localization at or between the dislocations, an electron and a hole form quite different polarons. The polaron which is localized at one misfit dislocation has very wide potential barriers for the tunneling. The carrier localized between the dislocations has the considerably narrower barriers for the tunneling. Therefore, for the last type of the polaron, the conductivity has to be considerably higher. So, such type of the functionalized nanowire can be applied for a rectifier construction.

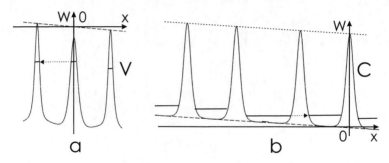

Figure 16. The band edges in an external electric field. a) A hole localization and the discrete polaron formation at a peak of the valence band top edge at one misfit dislocation.
b) An electron localization and the discrete polaron formation at the conduction band bottom edge between two misfit dislocations with wider interval. The resonant tunneling of a hole and an electron is shown by dotted arrows.

The following physical results have been obtained in this section.
(i) The periodic system of the misfit dislocations in the functionalizing molecular layer exists at the nanowire surface under certain conditions. There may be two types of the misfit dislocation: with extra or lack molecular row. They are analogies of bright or dark solitons. The periodic dislocations are divided by the wide commensurate regions.

(ii) A homogeneous distribution of the molecular dipoles creates a homogeneous shift of electric potential along the nanowire. The periodic misfit dislocations with extra or lack dipoles create a superlattice potential for the carrier inside the nanowire. A set of the quantum wells and the barriers arises, they modulate the band edges. The potential relief depends on the dipoles' orientation and type of the misfit dislocation. The topological invariant $sign(\sigma d) = \pm 1$ defines creation of the relatively narrow peaks or wells for electron (correspondingly the wells or peaks for hole).

(iii) The periodic modulation of the nanowire band structure by the misfit dislocations forms the superlattice carrier spectrum either for an electron or for a hole. The misfit dislocations create the same periodic relief of the valence and conduction bands edges. Considerable deviation of the bands edges relief from a sinusoidal function breaks this symmetry of the spectra. The narrow peaks are easily penetrated by a carrier that causes the narrow gap or quasi-gap in a corresponding band. In another band the narrow wells are divided by wide barriers; tunneling is rather difficult. It causes very narrow minibands or a set of the split quantum levels; localization is possible.

(iv) In its turn, the charge carrier attracts the molecular layer that makes the dislocation out of periodicity and the coherent regions expand. It destroys miniband structure and can lead to fast carrier localization. The carrier localization on a single peak causes an adjustment of the dislocation arrangement. At longitudinal quantization, effect of the walls of the potential well vanishes under the localization condition. The energy becomes independent on the nanowire length $2L$ however depends on the dislocation width λ_{0E} or dislocation period l_{xE} as it is shown in Figure 15 where $\lambda_{0E} \ll l_{xE} \ll L$. This scenario works only under presence of a carrier in the nanowire for a long time.

(v) In the nanowire, the spectra of holes and electrons are symmetric. The function-

Polarons in the Functionalized Nanowires 257

alizing molecules with the intrinsic electric dipoles breaks this symmetry, create opposite conditions for localization or tunneling along the dislocation superlattice in dependence on the charge sign, the dipoles orientation, and type of the dislocation.

(vi) The misfit dislocation system of the functionalized nanowire and the induced super-lattice spectra are extremely sensitive to variation of the molecule-molecule and molecule-substrate interactions. It may be caused by temperature, phase transitions, a filling of the functionalizing layers and impurity concentration. The high rigidity of the functionalizing molecular system leads to decrease of the dislocation mobility, and as sequence, to suppression of the localization. In these cases, the nanowire conductivity must grow significantly. Thus, the nanowire can be used as a sensor for the state of the molecular system.

(vii) Two types of the discrete polarons can be formed: localized at one misfit dislocation and between adjacent dislocations. The polarons' properties have to be very different, what can be used in rectifier devices.

6. Estimations

Charges, Dipoles and Energies

The linear density of the charge carrier can be evaluated as

$$\tau \sim \frac{e}{2L} \sim 10^{-13}\frac{\text{Cl}}{\text{m}}. \tag{88}$$

Here $2L \sim 1\mu\text{m}$ is the nanowire length. The electric dipole of a molecule and its changing Δd can be estimated as

$$\Delta d \sim d \sim |er_{at}| \sim 10^{-28}\text{Cl} \cdot \text{m}; \tag{89}$$

where $r_{at} \simeq 0.4\text{nm}$ is size of the dipole which is approximately equal to the interatomic distance. After the conformational transition according to (39), the carrier energy U and its decreasing ΔU in the functionalized nanowire can be estimated as

$$U \sim \Delta U = \left| -\frac{4\pi}{\varepsilon}n_s e \Delta d \right| \sim (10^{+1} \div 10^{-1})\text{eV}; \tag{90}$$

where the surface concentration can be estimated as $n_s \simeq S^{-1}$. The surface, occupied by one molecule, is $S \sim (10^2 \div 10^4) \cdot 10^{-20}\text{m}^2$. Let's keep in mind factor $\varepsilon_0 \sim 10^{-11}\text{F/m}$ in SI for the denominator. Thus, in experimental work [29], the surface concentration of the molecules was one per 100 elementary cells of the nanowire surface and $\Delta U \simeq 1.5\text{eV}$. We take this value for further estimations.

The estimations of $\tau \sim \tau_c$ and relations (38) allow evaluating the energies W of a dipole in the carrier's electric field, and the difference between the conformations W_c:

$$W \sim W_c = \frac{\tau_c \Delta d_r}{2\pi\varepsilon r'} \sim 10^{-3}\text{eV}. \tag{91}$$

The shorter nanowire ($2L \sim (10^{-7} \div 10^{-8})\text{m}$) then the greater critical conformational energy is: $W_c \sim (10^{-2} \div 10^{-1})\text{eV} \sim (100 \div 1000)\text{K}$. So, the shorter nanowire the more energy difference between the conformations can be overcome by the carrier electric field.

Relaxation to the Ground State

To obtain an analytical description we have applied the simplest models. Here we indicate factors that affect the obtained results. The relationship between the lifetime of the excited carrier state τ_e (determined by electron-phonon relaxation) and the relaxation time of the molecular system τ_M are necessary to take into account. For $\tau_M \sim \tau_e$ and $\tau_M \gg \tau_e$ (the adiabatic approximation) the charge carrier and the molecular-system relaxations should be considered simultaneously with phonons emission and filling a deep ground-state level.

The longitudinal quantization creates quantum levels which have energy distance in several orders less than for the transverse quantization. Therefore, a carrier can tunnel from a lead into the nanowire on a high level of energy (high number of nodes of the wave function), then several maxima are possible in initial stage of relaxation. The symmetry breaking gives an advantage to one quantum well only. It is right also for the conformational transition. If conditions (38) are satisfied in several space intervals at high level, CT goes independently; it is reason for a chaotic scenario. In the different wells, the conformational transition goes by independent jumps and wave function fluctuates along the nanowire going to a deep ground-state level. At low temperature, a more probable scenario is following: the first of all, the carrier goes into the ground quantum state and then the elastic deformation and the conformational domain are realized in one minimum. This scenario develops if the electron-phonon relaxation processes for the carrier energy are faster than molecules' relaxation: $\tau_e < \tau_M$. Both these scenarios are possible only if a wave functions of a high energy level $\psi_n(x)$ and of the ground state $\psi_1(x)$ exceed the critical value (38) for CT: $|\psi_1(x)|^2 \geq |\psi_n(x)|^2 \geq |\psi|^2$. The wave function of the ground state has only one maximum. Therefore, CT grows more or less chaotically from central point with maximum electric field to the region with decreasing wave function where conformational domain boundary reaches to the point $\psi_1(x) = \psi_c$. As the result, the conformational domain is formed closely to symmetric position. The systems with the misfit dislocations have already existing quantum wells: relaxation lowers the carrier to the deeper quantum levels, the molecular relaxation transforms the well shape.

7. Questions for Further Research

What are the conditions of existence of the ordered commensurate molecular structure at the nanowire (CNT) surface? What are the conditions of existence of the longitudinal and transverse incommensurate molecular structure at these surfaces?

What are the molecules with the most pronounced demonstration of only one soft degree of freedom (radial, conformational, incommensurate)? These molecular systems could be models for more complicated cases as DNA et al.

What are the other soft degrees of freedom in the functionalizing molecules that allow finding solutions exactly?

The polarons move until they reach the nanowire end. The question is how the edge polaronic states can be described?

What the physical effects are the most pronounced for each soft degree of freedom of the functionalizing molecules? What are the ranges of the parameters for realization?

What type of a functionalizing molecule is the most convenient for the polarons creation? What type of the polaron is better for the current rectifier? What type of the polaron is better for detecting of the functionalizing molecules' state as phase transitions, concentration, chemical reactions, and other responses to external influences?

Conclusion

The quantization, the localization, and the polaron formation have been investigated for an uncompensated charge carrier in the functionalized nanowire. The physical mechanism responsible for the specificity is pointed out: the linear and nonlinear interactions of the molecular dipole moments with the charge carrier. Organic molecules are too complicated; they contain polar groups with dipole moments. The molecule-substrate interaction gives to the molecular dipoles the same radial orientation (the nanowire radius). A homogeneous distribution of the molecular dipoles creates the homogeneous shift of electrical potential along the nanowire. Approximations of the long nanowire and thin molecular layer have been used. The following soft degrees of freedom in the functionalizing molecules are considered. (i) The radial elastic deformation of the molecules. (ii) The conformation of the molecules. (iii) The incommensurate molecular structure along the nanowire.

(i) The soft radial degrees of freedom of the functionalizing molecules. For the carrier, the longitudinal quantization has been described by the nonlinear Schrodinger equation and the analytical solutions obtained. For the heavier carriers and the softer molecules, effect of the nonlinear interaction and the polaron formation has to be more pronounced? It has been shown that the self-consistent polaron well is described by modified Poschl-Teller potential; its ground state wave function has quasi-soliton shape. *The elastic polaron* (the polaron with the soft radial degrees of freedom of the molecules) has almost the same properties for electrons and holes due to symmetry of the elastic deformation. The nearest analog of the polaron with the soft radial degrees of freedom is the Pekar's polaron [1].

(ii) The conformation in the functionalizing molecules. For the well length, the spectrum parameters, and critical charge density of the carrier, the self-consistent equation set has been derived. The system of four nonlinear transcendental equations then has been reduced to two equations and solved graphically. By criterion of the energy minimum, the stable solution has been pointed out. It has been shown that the rectangular polaron well (conformational domain) is to be sufficiently deep. The nearest analogs of the *conformational polaron* are the fluctuon [4], [8] and the fluctuon in ferroelectric-ferromagnetic systems [7].

(iii) The periodic misfit dislocations with extra or lack dipoles create a superlattice potential for the carrier inside the nanowire. The topological invariant $\text{sign}(\sigma d) = \pm 1$ defines creating of the relatively narrow peaks and wide wells for an electron (correspondingly the narrow wells and wide peaks for a hole). The narrow peaks are easily penetrated by the carrier that causes the narrow gap arising in the corresponding band. In another band, the narrow wells are divided by the wide barriers; the tunneling is rather difficult and causes very narrow miniband. In its turn, the charge carrier attracts the molecular layer and shifts the dislocations out of the periodicity expanding the coherent regions due to an inhomogeneous distribution of ψ. It destroys miniband structure and leads to the carrier localization. *The discrete polaron in the incommensurate structure* is formed at the self-consistently deformed one dislocation or one distance (gap) between adjacent dislocations in the functionalizing molecular layer.

The polaron mass has been found for the elastic polaron (polaron with the soft radial degrees of freedom) and the conformational polaron.

In the nanowire, the spectra of holes and electrons are symmetric. The functionalizing molecular layer breaks this symmetry in dependence on the charge sign, the electric dipoles orientation, and the type of the *dislocation or conformation*. The strong energy spectrum modification and the polaron formation are possible only for one sign of the carrier's charge. The carrier with another sign of charge feels the homogeneous potential along the nanowire. In these cases, the functionalized nanowire can be used as a semiconducting rectifier.

The functionalized nanowire and the induced spectra are extremely sensitive to variation of the molecule-molecule and molecule-substrate interactions. It may be caused by temperature, phase transitions, filling of the functionalizing layers, and an impurity concentration. The charge-carrier energy spectrum depends on the rigidity of the functionalizing molecular system. The high rigidity leads to suppression of the localization and the polaron formation. Under a phase transition, crystallization of the molecular film results in discontinuous growth of a rigidity and appropriate falling of the nonlinear interaction parameters, the longitudinal carrier localization disappears. In these cases, the nanowire conductivity must grow significantly. Thus, the functionalized nanowire can be used as a sensor for the state of the molecular system, chemisorption sensors. The considered processes should be taken into account in design of the CNT based chips, nanowires, and layered organic surface structures [22, 23].

Acknowledgments

This research is supported by FFI National Academy of Sciences of Ukraine, grant 4/17-H, Ministry of Science and Education of Ukraine under the Projects M0624, M05486.

References

[1] Pekar S. I. 1963. *Research in Electron Theory of Crystals*, English translation: AEC-tr-555, US Atomic Energy Commission (Pekar 1963).

[2] Lewin, Mathieu, and Rougerie, Nicolas 2013. "Derivation of Pekar's polarons from a microscopic model of quantum crystal." *SIAM Journal on Mathematical Analysis* 45: 1267-1301. https://doi.org/10.1137/110846312 (Lewin and Rougerie 2013).

[3] Davydov, Alexandr S. 1991. *Solitons in molecular systems*. Berlin: Springer-Verlag. (Davydov 1991).

[4] Krivoglaz M. A. 1974. "Fluctuon states of electrons" *Soviet Physics Uspekhi* 16: 856-877 DOI:10.1070/PU1974v016n06ABEH004095 (Krivoglaz 1974).

[5] Nagaev E. L. 1975 "Ferromagnetic and antiferromagnetic semiconductors" *Soviet Physics Uspekhi* 18:863-892; DOI:10.1070/PU1975v018n11ABEH005234 (Nagaev 1975).

[6] Nagaev, Eduard Leonovich 1983. *Physics of magnetic semiconductors.* Moscow: Mir Publishers. http://113.160.249.209:8080/xmlui/handle/123456789/12111 (Nagaev 1983).

[7] Yurkevich, V. E., Bystrov, V. S. and Rolov, B. N. 1989. "Fluctuons in ferroelectric-ferromagnetic systems." *Ferroelectrics* 89: 125-132 http://dx.doi.org/10.1080/00150198908017890 (Yurkevich et al. 1989).

[8] Brandt, N. B. and Kulbachinsky, V. A. 2005. *Quasiparticles in condensed matter physics.* Moscow: Fizmatlit. [in Russian] (Brandt and Kulbachinsky 2005).

[9] Lykah V. A. 1999. "Ferroelectric microdomain polaron in $YBa_2Cu_3O_{7-x}$". *Ferroelectrics* 233: 279-295. http://dx.doi.org/10.1080/00150199908018627 (Lykah 1999).

[10] Datta S. 1995. *Electronic transport in mesocopic systems.* Cambridge: Cambridge University Press. (Datta 1995).

[11] Ferry, David K., and Goodnick, Stephen M. 1997. *Transport in Nanostructures.* Cambridge: Cambridge University Press. (Ferry and Goodnick 1997).

[12] Yanson, Alex I., Yanson, Igor K., and van Ruitenbeek, Jan M. 2001. "Crossover from Electronic to Atomic Shell Structure in Alkali Metal Nanowires." *Physical review letters* 87: 216805 [4 pages]. (Yanson et al. 2001).

[13] Agraita, Nicolis, Yeyatib, Alfredo L., and van Ruitenbeek, Jan M. 2003. "Quantum properties of atomic-sized conductors." *Physics Reports* 377: 81-279. (Agraita et al. 2003, 81).

[14] Orlikowski, Daniel, Mehrez, Hatem, Taylor, Jeremy, Guo, Hong, Wang, Jian, and Roland, Christopher. 2001. "Resonant Transmission Through Finite-Sized Carbon Nanotubes." *Physical review B* 63: 155412 [12 pages]. (Orlikowski et al. 2001).

[15] Poncharal, Philippe, Berger, Claire, Yi, Yan, Wang, Z. L., and de Heer, Walt A. 2002. "Room temperature ballistic conduction in carbon nanotubes." *Journal of Physical Chemistry B* 106: 12104-12118. (Poncharal et al. 2002).

[16] Dekker, C. 1999. "Carbon nanotubes As Molecular Quantum Wires." *Physics Today* 52: 22-28. (Dekker 1999).

[17] McEuen, P. L. 2000. "Single-wall Carbon nanotubes." *Physical World* 13: 31-36. (McEuen 2000).

[18] Dai H. 2000. "Controlling nanotube growth." *Physical World* 13: 43-47 (Dai 2000).

[19] Rao, Daniel, S., Rao, Talasila P., Usha, Kota S., Naidu, Rani, S., Hea-Yeon, Lee, G.R.K., and Kawai, Tomoji. 2007. "A review of DNA functionalized/grafted carbon nanotubesand their characterization." *Sensors and Actuators B: Chemical* 122: 672-682. (Rao et al. 2007, 672).

[20] Ciraci, S., Dag, S., Yildirim, T., Gilseren, O., and Senger, R. T. 2004. "Functionalized carbon nanotubes and device applications." *Journal of Physics: Condensed Matter* 16: R901-R960. (Ciraci et al. 2004).

[21] Armitage, N. P., Gabriel, J.-C. P., and Gruner, G. 2004. "Quasi-Langmuir-Blodgett thin film deposition of carbon nanotubes." *Journal of Applied Physics* 95: 3228-3230 (Armitage et al. 2004, 3228). Accessed June 18, 2009 http://dx.doi.org/+10.1063/1.1646450.

[22] Buzaneva, E., Gorchynskyy, A., Popova, G., Karlash, A., Stogun, Y., Yakovkin, K., Zherebskiy, D., Matyshevska, O., Prylutskyy, Yu., Scharff, P. 2002. "Nanotechnology of DNA/nano-Si and DNA/carbonnanotubes/nano-Si chips." In *Frontiers of Multifunctional Nanosystems*, edited by E. Buzaneva, and P. Scharff. 191-212. NATO Science Series II: Mathematics, Physics and Chemistry Vol. 57. Kluwer. (Buzaneva et al. 2002).

[23] Neilands, O. 2002. "Organic compounds capable to form intermolecular hydrogen bonds for nanostructurs created on solid surface, aimed to sensor design." In *Molecular Low Dimensional and Nanostructured Materials for Advanced Applications*, edited by A. Graja, B. R. Bulka, amd F. Kajzar) 181-190. NATO Science Series II: Mathematics, Physics and Chemistry Vol. 59. Kluwer. (Neilands 2002).

[24] Courty, S., Mine, J., Tajbakhsh, A. R., and Terentjev, E. M. 2003. "Nematic elastomers with aligned carbon nanotubes: new electromechanical actuators." *EPL (Europhysics Letters)* 64: 654-660. (Courty et al. 2003).

[25] Zheng, M., Jagota, A., Strano, M. S., Santos, A. P., Barone, P., Chou, S. G., Diner, B. A., Dresselhaus, M. S., Mclean, R. S., Onoa, G. B., Samsonidze, G. G., Semke, E. D., Usrey, M., and Walls, D. J. 2003. "Structure-based carbon nanotube sorting by sequence-dependent DNA assembly." *Science* 302 (5650): 1545-1548. (Zheng et al. 2003).

[26] Somoza, A. M., Sagui, C., and Roland, C. 2001. "Liquid-crystal phases of capped carbon nanotubes." *Physical Review B* 63: 081403 [4 pages]. (Somoza et al. 2001).

[27] Lieber, C. M., and Wang, Zhong Lin. Guest Eds. 2007. "Functional Nanowires." *MRS Bulletin* 32: 99-108. (Lieber and Wang 2007).

[28] George, C. B., Ratner, M. A., and Lambert, J. B. 2009. "Strong Conductance Variation in Conformationally Constrained Oligosilane Tunnel Junctions." *Journal of Physical Chemistry A* 113: 3876-3880. (George et al. 2009).

[29] Canto, E. D., Flavin, K., Natali, M., Perova, T., Giordani, S. 2010. "Functionalization of Single Walled Carbon Nanotubes with Optically Switchable Spiropyrans." *Carbon* 48: 2815-2824.. doi: 10.1016/j.carbon. 2010.04.012 (Canto et al. 2010).

[30] Blinc, Robert, and Zeks B. 1974. *Soft modes in ferroelectrics and antiferroelectrics*. Amsterdam: North-Holland Publ. Co. (Blinc and Zeks 1974).

Polarons in the Functionalized Nanowires 263

[31] Sonin, A. S. 1983. *Introduction to Physics of Liquid Crystals.* Moscow: Nauka. [in Russian]. (Sonin 1983).

[32] Hiemenz, P. C., and Rajagopalan, R. (Eds.). 1997. *Principles of Colloid and Surface Chemistry, revised and expanded.* (Vol. 14). New York: Marcel Dekker. (Hiemenz and Rajagopalan 1997).

[33] Frolov, Y. G. 1988. *Course in Colloid Chemistry. Surface Phenomena and Disperse Systems.* Moscow: Chemistry. (Frolov 1988).

[34] Shchukin, E. D., Pertsov, A. V., Amelina, E. A., and Zelenev, A. S. 2001. *Colloid and surface chemistry* (Vol. 12). Elsevier. (Shchukin et al. 2001).

[35] Lykah, V. A., and Syrkin, E. S. 2004. "Soft polar molecular layers on charged nanowire." *Condensed Matter Physics* 7: 805-812. (Lykah and Syrkin 2004).

[36] Lykakh, Victor A., and Syrkin, Eugen S. 2005. "The effect of adsorbed molecules on the charge-carrier spectrum in a semiconductor nanowire." *Semiconductors* 39: 679-684. (Lykakh and Syrkin 2005, 679) Accessed June 01, 2005. doi: 10.1134/1.1944859.

[37] Lykah, Victor A., and Syrkin, Eugen S. 2012. "Carriers Spectra of Functionalized Semiconducting nanowires and Conformational Transition in Molecules." *Ukrainian Journal of Physics* 57: 711-717. (Lykah and Syrkin 2012, 711).

[38] Lykah, Victor A., and Syrkin, Eugen S. 2013. "Charge-carrier Spectra in Semi-conducting Nanowire Functionalized by Incommensurate Molecular Structures." *Advances in Optoelectronic Materials (AOM)* 1: 25-34. (Lykah and Syrkin 2013).

[39] Endo, Morinubo, Iijima, Sumio, and Dresselhaus, Mildred S., eds. 1996. *Carbon Nanotubes.* New York: Pergamon Press. (Endo et al. 1996).

[40] Landau, Lev D., and Lifshits, Eugen M. 1980. *Course of Theoretical Physics, Vol. 3: Quantum Mechanics: Non-Relativistic Theory.* New York: Pergamon. (Landau and Lifshits 1980).

[41] Landau, Lev D., and Lifshits, Eugen M. 1975. *Course of Theoretical Physics, Vol. 2: The Classical Theory of Fields.* New York: Pergamon. (Landau and Lifshits 1975).

[42] Dwight, Herbert B. 1961. *Tables of integrals and other mathematical data.* New York: Macmillan. (Dwight 1961).

[43] Prudnikov, A. P., Brychkov, Yu. A., Marichev, O. I. 1986. *Integrals and series.* New York: Gordon and Breach. (Prudnikov et al. 1986).

[44] Kovalyov, Alexandr S. 1978. "Frequency spectrum of monochromatic oscillations of one-dimensional nonlinear chain of finite length." *Theoretical and Mathwmatical Physics* 37: 135-144. [in Russian] (Kovalyov 1978).

[45] Blinov, L. M. 1983. *Electro-optical and magneto-optical properties of liquid crystals.* Chichester: J. Wiley. (Blinov 1983).

[46] Flugge, Siegfried. 1971. *Practical quantum mechanics*. Berlin: Springer-Verlag. (Flugge 1971).

[47] Janke, E., Emde, F., Losch, F. 1960. *Tafeln Hoherer Functionen*. Stuttgart: B.G.Taubner Verlagsgesellschaft. (Janke et al. 1960).

[48] Prudnikov, A. P., Brychkov, Yu. A., Marichev, O. I. 1989. *Integrals and series. Additional chapters*. Vol. 3. New York: Gordon and Breach. (Prudnikov et al. 1989).

[49] Abramowitz, M., Stegun, I. 1965. "Elliptic Integrals." In *Handbook of Mathematical Functions*. Edited by Milton Abramowitz and Irene A. Stegun. New York: Dover Publisher. 402-406. (Abramowitz and Stegun, 1965).

[50] Flygare, W. H. 1978. *Molecular structure and dynamics*. New Jersey: Prentice-Hall Inc. (Flygare 1978).

[51] Metzler, D. E. 2001. "Vitamins and Coenzymes." In *Encyclopedia of Physical Science and Technology [c]. Biochemistry*. Edited by Robert A. Meyers. 509-528 (3rd edition) New York: Academic Press. (Metzler 2001).

[52] Lines, M. E. and Glass, A. M. 1977. *Principles and Applications of Ferroelectrics and Related Materials*. Oxford: Oxford University Press. (Lines and Glass 1977). Published to Oxford Scholarship Online: February 2010. DOI:10.1093/acprof:oso/9780198507789.001.0001.

[53] Braun, Oleg M., and Kivshar, Yurii S. 2004. *The Frenkel-Kontorova Model. Concepts, Methods, and Applications*. Berlin, Heidelberg, New York: Springer. (Braun and Kivshar 2004).

In: Polarons: Recent Progress and Perspectives
Editor: Amel Laref

ISBN: 978-1-53613-935-8
© 2018 Nova Science Publishers, Inc.

Chapter 7

BOUND POLARONS AND EXCITON-PHONONS COUPLING IN SEMICONDUCTOR NANOSTRUCTURES

Abdelaziz El Moussaouy[*]
LDOM, Department of Physics, Faculty of Sciences,
Mohammed I University, Oujda, Morocco

ABSTRACT

In recent years, particular attention was paid to the polaron in physics of heterostructures of low-dimensionality such as quantum wells, quantum well wires, and quantum dots. Because of their reduction of dimensionality, the influences of optical phonons on confined charges carriers and excitons in nanostructures exhibit many new physical properties which are extremely interesting from the point of view of fundamental physics and also for their potential exploitations in microelectronic and optoelectronic device technology. This chapter offers an overview of the recent progress in the description of polaronic effects on confined impurities and excitons in semiconductor nanostructures and their relevance to improving electronic and excitonic characteristics in these systems. In the first step, the chapter presents the properties of semiconductor quantum dots and theory of polaron in these structures. Likewise, the chapter describes significant theoretical frameworks, currently developed in researches, considering the interactions between charge carriers and optical phonons in semiconductor of low-dimensional structures. Afterwards, the chapter explores and discusses the ground states binding energies and the photoluminescence's spectra by incorporating the polaronic contributions due to the optical phonon modes as function of some physical parameters. Furthermore, it investigates also the action of external perturbations such as hydrostatic pressure and temperature on these physical properties. The findings show the significant contributions of polarons and exciton-phonons coupling to some physical parameters and fosters better understanding of polaronic properties in such systems, which will favor more effective exploitations in future application devices.

---

[*] Corresponding Author address. Email: azize10@yahoo.fr.

Keywords: nanostructures, quantum dots, polarons, optical phonon modes, exciton-phonons, impurities, pressure, temperature

1. INTRODUCTION

Remarkable progress has been made over the last years regarding the properties of polaron and exciton-phonon interaction in semiconductors, in particular nanomaterial structures such as quantum well (QW), quantum well wire (QWW) and quantum dot (QD). The state of the art is described in detail in the present book. A large application-driven interest stems from the fact that nanostructures are good candidates in promising applications like QD lasers, single photon sources, and quantum information devices. The polaronic process has become a main research subject in the physics of nanosystems [1–8]. This is due to their role in scattering of charge carriers, which is interesting for the understanding of the experimental issues of the semiconductor optical spectra [1–3, 9, 10]. It is investigated that electrons and longitudinal optical (LO) phonons are in the strong coupling regime leading to polaronic influences [11–13]. This involves that their interaction can never be considered perturbatively and that, in fact, electrons and phonons form mixed states. A significant exception is when the energies of the LO phonons match the level separation [14], in this sense the exciton–phonon coupling can be remarkably strong. More recently, several works have been interested in this advancement and in its effect on various physical properties of polar semiconductor of low-dimensional structures [15]. Some recent optical measurements [16–18] of the photoluminescence (PL) spectra conducted on different QDs and QWs semiconductor structures, also reveal the LO phonon's influence on the PL line widths. Since II–VI materials have high ionicity, it is evident to expect the presence of electron and hole polaron states [19]. The comparison of polaronic contribution in different nanostructures has shown that the maximum polaron impact is in the QD [20].

This chapter discusses recent theoretical progress in the description of polaronic contribution to charge carriers states in semiconductor nanostructures along with their relevance to improve electronic and excitonic properties in these systems.

After a short introduction, some properties of semiconductor QD are presented in Sec. 2. The interaction between exciton and different phonon modes in QD are developed and the recent numerical findings are discussed in Sec. 3. Finally, the theoretical works of polaron in QD are highlighted in Sec. 4.

2. PROPERTIES OF SEMICONDUCTOR QUANTUM DOTS

QDs are semiconductor nanostructures, in which the charge carriers are confined in three spatial dimensions and have various unique properties and show noticeable phenomena, such as size emission wavelength, broad excitation and narrow emission peak [21-22]. The confinement of charge carriers in a QD leads to a transformation from the continuous optical spectrum of a bulk material to a size-tunable, atomic-like spectrum featuring a series of sharp peaks related with discrete electronic levels. Every peak can be related to a specific multi-particle complex, such as exciton, biexciton, trion, as well as other excitonic states attributed to higher energy levels in the nanostructure. The binding between electrons and holes via Coulombic interaction is improved in QDs with regard to other systems, which can permit to dramatic change of the exciton properties. Moreover, confinement diminishes the influences of external perturbations that can affect the radiative lifetime of excitonic resonances. These special electronic and optical properties of QDs have been considered for a wide range of various exploitations: integrating quantum information and quantum electronics, display technologies, optical communications, fluorescent labelling, solar energy harvesting, renewable energy, and environmental issues. The significant importance for many of these applications are the physical parameters identifying the optical transitions, such as the QD size, shape and material composition, which are not easy to manipulate during elaboration. The elaboration methods and fabrication procedures of the QD are inherently statistical processes that exhibit in an ensemble of QDs with a distribution of spectroscopic characteristics. Identifying and characterizing the relation between the form and the design of a QD and its electronic and optical properties is the major key for the effective integration of various QD-based exploitation devices.

The development of semiconductor growth techniques such as MBE and MOCVD [23, 24] permitted the realization of hetero-junction between various materials at the interfaces with novel electronic and optical characteristics. This significant advancement embarked a new regime of semiconductor quantum physics in which the effects of quantum confinements could be considered and controlled through spatial adjustment of the QD. Since the interesting work of Esaki and Tsu [25, 26] along with examining transport characteristics of a QW superlattice, the obtaining of energy quantization, and tunneling effects in GaAs-AlGaAs QWs by Dingle et al. [27], the physics of low-dimensional structures has been investigated-with growing momentum- and new observed quantum phenomena have been explored.

Historically, semiconductor lasers considering a QW as the active medium were achieved [28, 26], and localization and interaction influences on electrons and holes confined in one-dimensional QWWs were discussed [30]. In this regard, quantum size effects on the absorption spectrum of excitons confined in epitaxially-grown quasi-zero-dimensional quantum boxes and in colloidal crystals of low-dimensional systems were explored [31-33]. Based on these innovation studies, Reed et al. [34] offer the first indirect evidence of energy quantization, and identify the term QD to depict the zero-dimensional quantum boxes. Follow-up investigations by these authors [35] disclose discrete peaks in resonant tunnelling spectra, revealing direct evidence of the atomic-like density of states in QDs.

The above innovating researches permit to motivate the number of researchers focused on epitaxial-grown semiconductor QDs. Most works in the following decade have focused on identifying the fundamental electronic and optical properties of QDs and improving innovative methods of elaboration. The first experiments that investigated QDs were elaborated through electron beam lithography of GaAs-AlGaAs and InGaAs-InP [36, 37] QWs. The lowest energy optical transition of these nanostructures was indicated through ensemble photoluminescence measurements [14, 15]. The electronic and optical properties could be tuned through action of external fields and were identified by measuring Zeeman splitting of electronic states, resonant tunnel effects between QDs and QD charging with many electrons [38-42]. Investigation of the linear and nonlinear optical properties [43, 44], confinement influences on exciton creation and Coulomb interactions and the action of field leaded in the understanding of the experimental observations [45, 46]. However, considerable inhomogeneity of the QD properties due to size dimension makes it difficult to relate theory and experiments.

A particular result was attained when for the first time, the researches [47] exhibit μ-PL spectra from a GaAs-AlGaAs QD. As the QD lateral dimensions became similar to the exciton Bohr radius in bulk GaAs, discrete spectral lines observed in the photoluminescence spectrum, proving that are related to the atomic-like spectrum of an exciton in a single QD. Likewise, they developed the growth technique by instead interrupting the epitaxial growth process of a narrow QW, permitting the gallium and arsenide atoms to diffuse to nucleation sites, naturally obtaining monolayer interfacial islands, called as interfacial fluctuation QDs [48]. These studies pushed a revolution in single QD elaboration and characterization, permitting for better facilely-controlled, systematic works of size dependences on the QD properties. Afterwards, some authors have discussed the excited state spectrum of a single interfacial fluctuation QD, the temperature contribution of the exciton homogeneous line-width and the exchange interaction in QDs [49-51].

Similar to these advancements, important progress was achieved towards the elaboration of semiconductor QDs [52-60] considering various material organizations, integrating CdTe, ZnTe, InSb, GaSb, AlSb, and In(Ga)As, realized on a GaAs compound by employing the Stranski-Krastanow procedure [61, 62]. InGaAs/GaAs self-assembled QDs would form naturally to strain effect created in the InGaAs layer from the InGaAs/GaAs lattice coupling. InGaAs self-assembled QDs were identified via investigation of the electronic structure and carrier relaxation dynamics, excitonic localization, and dimensional effects on the exciton exchange interaction, biexciton binding energy and homogeneous line width [63-68].

Recently, electronic and optical properties of semiconductors QD have attracted significant attention in physics of low-dimensional structures. Optical properties of QD depend on both the dimension and composition of the dots. The linear and nonlinear optical properties have been investigated theoretically by different authors [69–73]. Wang and Guo [69] have studied excitonic effects on the third-harmonic generation coefficient for typical GaAs/AlGaAs parabolic QD. Khordad [70] has calculated the linear and nonlinear optical properties of T-shaped QWWs. Shao et al. [71] have developed third-harmonic generation coefficient in QDs in presence of electric field. Morales et al. [72] have discussed the donor-related optical absorption spectra for a double QW under hydrostatic pressure and electric field. Karabulut et al. [73] have studied the second and third-harmonic generation susceptibilities of spherical QDs with parabolic confinement subjected to an external electric field by considering the impurity.

2.1. Exciton States in Bulk of Semiconductor

2.1.1. Periodic Potential Lattice and Bloch Wave Functions

In a bulk crystal lattice, the Schrödinger equation for an electron is written as:

$$H.\psi(\vec{r}) = \left(-\frac{\hbar}{2m}\nabla^2 + U(\vec{r})\right).\psi(r) = E.\psi(\vec{r}), \qquad (1)$$

where $U(\vec{r})$ is the potential of periodic crystal lattice. Bloch's theorem then states that the energy eigenfunction for such a Hamiltonian, $\varphi_{n\vec{k}}(\vec{r})$, can be expressed as the product of a plane wave envelope function and a periodic Bloch function, $u_{n\vec{k}}(\vec{r})$, that has the same periodicity of the underlying atomic potential, which is equal to $u_{n\vec{k}}(\vec{r}).e^{n\vec{k}}$. $\vec{k}$, equivalent to the crystal momentum when multiplied by the reduced Planck's constant, is unique only up to a reciprocal lattice vector and therefore one only needs to consider $\vec{k}$ within the first Brillouin zone of the reciprocal lattice. The relating energy eigenvalues, $E_n(\vec{k})$, are also periodic with periodicity defined by the reciprocal lattice vector $\vec{K}$. The index n identifies an energy band for which the energy varies continuously with $\vec{k}$.

For direct gap III-V semiconductors, such as Ga(Al)As and In(Ga)As, the dispersion relation $E_n(\vec{k})$ is approximately quadratic at the centre of the Brillouin zone; thus an effective mass for the electron can be identified similarly as for a free electron plane wave, $m^* = \left(\frac{1}{h^2}\frac{d^2E}{d^2k^2}\right)^{-1}$.

The dispersion relation for an electron in the crystal is not the same for a free space electron, and thus m^* is generally different from the free-space electron mass. Until now, the Bloch theorem is simply a general statement of the wave function form in a periodic potential. Additional insight can be obtained by making specific supposition about the wave functions using the tight binding approximation [74, 75] in which the electrons are tightly-bound to the atom to which they associate and only weakly binding interact with electrons in neighbouring atoms.

When the QD confinement is considered, the envelope wave function must respect the boundary conditions imposed by the geometrical confining potential. In addition, other influences, such as Coulomb interactions responsible for generating exciton, complicate the picture. Therefore, in order to properly determine the electron and hole wave functions and energies, a more complete theory that considers these contributions is necessary, which will be incorporated in the forthcoming sections. In the following Section, excitonic effects on the optical properties in bulk materials are discussed.

2.1.2. Excitons in Semiconductors

In semiconductor, if energy of incident photon is similar to the band gap, then they can be absorbed by the electrons generating atomic bonds between neighbouring atoms, and so provide them with enough energy to be free and move around in the body of the crystal. This would be described as exciting an electron from the valence band across the band gap into the conduction band. If the energy of the photon is larger than the band gap, then a free electron is created and an empty state is left within the valence band. The empty state within the valence band acts like an air bubble in a liquid and rises to the top the lowest energy state. This hole acts as though it was positively charged and hence usually forms a bond with a conduction band electron.

The attractive Coulomb interaction between the electron and hole enables the formation of a bound state, known as an exciton. The exciton can be considered as a quasiparticle, and unbound electron-hole pairs are described as continuum states [76]. The concept of a quasiparticle is convenient for representing the most basic excitations arising from weak perturbations. Typically, excitons have large oscillator strengths and dominate the linear optical properties of semiconductors near the band edge. However, the residual Coulomb forces not accounted for in exciton formation leads to interactions between excitons and induce a nonlinear response [76].

Generally, the hole mass is very important, like the electron mass, and consequently the two-body system analogous to hydrogen atom, with the negatively charged electron revolving around the proton. The exciton is rather stable and can have a relatively long lifetime. Exciton recombination is a significant feature of low temperature photoluminescence, whereas as the binding energies are low enough, they tend to dissociate to free electron and hole at higher temperatures. Thus, the exciton binding energy and orbital radius can be well described by Bohr Theory, with the correction for the finite mass of the central charge [77, 78]. This is incorporated by exchanging the orbiting electron mass, with the reduced mass of the two-body system, in this case, the electron-hole pair. In the center of mass coordinates, the exciton has a reduced mass $\mu = \frac{m_e m_h}{m_e + m_h}$ and a total mass $M = m_e + m_h$, where $m_e (m_h)$ is respectively the electron (hole) masses.

Consequently, the binding energy reads:

$$E = -\frac{\mu e^4}{34\pi^2 \hbar^2 \varepsilon_r^2 \varepsilon_0^2},$$ (2)

and the Bohr radius is written as:

$$\lambda = \frac{4\pi \hbar^2 \varepsilon_r \varepsilon_0}{\mu e^2},$$ (3)

As an example using particular values of the bulk GaAs, the electron and heavy-hole effective masses, $m_e^* = 0.067 m_0$ and $m_{hh}^* = 0.62 m_0$, and reduced mass $\mu = 0.060\ m_0$. The static dielectric constant is $\varepsilon_r = 13.18$ [79], and then the exciton binding energy and Bohr radius are taken respectively: -4 meV and 115 nm. In the similar manner for CdTe the exciton binding energy and Bohr radius are taken respectively: -10.1 meV and 6.7 nm [79].

In the similar manner as in bulk material, the exciton in heterostructures can be constituted by the coupling of free electron and free hole pairs or through resonant excitation. While in bulk, the global energy of the exciton is absolutely the energy of the free electron and hole pair (band gap) added to the exciton binding energy, in a heterostructure there are additional terms due to the electron and hole confinement energies.

The Hamiltonian describing the coupling two-body electron-hole system can be considered to be the sum of three terms:

$$H = H_e + H_h + H_{e-h}.$$ (4)

The term $H_e (H_h)$ is the electron (hole) Hamiltonian appropriate to the conduction (valence bands) of a specific microstructure of interest. Whereas, for the exciton a constant effective mass along the growth axis must be considered, i.e., the Hamiltonians of electron and hole are expressed as:

$$H_e = -\frac{\hbar^2}{2m_e^*}\frac{\partial^2}{\partial z^2} + V_{\text{Conduction band}} \; ; \; H_h = -\frac{\hbar^2}{2m_h^*}\frac{\partial^2}{\partial z^2} + V_{\text{Valance band}}, \quad (5)$$

H_{e-h} describes the electron-hole coupling, and is itself constructed by two terms. One of these represents the kinetic energy of the relative motion of the electron and hole in the x-y plane, whereas the second describes the Coulombic potential energy, i.e.

$$H_{e-h} = \frac{P^2}{2\mu} - \frac{e^2}{4\pi\varepsilon r}, \quad (6)$$

where the parameter P is the quantum mechanical momentum operator for the in-plane component of the relative motion, and r is the electron-hole distance.

As the main interest is connected to the optical properties of excitons in QW systems, no term for the motion of the centre of mass of the exciton in the x-y plane in the Hamiltonian, i.e., the exciton is supposed to be at rest within the plane of the well.

In general, the problem is to research solution of Schrödinger equation to find the eigenfunctions and eigenvalues of energies describing the system:

$$H\psi = E\psi. \quad (7)$$

Following standard approaches, the two-body exciton wave function is chosen to have a trial form, as follows:

$$\psi = \psi_e(z_e)\psi_h(z_h)\psi_r, \quad (8)$$

where ψ_e and ψ_h are simply the eigenfunctions of the electron and hole Hamiltonians of the heterostructure, respectively. The term ψ_r describes the electron and hole relative motion. According to Hilton et al. [80, 81] and Harrison et al. [82, 83], ψ_r will be a variational wave function used to minimize the total energy of the system. The particular form of ψ_r will be developed later.

One of the main benefits of such model is that it is independent of the form of the one charge carrier Hamiltonians, and indeed calculations can be elaborated on any structure in which the standard electron and hole wave functions can be calculated [84-86].

By integrating over all space, the total exciton energy follows simply as the expectation value:

$$E = \frac{\langle\psi|H|\psi\rangle}{\langle\psi|\psi\rangle}. \quad (9)$$

2.2. Basic Concepts of Optical Phonons and Polarons

In semiconductor crystals, the massive atoms are all attached together by interactions via covalent bonds. These atoms are forever in a state of continual motion, which, because of the accurate crystal lattice structure, is vibrational around an equilibrium site. The atoms oscillate even at the hypothetical zero of absolute temperature. In few manners, the vibrations of these interattached atoms are similar to a classical system of a series of masses related by springs. There are essentially four separate modes of vibration each one of which is called a phonon. The acoustic phonon modes are described by the neighbouring atoms being in phase. In the longitudinal mode the atomic movements are in the same orientation as the direction of energy transfer, while in the transverse mode the atomic movements are perpendicular to this orientation. The terms of longitudinal and transverse also apply to the two kinds of optical phonon modes, though in this kind of lattice vibration the movements of neighbouring atoms are in opposite phase.

A conduction electron (hole) with its self-induced polarization in a polar semiconductor or an ionic crystal creates a quasiparticle, which is known by a polaron. The concept of polaron becomes nowadays more important, not only because it characterizes the specific physical properties of an electron in polar crystals and ionic semiconductors but also because it is an attractive field and theoretical model consisting of a fermion coupling with a scalar boson field. The early work on polarons was developed with general theoretical formulations and approximations, which now establish the fundamental theory with experiments on cyclotron resonance and transport properties. The bound polaron generates some novel physical properties and differ from those of the band-electron. Especially, the polaron is identified by its binding or (self-) energy, effective mass and by its response to external fields. The general polaron concept was firstly introduced by Landau [87]. Afterwards Pekar [88] investigated the self-energy and the effective mass of the polaron, for what was illustrated by Fröhlich [89] to correspond to the adiabatic or strong-coupling regime.

The initial work on polarons was interested to the coupling between the charge carriers and the long-wavelength optical phonon modes. The field-theoretical Hamiltonian considering this interaction, developed by Fröhlich, is expressed as:

$$H = \frac{p^2}{2m_b} + \sum_k \hbar\omega_{LO} a_k^+ a_k + \sum_k \left(V_k a_k e^{ik.r} + h.c. \right), \qquad (10)$$

where r is the coordinate operator describing the position of the electron with band mass m_b, p is its canonically conjugate momentum operator. The parameters a_k^+ and a_k are, respectively, the creation and the annihilation operators for longitudinal optical phonons of wave vector k and energy $\hbar\omega_{LO}$. The terms V_k represent Fourier components of the electron-phonon coupling:

$$V_k = -i \frac{\hbar\omega_{LO}}{k} \left(\frac{4\pi\alpha}{V} \right)^{\frac{1}{2}} \left(\frac{\hbar}{2m_b\omega_{LO}} \right)^{\frac{1}{4}}, \qquad (11)$$

with

$$\alpha = \frac{e^2}{\hbar c} \sqrt{\frac{m_b c^2}{2\hbar\omega_{LO}}} \left(\frac{1}{\varepsilon_\infty} - \frac{1}{\varepsilon_0} \right). \qquad (12)$$

α is the Fröhlich coupling constant. ε_∞ and ε_0 represent, respectively, the electronic and the static dielectric constant of the polar crystal.

It was assumed that the size extension of the polaron is considerable with regard to the solid lattice parameters, the spin and relativistic influences can be ignored, the band-electron has parabolic dispersion, and it is also supposed that the LO phonons of interest for the coupling, are the long-wavelength phonons with constant frequency ω_{LO}.

The concept of the polaron has been generalized over the years to consider polarization fields other than the LO-phonon field: the acoustical phonon field, the exciton field, etc. For some materials the continuum approximation is not suitable as far as the polarization is confined to a domain of the order of a unit cell; the so-called "small polaron" is a more pertinent quasiparticle in that case. The term "large polaron" or "Fröhlich-polaron" is habitually adopted for the quasiparticle considering the charge carriers (electron or hole) and the polarization induced by the LO-phonons modes. For more reviews on polaron and exciton-phonon interaction see Refs. [90-95].

3. EXCITON-PHONONS STATES IN QUANTUM DOTS

3.1. Exciton States in Quantum Dot

To treat theoretically the exciton in QD, the Hamiltonian variational calculation is proved to be an important method frequently used in recent studies. The model of QD structure presented here is assumed to have a cylindrical form of radius R and height H = 2d, whose prototypical scheme and potential energy profile are those shown in figure 1.

From the methodological point of view the basic Hamiltonian for exciton states in cylindrical QD within the framework of the effective-mass and non-degenerate band approximation, can be expressed as follows:

$$H = -\frac{\hbar^2}{2m_e^*}\nabla_e^2 - \frac{\hbar^2}{2m_h^*}\nabla_h^2 - \frac{e^2}{\varepsilon_0|r_e - r_h|} + V_w^e(r_e) + V_w^h(r_h), \qquad (13)$$

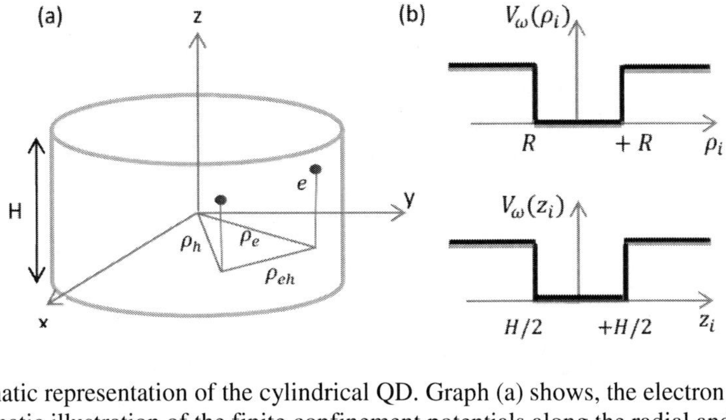

Figure 1. Schematic representation of the cylindrical QD. Graph (a) shows, the electron and hole positions. Schematic illustration of the finite confinement potentials along the radial and axial directions are depicted in Graphs (b).

where m_e^* and m_h^* are the effective masses of the electron and hole, respectively. $r_e = (\rho_e, z_e)$ and $r_h = (\rho_h, z_h)$ are the spatial coordinates of the electron and hole in cylindrical frame, respectively.

$-\dfrac{e^2}{\varepsilon_0(T)|r_e - r_h|}$ is the Coulomb potential, where ε_0 is the static dielectric constant. $V_w^e(r_e)$ ($V_w^h(r_h)$) is the corresponding electron (hole) confining potential written as follows:

$$V_w(r) = \begin{cases} 0 & \text{if } \rho_i \leq R \text{ and } |z_i| \leq d \\ V_i & \text{otherwise.} \end{cases} \tag{14}$$

The expressions of V_e and V_h depend on the type of semiconductor materials used in the calculations [96].

The effective Hamiltonian, in the atomic units system ($a_{ex}^* = \dfrac{\varepsilon_0 \hbar^2}{\mu e^2}$ excitonic Bohr radius and Rydberg energy $R_{ex}^* = \dfrac{\mu e^2}{2\varepsilon_0^2 \hbar^2}$), reads:

$$H_{eff} = -\frac{1}{1+\sigma}\left[\frac{\partial^2}{\partial \rho_e^2} + \frac{1}{\rho_e}\frac{\rho_{eh}^2 + \rho_e^2 - \rho_h^2}{\rho_e \rho_{eh}}\frac{\partial^2}{\partial \rho_e \partial \rho_{eh}} + \frac{\partial^2}{\partial z_e^2}\right]$$
$$-\frac{\sigma}{1+\sigma}\left[\frac{\partial^2}{\partial \rho_h^2} + \frac{1}{\rho_h}\frac{\rho_{eh}^2 + \rho_h^2 - \rho_e^2}{\rho_h \rho_{eh}}\frac{\partial^2}{\partial \rho_{eh} \partial \rho_{eh}} + \frac{\partial^2}{\partial z_h^2}\right] - \left[\frac{\partial^2}{\partial \rho_{eh}^2} + \frac{1}{\rho_{eh}}\frac{\partial}{\partial \rho_{eh}}\right]$$
$$-\frac{2}{\sqrt{\rho_{eh}^2 + (z_e - z_h)^2}} + V_w^e(z_e) + V_w^h(z_h), \tag{15}$$

where $\sigma = \dfrac{m_e^*}{m_h^*}$.

The trial exciton wave function is chosen to be [96]:

$$\psi_{ex}(\rho_e, \rho_h, z_e, z_h) = F_e(\rho_e, z_e) F_h(\rho_h, z_h) F_{eh}(\rho_{eh}, |z_e - z_h|), \tag{16}$$

with

$$F_{eh}(\rho_{eh}, |z_e - z_h|) = \exp(-\alpha \rho_{eh}) \exp(-\gamma(z_e - z_h)^2), \tag{17}$$

and

$$F_i(\rho_i, z_i) = f_i(\rho_i) g_i(z_i), \qquad (i = e, h). \tag{18}$$

Respectively, the corresponding 2D (lateral direction) and 1D (longitudinal direction) effective mass Schrödinger equations are:

$$\left\{-\frac{\hbar^2}{2m_i^*}\nabla_i^2 + V_w^i(\rho_i)\right\} f_i(\rho_i) = E_i(\rho_i) f_i(\rho_i), \qquad (i = e, h), \tag{19}$$

$$\left\{-\frac{\hbar^2}{2m_i^*}\nabla_i^2 + V_w^i(z_i)\right\} g_i(z_i) = E_i(z_i) g_i(z_i), \qquad (i = e, h), \tag{20}$$

Bound Polarons and Exciton-Phonons Coupling in Semiconductor Nanostructures 275

with solutions of the form:

$$f_i(\rho_i, T) = \begin{cases} J_0(\theta_i \frac{\rho_i}{R}) & \text{for } \rho_i \leq R, \\ A_i K_0(\beta_i \rho_i) & \text{for } \rho_i > R \quad (i = e, h), \end{cases} \quad (21)$$

$$g_i(z_i, T) = \begin{cases} \cos(\pi_i \frac{z_i}{2d}) & \text{for } |z_i| \leq d \\ B_i \exp(k_i |z_i|) & \text{for } |z_i| > d \quad (i = e, h). \end{cases} \quad (22)$$

J_0 and K_0 are the modified Bessel functions of 0th order. $\theta_i, \beta_i, \pi_i, k_i, A_i$ and B_i are determined from the boundary conditions at the interfaces $\rho_i = R$ and $|z_i| = d$.

To calculate the exciton ground state energy, the expectation value of the effective Hamiltonian, Eq. (15) is minimized with respect to the variational parameters α and γ. The exciton binding energy is given with respect to the subband energy and the self energies of the electron and hole [96].

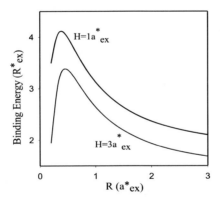

Figure 2. Binding energy of an exciton in a QD as a function of cylinder radius for two values of the height and zinc concentration x = 0.3.

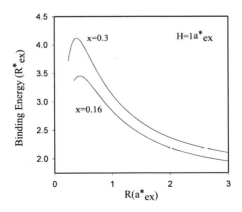

Figure 3. Binding energy of an exciton in a QD of height $H = 1a_{ex}^*$ as a function of the cylinder radius for two values of zinc concentration.

The binding energy of an exciton in a QD requires a more clarified definition. Generally, this physical parameter relates to the amount of energy needed to break the bound state of the e–h pair and to situate the component particles in their one-particle levels. Similar transition is shown in bulk material, in a QW, and even in a QWW, where the electron-hole system can be dimensionally separated in at least one free size. Though, in three spatial confinements, where the exciton is confined inside the dot, the electron and the hole cannot be considered in uncorrelated one-particle levels. Therefore, the quantity is related to the Coulomb correlation energy of the exciton. Consequently, the term "exciton binding energy" must be interpreted within this situation.

Figure 2 plots the variation of the binding energy of an exciton in a QD, made of CdTe embedded in $Cd_{1-x}Zn_xTe$ material, as a function of the cylinder radius R for two values of the dot heights. The Zn concentration is chosen to be x = 0.3. the physical parameters corresponding to the (well) crystal are [96]: $\varepsilon_0 = 9.6$, $m_e^* = 0.098\, m_0$, $m_h^* = 0.11\, m_0$. m_e^* and m_h^* are respectively, the electron and the hole effective masses. The curves are presented in units of the effective Rydberg energy $R_{ex}^* = 7.62\, meV$ and of the effective Bohr radius $a_{ex}^* = 98.17\, Å$. It can be seen from the curves that, for each value of the dot height, the binding energy increases as the dot radius reduces, reach a maximum for a critical value of the radius and then decreases. This behavior is explained via the finite confinement potential model considered in the calculation. The exciton wave function is compressed when the dot radius is reduced and consequently the binding energy is largely affected. It is also shown that, because of the tunneling effect, discrete levels are absent inside the dot, and the charge carriers wave functions are distributed outside the QD. When the radius is increased, the exciton levels fall from the continuum spectrum into the well.

The effect of an increasing barrier height, by augmenting zinc concentration x, is shown in figure 3, where the later plots the binding energy as a function of two value of zinc concentration. It can be seen that, the binding energy increases as the radius reduces and present a maximum value for a critical radius Rc. This critical radius depends on the zinc concentration and this effect is more significant for the smaller dot radius.

3.2. Exciton-Phonon Coupling in Quantum Dot

Recently, there has been a more significant interest in investigating the electronic and optical properties of QDs of semiconductors due to their remarkable exploitations in optoelectronic devices and many other applications. QDs based on GaAs, CdTe, ZnSe, ZnO, MgO, CuCl and CdS [97-104] have been elaborated and their properties are examined in more detail. It has been suggested that the lasing process in these systems is excitonic in nature and therefore can lead to lower values of the threshold current densities. Consequently, a better understanding of the exciton related phenomena in these confined structures is of particular importance. Of significant interests are the current works of polaronic contributions to low-dimensionally confined excitons [105–111]. More interestingly, summaries on the size dependence of the exciton–LO-phonon coupling in a QD structure are even contradictory. Some works point out that Fröhlich type exciton–LO-phonon interaction should disappear in small nanocrystals. However, in other studies, using a donor-like exciton model and adiabatic approximation it has been achieved that exciton– LO-phonon coupling strength is unattached to dot radius. This was contradicted in another work, which claims that polaronic effects in a donor like exciton should increase with decreasing dot size (for further discussion see Ref. [110] and the references cited therein). Regarding these divergent achievements, the problem of exciton-phonon interaction in a QDs requests more attention.

The excitonic properties of polar semiconductor QDs are importantly modified due to the effect of the exciton–optical-phonon coupling. For example, exciton binding energy is improved, controlled by the strength of this interaction. Consequently, a proper description of the nature of this interaction among electrons, holes, and phonons should not be oversimplified. This influence must be appropriately described by an effective electron–hole coupling and not by simply considering that these charge carrier band masses are reduced to their respective polaron masses. Indeed, the polaron-like coupling among electrons and holes is related to their mutual distance in such a way that when this is much larger than their polaron radii, the electron and the hole interact like polarons through the statically screened Coulomb potential. However, when their distance is less than their polaron radii, the two polarized virtual phonon clouds about each particle overlap and partially cancels out their renormalization effects, so that the electron–hole interaction approaches the dynamically screened Coulomb coupling. Similar representation of the effective electron–hole coupling, considering the above mentioned limiting cases has previously been given across some effective potential [111–116].

The theoretical model adopted in investigating the properties of exciton-phonon in QD system is also treated by a variational procedure. In the similar model of QD as mentioned above, the basic Hamiltonian of an exciton in such QD coupling to the different optical phonons, can be expressed within the framework of the effective-mass and nondegenerate-band approximations, as follows [8]:

$$H_{exciton-phonons} = H_{ex} + W_{phonons}, \tag{23}$$

where H_{ex} is the excitonic Hamiltonian which, in the absence of phonon modes is expressed as Eq.13. However, when these phonons are incorporated in the calculation the Coulomb potential will be expressed as $-\frac{e^2}{\varepsilon_\infty |r_e - r_h|}$, where ϵ_∞ is the high frequency dielectric constant.

$W_{phonons}$ is the Hamiltonian describing the contribution of different optical phonons modes which is written as:

$$H_{phonon} = H_{LO} + H_{TSO} + H_{SSO} + H_{ex-LO} + H_{ex-TSO} + H_{ex-SSO}, \tag{24}$$

where H_{LO}, H_{TSO} and H_{SSO} describe the Hamiltonian operators for confined LO, TSO, SSO phonon modes, respectively, which can be expressed as:

$$H_{LO} = \sum_{l,n1} \hbar\omega_{LO} a^+_{l,n1} a_{l,n1}, \tag{25}$$

where the parameter $a^+_{l,n1}(a_{l,n1})$ is the creation (the annihilation) operator for the LO-phonon of the (l, ln1)th mode, with frequency $\hbar\omega_{LO}$ and wave vector $(K_p = \frac{\chi_{n1}}{R}, K_z = \frac{l\pi}{2d})$ where χ_{n1} is the n1-th root of the zero-order Bessel function.

In Eq. 24 H_{TSO} and H_{TSO} are, respectively, the Hamiltonian operators for TSO and SSO phonons:

$$H_{TSO} = \sum_{n2,m,p} \hbar\omega_p b_{n2,m,p}^+ b_{n2,m,p},$$ (26)

and

$$H_{SSO} = \sum_{n3,m,p} \hbar\omega_{SS} C_{n3,m,p}^+ C_{n3,m,p}.$$ (27)

$b_{n2,m,p}^+(b_{n2,m,p})$ and $C_{n3,m,p}^+(C_{n3,m,p})$ are, respectively, the creation (the annihilation) operators of TSO and SSO phonons with frequency $\omega_p(p = +, -)$ and ω_{SS} of the (n2, m)th and (n3, m)th modes. The frequencies of different phonon modes are expressed as:

$$\omega_{LO}^2 = \frac{\varepsilon_0}{\varepsilon_\infty} \omega_{TSO}^2,$$ (28)

$$\omega_{TSO\pm}^2 = \frac{[\varepsilon_0 + \varepsilon_b] \mp [\varepsilon_0 - \varepsilon_b]e^{(-2q\pm d)}}{[\varepsilon_\infty + \varepsilon_b] \mp [\varepsilon_\infty - \varepsilon_b]e^{(-2q\pm d)}} \omega_{TO}^2,$$ (29)

$$\omega_{SSO}^2 = \left(1 - \frac{\varepsilon_0 + \varepsilon_\infty}{\varepsilon_0 + \varepsilon_b \frac{I_0(k_{n3}R)K_1(k_{n3}R)}{I_1(k_{n3}R)K_0(k_{n3}R)}}\right).$$ (30)

The parameters I_0, K_0, I_1 and K_0 represent the modified Bessel functions of zero and first order, $k_{n3} = \frac{n3\pi}{2d}$ describes the wave vector of the SSO phonon mode, ω_{TSO} the transverse optical phonon frequency and ε_b is the dielectric constant of the barrier material. The terms $q_{\pm,n2}$ are the wave vectors of TSO, which can be acquired by obtaining the n2th roots of the following equations:

$$\pi q_{+,n2} R J_1(q_{+,n2}R)J_0(q_{+,n2}R) = 1 - e^{(-2q_{+,n2}d)},$$ (31)

$$\pi q_{-,n2} R J_1(q_{-,n2}R)J_0(q_{-,n2}R) = 1 - e^{(-2q_{-,n2}d)}.$$ (32)

The Hamiltonians in Eq. (24), H_{ex-LO}, H_{ex-TSO} and H_{ex-SSO} describe respectively, the interaction operators between the exciton and different phonon modes (LO, TSO, SSO). Thus:

$$H_{ex-LO} = H_{e-LO} + H_{h-LO},$$ (33)

where

$$H_{i-LO} = -\sum_{n1} J_0\left(\frac{\chi_{n1}}{R}\rho_i\right) \times \left[\begin{array}{l} \sum_{l=1,3..} V_{ln1}(P,T) \cos\left(\frac{l\pi_i}{2d}z_i\right)(a_{ln1} + a_{ln1}^+) + \\ \sum_{l=2,4..} V_{ln1}(P,T) \sin\left(\frac{l\pi_i}{2d}z_i\right)(a_{ln1} + a_{ln1}^+) \end{array}\right]$$

$$(i = e, h), \quad (34)$$

and

$$V_{ln1}^2 = \frac{1}{V}\frac{4\pi e^2 \hbar\omega_{LO}}{\left(\frac{\chi_{n1}}{R}\right)^2 J_1^2(\chi_{n1})\left[1 + \left(\frac{l\pi R}{2d\chi_{n1}}\right)^2\right]}\left(\frac{1}{\varepsilon_\infty} - \frac{1}{\varepsilon_0}\right). \quad (35)$$

In a similar manner, the Hamiltonian describing the exciton-TSO interaction can be written as:

$$H_{ex-TSO} = H_{e-TSO}(P,T) + H_{h-TSO}, \quad (36)$$

with

$$H_{i-TSO} = -\sum_{n2,+} V_{n2+} J_0(q_{n2+}\rho_i) \cosh(q_{n2+}z_i)(b_{n2+} + a_{n2+}^+)$$
$$- \sum_{n2,-} V_{n2-} J_0(q_{n2-}\rho_i) \cosh(q_{n2-}z_i)(b_{n2-} + a_{n2-}^+), \quad (i = e, h), \quad (37)$$

$$V_{n2+}(P,T) = \frac{1}{S}\frac{4\pi e^2 \hbar\omega_\pm}{[A+B]}\left(\frac{1}{\varepsilon(\omega_\pm) - \varepsilon_0} - \frac{1}{\varepsilon(\omega_\pm) - \varepsilon_\infty}\right), \quad (38)$$

where A and B are written as

$$A = [snih(2q_{n2\pm}d) - 2q_{n2\pm}d][J_1^2(q_{n2\pm}R) - J_0(q_{n2\pm}R)J_2(q_{n2\pm}R], \quad (39)$$

$$B = [snih(2q_{n2\pm}d) - 2q_{n2\pm}d][J_1^2(q_{n2\pm}R) + J_1^2(q_{n2\pm}R)], \quad (40)$$

$$\varepsilon(\omega_+) = -\varepsilon_b \coth(q_{n2+}d), \quad (41)$$

$$\varepsilon(\omega_-) = -\varepsilon_b \tanh(q_{n2+}d), \quad (42)$$

and

$$H_{ex-SSO} = H_{e-SSO} + H_{h-SSO} \quad (43)$$

$$H_{ex-SSO} = -\sum_{n3=2,4..} \Gamma_{n3+} I_0\left(\frac{n2\pi}{2d}\rho_i\right) \cosh\left(\frac{n3\pi}{2d}z_i\right)(c_{n3+} + c_{n3+}^+) -$$
$$\sum_{n3=1,3..} \Gamma_{n3-} I_0\left(\frac{n2\pi}{2d}\rho_i\right) \sinh\left(\frac{n3\pi}{2d}z_i\right)(c_{n3-} + c_{n3-}^+),$$
$$(i = e, h) \quad (44)$$

where

$$\Gamma_{n_{3+}}^2 = \frac{1}{S} \frac{4\pi e^2 \hbar \omega_{ss}}{dk_{n3}^2 [C+D]} \left(\frac{1}{\varepsilon(\omega_{ss}) - \varepsilon_0} - \frac{1}{\varepsilon(\omega_{ss}) - \varepsilon_\infty} \right), \tag{45}$$

where $C = I_0^2 k_{n3} R$, $D = I_0(k_{n3}R)I_2(k_{n3}R)$ and $S = \pi R^2$.

The effective Hamiltonian by adopting variational treatment for quasi-zero-dimensional systems considering different phonon modes, in the atomic units system, excitonic Bohr radius and Rydberg energy, becomes:

$$
\begin{aligned}
H_{eff} = & -\frac{1}{1+\sigma} \left[\frac{\partial^2}{\partial \rho_e^2} + \frac{1}{\rho_e} \frac{\partial}{\partial \rho_e} + \frac{\rho_{eh}^2 + \rho_e^2 + \rho_h^2}{\rho_e \rho_{eh}} \frac{\partial^2}{\partial \rho_e \partial \rho_{eh}} + \frac{\partial^2}{\partial z_e^2} \right] \\
& -\frac{1}{1+\sigma} \left[\frac{\partial^2}{\partial \rho_h^2} + \frac{1}{\rho_h} \frac{\partial}{\partial \rho_h} + \frac{\rho_{eh}^2 + \rho_h^2 + \rho_e^2}{\rho_h \rho_{eh}} \frac{\partial^2}{\partial \rho_h \partial \rho_{eh}} + \frac{\partial^2}{\partial z_h^2} \right] - \left[\frac{\partial^2}{\partial \rho_{eh}^2} + \frac{1}{\rho_{eh}} \frac{\partial}{\partial \rho_{eh}} \right] - \\
& \frac{\varepsilon_0}{\varepsilon_\infty} \frac{2}{\sqrt{\rho_{eh}^2 + (z_e - z_h)^2}} + V_w^e(z_e) + V_w^h(z_h) + V_{e-LO}(\rho_e, z_e) + V_{h-LO}(\rho_h, z_h) + \\
& V_{e-TSO}(\rho_e, z_{e'}) + V_{h-TSO}(\rho_h, z_h) + V_{e-SSO}(\rho_e, z_e) + V_{h-SSO}(\rho_h, z_h) + \\
& V_{e-LO-h}^{exc}(\rho_e, z_e, \rho_h, z_h) + V_{e-TSO-h}^{exc}(\rho_e, z_e, \rho_h, z_h) + \\
& V_{e-SSO-h}^{exc}(\rho_e, z_e, \rho_h, z_h).
\end{aligned}
\tag{46}
$$

The effective potentials induced by the interaction between the electron (hole) with the confined LO, the TSO and SSO phonon modes are expressed as:

$$V_{i-LO}(\rho_i, z_i) = -\sum_{n1} \frac{J_0^2\left(\frac{x_{n1}}{R}\rho_i\right)}{\hbar \omega_{LO}} \left[\sum_{l=1,3..} V_{ln1}^2 \cos^2\left(\frac{l\pi_i}{2d}z_i\right) + \sum_{l=2,4..} V_{ln1}^2 \sin^2\left(\frac{l\pi_i}{2d}z_i\right) \right],$$
$$(i = e, h) \tag{47}$$

$$
\begin{aligned}
V_{i-TSO}(\rho_i, z_i) = & -\sum_{n2,+} \frac{V_{n2,+}^2}{\hbar \omega_+} J_0^2(q_{n2,+}\rho_i) \cosh^2(q_{n2,+}z_i) \\
& -\sum_{n2,-} \frac{V_{n2,-}^2}{\hbar \omega_-} J_0^2(q_{n2,-}\rho_i) \sinh^2(q_{n2,-}z_i), (i = e, h)
\end{aligned}
\tag{48}
$$

$$
\begin{aligned}
V_{i-SSO}(\rho_i, z_i) = & -\sum_{n3=2,4..} \frac{\Gamma_{n3,+}^2}{\hbar \omega_{ss}} I_0^2\left(\frac{n_2\pi}{2d}\rho_i\right) \cos^2\left(\frac{n_3\pi}{2d}z_i\right) \\
& -\sum_{n3=1,3..} \frac{\Gamma_{n3,-}^2}{\hbar \omega_{ss}} I_0^2\left(\frac{n_2\pi}{2d}\rho_i\right) \sin^2\left(\frac{n_3\pi}{2d}z_i\right), (i = e, h).
\end{aligned}
\tag{49}
$$

The exchange potentials electron–hole via different phonon modes are given by:

$$V_{e-LO-h}^{exc}(\rho_e, z_e, \rho_h, z_h)$$
$$= \sum_{n1} \frac{2J_0\left(\frac{\chi_{n1}}{R}\rho_e\right)J_0\left(\frac{\chi_{n1}}{R}\rho_h\right)}{\hbar\omega_{LO}} \left[\sum_{l=1,3..} V_{ln1}^2 \cos\left(\frac{l\pi_h}{2d}z_h\right)\cos\left(\frac{l\pi_e}{2d}z_e\right) \right.$$
$$\left. + \sum_{l=2,4..} V_{ln1}^2 \sin\left(\frac{l\pi_e}{2d}z_e\right)\sin\left(\frac{l\pi_h}{2d}z_h\right) \right], \qquad (50)$$

$$V_{e-TSO-h}^{exc}(\rho_e, z_e, \rho_h, z_h)$$
$$= \sum_{n2,+} \frac{V_{n2,+}^2}{\hbar\omega_+} J_0(q_{n2,+}\rho_e)J_0(q_{n2,+}\rho_h)\cosh(q_{n2,+}z_e)\cosh(q_{n2,+}z_h)$$
$$+ \sum_{n2,-} \frac{V_{n2,-}^2(P,T)}{\hbar\omega_-} J_0(q_{n2,-}\rho_e)J_0(q_{n2,-}\rho_e)\sinh(q_{n2,-}z_e)\sinh(q_{n2,-}z_h), \qquad (51)$$

$$V_{e-SSO-h}(\rho_i, z_i)$$
$$= \sum_{n3=2,4..} \frac{\Gamma_{n3,+}^2}{\hbar\omega_{ss}} I_0\left(\frac{n_2\pi}{2d}\rho_e\right)I_0\left(\frac{n_2\pi}{2d}\rho_h\right)\cos\left(\frac{n_3\pi}{2d}z_e\right)\cos\left(\frac{n_3\pi}{2d}z_h\right)$$
$$- \sum_{n3=1,3..} \frac{\Gamma_{n3,-}^2}{\hbar\omega_{ss}} I_0\left(\frac{n_2\pi}{2d}\rho_e\right)I_0\left(\frac{n_2\pi}{2d}\rho_h\right)\sin\left(\frac{n_3\pi}{2d}z_e\right)\sin\left(\frac{n_3\pi}{2d}z_h\right). \qquad (52)$$

Figure 4 displays the binding energy with and without LO phonons as a function of the dot radius for two potential barriers and fixed dot height $H = 1a_{ex}^*$. The physical parameters for $CdTe/Cd_{1-x}Zn_{1-x}Te$ considered are similar to that used in figure 2. It is understood that the variation of the binding energy with the zinc concentration is significant and the shape of the curves with and without confined LO phonon modes are similar. As highlighted by the curves, the contribution of the phonons is more important for smaller QD. Consequently, in the range of strong confinement regime, the exciton states will be most stable.

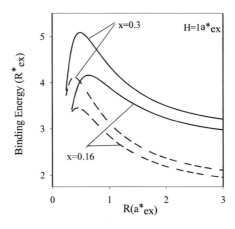

Figure 4. Binding energy of an exciton in a QD as a function of cylinder radius for two values of zinc concentration and for height $H = 1a_{ex}^*$ with (solid line) and without (dashed line) the LO phonon correction.

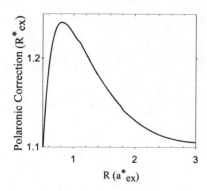

Figure 5. Polaronic correction to the exciton binding energy as a function of the QD radius.

To understand more the contribution of LO phonons to the binding energy, figure 5 displays the results of calculation of polaronic correction to the exciton binding energy in a QD of height $H = 1a^*_{ex}$ as a function of the radius for zinc concentration $x = 0.3$. The behavior of the polaronic correction we observe is similar to that in the case of the variation of the binding energy with radius; it increases as the dot radius decreases, attains a maximum for a critical value of the radius and then decreases. It can be seen also that the major contribution of the polaronic effect to the binding energy appears for the smaller QD.

Let us mention that the cylindrical QD can also present information about the transition between different confinement limits corresponding to quasi-zero-dimensional, quasi-one-dimensional, quasi-two-dimensional and quasi-three-dimensional. Consequentlly, the QD can be the most appropriate structure for practical applications because the exciton in this system becomes more stable.

3.3. Pressure and Temperature Effects

Consider now the influence of the simultaneous contribution of the pressure and temperature on the binding energy in cylindrical QD. When introducing these effects the Hamiltonian of the system and some parameters depend on the pressure and temperature, i.e., the effective masses, the static and the high frequency dielectric constants, the band gaps and the confining potentials.

The effective mass versus pressure and temperature for the electron is taken as:

$$m_e^*(P,T) = \frac{m_0}{1 + 7.51\left(\frac{2}{E_g(P,T)} + \frac{2}{E_g(P,T)+0.341}\right)}, \quad (53)$$

where m_0 represents the free electron mass and $E_g(P,T)$ is the pressure and temperature dependent energy for the GaAs QW in units of eV is given by [8]:

$$E_g(P,T) = E_g(0,T) + 1.26 \times 10^{-2}P - 3.77 \times 10^{-5}P^2, \quad (54)$$

where,

Bound Polarons and Exciton-Phonons Coupling in Semiconductor Nanostructures 283

$$E_g(0,T) = 1.519 - \frac{5.405 \times 10^{-4}T^2}{T+204}. \tag{55}$$

The effective mass versus pressure and temperature for the hole is taken as [8]:

$$m_h^*(P,T) = m_0(0.09 - 0.20 \times 10^{-3}P - 3.55 \times 10^{-5}T). \tag{56}$$

The pressure and temperature dependent static dielectric constant is expressed as [8]:

$$\varepsilon(P,T) = \varepsilon_a e^{\alpha_1 T + \alpha_2 P}, \tag{57}$$

where for $T < 200\ K$, $\varepsilon_a = 12.649812$, $\alpha_1 = 9.4 \times 10^{-5}K^{-1}$ and $\alpha_1 = -1.67 \times 10^{-3}kbar^{-1}$, whereas for $T \geq 200\ K$, $\alpha_1 = 20.4 \times 10^{-5}K^{-1}$ and $\alpha_1 = -1.73 \times 10^{-3}kbar^{-1}$. for $T \geq 200\ K$, the quantity of ε_a is calculated for continuity of ε_0 at $T = 200\ K$ and for a given value of the pressure.

The electron and hole confining potentials at pressure P and temperature T are given by:

$$\begin{cases} V_e(x,P,T) = 0.658\ \Delta E_g(x,P,T), \\ V_h(x,P,T) = 0.342\ \Delta E_g(x,P,T), \end{cases} \tag{58}$$

where

$$\Delta E_g(x,P,T) = \Delta E_g(x) - P(1.3 \times 10^{-3})x - T(1.11 \times 10^{-4})x, \tag{59}$$

with

$$\Delta E_g(x) = E_{gG_a(1-x)Al(x)As}(x) - E_{gG_aAs}(x) = 1.155x + 0.37x^2. \tag{60}$$

The high frequency dielectric constant at pressure P and temperature T is given by [8]:

$$\varepsilon_\infty(P,T) = 10.89 \exp(-1.4 \times 10^{-3}P) \exp(11.4 \times 10^{-5}(T-300)). \tag{61}$$

The variation of the dot height with pressure is considered via the equation $\frac{da}{dP} = -2.6694\ 10^{-4}a_0$, where a_0 is the lattice constant [8].

The wave function of the system can be written as:

$$|\psi(\rho,z,k)\rangle = |\psi_{ex}(\rho,z)\rangle|N_{l,n1}\rangle|N_{p,n2}\rangle|N_{p,n3}\rangle, \tag{62}$$

where $|N_{l,n1}\rangle$, $|N_{p,n2}\rangle$ and $|N_{p,n3}\rangle$ are respectively, the wave functions of the LO phonon field, the TSO phonon field and the SSO phonon field in the particle number representation.

Considering these modifications and using variational treatment, the effective Hamiltonian by incorporating the combined effect of both the pressure and temperature, in the atomic units system, excitonic Bohr radius and Rydberg energy, becomes:

$$H_{eff}$$

$$= -\frac{1}{1+\sigma}\frac{m_e^*(0)}{m_h^*(P,T)}\left[\frac{\partial^2}{\partial\rho_e^2}+\frac{1}{\rho_e}\frac{\partial}{\partial\rho_e}+\frac{\rho_{eh}^2+\rho_e^2+\rho_h^2}{\rho_e\rho_{eh}}\frac{\partial^2}{\partial\rho_e\,\partial\rho_{eh}}+\frac{\partial^2}{\partial z_e^2}\right]$$

$$-\frac{1}{1+\sigma}\frac{m_h^*(0)}{m_h^*(P,T)}\left[\frac{\partial^2}{\partial\rho_h^2}+\frac{1}{\rho_h}\frac{\partial}{\partial\rho_h}+\frac{\rho_{eh}^2+\rho_h^2+\rho_e^2}{\rho_h\rho_{eh}}\frac{\partial^2}{\partial\rho_h\,\partial\rho_{eh}}+\frac{\partial^2}{\partial z_h^2}\right]-\left[\frac{\partial^2}{\partial\rho_{eh}^2}+\frac{1}{\rho_{eh}}\frac{\partial}{\partial\rho_{eh}}\right]$$

$$-\frac{\varepsilon_0(0)}{\varepsilon_\infty(P,T)}\frac{2}{\sqrt{\rho_{eh}^2+(z_e-z_h)^2}}+V_w^e(z_e,P,T)+V_w^h(z_h,P,T)+V_{e-LO}(\rho_e,z_e,P,T)$$

$$+V_{h-LO}(\rho_h,z_h,P,T)+V_{e-TSO}(\rho_e,z_e,P,T)+V_{h-TSO}(\rho_h,z_h,P,T)+V_{e-SSO}(\rho_e,z_e,P,T)$$

$$+V_{h-SSO}(\rho_h,z_h,P,T)+V_{e-LO-h}^{exc}(\rho_e,z_e,\rho_h,z_h,P,T)+V_{e-TSO-h}^{exc}(\rho_e,z_e,\rho_h,z_h,P,T)$$

$$+V_{e-SSO-h}^{exc}(\rho_e,z_e,\rho_h,z_h,P,T)$$

$$+\sum_{n1}\hbar\omega_{LO}\frac{1}{\exp\left(\frac{\hbar\omega_{LO}}{KT}\right)-1}+\sum_{n2}\hbar\omega_{\pm}\frac{1}{\exp\left(\frac{\hbar\omega_{\pm}}{KT}\right)-1}$$

$$+\sum_{n3}\hbar\omega_{ss}\frac{1}{\exp\left(\frac{\hbar\omega_{ss}}{KT}\right)-1}. \tag{63}$$

The first illustration of the results of calculation of the effect of both pressure and temperature are shown in figure 6 (a and b). This figure shows the dependence of the binding energy in a QD of a radius $R = 1a_{ex}^*$, made of GaAs surrounded by $Ga_{1-x}Al_xAs$ barriers, on the cylinder height. This is calculated for two values of the pressure along the z-direction at low and room temperature with and without considering LO phonon modes corrections. The curves are presented in units of the effective Rydberg energy $R_{ex}^* = 3$ meV and of the effective Bohr radius $a_{ex}^* = 183.431$ Å. The behavior of the curve for each case shows that the binding energy is enhanced as the dot height is decreased, reached a maximum and then reduced. It can be seen also that for a given dot height the binding energy of exciton is sensitive to both pressure and temperature; it increases with pressure and decreases with temperature. The effect of pressure is more pronounced compared to the influence of temperature. Consequently, the exciton in QD remains stable even at room temperature when considering the pressure effect. The behavior of the effect of pressure in QD explains the fact that this influence generates an additive electronic confinement in the system which will be added to the existing geometrical one. From this figure, it can be seen also that the contribution of LO phonon modes on the exciton binding energy is significant, especially in the presence of pressure. Thus, the stability of the exciton is enhanced when including pressure effect, at low-temperature, and when considering the polaronic interaction.

To facilitate the comprehension of pressure and temperature effect, Figure 7 (a and b) shows the results of calculations of the variation of the exciton binding energy, in the presence and the absence of LO phonons contribution as a function of pressure and temperature. All curves show an interesting finding: the binding energy increases linearly with pressure whatever the temperature and LO phonons contribution, so the pressure acts like an extra-confinement. It is clearly seen that the slope of the curve depends on the absence (presence) of LO phonon modes and the temperature value, and the maximum is attained when including LO phonon modes at low temperature. As seen in the figure, for a fixed pressure value, the binding energy decreases significantly with increasing temperature. It is shown also that the exciton binding energy diminishes with enhancing temperature and decreasing pressure. This behavior is due to the fact that the exciton loses its stability when decreasing temperature and diminishing stress and the exciton tends to decrease the electron–hole interaction. Furthermore, the binding energy becomes important by considering LO phonon modes and is more significant at low temperature when including pressure effect. On can conclude that, in the practice, the technologic interests of these antagonistic effects is that they can be employed to improve the physical properties connected to the presence of excitons at the working temperature without manipulating the dot sizes but simply by adjusting the applied pressure.

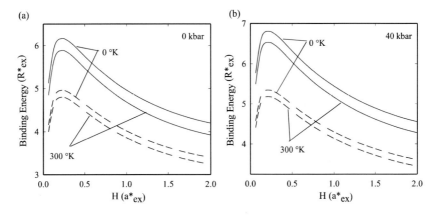

Figure 6. Variation of the binding energy of an exciton in a QD as a function the height H with (solid lines) and without (dashed lines) LO phonon correction at low and room temperature (a) in the absence and (b) the presence of pressure for x = 0.3.

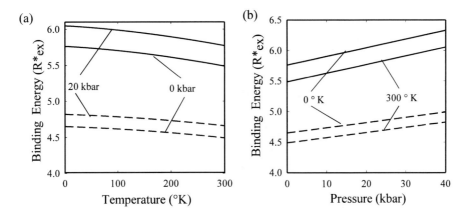

Figure 7. Binding energy of an exciton in a QD of a radius $R = 1a^*_{ex}$ and height $H = 0.5\ a^*_{ex}$ with (solid lines) and without (dashed lines) LO phonon correction as a function of (a) temperature for two value of pressure (b) pressure at low and room temperature for x = 0.3.

To support the comprehension of the polaronic contribution on exciton states in QD, let us examine the LO, TSO and SSO phonon modes influences on the behavior of the binding energy when considering the temperature and pressure effects. Figure 8 (a and b) shows the variation of the polaronic contribution, due to LO phonon modes, to the binding energy of exciton as a function of the dot height with a fixed radius $R = 1a^*_{ex}$ and aluminum concentaraion $x = 0.3$ at low and room temperature in the absence and presence of pressure effect. The similar behavior is observed as shown in figure 6. For each pressure and temperature values, it is seen that the polaronic contribution increases from its quasi-QWW value in GaAs as the cylinder height is diminished, achieves a maximum, and then decreases. Furthermore, when increasing pressure and decreasing temperature, the exciton binding energy is enhanced; i.e., for the significant pressure value, the exciton and LO phonon modes are more confined and the polaronic contribution to the binding energy enhances. These important findings have clarified the contribution of various parameters in QD system such as the cylinder height, the LO phonon modes, the temperature and the hydrostatic pressure. It is important to note here also that the presence of TSO and SSO phonon modes can affect the binding energy. Figure 8 (c and d) illustrates the results of calculation of the dependence of the polaronic corrections due to these surface phonon modes on the cylinder height with a fixed radius $R = 1a_{ex}^*$ and aluminum concentration $x = 0.3$ in absence and the presence of hydrostatic pressure. The same behavior as in figure 8 (a and b) is observed. Decreasing dot height implies an enhancement of the TSO and SSO phonon modes correction to the binding energy until reaching a maximum and then diminishes. The large contribution of polaronic effect is found at low temperature and the important applied pressure. It is interesting to mention that the LO phonon mode contribution is significant than the optical surface phonon ones, and therefore the important part of the polaronic shift is based on the LO phonons. In the presence of applied pressure the surface phonons becomes significant and it is important to consider in investigations.

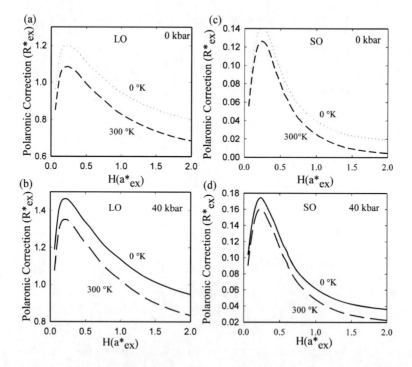

Figure 8. Polaronic correction at low and room temperature induced by the (a and b) LO phonon modes with and without pressure (c and d)) surface phonon modes with and without pressure, in a QD as a function of the dot height.

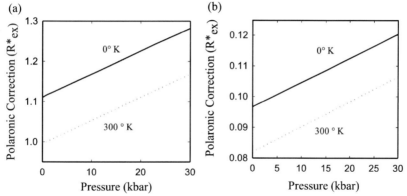

Figure 9. Polaronic correction induced by (a) LO phonon modes and (b) surface phonon modes, in a QD of $R = 1\, a^*_{ex}$ and $H = 0.5\, a^*_{ex}$ as a function of the pressure at low and room temperature.

The previous descriptions can be supported by analyzing the variation of the polaronic contribution to the binding energy as a function of pressure. Figure 9 (a and b), shows the polaronic correction due to the LO and to TSO and SSO phonon modes, respectively, in QD of fixed radius and height as a function of the applied pressure at two temperature values and luminium concentration $x = 0.3$. For every type of phonon modes the polaronic correction to the binding energy follows a linear behavior with increasing pressure. The influence of the surface phonon modes on the binding energy is less significant with regard to that of the LO phonon mode. The competing contribution of these effects to the binding energy permits the existence of excitons at a given temperature without changing the sizes of the QD.

To further facilitate the comprehension of the applied temperature on the polaronic effect, it is plotted in figure 10 (a and b) the dependence of the polaronic contribution to the binding energy on the temperature. This figure shows the variation of the polaronic correction due to the LO and to the surfaces phonon modes, in a dot of a fixed radius and height as a function of the temperature for $x = 0.3$. For all cases, the polaronic correction to the binding energy decreases significantly with increasing temperature. At high temperature the stability of exciton state depends on the competition between the pressure, the temperature and the strength exciton–phonon interaction.

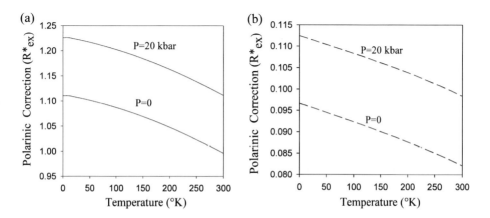

Figure 10. Polaronic correction induced by (a) LO phonon modes and (b) surface phonon modes, in a QD of $R = 1\, a^*_{ex}$ and $H = 0.5\, a^*_{ex}$ as a function of the temperature for two values of pressure.

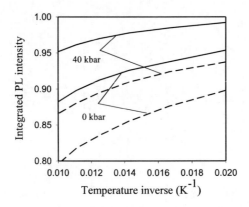

Figure 11. Arrhenius plot of the integrated PL intensity in a QD of R = 1 a^*_{ex} and H = 0.5 a^*_{ex} as a function of the temperature inverse with (solid lines) and without (dashed lines) all type of the phonon modes for two values of the pressure.

More advantage, in the field of QDs devices, has been interested on the study of the role of charge carriers coupling not only on the electronic point of view but also on the optical ones. The emission spectra in such system characterized by the photoluminescence peak which is related to the exciton-phonon coupling energy in presence of pressure and temperature effects can generate further information and allows for better understanding of the optical properties of such QDs. To facilitate the comprehension of the optical properties of these nanostructures, integrated Photoluminescence (PL) intensity is an important parameter able to present a proper illustration of these characteristics. In line with several works in literature, the Arrhenius model can be adopted to study the temperature dependence of the integrated PL intensity at a given pressure P and temperature T, which is expressed as [8]:

$$\frac{I(T)}{I(0)} = \frac{1}{1 + C_1 \exp\left(-\frac{E_1}{KT}\right) + C_2 \exp\left(-\frac{E_2}{KT}\right)}, \quad (64)$$

$I(T)(I(0))$ is the integrated PL intensity emission at temperature T (T=0 K). E_1 and E_2 are activation energies, and C_1 and C_2 are constants characterizing the ratio of non-radiative to radiative recombination rates.

$$E_1 = E_e + E_h - min_{\alpha,\gamma}\langle\psi_{ex}|H_{eff}|\psi_{ex}\rangle, \quad (65)$$

$$E_2 = E_e + E_h + E_{e-LO} + E_{h-LO} + E_{e-TSO} + E_{h-TSO} + E_{e-SSO} + E_{h-SSO}. \quad (66)$$

$E_e(E_h)$ is the electron (hole) energy at pressure P and temperature T. $E_{e-LO}(E_{h-LO})$, $E_{e-TSO}(E_{h-TSO})$ and $E_{e-SSO}(E_{h-SSO})$ are the interaction energies between electron (hole) and different type of phonon modes.

It is important to note that that the contribution of pressure and temperature to the exciton–phonon interaction appears only in the activation energies and thus the PL parameter. Therefore, it now remains to understand the combined effects of pressure and temperature on the integrated PL intensity. To reach this objective and in the interests of simplicity to get

Bound Polarons and Exciton-Phonons Coupling in Semiconductor Nanostructures   289

significant information about this physical parameter, the constants C_1 and C_2 are chosen equal to unit. Figure 11 shows the variation of the integrated PL intensity as a function of the temperature inverse, in presence and absence of polaronic contribution due to different phonon modes. It can be seen that the influence of the polaronic correction to the binding energy on the integrated PL intensity seems at high temperature. The polaronic correction increases the integrated PL intensity. Moreover, it is important to note that, for every temperature value, in the high temperature regime, the integrated PL intensity enhances with augmenting pressure.

4. POLARON STATES IN QUANTUM DOTS

4.1. Impurities in Quantum Dot

When the impurity donor is located within a QD system, the situation is significantly more complex than in the bulk material, due to three degrees of freedom. First, the binding energy depends upon the confining potential due to the QD structure. Second, the donor binding energy and the wave function depend also on the donor location within the nanostructure. The binding energy is, of course, different for a donor at the QD centre than it is for a donor at the edge of a QD.

As in the case of exciton, any theoretical investigation of the properties of donor impurity in nanostructures requires solving the standard Schrödinger equation for the particular system with incorporation of the additional Coulomb interaction.

Within the envelope function and effective mass approximations, the Hamiltonian for an electron confined in cylindrical QD and in the presence of a shallow donor impurity is merely the standard Hamiltonian of earlier, plus a further term due to the Coulombic interaction as follows:

$$H_e(r, P) = -\frac{\hbar^2}{2m^*} \nabla^2 - \frac{e^2}{\varepsilon_0 r} + V_b(r), \qquad (67)$$

where $r = [\rho^2 + (z - z_i)^2]^{\frac{1}{2}}$. z_i is the position of the donor impurity along the growth z-direction. Note that for now the effective mass m^* has been assumed to be constant when any external field effect is considered.

The potential $V_b(r)$ describes the conduction-band-edge potential of the QD, given by:

$$V_b(r) = \begin{cases} 0 & \text{if } \rho_i \leq R \text{ and } |z| \leq d \\ V_0 & \text{otherwise} \end{cases}. \qquad (68)$$

The expression of $V_0(x)$ is written as follows:

$$V_0(x, P) = 0.658\Delta E_g(x), \qquad (69)$$

with

$$\Delta E_g(x) = E_{gG_a(1-x)Al(x)As}(x) - E_{gG_aAs}(x) = 1.155x + 0.37x^2. \qquad (70)$$

The effective Hamiltonian, in the atomic units system ($a^* = \frac{\varepsilon_0 \hbar^2}{m^* e^2}$ excitonic Bohr radius and Rydberg energy $R^* = \frac{m^* e^2}{2\varepsilon_0^2 \hbar^2}$), becomes:

$$H_{eff} = -\left[\frac{\partial^2}{\partial \rho^2} + \frac{1}{\rho}\frac{\partial}{\partial \rho} + \frac{1}{\rho^2}\frac{\partial^2}{\partial \varphi^2} + \frac{\partial^2}{\partial z^2}\right] - \frac{2}{\sqrt{\rho^2 + (z - z_i)^2}}$$
$$+ V_b(r) \tag{71}$$

Generally, the procedures of solution of the Schrödinger equation have been based on two principal approaches. The first of these concerns developing the electron wave function as a linear combination of Gaussian functions [117]. While this procedure has been fruitful in calculating the properties of donors in QW structures, the generalization to more complex structures, including graded gap materials and systems where piezo electric fields are present, is non-trivial.

As shown in the case for exciton, the other approach centers on the variational procedure. In this procedure, a trial wave function is picked whose functional form may include one or more unknown parameters. These parameters are changed systematically and the expectation value of the energy calculated for each set. The variational procedure [117] shows that the lowest energy obtained is the closest approximation to the true state of the system. The achievement of variational method is based on the general choice of the trial wave function.

The trial wave function for an electron confined in a QD of the cylindrical geometry in the presence of a positively charged donor, hydroginic impurity, can be chosen as follows [118]:

$$\phi(\rho, z, P, T) = Nf(\rho, P, T)g_i(z, P, T)\exp(-\alpha r), \tag{72}$$

where N and α are, respectively, the normalization constant and the variational parameter. The exponential factor describes the correlation effect between the electron and the positive charge donor (ion). f and g are, respectively, the ground state solution of the Schrödinger equations in the lateral plane and z-axis.

Respectively, the corresponding 2D (lateral direction) and 1D (longitudinal direction) effective mass Schrödinger equations are:

$$\left\{-\frac{\hbar^2}{2m^*}\nabla^2 + V_b(\rho)\right\}f(\rho) = E(\rho)f(\rho), \tag{73}$$

$$\left\{-\frac{\hbar^2}{2m^*}\nabla^2 + V_b(z)\right\}g(z) = E_z(z)f(z). \tag{74}$$

The solutions are of the form:

$$f_i(\rho) = \begin{cases} J_0\left(\theta\frac{\rho}{R}\right) & \text{for } \rho \leq R \\ AK_0(\beta\rho_i) & \text{for } \rho > R, \end{cases} \tag{75}$$

$$g(z) = \begin{cases} \cos(\pi\dfrac{z}{2d}) & \text{for } |z| \leq d \\ B\exp(k|z|) & \text{for } |z| > d \ . \end{cases} \tag{76}$$

The ground-state energy of shallow donor impurity is obtained by:

$$E_G(P) = \min_\alpha \langle \psi | H_{eff} | \psi \rangle. \tag{77}$$

4.2. Polaronic Effect

The basic Hamiltonian of a shallow hydrogenic impurity in such QD interacting with the optical phonon modes (LO, TSO and SSO), can be expressed within the framework of the effective-mass and nondegenerate-band approximations, as follows [6]:

$$H = H_e + H_{LO} + H_{TSO} + H_{SSO} + H_{e-LO} + H_{e-TSO} + H_{e-SSO} + H_{ion-LO} + H_{ion-TSO} + H_{ion-SSO}. \tag{78}$$

In the presence of different phonon modes, the effective Hamiltonian, in the atomic units system Bohr radius and Rydberg energy, reads:

$$H_{eff} = -\left[\frac{\partial^2}{\partial\rho^2} + \frac{1}{\rho}\frac{\partial}{\partial\rho} + \frac{1}{\rho^2}\frac{\partial^2}{\partial\varphi^2} + \frac{\partial^2}{\partial z^2}\right] - \frac{\varepsilon_0}{\varepsilon_\infty}\frac{2}{\sqrt{\rho^2 + (z-z_i)^2}} + V_b(r) + V_{e-LO}(\rho,z)$$
$$+ V_{ion-LO}(\rho,z) + V_{e-LO-ion}^{exc}(\rho,z) + V_{e-TSO}(\rho,z) + V_{ion-TSO}(\rho,z)$$
$$+ V_{e-TSO-ion}^{exc}(\rho,z) + V_{e-SSO}(\rho,z) + V_{ion-SSO}(\rho,z)$$
$$+ V_{e-SSO-ion}^{exc}(\rho,z). \tag{79}$$

The expressions of effective potentials induced by the interactions between electron (ion) and different phonon modes are written as:

$$V_{e-LO}(\rho,z)$$
$$= -\sum_{n1}\frac{J_0^2\left(\frac{\chi_{n1}}{R}\rho\right)}{\hbar\omega_{LO}}\left[\sum_{l=1,3..}V_{ln1}^2(P)\cos^2\left(\frac{l\pi}{2d}z\right)\right.$$
$$\left. + \sum_{l=2,4..}V_{ln1}^2(P)\sin^2\left(\frac{l\pi}{2d}z\right)\right], \tag{80}$$

$$V_{ion-LO} = -\sum_{n1}\sum_{l=1,3..}\frac{V_{ln1}^2}{\hbar\omega_{LO}}, \tag{81}$$

$$V_{e-LO-ion}^{exc}(\rho,z) = \sum_{n1}\sum_{l=1,3..}\frac{2V_{ln1}^2(P)}{\hbar\omega_{LO}(P)}J_0\left(\frac{\chi_{n1}}{R}\rho\right)\cos\left(\frac{l\pi}{2d}z\right), \tag{82}$$

$$V_{e-TSO}(\rho,z)$$

$$= -\sum_{n2,+} \frac{V_{n2,+}^2(P)}{\hbar\omega_+} J_0^2(q_{n_{2,+}}\rho)\cosh^2(q_{n_{2,+}}z)$$

$$- \sum_{n2,-} \frac{V_{n2,-}^2(P)}{\hbar\omega_-} J_0^2(q_{n_{2,-}}\rho)\sinh^2(q_{n_{2,-}}z), \tag{83}$$

$$V_{ion-TSO}(\rho,z) = -\sum_{n2,+} \frac{V_{n2,+}^2}{\hbar\omega_+}, \tag{84}$$

$$V_{e-TSO-ion}^{exc}(\rho,z) = \sum_{n2,+} \frac{2V_{n2,+}^2}{\hbar\omega_+} J_0(q_{n_{2,+}}\rho)\cosh(q_{n_{2,+}}z), \tag{85}$$

$$V_{e-SSO} = -\sum_{n3=2,4..} \frac{\Gamma_{n3,+}^2}{\hbar\omega_{ss}} I_0^2\left(\frac{n_2\pi}{2d}\rho\right)\cos^2\left(\frac{n_3\pi}{2d}z\right)$$

$$- \sum_{n3=1,3..} \frac{\Gamma_{n3,-}^2(P)}{\hbar\omega_{ss}} I_0^2\left(\frac{n_2\pi}{2d}\rho\right)\sin^2\left(\frac{n_3\pi}{2d}z\right), \tag{86}$$

$$V_{ion-SSO} = -\sum_{n3=2,4..} \frac{\Gamma_{n3,+}^2}{\hbar\omega_{ss}} \tag{87}$$

and

$$V_{e-SSO-ion}^{exc}(\rho,z) = \sum_{n3=2,4..} \frac{2\Gamma_{n3,+}^2(P)}{\hbar\omega_{ss}(P)} I_0\left(\frac{n_3\pi}{2d}\rho\right)\cos\left(\frac{n_3\pi}{2d}z\right) \tag{88}$$

4.3. Combined Effect of Pressure and Temperature

In order to describe the effect of combined applied pressure and temperature on the donor impurity states by considering phonon modes, the similar modifications are adopted on some parameters as down in the case of exciton. In the presence of the confined LO phonon modes and using variational treatment, the effective Hamiltonian by introducing the combined effect of both the pressure and temperature, in the atomic units system Bohr radius and Rydberg energy, becomes:

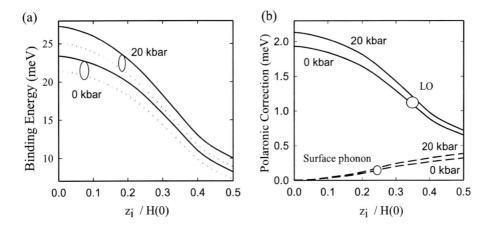

Figure 12. (a) Binding energy of shallow donor impurity and (b) polaronic correction, as a function of the donor impurity position along the z-axis for two values of pressure and for radius 100 Å and height 100 Å.

$$H_{eff} = -\frac{m^*(0)}{m^*(P,T)}\left[\frac{\partial^2}{\partial \rho^2} + \frac{1}{\rho}\frac{\partial}{\partial \rho} + \frac{1}{\rho^2}\frac{\partial^2}{\partial \varphi^2} + \frac{\partial^2}{\partial z^2}\right] - \frac{\varepsilon_0(0)}{\varepsilon_\infty(P,T)}\frac{2}{\sqrt{\rho^2 + (z-z_i)^2}}$$
$$+ V_b(r,P,T) + V_{ion-LO}(\rho,z,P,T) + V_{ion-LO}(\rho,z,P,T)$$
$$+ V^{exc}_{e-LO-ion}(\rho,z,P,T) + V_{e-TSO}(\rho,z,P,T) + V_{ion-TSO}(\rho,z,P,T)$$
$$+ V^{exc}_{e-TSO-ion}(\rho,z,P,T)$$
$$+ V_{e-SSO}(\rho,z,P,T) + V_{ion-SSO}(\rho,z,P,T) + V^{exc}_{e-SSO-ion}(\rho,z,P,T)$$
$$+ \sum_{n1}\hbar\omega_{LO}\frac{1}{\exp\left(\frac{\hbar\omega_{LO}}{KT}\right)-1} + \sum_{n2}\hbar\omega_{\pm}\frac{1}{\exp\left(\frac{\hbar\omega_{\pm}}{KT}\right)-1}$$
$$+ \sum_{n3}\hbar\omega_{ss}\frac{1}{\exp\left(\frac{\hbar\omega_{ss}}{KT}\right)-1} \tag{89}$$

For the numerical results, the energies and the sizes are presented in units meV and Å, respectively. Also, the aluminium concentration in the barrier material is taken $x = 0.3$.

Figure 12a shows the results of the numerical calculations for the binding energy of the shallow donor impurity in a QD of a fixed radius 100 Å and height 100 Å as a function of the donor position along the z-axis when hydrostatic pressure is applied. This is depicted with and without the contribution of different phonon modes. First, it can be observed, in every case, that the binding energy of the shallow donor impurity decreases as the impurity moves away from the centre to the edge of the QD; this behavior is similar to the results, in the literature, published previously [119, 120] without phonon contribution. The total quantity of the phonon correction to the binding energy decreases as the impurity displaces from the centre to the boundary of the QD. Also, depending on the position of the donor interior the QD, the binding energy enhances with augmenting pressure; this is because the pressure generates an important electronic confinement in the QD. It can be also observed that when the impurity approaches the edge, the effect of pressures becomes less important. Second, it can be seen that the polaronic correction to the binding energy is significant when the values of the donor position decays and achieve a maximum value when the donor impurity is located in the origin of the cylinder. A more illustrated behavior for the phonon correction to the binding energy is shown in figure 12b, in which the contributions of different phonon modes are singly presented. It can be seen that the contribution of the LO phonon decreases while that of the surface optical phonon enhances as the donor shifts away from the QD origin. However, the contribution of the surface optical phonon modes to the binding energy is quite weak and has a small role when considering the binding energy of a donor located in the origin of the dot. Also, for both LO phonon and surface phonons the polaronic contribution to the binding energy augments with enhancing pressure. The influence of stress is very significant for LO phonon in the case of the donor on the QD origin. The pressure action on the surface phonon becomes significant when the impurity is situated at the QD edge.

In what follows, the results are given for the donor impurity placed in the origin of the cylinder. In figure 13a, are presented the numerical results for the binding energy of the shallow donor impurity in a dot of a fixed radius, as a function of the height of the cylinder in view of different values of the pressure along the z-axis of the cylinder with and without the overall phonon effects. It can be shown that the binding energy decreases as the height augments. This behavior can be interpreted as that the electron wave function is more free to move along the z-axis when the potential barriers are far away. Furthermore, it is expected that at each value of the height, the binding energy enhances with augmenting the compression due to the applied pressure; this action is larger when the height is decreased and when considering phonon effects. This behavior can be understood as that the introduction of the pressure creates an additive confinement which will be added to the geometrical one, and changes the motion of the carrier and the nature of the relating wave function. The overall polaronic contribution to the binding energy in a QD of a fixed radius when applying stress as a function of the dot height is posted in figure 13b. Here also it can be shown that the phonon contribution to the binding energy decays as the height increases to the value of quasi-QWW limit. Additionally, it can be seen that the effect of the hydrostatic stress is significant for smaller values of the dot height. This behavior is similar to those shown aforementioned in the case of excitons.

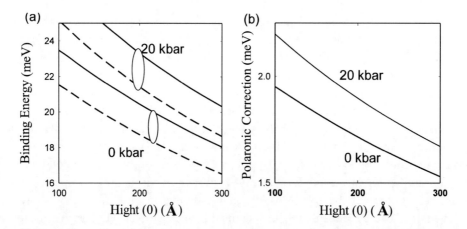

Figure 13. (a) Binding energy of shallow donor impurity as a function of the height with (solid lines) and without (dashed lines) the total phonon modes; (b) polaronic correction as a function of the height. The dot radius is fixed at 100 Å.

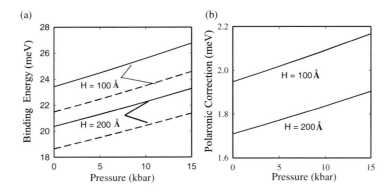

Figure 14. (a) Binding energy of shallow donor impurity in a QD as a function of the pressure with (solid lines) and without (dashed lines) the total phonon modes; (b) polaronic correction as a function of the pressure. The dot radius is fixed at 100 Å.

The influence of pressure in such structure is highlighted by the plot, in figure 14a, of the magnitude of the binding energy of shallow donor impurity, as a function of the action of stress with fixed dot radius and two height values, with and without phonon contribution. Clearly, the behavior of the binding energy follows a nearly linear function and it increases with the augmentation of pressure. In addition, it is apparently observed that as the height of the dot enhances the slope of the curve decreases. For the comprehension of this behavior it can be mention that the electron wave function does not feel the small compression in the dot system when the dimension is large. The role of the pressure becomes very significant by considering phonon modes. The polaronic influence on the binding energy of the shallow donor impurity in a QD of a fixed radius, as a function of the applied pressure for different height is presented in figure 14b. Here also, the variation of the polaronic correction follows a linear function of the pressure. Furthermore, as the height of the dot augments the slope of the curve decays. The similar interpretation as mentioned above can be adopted; the phonon mode contributions decays and they do not feel the lower compression in the dot for the large size.

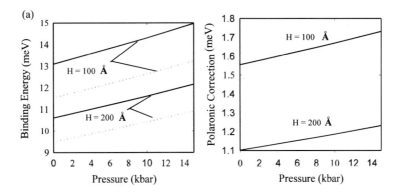

Figure 15. (a) Binding energy of shallow donor impurity in a quasi-QW as a function of the pressure with (solid lines) and without (dashed lines) the total phonon modes; (b) polaronic correction as a function of the pressure. The dot radius is fixed at 1000 Å.

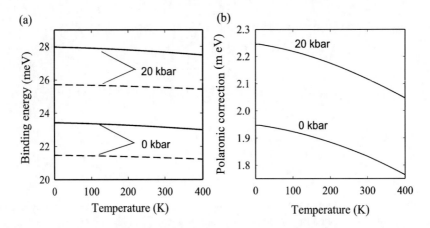

Figure 16. (a) Binding energy of shallow donor impurity in a QD as a function of the temperature with (solid lines) and without (dashed lines) the total phonon modes; (b) polaronic correction as a function of the temperature.

The possibility to control dimensions of the QD, offers by the cylindrical geometry, permits to obtain numerical results of different limiting cases; QWW, quantum disc, and QW. Figure 15a shows the binding energy of the shallow donor impurity as a function of hydrostatic pressure for a QD of radii R = 1000 Å and two values of the height with and without considering the total quantity of phonon contribution. The large value of the radius here considered permits to claim that the structure is almost in the QW limit. Here again, the binding energy is relatively a linear function of the pressure. An analogous behavior is observed when introducing phonon contribution. In figure 15b, it is plotted the polaronic influence on the binding energy in a cylindrical QW of radius R = 1000 Å, as a function of the pressure for different QW thicknesses. Similarly, it can be seen that the polaronic correction increases linearly with increasing pressure. This figure also shows that as the thickness of the dot enhances the slope of the curve decreases. In addition, it can be clearly seen that the polaronic contribution due to the phonon modes to the binding energy under hydrostatic stress is more significant as compared with the one and two-dimensional cases.

The combined effect of the pressure and temperature on the binding energy of donor impurity in QD can affects the characteristics of polaron structure and will be an important issue in future exploitation. When the effects of the hydrostatic stress and temperature are introduced, the following influences are shown: (i) the value of the electron effective mass increases with pressure and decreases with temperature, (ii) the GaAs static dielectric constant decreases with pressure and increases with temperature, (iii) the effective Bohr radius decreases with pressure and increases with the temperature, and (iv) the value of the effective Rydberg increases with pressure and decreases with temperature. In order to draw some properties of the binding energy of the donor in presence and absence of phonon modes, it is plotted in figure 16a the binding energy of the shallow donor impurity in a QD of a fixed radius and height, as a function of the temperature by considering different values of the applied pressure with and without the confined LO phonon modes corrections. It can be seen from the figure that, for a given stress value, the binding energy decays sensitively with enhancing temperature. This behavior is in good agreement with that illustrated in the case of exciton confined in a QD. It becomes important to mention that an increasing of the applied compression indicates a weakening of the temperature effect on the binding energy. At room temperature and for GaAs material, the binding energy of the donor impurity in a QD remains significant, and it empowers obtaining the electronic and optical properties even at operating temperature which can be attractive for their potential exploitations in low-dimensional structure device technology.

To more understand the combined effect of the hydrostatic stress and the temperature on the binding energy, it is plotted in figure 16b the behavior of the polaronic correction to the binding energy of the shallow donor impurity as a function of the temperature in a QD of fixed radius and height at two pressure values. It is clearly seen that, for a given stress value, the polaronic correction to the binding energy diminishes as the temperature augments. Also, the pressure effect on the polaronic correction becomes more significant by decreasing temperature.

REFERENCES

[1] Gopal, A. V., Kumar, R., Vengurlekar, A. S., Bosacchi, A., Franchi, S., and Pfeiffer, L. N. 2000. "Photoluminescence study of exciton–optical phonon scattering in bulk GaAs and GaAs quantum wells." *Journal of Applied Physics* 87: 1858-1862.

[2] Devreese, J. T., Fomin, V. M., Gladilin, V. N., & Klimin, S. N. 2001. "Photoluminescence Spectra of Quantum Dots: Enhanced Efficiency of the Electron– Phonon Interaction." *Physica status solidi* (b) 224: 609-12.

[3] Zheng, R., Matsuura, M., and Taguchi, T. 2000. "Exciton–LO-phonon interaction in zinc-compound quantum wells." *Physical Review B* 61: 9960-963.

[4] Ipatova, I. P., Maslov, A. Y., and Proshina, O. V. 1995. "Multiphonon processes involved in optical transitions in quantum nanostructures". *Physics of the Solid State* 37: 991-994.

[5] Ipatova, I. P., Maslov, A. Y., and Proshina, O. V. 1999. "Polar state of a particle with a degenerate band spectrum in a quantum dot." *Semiconductors* 33: 765-70.

[6] El Moussaouy, A., Ouchani, N., El Hassouani, Y., and Benami, A. 2014. "Bound polaron states in GaAs/Ga(1–x)Al(x)As cylindrical quantum dot under hydrostatic pressure effect." *Surface Science* 624: 95-102.

[7] El Moussaouy, A., and Ouchani, N. 2014. "The role of hydrostatic pressure and temperature on bound polaron in semiconductor quantum dot." *Physica B: Condensed Matter* 436: 26-32.

[8] El Moussaouy, A., Ouchani, N., El Hassouani, Y., and Abouelaoualim, D. 2014. "Temperature and hydrostatic pressure effects on exciton–phonon coupled states in semiconductor quantum dot." *Superlattices and Microstructures* 73: 22-37.

[9] Ham, H., and Spector, H. N. 2001. "Exciton linewidth due to scattering by polar optical phonons in semiconducting cylindrical quantum wire structures." *Physical Review* B 62: 13599-3603.

[10] Devreese, J. T. 1996."Polarons." *Digital Encyclopedia of Applied Physics.* doi:10.1002/ 3527600434.eap347.

[11] Verzelen, O., Ferreira, R., and Bastard, G. 2000. "Polaron lifetime and energy relaxation in semiconductor quantum dots." *Physical Review B* 62: R4809(R).

[12] Verzelen, O., Ferreira, R., and Bastard, G. 2002. "Excitonic Polarons in Semiconductor Quantum Dots." *Physical Review Letters* 88: 146803.

[13] Hameau, S., Guldner, Y., Verzelen, O., Ferreira, R., Bastard, G., Zeman, J., ... and Gérard, J. M. 1999. "Strong Electron-Phonon Coupling Regime in Quantum Dots: Evidence for Everlasting Resonant Polarons." *Physical Review Letters* 83: 4152-4155.

[14] Tasai, T., and Eto, M. 2003."Transport properties of quantum dots with strong electron–phonon interaction." *Physica E: Low-dimensional Systems and Nanostructures* 17: 139-42.

[15] Zaitsev, V. V., Bagaev, V. S., and Onishchenko, E. E. 1999. "Specific features in the temperature dependence of photoluminescence from CdTe/ZnTe size-quantized islands and ultrathin quantum wells." *Physics of the Solid State* 41: 647-53.

[16] Bouhassoune, M., Charrour, R., Fliyou, M., Bria, D., and Nougaoui, A. 2002. "Polaronic and magnetic field effects on the binding energy of an exciton in a quantum well wire." *Journal of Applied Physics* 91: 232.

[17] Dai, N., Brown, F., Doezema, R. E., Chung, S. J., and Santos, M. B. 2001. "Temperature dependence of exciton linewidths in InSb quantum wells." *Physical Review B* 63: 115321.

[18] Masumoto, Y., Ikezawa, M., Hyun, B. R., Takemoto, K., and Furuya, M. 2001. "Homogeneous width of confined excitons in quantum dots at very low temperatures." *Physica status solidi* (b), 224: 613-619.

[19] Pekar, S. I. 1954. *"Untersuchungen über die Elektronentheorie akr Kristalle."* [Investigations on the electron theory ark crystals] Akademie-Verlag, Berlin.

[20] Ipatova, I. P., Maslov, A. Y., and Proshina, O. V. 2001."Multi-phonon transitions in II-VI quantum dot." *Europhysics Letters (EPL)* 53: 769-75.

[21] Ashoori, R. C. 1996. "Electrons in artificial atoms." *Nature* 379, no. 6564 (1996): 413-419.

[22] Alivisatos, A. P. 1996. "Semiconductor Clusters, Nanocrystals, and Quantum Dots." *Science* 271: 933-937.

[23] Manasevit, H. M. 1968. "Single-Crystal Gallium Arsenide on Insulating Substrates." *Applied Physics Letters* 12: 156-159.

[24] Cho, A. Y., and Arthur, J. R. 1975. "Molecular beam epitaxy." *Progress in Solid State Chemistry* 10: 157-191.

[25] Esaki, L., and Tsu, R. 1970. "Superlattice and Negative Differential Conductivity in Semiconductors." *IBM Journal of Research and Development* 14: 61-65.

[26] Tsu, R., and Esaki, L. 1973. "Tunneling in a finite superlattice." *Applied Physics Letters* 22: 562-564.

[27] Dingle, R., Gossard, A. C., and Wiegmann, W. 1975. "Direct Observation of Superlattice Formation in a Semiconductor Heterostructure." *Physical Review Letters* 34: 1327-1330.

[28] Chin, R., Holonyak Jr, N., Vojak, B. A., Hess, K., Dupuis, R. D., & Dapkus, P. D. 1980. "Temperature dependence of threshold current for quantum-well Al x Ga1− x AsxGaAs heterostructure laser diodes." *Applied Physics Letters* 36: 19-21.

[29] Arakawa, Y., and Sakaki, H. 1982. "Multidimensional quantum well laser and temperature dependence of its threshold current." *Applied Physics Letters* 40: 939-941.

Bound Polarons and Exciton-Phonons Coupling in Semiconductor Nanostructures 299

[30] Skocpol, W. J., Jackel, L. D., Hu, E. L., Howard, R. E., and Fetter, L. A. 1982. "One-dimensional localization and interaction effects in narrow (0.1-μm) silicon inversion layers." *Physical Review Letters* 49: 951-955.

[31] Ekimov, A. I., and Onushchenko, A. A. 1981. "Quantum size effect in three-dimensional microscopic semiconductor crystals." *Jetp Lett* 34: 345-349.

[32] Brus, L. E. 1983. "A simple model for the ionization potential, electron affinity, and aqueous redox potentials of small semiconductor crystallites." *The Journal of Chemical Physics* 79: 5566-5571.

[33] Brus, L. E. 1984. "Electron–electron and electron-hole interactions in small semiconductor crystallites: The size dependence of the lowest excited electronic state." *The Journal of Chemical Physics* 80: 4403-4409.

[34] Reed, M. A. 1986. "Spatial quantization in GaAs–AlGaAs multiple quantum dots." *Journal of Vacuum Science & Technology B: Microelectronics and Nanometer Structures* 4: 358.

[35] Reed, M., J. Randall, R. Aggarwal, R. Matyi, T. Moore, and A. Wetsel. 1988. "Observation of discrete electronic states in a zero-dimensional semiconductor nanostructure." *Physical Review Letters* 60: 535-537.

[36] Kash, K., A. Scherer, J. M. Worlock, H. G. Craighead, and M. C. Tamargo. 1986. "Optical spectroscopy of ultrasmall structures etched from quantum wells." *Applied Physics Letters* 49: 1043-1045.

[37] Temkin, H., Dolan, G. J., Panish, M. B., and Chu, S. N. G. 1987. "Low-temperature photoluminescence from InGaAs/InP quantum wires and boxes." *Applied Physics Letters* 50: 413-415.

[38] Hansen, W., Smith III, T. P., Lee, K. Y., Brum, J. A., Knoedler, C. M., Hong, J. M., & Kern, D. P. 1989." Zeeman bifurcation of quantum-dot spectra." *Physical review letters,* 62: 2168.

[39] Tewordt, M., Asahi, H., Law, V. J., Syme, R. T., Kelly, M. J., Ritchie, D. A., and Jones, G. A. C. 1992. "Resonant tunneling in coupled quantum dots." *Applied Physics Letters* 60: 595-597.

[40] Meurer, B., Heitmann, D., and Ploog, K. 1992."Single-electron charging of quantum-dot atoms." *Physical Review Letters* 68: 1371-1374.

[41] Field, M., Smith, C. G., Pepper, M., Ritchie, D. A., Frost, J. E. F., Jones, G. A. C., and Hasko, D. G. 1993. "Measurements of Coulomb blockade with a noninvasive voltage probe." *Physical Review Letters* 70: 1311.

[42] Ashoori, R. C., Stormer, H. L., Weiner, J. S., Pfeiffer, L. N., Baldwin, K. W., and West, K. W. 1993. "N-electron ground state energies of a quantum dot in magnetic field." *Physical Review Letters* 71: 613.

[43] Schmitt-Rink, S. D. A. B. M., Miller, D. A. B., and Chemla, D. S. 1987. "Theory of the linear and nonlinear optical properties of semiconductor microcrystallites." *Physical Review B* 35: 8113-8125.

[44] Wang, L. W., and Zunger, A. 1994. "Electronic Structure Pseudopotential Calculations of Large (.apprx.1000 Atoms) Si Quantum Dots." *The Journal of Physical Chemistry* 98: 2158-2165.

[45] Bryant, G. W. 1988. "Excitons in quantum boxes : Correlation effects and quantum confinement." *Physical Review B* 37: 8763-8772.

[46]   Kumar, A., Laux, S. E., and Stern, F. 1990. "Electron states in a GaAs quantum dot in a magnetic field." *Physical Review B* 42: 5166-5175.

[47]   Brunner, K., Bockelmann, U., Abstreiter, G., Walther, M., Böhm, G., Tränkle, G., and Weimann, G. 1992. "Photoluminescence from a single GaAs/AlGaAs quantum dot." *Physical Review Letters* 69: 3216-3219.

[48]   Brunner, K., Abstreiter, G., Böhm, G., Tränkle, G., & Weimann, G. 1994. "Sharp-Line Photoluminescence and Two-Photon Absorption of Zero-Dimensional Biexcitons in a GaAs/AlGaAs Structure." *Physical Review Letters* 73: 1138-1141.

[49]   Gammon, D., Snow, E. S., and Katzer, D. S. 1995. "Excited state spectroscopy of excitons in single quantum dots." *Applied Physics Letters* 67: 2391-2393.

[50]   Gammon, D., Snow, E. S., Shanabrook, B. V., Katzer, D. S., and Park, D. 1996. "Homogeneous Linewidths in the Optical Spectrum of a Single Gallium Arsenide Quantum Dot." *Science* 273: 87-90.

[51]   Gammon, D., Snow, E. S., Shanabrook, B. V., Katzer, D. S., and Park, D. 1996. "Fine Structure Splitting in the Optical Spectra of Single GaAs Quantum Dots." *Physical Review Letters* 76: 3005-3008.

[52]   Glaser, E. R., Bennett, B. R., Shanabrook, B. V., and Magno, R. 1996. "Photoluminescence studies of self-assembled InSb, GaSb, and AlSb quantum dot heterostructures." *Applied Physics Letters* 68: 3614-3616.

[53]   Esch, V., Fluegel, B., Khitrova, G., Gibbs, H. M., Jiajin, X., Kang, K., ... and Peyghambarian, N. 1990. "State filling, Coulomb, and trapping effects in the optical nonlinearity of CdTe quantum dots in glass." *Physical Review B* 42: 7450-7453.

[54]   Rajh, T., Micic, O. I., and Nozik, A. J. 1993. "Synthesis and characterization of surface-modified colloidal cadmium telluride quantum dots." *The Journal of Physical Chemistry* 97: 11999-12003.

[55]   Gourgon, C., Eriksson, B., Dang, L. S., Mariette, H., and Vieu, C. 1994. "Photoluminescence of CdTe / ZnTe semiconductor wires and dots." *Journal of Crystal Growth* 138: 590-594.

[56]   Magnea, N. 1994. "ZnTe fractional monolayers and dots in a CdTe matrix." *Journal of Crystal Growth* 138: 550-558.

[57]   Calvo, V., Lefebvre, P., Allègre, J., Bellabchara, A., Mathieu, H., Zhao, Q. X., and Magnea, N. 1996. "Evidence of the ordered growth of monomolecular ZnTe islands in CdTe/ (Cd,Zn)Te quantum wells on a nominal (001) surface." *Physical Review* B 53.

[58]   Marzin, J. Y., Gérard, J. M., Izraël, A., Barrier, D., & Bastard, G. 1994. "Photoluminescence of Single InAs Quantum Dots Obtained by Self-Organized Growth on GaAs." *Physical Review Letters* 73: 716-719.

[59]   Nabetani, Y., Ishikawa, T., Noda, S., and Sasaki, A. 1994. "Initial growth stage and optical properties of a three-dimensional InAs structure on GaAs." *Journal of Applied Physics* 76: 347-351.

[60]   Seifert, W., Carlsson, N., Miller, M., Pistol, M. E., Samuelson, L., and Wallenberg, L. R. 1996. "In-situ growth of quantum dot structures by the Stranski-Krastanow growth mode." *Progress in Crystal Growth and Characterization of Materials* 33: 423-471.

[61]   Stranski, I. N., and Krastanow, L. 1937. "Zur Theorie der orientierten Ausscheidung von Ionenkristallen aufeinander." *Monatshefte für Chemie/Chemical Monthly*, 71: 351-364.

Bound Polarons and Exciton-Phonons Coupling in Semiconductor Nanostructures 301

[62] Goldstein, L., Glas, F., Marzin, J. Y., Charasse, M. N., and Le Roux, G. 1985. "Growth by molecular beam epitaxy and characterization of InAs/GaAs strained layer superlattices." *Applied Physics Letters* 47: 1099-1101.

[63] Schmidt, K. H., Medeiros-Ribeiro, G., Oestreich, M., Petroff, P. M., and Döhler, G. H. 1996. "Carrier relaxation and electronic structure in InAs self-assembled quantum dots." *Physical Review B* 54: 11346-11353.

[64] Tarucha, S., Austing, D. G., Honda, T., Van der Hage, R. J., and Kouwenhoven, L. P. 1996."Shell Filling and Spin Effects in a Few Electron Quantum Dot." *Physical Review Letters* 77: 3613-3616.

[65] Lubyshev, D. I., González-Borrero, P. P., Marega Jr, E., Petitprez, E., La Scala Jr, N., and Basmaji, P. 1996."Exciton localization and temperature stability in self-organized InAs quantum dots." *Applied Physics Letters* 68: 205-207.

[66] Bayer, M., Ortner, G., Stern, O., Kuther, A., Gorbunov, A. A., Forchel, A., ... and Walck, S. N. 2002."Fine structure of neutral and charged excitons in self-assembled In(Ga)As/(Al)GaAs quantum dots." *Physical Review B* 65: 195315.

[67] Langbein, W., Borri, P., Woggon, U., Stavarache, V., Reuter, D., and Wieck, A. D. 2004. "Control of fine-structure splitting and biexciton binding inInxGa1−xAsquantum dots by annealing." *Physical Review B* 69: 161301.

[68] Borri, P., Langbein, W., Schneider, S., Woggon, U., Sellin, R. L., Ouyang, D., and Bimberg, D. 2001. "Ultralong Dephasing Time in InGaAs Quantum Dots." *Physical Review Letters* 87: 157401. doi:10.1103/physrevlett.87.157401.

[69] Wang, G. and Guo, Q. 2008. "Third-harmonic generation in cylindrical parabolic quantum wires with static magnetic fields." *Physica B* 403: 37-43

[70] Khordad, R. 2014. "Refractive index change and absorption coefficient of T shaped quantum wires: comparing with experimental results." *Opt. Quant. Electron.* 46, 283

[71] Shao, S., Guo, K. X., Zhang, Z. H., Li, N. and Peng, C. 2010. "Studies on the third-harmonic generations in cylindrical quantum dots with an applied electric field." *Superlatt. Microstruct.* 48, 541.

[72] Morales, A. L., Raigoza, N. and Duque, C. A. 2006. "Donor-Related Optical Absorption Spectra for a GaAs-Ga(0,7)Al(0,3)As Double Quantum Well under Hydrostatic Pressure and Applied Electric Field Effects." *Braz. J. Phys.* 36, 862.

[73] Karabulut, I. and Baskoutas, S. 2009. "Second and Third Harmonic Generation Susceptibilities of Spherical Quantum Dots: Effects of Impurities, Electric Field and Size. " *J. Comput. Theoret. Nanosci.* 6, 153.

[74] Ashcroft, N. W., and Mermin, N. D. 1976. *Solid state physics,* 1st edn. Brooks Cole.

[75] Slater, J. C., and Koster, G. F. 1954. "Simplified LCAO Method for the Periodic Potential Problem." *Physical Review* 94: 1498-1524.

[76] Chemla, D. S., and Shah, J. 2001. "Many-body and correlation effects in semiconductors." *Nature* 411: 549-557.

[77] R., Lea, S. M., and Burke, J. R. 1961. *Modern Physics* (pp. 30-35). New York: John Wiley and Sons.

[78] Weidner, R. R., and Sells, R. L. 1980. *Elementary modern physics.* Allyn and Bacon, Boston, Third edition.

[79] Adachi, S. 1994. *GaAs and related materials: bulk semiconducting and superlattice properties.* World Scientific.

[70]   Hilton, C. P., Hagston, W. E., and Nicholls, J. E. 1992."Variational methods for calculating exciton binding energies in quantum well structures." *Journal of Physics A: Mathematical and General* 25: 2395-2401.

[81]   Hilton, P., Goodwin, J., Harrison, P., and Hagston, W. E. 1992. "Theory of exciton energy levels in multiply periodic systems." *Journal of Physics A: Mathematical and General* 25: 5365-5372.

[82]   Harrison, P., Goodwin, J., and Hagston, W. E. 1992. "Exciton energies and band-offset determination in magnetic superlattices." *Physical Review B* 46: 12377-12383.

[83]   Harrison, P., Piorek, T., Hagston, W. E., and Stirner, T. 1996."The symmetry of the relative motion of excitons in semiconductor heterostructures." *Superlattices and Microstructures* 20: 45-57. doi:10.1006/spmi.1996.0048.

[84]   Harrison, P., Hagston, W. E., and Stirner, T. 1993. "Excitons in diffused quantum wells." *Physical Review B* 47: 16404-16409.

[85]   Roberts, R. G., Harrison, P., Stirner, T., and Hagston, W. E. 1993. "Stark ladders in strongly coupled finite superlattices." *Le Journal de Physique* IV 03: 203-206.

[86]   Harrison, P., and Hagston, W. E. 1996. "The effect of linear and non-linear diffusion on exciton energies in quantum wells." *Journal of Applied Physics* 79: 8451-8455.

[87]   Landau, L. D. 1933, *Phys. Z. Sovjet.* 3, 664 [English translation: (1965), Collected Papers, New York: Gordon and Breach, p. 67 - 68].

[88]   Pekar, S. I. 1951. *Research in Electron Theory of Crystals, Moscow: Gostekhizdat [German translation:* (1954), Untersuchungenuber die Elektronentheorie der Kristalle, Berlin: Akademie Verlag; English translation: (1963), *Research in Electron Theory of Crystals,* US AEC Report AEC-tr-5575].

[89]   Fröhlich, H. 1954. "Electrons in lattice fields." *Advances in Physics 3*: 325-61.

[70]   Doniach, S., and Kuper, W. 1963. *Polarons and Excitons. Ed.: CG Kuper, and GD Whitfield.* Edinburgh: Oliver and Boyd, 191.

[91]   Appel, J. 1968. "Polarons." *Solid State Physics,* 193-391. doi:10.1016/s0081-1947(08) 60741-9.

[92]   Devreese, J. T. (Ed.). 1972. *Polarons in ionic crystals and polar semiconductors: Antwerp Advanced Study Institute 1971 on Fröhlich polarons and electron-phonon interaction in polar semiconductors.* North-Holland.

[92]   Devreese, J. T., Peeters, F. M. (Eds.) 1987. *The Physics of the Two-Dimensional Electron Gas,* New York: Plenum.

[94]   Senger, R. T., and Bajaj, K. K. 2003."Polaronic exciton in a parabolic quantum dot." *Physica status solidi* (b) 236: 82-89.

[95]   El Moussaouy, A. 2011. *Etude des états d'excitons dans les nanostructures de semiconducteurs.* Edition universitaires européennes. [Study of exciton states in semiconductor nanostructures. European University Publishing]

[96]   El Moussaouy, A., Bria, D., Nougauoi, A., Charrour, R., and Bouhassoune, M. 2003. "Exciton–phonon coupled states in CdTe/Cd1−xZnxTe quantum dots." *Journal of Applied Physics* 93: 2906-2911.

[97]   Ramvall, P., Riblet, P., Nomura, S., Aoyagi, Y., and Tanaka, S. 2000. "Optical properties of GaN quantum dots." *Journal of Applied Physics* 87: 3883-3890.

[98]   Tawara, T., Tanaka, S., Kumano, H., and Suemune, I. 1999. "Growth and luminescence properties of self-organized ZnSe quantum dots." *Applied Physics Letters* 75: 235-237.

Bound Polarons and Exciton-Phonons Coupling in Semiconductor Nanostructures 303

[99]   Tang, Z. K., Wong, G. K., Yu, P., Kawasaki, M., Ohtomo, A., Koinuma, H., adn Segawa, Y. 1998. "Room-temperature ultraviolet laser emission from self-assembled ZnO microcrystallite thin films." *Applied Physics Letters* 72: 3270-3272.

[100] Lu, M., Yang, X. J., Perry, S. S., and Rabalais, J. W. 2002."Self-organized nanodot formation on MgO (100) by ion bombardment at high temperatures." *Applied Physics Letters* 80: 2096-098.

[101] Yano, S., Goto, T., and Itoh, T. 1996. "Excitonic optical nonlinearity of CuCl microcrystals in a NaCl matrix." *Journal of Applied Physics* 79: 8216-8222.

[102] Nandakumar, P., Vijayan, C., and Murti, Y. V. G. S. 2002."Optical absorption and photoluminescence studies on CdS quantum dots in Nafion*." Journal of Applied Physics* 91: 1509-1514.

[103] El Moussaouy, A., Bria, D., and Nougaoui, A. 2005."Hydrostatic stress dependence of the exciton–phonon coupled states in cylindrical quantum dots." *Physica B: Condensed Matter* 370: 178-185.

[104] El Moussaouy, A., Bria, D., & Nougaoui, A. 2006."Thermal effect on bound exciton in CdTe/Cd1−xZnxTe cylindrical quantum dots." *Solar Energy Materials and Solar Cells* 90: 1403-1412.

[105] Ramvall, P., Riblet, P., Nomura, S., Aoyagi, Y., and Tanaka, S. 2000. "Optical properties of GaN quantum dots." *Journal of Applied Physics* 87: 3883-3890.

[106] Nomura, S., and Kobayashi, T. 1992. "Exciton–LO-phonon couplings in spherical semiconductor microcrystallites." *Physical Review B* 45: 1305-316.

[107] Schmitt-Rink, S. D. A. B. M., Miller, D. A. B., and Chemla, D. S. 1987. "Theory of the linear and nonlinear optical properties of semiconductor microcrystallites." *Physical Review B* 35: 8113-8125.

[108] Klein, M. C., Hache, F., Ricard, D., and Flytzanis, C. 1990. "Size dependence of electron-phonon coupling in semiconductor nanospheres: The case of CdSe." *Physical Review B* 42: 11123-11132.

[109] Marini, J. C., Stebe, B., and Kartheuser, E. 1994. "Exciton-phonon interaction in CdSe and CuCl polar semiconductor nanospheres." *Physical Review B* 50, no. 19: 14302-14308.

[110] Oshiro, K., Akai, K., and Matsuura, M. 1999. "Size dependence of polaronic effects on an exciton in a spherical quantum dot." *Physical Review B* 59: 10850-10855.

[111] Ajiki, H. 2001. "Exciton–Phonon Interaction in a Spherical Quantum Dot: Effect of Electron–Hole Exchange Interaction." *Physica status solidi* (b) 224: 633-637.

[112] Haken, H. 1956. "Zur Quantentheorie des Mehrelektronensystems im schwingenden Gitter. I." [To the quantum theory of multi-electron systems in the vibrating grid. I.] *Zeitschrift fur Physik* 146: 527-554. doi:10.1007/bf01333179.

[113] Pollmann, J., and Büttner, H. 1975."Upper bounds for the ground-state energy of the exciton-phonon system." *Solid State Communications* 17: 1171-1174.

[114] Pollmann, J., and Büttner, H. 1977. "Effective Hamiltonians and bindings energies of Wannier excitons in polar semiconductors." *Physical Review B* 16: 4480-4490.

[115] Bednarek, S., Adamowski, J., and Suffczyński, M. 1977. "Effective Hamiltonian for few-particle systems in polar semiconductors." *Solid State Communications* 21: 1-3.

[116] Kane, E. O. 1978. "Pollmann-Büttner variational method for excitonic polarons" *Phys. Rev. B* 18: 6849.

[117] Harrison, P. 2005. *Quantum Wells, Wires and Dots.* Willey-Interscience, England.

[118] El Moussaouy, A., and Ouchani, N. 2014. "The role of hydrostatic pressure and temperature on bound polaron in semiconductor quantum dot." *Physica B: Condensed Matter* 436: 26-32.

[119] Silva-Valencia, J., and Porras-Montenegro, N. 1997. "Optical-absorption spectra associated with shallow donor impurities in spherical infinite-well GaAs quantum dots." *Journal of Applied Physics* 81: 901-904.

[120] Movilla, J. L., and Planelles, J. 2005."Off-centering of hydrogenic impurities in quantum dots." *Physical Review* B 71: 075319.

In: Polarons: Recent Progress and Perspectives
Editor: Amel Laref

ISBN: 978-1-53613-935-8
© 2018 Nova Science Publishers, Inc.

Chapter 8

THE ELECTRICAL AND STRUCTURAL STUDY OF COMPOUNDS WITH A MODULATED SCHEELITE-TYPE STRUCTURE AT HIGH TEMPERATURE

C. González-Silgo[1], M. E. Torres[1], N. P. Sabalisck[1], I. T. Martín-Mateos[2], E. Zanardi[1], A. Mujica[1], F. Lahoz[1], J. López-Solano[3] and C. Guzmán-Afonso[4]

[1]Departamento de Física, Universidad de La Laguna, La Laguna, Spain
[2]Departamento de Ingeniería Industrial, Universidad de La Laguna, La Laguna, Spain
[3]Izaña Atmospheric Research Center, AEMET, Santa Cruz de Tenerife, Spain
[4]Advanced Solid-State NMR Unit. RIKEN CLST-JEOL, Yokohama Kanagawa, Japan

ABSTRACT

The structural flexibility of scheelites and related compounds (SRCs) makes it possible for this family to exhibit phenomena as diverse as photoluminescence, catalytic activity, mixed ionic-electronic conduction, etc, with the crystal and electronic structures determining the material's properties (the electrical ones having been much less studied than the optical ones). Here we review the ac conductivity at high temperature in SRCs and discuss Jonscher's Universal Power Law for Eu, Sm and Nd molybdates. Polaronic mechanisms can be explained and correlated with the thermal dependence of the crystal structure.

Keywords: scheelites and related compounds, SRCs, ac conductivity, Jonscher Universal Power Law, Overlapping Large Polaron, modulated structures

1. INTRODUCTION

1.1. Scheelite and Related Compounds: Correlation Crystal Structure-Physical Properties and Technological Applications

It is a truth universally acknowledged that practical applications drive modern research in materials science, where the focus is on the development, by design or serendipitous discovery, of novel materials with technologically-wise interesting properties. The new and improved functionalities to which the research in the field aims can only be fulfilled provided that a thorough understanding of the relationship between structure and properties of the materials under research has been previously achieved. On the other hand, materials properties are not uniquely determined by the composition and structure of the ideal host material but often extrinsic and/or intrinsic defects play a significant role on their enhancement. As a matter of fact, many physical and chemical properties depend on the delicate interplay that is established among a specific ordering of defects with subtle and often competing structural distortions and electronic instabilities (Batuk et al., 2015). There have been a number of reports concerning the study and development of interesting electric transport and luminescence properties, which can be potentially further controlled, improved and functionally tuned by structurally altering the host lattice through either cation or anion total or partial substitution (recently: Fabbri et al., 2012, Morozov et al., 2013, Han et al., 2015, and Sharp et al., 2017).

The scheelite related compounds (SRCs), which comprise scheelites properly, as well as wolframites, fergusonite-types, and modulated scheelites, with the general formula $A_n(XO_4)_m$, belong to a broad spectrum of compositions, with A representing either a single element or combination of up to three cations (which can be Ca, Sr, Pb, Ba, Zn, Cd, In, Ga, Tl, Ln, Y and Bi); while X stands for either Mo, W, Nb, V, Ta or, possibly, a combination thereof (Arakcheeva et al., 2012). Their high stability, easiness of preparation and promising optical and electrical properties makes them ubiquitous for many technological applications (Pang et al., 2011). Such physical properties are primarily associated to their cluster constituents: initially, anionic tetrahedral XO_4 and AO_8 polyhedral, which can be deformed, respectively, into XO_6 and either from AO_6 to AO_{12} polyhedrals. Distorted clusters due to lattice defects yield a local lattice distortion that propagates through the material, pushing the surrounding clusters away from their ideal positions and changing the electronic distribution along the network (Gracia et al., 2011). As a result, the distorted electronic structure affects both the optical and electrical transport properties. The technological applications of this family of materials are various and many, and in what follows we will highlight some of the research in the field, with a focus in progress made during the last two decades. We will distinguish between: 1) optical devices for detectors and sources: scintillators, Raman shifters, stoichiometric or doped with rare earth efficient phosphors, and laser hosts; and 2) electro chemical devices for energy storages: electrolytes, and electrodes in fuel cells and batteries, supercapacitors and photocatalysts.

1.1.1. Optical Properties

One of the main optical applications of these materials is as scintillators. In particular, the mineral with chemical composition $CaWO_4$ (scheelite) was the first material found for efficient conversion from X-ray to visible light (Röntgen, 1896). During most of the twentieth century, $CaWO_4$ was widely used as the reference scintillator until its dominant position was overtaken by rare-earth phosphors in recent years. Nowadays another compound with scheelite structure, lead tungstate ($PbWO_4$), is being considered the most attractive material for high-energy physics applications because of its high density, short decay time and high irradiation damage resistance (Kobayashi et al., 1998, Hara et al., 1998). Annenkov et al. (2002) have explained how only shallow polaronic and distorted regular centers contribute to the scintillation processes through relatively fast electron exchange with the conducting zone. The shallowest of them occurs in all crystals since it is an intrinsic defect: an additional electron auto-localized at an anionic WO_4^{2-} complex, via a Jahn-Teller distortion, creating a WO_4^{3-} polaronic center. The released electrons partly recombine radiatively, and partly are caught by deeper traps. Anion vacancies or their associations, contribute to slow components in scintillations, phosphorescence and additional transient or optical absorption induced by irradiation. Further, the systematic deficiency of lead in the crystals introduces additional oxygen vacancies leading to irregular WO_3 groups where the luminescence is shifted to the green region, while regular WO_4^{2-} groups show blue luminescence (Abraham et al., 2001). The theoretical calculations indicate that locally disordered Pb clusters are more favorable for the PL effect generating localized states and inhomogeneous charge distribution in the cell, thus allowing the trapping of electrons and holes before excitation (Anicete-Santos et al., 2007).

Another interesting application of SRCs is as Raman lasers. In these kind of luminescent materials, the fundamental light amplification is stimulated by Raman scattering, depending on spectroscopic parameters such as the integral Raman scattering cross-sections and linewidth. Raman spectra of tungstate and molybdate single crystals consists of intensive and narrow lines corresponding to symmetrical vibrations of $[WO_4]^{2-}$ and $[MoO_4]^{2-}$ groups making this type of compounds as promising Raman active crystals (Basiev et al., 2000, Zverev, 2012).

Rare earth (RE) tungstates and molybdates, where the RE replaces the A atom through stoichiometric substitution in the general formula $A_n(XO_4)_m$, or by doping, are widely studied both for their important applications in optics and electronics, as well as for the fundamental physics underlying those processes (Gai et al., 2014). The optical properties of RE ions are basically influenced by the spin- and parity-forbidden character of the intra-configurational 4f $\rightarrow$ 4f electronic transitions, which produce relatively small absorption cross-sections and long radiative lifetime decays (Ofelt, 1963). In particular, tungstates and molybdates have demonstrated to be exceptional hosts for RE ions in related scheelites, due to its large ionic radius. For instance, a study in ytterbium- and erbium-co-doped lanthanum molybdates revealed that these materials have an efficient near-IR to visible light up-conversion, where ytterbium acts as an absorber and erbium as an emitter in the crystal lattice (Yi et al., 2002). More recently, a wide range of luminescence temporal behaviour from nanoseconds to hundreds of microseconds was found in $RE_2(WO_4)_3$ crystals (Lahoz et al., 2015); and even the excited state RE lifetime can be modulated taking advantage of energy transfer processes (Sabalisck et al., 2017a).

The compounds with scheelite and scheelite-like structures with formula $ARE(XO_4)_2$ (A = Li, Na, K; RE = lanthanide and X = W, Mo) have been studied because of their potential as solid-state lasers. They show high quantum efficiency, broad absorption and emission bands, as well as relatively long upper-level lifetimes. As a consequence of the possibility to have a locally variable position of the active ion in the crystal structure, the linewidths of the electronic transitions for the rare earth elements are found to be broader in disordered than in ordered crystals which could be suitable in the generation of ultrashort laser pulses (Méndez-Blas et al., 2007). The polymorphism and modulated structures of these compounds have been studied in order to know the degree of ordering for the rare earth which is closely related to the symmetry of its crystal structure (Morozov et al., 2006).

1.1.2. Electrochemical Properties

In the near future, new and more efficient electrochemical energy storage devices (fuel cells, batteries, and supercapacitors) will be increasingly needed to cope with the exhaustion of fossil-based energy sources and the challenges imposed by the changes in weather at a global scale (Liu et al., 2010). In what pertains to compounds within the scheelite family, the first steps were aimed at their usage as ionic superconductors. Both anionic and cationic conductors were found in solid solutions, for instance, oxide ion conductors based on $Pb_{1-x}RE_{3+x}WO_{4+\delta}$ (RE = La, Sm, Pr, Tb) and lithium ion conductors based on $A_{1-x}Li_{2x}XO_4$ (A = Ca, Sr, Ba, X = W, Mo) (Esaka, 2000).

Also, binary transition metal oxides which naturally exhibit high structural stability, high reversible capacity, cyclability and electronic conductivity, have been widely studied with a view on the capacity requirements of next-generation devices. Several tungstates have been proposed as alternatives for graphite anode in lithium-ion batteries material, as for instance $ZnWO_4$ and $FeWO_4$, which exhibit excellent electrochemical performances (Shim et al., 2011, Gong et al., 2013). Synthesis methods for novel electrode materials ($NiMoO_4$ and $CoMoO_4$) used in supercapacitors have also been described (Zhang et al., 2015). In order to improve the functionality of this type of transition-metal tungstates and molybdates, their bulk properties and defect physics must be understood first. The study of $FeWO_4$ and $MnWO_4$ via first-principles total energy methods, using screened hybrid density-functionals, has shown that their electronic structures near the band edges are determined by the highly localized transition-metal d states, which allow for the formation of both hole polarons at the Fe and Mn sites and electron polarons at the W sites. However, the point defects in both compounds due to the synthesis conditions are hole polarons or/and cation vacancies at the Fe and Mn sites. The presence of low-energy and highly mobile polarons provides an explanation for the good p-type conductivity observed in the experiments and the ability of these materials to store energy via a pseudocapacitive mechanism (Hoang, 2017).

As alternatives to perovskite electrolytes, other oxides have recently been reported to display proton conductivity together with rather good chemical stability, opening thus new perspectives. However, at present, the proton conductivity achieved for these prospective materials is still small for practical solid oxide fuel cell applications. It is worth mentioning here the class of compounds described as $RE_{(1-x)}A_xXO_4$, (RE = La, Nd, Gd, Er; A = Ca, Sr, Ba is the dope; X = Ta or Nb and x varies between 0.01 and 0.05). The largest conductivity within this class of compounds was observed for $LaNbO_4$. Besides its low conductivity this compound shows a fergusonite structure at ambient temperature and undergoes a phase transition to the scheelite structure at around 500 °C, and this structural change affects its thermomechanical and conduction properties (Fabbri et al., 2012). Theoretical simulations have been used to investigate the energetics of defect formation, dopant solution, water incorporation and defect clustering in the high-temperature proton conductor $LaNbO_4$ (Mather et al., 2010).

On the other hand, many efforts have been made to find appropriate materials and systems that can utilize solar energy to produce chemical fuels through photocatalytic processes. Photocatalysis is a chemical reaction that uses light to activate a substance, the photocatalyst, which generates or accelerates a reaction by creating free radicals that reduce the activation energy of the reaction, without altering it. One of the most viable options is the construction of a photoelectrochemical cell that can produce H_2 from water. Among currently used semiconductor photoanodes, the ms-$BiVO_4$ (monoclinic phase) has recently been identified as a suitable material for the study of photoanodes, in which water is oxidized into O_2 (Park et al., 2013). A comprehensive approach for a deep understanding of the electronic structure of ms-$BiVO_4$, including both valence band (VB) and conduction band (CB) orbital character, has been presented by Cooper et al. (2014). Density functional theory (DFT) calculations have been performed to obtain the total and partial density of states (DOS) near the bandgap, and the results have been successfully compared to experimental data (Kweon, et al., 2015). Electron and hole self-trapping via small polaron formation in some semiconductors, especially in transition-metal oxides, can severely reduce mobilities, thereby limiting the efficiency of the photoelectrochemical cell (Peng and Lany, 2012). In spite of this, because small polarons exist as deep level states within the band gap, they can also introduce additional recombination channels. Thus, identifying polaronic conduction and understanding its implications is essential for the evaluation of materials and development of approaches to improve their efficiencies (Sharp et al., 2017).

Finally, SRCs ceramic materials are also utilized in microwave applications (wireless communication devices) because of their moderate dielectric constants and high quality factors Q_f. These applications also require ceramics of zero temperature coefficients at the resonant frequency τ_f, which can be achieved by ionic substitution (Chen et al., 2017). For instance, Pang et al. (2017) found that the $(Na_{0.5}Bi_{0.5})(Mo_{1-x}W_x)O_4$ ceramic is a good candidate for low temperature co-fired ceramics. With the increase of the W content, the permittivity decreased from 34.4 to 25.7 and the Q_f value increased from 12300 GHz to 17500 GHz while τ_f shifted from +43 to −18 ppm/°C, which indicates that τ_f is strongly dependent on the $[XO_4]$ tetrahedron in the scheelite structure. Although the deformation of the tetrahedra influences the dielectric properties, in the range of microwaves, we are not aware that, for instance, polaron conduction loss has been studied for SRCs.

1.2. Polarons, Dielectric Spectroscopy and Crystal Structure

As a common feature, more or less disordered defects, mostly acceptor vacancies, and their subsequent lattice distortions, form small polarons. These quasiparticles play a dominant role in scintillation, phosphors emission, energy storage and transport for scheelite-related compounds. The presence of polarons in transition metal oxides and their optical and transport properties, have been reviewed by Holstein (1959a, 1959b), Friedman and Holstein (1963), Austin and Mott (1969), Emin (1993), and Alexandrov and Kornilovitch (1999) among others. Usually small polarons can be well distinguished from large polarons. Small polarons can form when a short-range electron-lattice interaction, such as the deformation-potential interaction, is dominant. In contrast, polarons whose electronic carriers are not self-trapped, i.e., which are weakly coupled, are called large polarons. Traditionally, polarons have been detected using absorption spectroscopy in which the absorption bands of large and small polaronic carriers have qualitatively different shapes and temperature dependences. Moreover, the decrease of the negative (positive) Seebeck coefficient, with increasing temperature, is due to large electron (hole) polarons, while the decrease of the negative (positive) Seebeck coefficient is attributed to small electron (hole) polarons (Emin, 1999). In Hall effect experiments, conventional carriers have very similar Hall and drift mobilities. However, for small polarons Hall mobility measurements can help distinguish between small-polaron's intrinsic hopping and other carrier's hopping among defects, impurities or dopants (Mott, 2001), because both mobilities are quite different.

The possible coexistence of both types of polarons together with other conventional carriers can be inferred using the mentioned techniques, but it is a complex task. Thermally activated conductivity is a result of the temperature dependence of the carrier concentration and/or mobility and the particular dependence of different carrier features at high temperature can be used to distinguish among different conduction mechanisms. For wide band gap oxides, the density of electronic charge carriers may not be significant until a high temperature is reached, from which different contributions and transport and/or relaxing mechanisms can be distinguished: electronic, polaronic, ionic and tunnelling or hopping. Hence, performing dielectric measurements over a wide range of temperatures provides the values of the energy barrier or activation energy associated with each kind of relaxing or conducting species. The dielectric response of a given material strongly depends on the nature of its chemical bonds, as well as on the compositional and structural order at short, medium and long range. Therefore, dielectric spectroscopy provides a tool to study interatomic interactions in materials, not necessarily restricted to electrical applications (Macdonald and Johnson 2005). A better understanding of the dielectric response and its relationship to microscopic features in solids resulted from the work of Jonscher (1977), Dissado and Hill (1984) and León et al. (2001). They introduced the universal dielectric response explained in a many-body theory framework involving cooperative contributions from interacting electric species (charges or dipoles) during relaxation. From the dielectric spectroscopy measurements, one can obtain the complex dielectric permittivity, with the real part of the ac conductivity being proportional to the imaginary part of the permittivity. A fascinating aspect of ac conductivity in solids is that, at high frequencies, it is quite similar, whether they are glasses, polycrystalline semiconductors, polymers, transition metal oxide or even organic-inorganic composites (Overhof, 1998). Most materials display a similar conductivity-frequency dependence as proposed by Jonscher, which is known as the Universal Power Law (Jonscher, 1977). Different theories for the ac conduction have been proposed to account for the frequency and temperature dependence of the ac conductivity (Long, 1982, Elliot, 1987). It is commonly assumed that the dielectric loss occurs because the carrier motion is considered to be localized within pairs of sites. In essence, two distinct processes are considered for the relaxation and conduction mechanism, namely quantum mechanical tunnelling through the barrier separating two equilibrium positions and classical hopping of a carrier over the barrier. Moreover, it is assumed that electrons, polarons or ions are the responsible carriers. As extreme cases, while the ionic conductivity arises by a barrier crossing process, the electronic conduction occurs via quantum mechanical tunnelling between localized states. In Section 4 we will present models for small and large polaron dynamics.

One must not forget the main differences between free-charge carriers versus dynamic dipoles. The first are responsible for dc-type processes without implying any polarization process. Meanwhile, the dipoles are localized without contributing to the dc conductivity. The appearance of the localization or restriction of the free motion in a solid is closely related to the nature of their chemical bonds and to disorder, such as structural or compositional defects. In the field of energy storage and electric transport, the materials are grouped on the basis of composition as well as, especially, in terms of their major structural types. A detailed structural characterization can be usually found at room temperature. Nevertheless, there are far fewer studies on the thermal dependence of the crystalline structure and dielectric properties, excepting studies on polymorphs, phase transitions and negative thermal expansion (e.g., Sabalisck et al., 2017b).

Determining or "simulating" unambiguously the "atomic structure" of structurally disordered materials (without long-range periodic symmetry) is unfortunately a nontrivial task, however it must be carried out in order to get a better understanding. In theoretical calculations it requires using large simulation cells, which is very time consuming. On the other hand, owing to the advances in X-ray and neutron diffraction instrumentation, as well as in transmission electron microscopy (TEM), much progress has been made in the knowledge of defective structures. In particular, in many functional materials, the interplay between interatomic interaction and lattice strains forces point defects to get as homogeneously distributed as possible and ordered into modulated uniform patterns (Batuk et al., 2015). In an interesting way, the correlation between the crystal structure and the luminescence parameters is being recently studied in modulated scheelites, where the relative amount of rare earth dimers, detected in different modulated structures are well correlated with the overall and intrinsic quantum yields and the observed lifetimes (Arakcheeva et al., 2012). All these electronic and optical properties of modulated scheelite or related compounds and, in particular, the interesting crystal structure-optical correlations found in modulated scheelite, have motivated us to undertake its detailed electrical study. As previously mentioned, these compounds exhibit high stability, easiness in their preparation and some advantages in their physical properties, due to the ordering of their stoichiometric cation vacancies, regarding other materials with related-scheelite structure. Moreover, the compounds with stoichiometric RE content are more suitable for this purpose because the RE ion position can be reliably defined from crystal structure analysis (Lavín et al., 2002). In addition, the creation of cationic vacancies in the crystal lattice and the ordering of these cations and vacancies may be a new factor in controlling their structural and physical properties (Morozov et al., 2012).

Finally, in contrast with other techniques, both impedance bridges and diffraction equipments are increasingly available techniques (with access to large facilities) and can be applied in the same samples, following similar heating and cooling programs. Therefore, one can be able to correlate the thermal dependence of the crystal structure with its dielectric response in the same temperature range.

1.3. How This Chapter Is Organized

This chapter is organized in six sections, with this Introduction being the first one. In the next section we will describe the crystalline structures of the pure scheelite, the related-types and the modulated structures based on scheelite, differentiating between stoichiometric and non-stoichiometric vacancies and their possible orderings. We will give relevance to those structural aspects that can be related to the electronic structure of the material, and hence to its electrical and optical properties. At the end of that section we will focus on two particular scheelite structures belonging to the trimolybdates family with chemical formula $RE_2(MoO_4)_3$: those of the α-phase and the $La_2(MoO_4)_3$-phase. We will account also for other polymorphic phases that can coexist, and whose formation depends on the ionic radius of the rare earth and the method of synthesis. In Section 3 we will give the experimental results of the measurements by conventional X-ray diffraction, at different temperatures, of three molybdates: α-$Eu_2(MoO_4)_3$, α-$Sm_2(MoO_4)_3$ and $Nd_2(MoO_4)_3$, with the same structure as of $La_2(MoO_4)_3$. We will discuss the thermal dependence of the cell parameters, making connection with the distortions of the MoO_4^{2-} and REO_8 groups and their shortening-lengthening oxygen bridges A/X...O...X/A. In Section 4 we will give the bases for the study of the ac conductivity from the fitting of Jonscher´s power law, including the macroscopic models that explain the different conduction mechanisms, emphasizing those of polaronic type. After that, we will review all (to the best of our knowledge) the SRCs in which the ac conductivity has been well studied and for which the mechanisms of relaxation and conduction (including different charge carriers) have been explained in the previous terms. In Section 5, we will present our results on dielectric measures in the three trimolybdates and will discuss the possible charge transport mechanisms. We will focus on the intermediate temperature region where we observe a polaronic conductivity. Finally, in Section 6 we make connections between our structural results described in Section 3 and those obtained for the ac conductivity in Section 5, that will allow to analyse similarities with other compounds that, although they do not have modulated or related scheelite structures, undergo structural phase transitions. We will end this chapter giving some conclusions.

2. REVIEW OF THE CRYSTALLINE AND STRUCTURE OF THE SCHEELITE FAMILY OF COMPOUNDS

Compounds with the scheelite or related structures can be considered in terms of the packing of its constituent clusters as isolated anionic tetrahedral $[XO_4]$ and bisdisphenoid $[AO_8]$ polyhedra (X and A are different metals) into the crystal lattice, with all polyhedra linked by oxygen bridges $[X/A...O...X/A]$ between neighbouring clusters. Symmetry-breaking processes on these clusters, such as distortions, breathings and tilts, create a huge number of different crystal structures and subsequently different functional properties. *Distorted clusters yield a local lattice distortion that is propagated along the overall material, pushing the surrounding clusters away from their ideal positions* (Gracia et al., 2011). Therefore, distorted clusters must move for these properties to occur, changing the electronic distribution along the network of these clusters and this electronic structure dictates both the optical and charge-transport properties. Polyhedral distortions by different cations

The Electrical and Structural Study of Compounds ... 313

substitutions, dopes, and by imposing non-environment conditions are being studied in order to understand the origin and limitations to get enhancements and control of such properties.

In this section, we will review the crystalline structure of the family, ranging from pure scheelite to modulated scheelites, in which the compounds of our interest are found. We will give some details of their crystalline and, in some cases, also of their electronic structure in order to determine the limiting factor of their luminescence or charge transport, for which the metal-metal interactions play an important role. We will try to get a better understanding of the transport mechanisms for different structural types, as explained in section 4. We also include some keys that provide the experimental studies at high pressure, which provide further insight into the short-range (Raman spectroscopy) and long-range (diffraction) evolution of the structural motifs under pressure and the possibility of monitoring them via theoretical calculations. Moreover, we will review other phases, not related by symmetry, that occur in polymorphs at other temperatures and that are stable, even at room temperature. These less dense phases have higher electrical conductivity and the thermal dependence of their crystal structure has received more attention because of their anomalous behaviours (e. g. compounds with negative thermal expansion) or because they give rise to paraelectric-ferroelectric phase transition.

2.1. The Parent Structure: AXO_4 with Scheelite-Type Structure

Compounds with scheelite-type structure can be expressed with the chemical formula AXO_4 where A = K, Ca, Sr, Ba, Pb and X = Mo, W, among others. A great number of double tungstates and molybdates with chemical formula $AA'(XO_4)_2$, where A is an alkaline metal and A' = rare earth or trivalent metal, and X = Mo and W, show this crystal structure too. Another oxide belonging to this family is the orthovanadate ts-$BiVO_4$ (Sleight et al.,1979). Scheelite is the name of the $CaWO_4$ mineral (SG: $I4_1/a$, Z = 4) (Zalkin and Templeton, 1964), this "parent structure" is made up of AO_8 dodecahedra and XO_4 tetrahedra connected via common vertices (with X/A...O...X/A connections) (Figure 2.1). Each dodecahedra is joined to four adjacent dodecahedra, two along the directions $[10\bar{1}]$ and $[\bar{1}0\bar{1}]$ and other two along the directions $[01\bar{1}]$ and $[0\bar{1}\bar{1}]$, via one common edge and forming a 3D framework; the A...A distances being about 4±0.25Å. The next neighbours (NN) are at distance equal to the a or b cell parameter (Figure 2.1). There are six dodecahedra joined in this way along the length of the c cell parameter in each unit cell. The tetrahedra XO_4 do not share edges or vertices but are linked in the same way with X...X distances equal the A...A distances. This is because the X and A atoms are in equivalent crystallographic positions. Therefore, three tetrahedrons are arranged in planes ab occupying three non consecutive parts of six, so that in the plane immediately below (at $c/4$) the tetrahedrons occupy the other three parts (Figure 2.1). A more dense and complex network is formed by the X...A cation which are connected in a similar way along the directions with distances about 4±0.25Å, in the directions $[10\bar{1}]$, $[\bar{1}0\bar{1}]$, $[01\bar{1}]$ and $[0\bar{1}\bar{1}]$; and a shorter contact about 3.8±0.25Å along the directions $[100]$, $[\bar{1}00$, $[010]$ and $[0\bar{1}0]$. Therefore, each atom A (or X) is connected to four atoms X (respectively, A) in the ab plane (Figure 2.1). The lattice periods $a \cong b \cong 5.5 \pm 0.5$ Å and $c \cong 2a$; and angles, close to 90°, are characteristics of the scheelite structure.

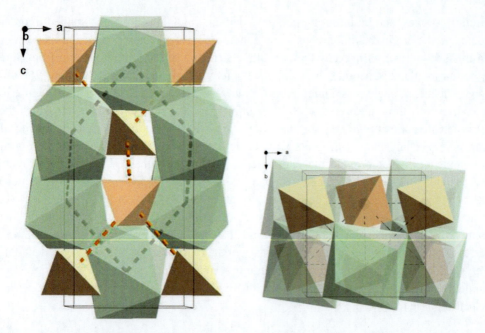

Figure 2.1. Scheelite structure: (left) view along the *b* axis of the scheelite structure. Nearest-neighbour (NN) jumps (X… X and A…A) are marked by orange and green dashed lines, proceeding along zig-zag chains in the *ac* and *bc* plane; (right) view along the c axis NN jumps (A…X) are marked by black dashed lines.

The physical origin of the transport anisotropy can be explained on the basis of the characteristics of scheelite-type crystal structure and taking into account the A…A and X…X cation distances (Rettie et al., 2015).

2.2. Related Scheelite Types: Wolframite and Fergusonite

Symmetry breaking processes undergone by these polyhedra, such as distortions, breathings and tilts, can create a huge number of different structures and, subsequently, different material properties. The crystalline and electronic structures of crystals with symmetry related to the scheelite type have been well studied. For instance, the fergusonite structure (SG: I2/b, Z = 4) is a monoclinic distortion of the scheelite structure, which transforms to the scheelite structure at high temperatures. A particular example is the *ms*-BiVO$_4$ phase (Stoltzfus et al., 2007), a subtle distortion of the *ts*-BiVO$_4$ phase, whose structural differences have been studied in detail, as explained in Section 1. In the fergusonite polymorph, A and X-sites are coordinated similarly to the scheelite form, with two additional near-oxygen sites at each X-site cation (Figure 2.2), however, the other X ... X contacts elongate. Several binary tungstates has been revised by López-Solano et al., in 2006 and López-Solano et al., in 2007, showing other two fergusonite-related structures: the structure of the isomorphous phases PbWO$_4$-III and BaWO$_4$-II (SG: P2$_1$/n, Z = 8) and the BaMnF$_4$-type structure (SG: Cmc2$_1$, Z = 4). On the other hand, compounds with small divalent cations can take the wolframite structure (SG: P2/c, Z = 2). In the wolframite-type both the A and X cations have octahedral oxygen coordination and each octahedron AO$_6$ shares two corners with its neighbours XO$_6$ and *vice versa* while AO$_6$(XO$_6$) shares one edge with the consecutive

$AO_6(XO_6)$ octahedron, forming polymeric octahedral ions along the directions [01$\bar{1}$] and [0$\bar{1}\bar{1}$]. The $AO_6(XO_6)$ octahedra are distorted, with two of the A(X)-O distances being shorter than the other four (Lacomba-Perales et al., 2009). Distances A/X...A/X for NN neighbours are about 3.25 Å (Figure 2.2), shorter than in the fergusonite structure. Other structures related to the wolframite-type are: the $CuWO_4$-type (SG: P-1, Z = 2), the orthorhombic $Ni_2(VO_4)_3$-type (SG: Cmca, Z = 4) and the monoclinic $MnMoO_4$ (SG: C2/m, Z = 8) (Errandonea and Manjón, 2008). Double tungstates of wolframite and fergusonite-type are also possible, depending on the A and A' radii. Several other space groups are described for the monoclinic distorted scheelites: α-Hg_2WO_4-type or α-Hg_2MoO_4-type (SG: C2/c, Z = 4) and β-Hg_2MoO_4-type (SG: $P2_1/c$, Z = 4) (Mormann and Jeitschko, 2000). Interestingly, for second harmonic generation applications, a non centrosymmetric phase (SG: I-4, Z = 2) whose structure is a slightly distorted scheelite has been found by Cascales et al. in 2006.

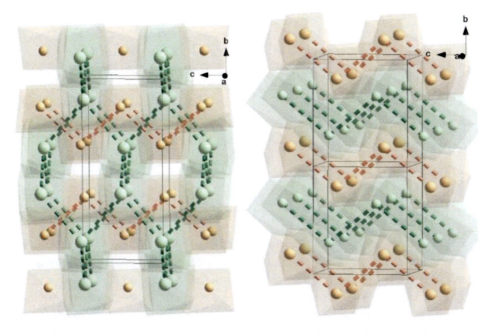

Figure 2.2. View along the *a* axis of the fergusonite structure where NN (A...A and X...X) jumps are marked by green and orange dashed lines (left). View along the *a* axis of the wolframite structure where NN (A...A and X...X) jumps are marked by green and orange dashed lines (right).

2.3. Scheelites or Related Types with Non-Stoichiometric Vacancies, Interstitials Ions

Nano-scale doping and substitution can achieve improved electrical, polaronic or ionic conductivities (Esaka, 2000) while the luminescence can be degraded. For instance, centres of non-stoichometric cations vacancies (A or X) due to metal dopes are characterized by symmetry-breaking holes located at one of several equivalent ligands. The localization is caused predominantly by the short-range acceptor-ligand bond distortions, stabilizing the holes as small polarons. The corresponding energy lowering of the hole increases the energy necessary to excite it thermally to the valence band maximum (Schirmer, 2011). Particularly, the effect of RE doping on the scintillating properties of $PbWO_4$ single crystal has been intensively investigated and was found to be dependent on the doping concentration. Doping at a low concentration level, the trivalent ion La^{3+} may substitute the Pb^{2+} ions, with the excess charge compensated by intrinsic Pb vacancies, showing a lower luminescence intensity compared with the pure one. However, heavy La^{3+} doping might also cause La^{3+} self-compensation by substitution in the W-site, at which a new effective recombination takes place, and thus severely degrades the luminescence (Huang et al., 2003a). The explanation on the ac conductivity of $La:PWO_4$ will be dealt with in Section 4 (Huang et al. 2003b).

On the other hand, if the constituent oxides of a compound have different vapour pressures at the operating temperatures, stoichiometric deviations in the grown crystal can occur. For instance, $CaMoO_4$ annealed at $1100°C$ in the presence of CaO is a mixed ionic/electronic conductor while $CaMoO_4$ annealed in MoO_3 vapor at $1100°C$ is an electronic conductor (Petrov and Kostadf, 1979). PbO has a high vapour pressure at the growth temperature, leading to significant stoichiometric deviations. Thus, lead and oxygen vacancies are generally created due to Pb and O deficiency in the $PbWO_4$ crystal (Ganesamoorthy et al., 2004). Moreover, the systematic deficiency of lead in the crystal introduces additional peculiarities to the crystal structure. A superstructure created by cation vacancies has been identified (Moreau et al., 1999); although the structural motive of the crystal remains the same, a final compound composition of $Pb_{7.5}W_8O_{32}$ ($Z = 4$) is found to be in agreement with both X-ray and neutron diffraction measurements. The ordering of the vacancies is compensated by the distortion of the tungstate anionic polyhedral and there are four unequivalent positions of Pb in such a structure; the occupancy factor of Pb(4) is 0.5.

2.4. Modulated Scheelites with Stoichiometric Vacancies

The possibility to create cationic vacancies in the crystal lattice and the ordering of these cations and vacancies is a new factor in controlling the structural and physical properties of this type of materials. The modulated scheelite structure is a cation deficient structure with formula (A, A')n[(X, X')O4]m; when (A, A)/(X',X) = n/m = 1, there are no possibilities for vacancies in the A' A" subset. Only charge neutrality limits the m/n ratio in the modulated structure model where the vacancies can be ordered or randomly distributed on the A positions, affecting the symmetry and the unit-cell dimensions in the ab plane. Thus, the symmetry of the modulated structures covers a large spectrum from the tetragonal to the monoclinic system (Arakcheeva and Chapuis, 2008). The modulated scheelites exhibit an occupational modulation wave propagating in the ab plane of the tetragonal scheelite subcell with a modulation vector $q = \alpha a^* + \beta b^*$ (giving additional satellite reflections in the lower angle of the XRD pattern). If α or/and β are rational numbers the modulation is commensurable, conversely, if α or/and β are irrational numbers the modulation is incommensurable. The modulation of the periodicities should be achievable through the variation of the chemical composition. The symmetry of these modulated structures is described in the superspace group: $I2/b(\alpha\beta0)00$. The (3 + 1)-dimensional symmetry concept (Schreiber, 2008) shows that the entire family of compositionally different scheelite related compounds could be described from a unique model of modulated structures.

This directly affects the placement of the luminescence centres in these modulated structures and it has been proposed that there is a relation between the incommensurate modulation in the scheelite structure and its luminescent properties. For several years, scientists have vainly attempted to correlate the luminescence efficiency, in particular by the Na/Eu content (van Vliet et al., 1988). A tentative relation between the quantum yield of the luminescence and a Eu^{3+} association into "dimers" has been postulated (Arakcheeva et al., 2012). This study pointed to the importance of considering the superspace formalisms without which no direct correlation could be found between composition and luminescent properties. Once the superspace model is established, it is straightforward to explore the best composition with larger numbers of A dimers (or other clusters) in order to maximize the luminescence or other physical property (for instance, the electric conductivity) by material engineering.

2.5. Rare Earth Molybdates and Tungstates with Modulated Scheelite Structure

The first rare earth molybdates represented by the formula $RE_2(MoO_4)_3$ were prepared by Hitchcock in 1895. In particular, Hitchcock synthesized the lanthanum, praseodymium and neodymium molybdates. However, the study of RE molybdates did not show any progress until almost a century later, when Borchardt and Bierstedt (1966) prepared $Gd_2(MoO_4)_3$ and found that this compound is ferroelectric. This discovery promoted the interest on the studies of the $RE_2(MoO_4)_3$ series until the present day.

The series of RE molybdates $[RE_2(MoO_4)_3]$ can crystallize in three structural groups depending on the RE cation size and the synthesis conditions. These three structural types are known as: 1) the modulated scheelite-type structures, 2) the β'-phase and 3) the γ-phase, without any simple symmetry relation among them. This rich polymorphism of RE molybdates is also made evident under changes of pressure and temperature, i.e., nine different crystal structures are known for these compounds from room temperature to their respective melting points (Brixner et al., 1979), and from ambient pressure (AP, 1 atm) to their respective pressure-induced amorphization (Maczka et al., 2012). Furthermore, the family of RE tungstates $[RE_2(WO_4)_3]$ crystalize in two crystal groups, depending on the ionic radii of the RE and the synthesis conditions with modulated scheelite-type structure and the γ-phase (Nassau et al., 1971, Brixner and Sleight, 1973). This variety of crystal structures makes them appropriate to develop new concepts about the physics of thermal anomalies, phase transitions and amorphization processes under high pressure, as well as related physical properties and their possible applications.

One can distinguish two substructures in RE molybdates with different ordering of the vacancies depending on the RE ion. The first substructure appears for the following ions: La^{3+}, Ce^{3+}, Pr^{3+}, and Nd^{3+}. This structure, denominated $La_2(MoO_4)_3$-type, is monoclinic with SG C2/c and Z = 12 formula units per conventional cell (Jeitschko, 1973). Its volume is nine times the scheelite volume. Its asymmetric unit consist of 23 atoms with non-equivalent positions, all of them in general positions 8f, except one Mo in a 4e Wyckoff position with the site symmetry C2. There are two RE atoms and one stoichiometric vacancy for every three Mo atoms. A view of the structure along the b axis, the scheelite cell, the supercell and the RE^{3+} and vacancies ordering is shown in Figure 3.5 (Section 3). The second substructure, known as the α-phase, occurs for the ions: Sm^{3+}, Eu^{3+}, Gd^{3+}, Tb^{3+} and Dy^{3+}. This crystal structure with SG C2/c and Z = 4 comprises three scheelite-like subcells and its named as the α-EuWO$_4$-phase (Templeton and Zalkin,1963). Its asymmetric unit consists of 9 atoms with non-equivalent positions, all of them in general positions 8f, except for one Mo occupying a Wyckoff position 4e with site symmetry C2. There are two RE atoms and one stoichiometric vacancy for every three Mo atoms. A view of the structure along the b axis, the scheelite cell, the supercell and the RE^{3+} and vacancies ordering is shown in Figure 3.5 (Section 3). Among the molybdates featuring the α-phase, only the structure of Eu$_2$(MoO$_4$)$_3$ has been determined (Boulahya et al., 2005). Lighter RE and bismuth tungstates (RE = La, Ce, Pr, Nd, Sm, Eu, Gd, Tb, Dy and Ho) crystalize in the α-phase or Eu$_2$WO$_4$-type (Brixner and Sleight, 1973). A special case is Bi$_2$(MoO$_4$)$_3$ (SG: P2$_1$/c and Z = 4) whose Bi^{3+} vacancies are ordered in a different way than the α-phase (Van den Elzen and Rieck, 1973). The bismuth molybdate, tungstate, and vanadate with the scheelite-type structure have traditionally been investigated for their catalytic properties (Sleight et al., 1975). In addition, a new possible cation ordering has been found for Pr$_2$(MoO$_4$)$_3$ which has been described as an incommensurable structure in the superspace group I2/b(αβ0)00 and Z = 7 (Logvinovich et al., 2010).

2.6. Non-Environmental Conditions and Other Phases Unrelated by Symmetry

Since rare earth molybdates, tungtates, and vanadates (A$_n$(XO$_4$)$_m$ materials, A = RE, X = Mo, W, V) with scheelite structures constitute an important class of materials that exhibits various functional properties, which depend on the structure, its evolution under pressure has attracted renewed interest in the last years (Errandonea and Manjón, 2008 and Maczka et al., 2012). However, much less work has been developed for RE$_2$(MoO$_4$)$_3$ compounds with a modulated scheelite-structure. No pressure-induced phase transitions have been observed for any of the studied compounds: high pressure Raman spectroscopy in Nd$_2$(MoO$_4$)$_3$, Tb$_2$(MoO$_4$)$_3$ (Jayaraman et al., 1997), and Eu$_2$(MoO$_4$)$_3$ (Le Bacq et al., 2011), and high pressure XRD in Tb$_2$(MoO$_4$)$_3$ (Guzmán-Afonso et al., 2014) and Eu$_2$(MoO$_4$)$_3$ (Guzmán-Afonso et al., 2015). However, La$_2$(WO$_4$)$_3$, the only tungstate studied using both Raman spectroscopy and X ray diffraction, has been found to undergo two distinguishable phase transitions at high pressure compatible with the results of ab $initio$ calculations (Sabalisck et al., 2014). This study provides a clear example of the potential of the theoretical calculations to explain the appearance of a mixture of phases, at lower pressure, which could be compatible with a new superstructure in the family of modulated scheelites. At higher pressure, the candidate structures have in common the increased coordination of both cations A and X, as it occurs in the better known AXO$_4$ compounds. On the other hand, although rare earth vanadates crystalize in the zircon-phase (ZrSiO$_4$, with SG I4$_1$/amd, Z = 4), an in $situ$ high pressure synchrotron X-ray study of CeVO$_4$ and TbVO$_4$ up to 50 GPa shows that TbVO$_4$ undergoes a transition to a scheelite-type phase at 6.4 GPa and to a fergusonite-type phase at 33.9 GPa (Errandonea et al., 2011).

On the other hand, RE$_2$(MoO$_4$)$_3$ (RE = Sm-Ho) undergoes a phase transition at ~700°C to the β-Gd$_2$(MoO$_4$)$_3$ phase, a tetragonal structure with SG P$\overline{4}$2$_1$m and Z = 2 (Brixner et al., 1979). At low temperature, the α-phase can be recovered performing a slow cooling rate, but the β'-Gd$_2$(MoO$_4$)$_3$ phase (SG Pba2, Z = 4) can be obtained by quenching or using a faster cooling rate. This ferroelectric phase (Borchardt et al., 1966 and Jeitschko, 1972) is particular of this family of molybdates and has not been found in any other similar tungstates, chromates, phosphates or vanadates. The main properties of molybdates with β'-phase is its ferroelectricity and ferroelasticity due to the possibility to develop a spontaneous polarization in the direction of the polar axis, parallel to the c axis. They may also exhibit a spontaneous strain. Both spontaneous polarization and strain disappear at the β'-β phase transition (Keve, 1971 and Jeitschko, 1972). In addition, these compounds are non-centrosymmetric with potential non-linear properties, specially optical ones. In particular, $Gd_2(MoO_4)_3$ is an efficient frequency doubling medium for laser diode pumping because it displays second order harmonic generation (Kaminskii et al., 1997). The optical properties of this phase have been also analysed as phosphor hosts because of their RE components. For instance, Bubb et al. (2005) studied the near-IR to visible light up-conversion of $LaEr(MoO_4)_3$. From the reviewed bibliography, we are only aware that dc electrical conductivity studies have been done in the β'-phases of $Gd_2(MoO_4)_3$ and $Tb_2(MoO_4)_3$ (Tripathi and Lal, 1980) for the ferroelectric molybdates.

Heavy molybdates and tungstates with RE = Y, Ho-Lu can crystallize in the orthorhombic SG Pbcn with Z = 4, known as the γ-phase. This is isostructural with $Sc_2(WO_4)_3$ (Abrahams and Bernstein, 1966). There is no simple symmetry-group relation with the $La_2(MoO_4)_3$-phase, α-phase, or the β'-phase. However, at high temperature some lighter rare earth molybdates (with β'-phase) and tungstates (with α-phase) undergo a transition to the γ-phase in a reversible process (Brixner et al., 1979 and González-Silgo et al., 2013). The γ-phase is hydrated at room conditions, which complicates studies at low temperature. One of the most interesting properties of the compounds with the γ-phase is that they exhibit negative thermal expansion (Evans et al., 1998). Tungstates usually exhibit a wider temperature range with negative thermal expansion than molybdates, while molybdates have more negative coefficients (Sumithra and Umarji, 2004). Moreover, as the RE cation size decreases, the thermal expansion coefficient also becomes less negative (Sumithra and Umarji 2006). Although, in general, structural flexibility seems to be a key factor for negative thermal expansion, different mechanisms have been proposed to explain it in this type of compounds (Marinkovic et al., 2009 and Guzmán-Afonso et al., 2011). The interest of the γ-$Sc_2(WO_4)_3$ phase has grown in the last decade because its large trivalent ion conductivity, where the ionic conductivities of the molybdate series become considerably higher in comparison to those of the tungstate series (Imanaka et al., 1998 and 2000). However, there are no studies of the ac conductivity, and in particular, the relationship between the ac conductivity and the contraction of the structure as the temperature increases has not been analysed.

3. THERMAL DEPENDENCE OF CRYSTAL STRUCTURE FOR MODULATED SCHEELITES: $ND_2(MOO_4)_3$, $SM_2(MOO_4)_3$ AND $EU_2(MOO_4)_3$

In this section we review the evolution of the structure of the α-phases for samarium and europium molybdates (Guzmán-Afonso et al., 2011 and 2013) upon the results of two experiments in which the thermal dependence of the cell parameters was determined. Higher resolution data were collected at three temperatures in order to refine the crystal structure with respect to the scheelite-structure. We were interested in making a comparison of these results with those of neodymium molybdate that has another structure also related to the scheelite-type. For that reason, we decided to refine the three structures using the modulated-phase method. We will describe the experimental conditions for the three compounds and will discuss the results comparing both structural types. Finally, we will evaluate the results provided by the Bond Valence Model to investigate mechanisms and driving paths in scheelite or related compounds.

3.1. Le Bail and Rietveld Refinement by Using Amplimodes

To be successful in our analysis, we need to model the diffraction pattern starting from a set of initial parameters of position, intensity and profile (shape and broadening) as close as possible to the final values obtained after a fitting procedure. The powerful method of Rietveld (Young, 1993) is still used for refining the crystal structure and the profile parameters by fitting to the diffraction patterns. In the last decades new software has been introduced to address specific structural and microstructural problems. All crystallographic information (cell unit, space group and atomic positions...) can be found in the ICSD database so that peak positions (Bragg reflexions) and intensities can be refined. When the diffractograms have not enough resolution, the Le Bail refinement (Le Bail, 2005), based on the Rietveld refinement, may be used. In this case, the atomic positions are not needed.

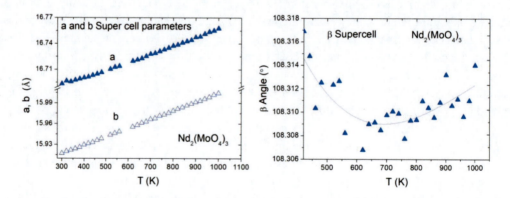

Figure 3.1. Thermal dependence of the lattice parameters a, c, and β angle of $Nd_2(MoO_4)_3$. The data were obtained by XRD in a polycrystalline sample and using Le Bail refinement.

An alternative to refine the atomic positions in order to calculate the peak intensities is the refinement of the amplitudes of the atomic displacements with respect to a known structure. AMPLIMODES (Orobengoa et al., 2009) is the tool that allows calculating these new parameters, called "amplimodes". This software, available at the Bilbao Crystallographic Server (http://www.cryst.ehu.es/), calculates the global structural distortion that relates the high- and low-symmetry phases, carrying out a symmetry-mode analysis. Then, the symmetry modes (amplitudes and polarization vectors) compatible with the symmetry break are calculated. The set of structural parameters used in a mode description of a distorted phase will be, in general, better adapted for a controlled refinement of the structure. The

AMPLIMODES program is currently being used (in this decade) to describe the transition between two different phases regarding the reversible symmetry-allowed distortions. There are still very few studies that relate these distortions, which are described hierarchically, to the formation of polarons. For instance, the influence in the structural distortions introduced by Li^+ and Na^+ intercalation in $Fe_2(MoO_4)_3$ (with the Anti-NASICON crystal structure) would be related to polaronic hopping of the electrons (Zhou et al., 2016). On the other hand, a first-principles theoretical calculation in WO_{3-x} reported the distortion of several phases with respect to the more symmetric cubic phase. In this case, the polaronic electron transport is a characteristic property of this material (Hamdi et al., 2016).

The thermal dependence of the cell parameters was obtained from the Le Bail refinement for the α-$Sm_2(MoO_4)_3$ and α-$Eu_2(MoO_4)_3$ diffractograms, from 300K to 1000K (Guzmán-Afonso et al., 2011 and Guzmán-Afonso et al., 2013), and both results showed a similar behaviour. With increasing temperature, the lattice parameters b and c expanded monotonously over the entire range of temperatures. The evolution of the parameter a showed three different regions compatibles with the behaviour of the dc conductivity (as we will see in section 5). Firstly, it undergoes a contraction until 600; secondly, this parameter remains constant from 600 to 700 K (or until 800 K for the case of europium molybdate), and finally, it increases. The monoclinic angle decreases, and the volume increases for both samples. The anomalous behaviour of the parameter a is well correlated to the thermal dependence of two parameters obtained from the ac conductivity leading us to conduct a more in-depth structural study using high-resolution diffractograms at three temperatures taken in each region. We performed a refinement using AMPLIMODES, where the scheelite structure was the higher-symmetry structure and the α-phase was the lower one. We considered that the higher temperature structure was closer to the scheelite-type. Comparing the distortions at different temperatures, we explained the unusual behaviour of the cell parameter a as due to the transversal displacements of the two oxygen atoms in the Sm(Eu)-O(1, 2)-Mo(2) bridges [Mo(2) is the more symmetric molybdenum] that reduce the Sm(Eu)-Mo(2) distances, and also the elongation of the tetrahedra around Mo(1) [Mo(1) is the less symmetric molybdenum] in the direction perpendicular to the a axis.

After the above study, diffractograms of $Nd_2(MoO_4)_3$ were refined by the Le Bail method. However, the lattice parameter a did not show similar anomalies as in the α-phase, while the β angle showed a minimum around 650 K (see Figure 3.1). We then decided to perform a new refinement that would allow the comparison of the α- and $La_2(MoO_4)_3$-phases, both transformed in the same average crystal structure.

3.2. Experimental Method: Solid-State Synthesis and Thermodiffractometry

Three samples were prepared by conventional solid-state synthesis, starting with MoO_3 and RE_2O_3 (RE = Nd, Sm, Eu) powder (Aldrich, 99.99%). The starting reactants were pre-heated at 923 and 1173 K, respectively, for 10 hours. These pre-heated powders were weighed in stoichiometric amounts, mixed and homogenized in an agate mortar. The resulting powder was pressed (3 tn) into cylindrical pellets of 13 mm in diameter and 1 mm in thickness, approximately, and finally synthesized in an air atmosphere at 1123 K for 48 hours in a platinum crucible. By using X-ray diffraction, the α-phase was confirmed for

Eu$_2$(MoO$_4$)$_3$ and Sm$_2$(MoO$_4$)$_3$, and the La$_2$(MoO$_4$)$_3$-phase for Nd$_2$(MoO$_4$)$_3$, to be without impurities or satellite peaks. The powder diffraction data were collected at the CAI facility (Universidad Complutense de Madrid) in a PANalytical X'Pert PRO diffractometer (Bragg-Brentano mode) with an X'Celerator detector using Cu Kα radiation and a Ni β-filter. All measurements were carried out during a heating process under a still air atmosphere, using an Anton Paar HTK-2000 camera from 300 to 1000K. The angular range 5° < 2θ < 80° was scanned at a step size of 0.017° and 100 s step time. Divergence and anti-scatter slits were set at 0.5 and 0.5°, respectively, and the detector was used in a scanning mode with an active length of 2.12°. 0.02 rad incident and diffracted Soller slits were used. These experimental settings are routine and standardized procedures.

We performed a "modulated Rietveld refinement" using the JANA2006 program package. A pseudo-Voight peak-shape function with axial divergence was used for the profile fittings. The background was modelled with a Legendre polynomial of 10 coefficients. The refinements were performed from the commensurate modulated structure of α-Eu$_2$(MoO$_4$)$_3$ (Martínez-García et al., 2009) and La$_2$(MoO$_4$)$_3$ (Logvinovich et al., 2010) in the superspace group I2/b(αβ0)00, with the modulation vector $q = 2/3a^* + 2/3b^*$, for samarium and europium molybdates, and $q = 2/3a^* + 8/9b^*$, for neodymium molybdate. The distribution of the RE (Nd, Sm, Eu) was modelled by a Crenel function with the centre of the RE-atomic domain located at $x_4 = 0.5$ and the occupation parameter was $\Delta RE = 2/3$. Cell parameters, atomic coordinates (isotropically) and Fourier amplitudes of the positional modulation (simulated with one harmonic function) were refined. The results obtained for the three samples were very similar (see Figure 3.2). For these experimental data, the resolution is not enough to refine the atomic coordinates using AMPLIMODES, although it is possible to obtain the distortion modes by comparing the obtained structures.

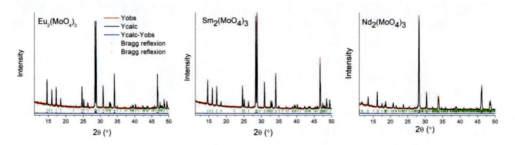

Figure 3.2. Modulated Rietveld refinement for RE$_2$(MoO$_4$)$_3$. The red plot indicates the observed intensities, and the black one shows the profile fits. The difference between observed and calculated intensities is plotted at the bottom in blue, and the vertical bars represent the expected positions of main diffraction peaks (green) and satellite peaks (olive).

3.3. Comparative Study: Temperature Dependence of the α-Phase and the La$_2$(MoO$_4$)$_3$-Phases

The thermal dependence of the average structure is shown in Figure 3.3. The parameters a_s and b_s increase when the temperature increases and they coincide throughout the whole temperature range for the α-phase (samarium and europium molybdates). On the other hand, the γ_s angle decreases with increasing temperature. For the La$_2$(MoO$_4$)$_3$-phase of neodymium, the a_s and b_s parameters are well differentiated, both increasing when the temperature increases, and the γ_s angle slightly increases with temperature, although it remains very close to 90°. The behaviour of the c_s parameter, which coincides with the b_α parameter (for the α-phase) and the b_{LaMo} parameter [for the La$_2$(MoO$_4$)$_3$-phase] in both supercells, is the same in all three compounds; along this axis, the thermal expansion is always linear, monotonous and without anomalies.

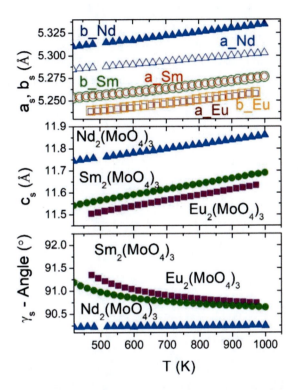

Figure 3.3. Thermal dependence of the lattice parameters a_s, b_s, c_s, and the γ_s angle for Eu$_2$(MoO$_4$)$_3$, Sm$_2$(MoO$_4$)$_3$ and Nd$_2$(MoO$_4$)$_3$. The data were obtained by XRD in polycrystalline samples and analyzed by "modulated Rietveld refinement" using the JANA2006 program package.

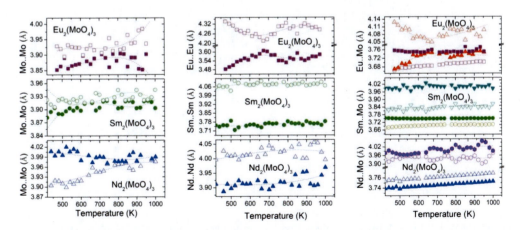

Figure 3.4. From left to right, the thermal dependence of RE…RE, RE…Mo and Mo…Mo distances, for Eu$_2$(MoO$_4$)$_3$, Sm$_2$(MoO$_4$)$_3$ and Nd$_2$(MoO$_4$)$_3$.

Although the oxygen atoms are not very well determined, we show the contacts between metals to analyse the polyhedral distortion (see Figure 3.4). Note that the RE sites are an average between the occupied sites and the vacancies, therefore their thermal evolution is more difficult to interpret. It can be observed that the length of the bonds is similar to what we described in section 2 for the nearest neighbours of the scheelite–type crystal structure. This average structure is very similar to the ferguson structure described in section 2, with two different distances for the first neighbours Mo…Mo and RE…RE, and four for the RE…Mo contacts.

The differences between the shorter and longer Mo…Mo distances decrease with temperature and reaches a minimum between 650 and 750K. In order to interpret this behaviour, by connecting with the real structures, we have transformed all the cell parameters to the supercell ones using the following equations:

$$a_\alpha = \sqrt{a_s^2 + b_s^2 - 2a_s b_s \cos\gamma_s} \tag{3.1}$$

$$c_\alpha = \sqrt{4a_s^2 + b_s^2 + 4a_s b_s \cos\gamma_s} \tag{3.2}$$

$$\beta_\alpha = A\cos\left(\frac{-2b_s^2 + a_s^2 + 2a_s b_s \cos\gamma_s}{a_\alpha c_\alpha}\right) \tag{3.3}$$

$$a_{LaMo} = \sqrt{9a_s^2 + b_s^2 + 6a_s b_s \cos\gamma_s} \tag{3.4}$$

$$c_{LaMo} = 3c_s \tag{3.5}$$

$$\beta_{LaMo} = A\cos\left(\frac{-9b_s a_s + 2a_s b_s \cos\gamma_s - 3b_s^2}{a_{LaMo} c_{LaMo}}\right) \tag{3.6}$$

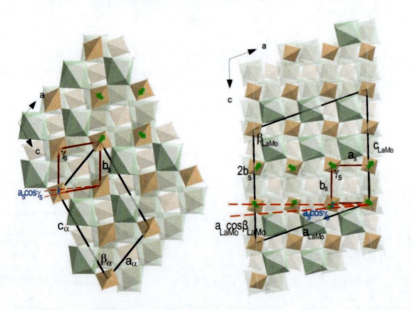

Figure 3.5. From left to right, the distribution of vacancies in the α- and $La_2(MoO_4)_3$-phases viewed along the b axis. The green arrows represent the displacements of the tetrahedra (molybdates) around the vacancies.

We plotted one layer, the darker one, perpendicular to the monoclinic axis b for both structure types (see Figure 3.5) to visualize more clearly the distribution of vacancies in each structure.

In this analysis, we cannot observe the distortions of the tetrahedra nor the shortening and lengthening of the RE(Mo)-O-Mo(RE) bridges but, in general terms, we can observe differences between the particular shape of vacancies and clusters of rare earth ions that could subtly affect their electrical properties. In Figure 3.5, the green arrows represent a tentative image of the displacements of the tetrahedra (molybdates), around the vacancies, according to the direction of increase or decrease of the monoclinic angles. Although the interaction between tetrahedra is from neighbours next to the closest, the behaviour must be similar to that of the nearest neighbours, as the longer Mo...Mo distances shorten while the shorter Mo...Mo distances get longer, as seen in Figure 3.5.

The tendency to make the Mo...Mo lengths almost equal seems the key for the contraction of the a_α parameter in $Eu_2(MoO_4)_3$ and $Sm_2(MoO_4)_3$ structures; as well as for the contraction of the γ_{LaMo} angle in $Nd_2(MoO_4)_3$. However, we need to find a common structural parameter for both structural types that allows comparing this contraction with possible differences in the behaviour of their ac conductivity (see section 5). For this, we will define the parameters p_α and p_{LaMo}. These parameters must have the same thermal dependence as the a_α parameter of the α-phase and the β_{LaMo} parameter of the $La_2(MoO_4)_3$-phase. In Figure 3.5, it can be seen how these parameters have been defined and in Figure 3.6 their thermal dependence is plotted. Note that a_s and b_s are almost equal in the α-structure and for the $La_2(MoO_4)_3$-structure: $3c_s = c_{LaMo} \approx a_{LaMo}\cos\beta_{LaMo}$.

$$p_\alpha = a_s\sqrt{(1-\cos\gamma_s)} = b_s\sqrt{(1-\cos\gamma_s)} \qquad (3.7)$$

$$p_{LaMo} = -3a_{Lamo}\cos\beta_{Lamo} - \frac{2}{3}c_{Lamo} = -3a_{Lamo}\cos\beta_{Lamo} - 2b_s = a_s +$$
$$+ 9b_s\cos\gamma_s \approx a_s\sqrt{(1+18\cos\gamma_s)} \approx b_s\sqrt{(1+18\cos\gamma_s)} \qquad (3.8)$$

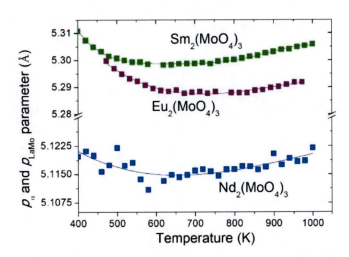

Figure 3.6. p_α p_{LaMo} parameters for $RE_2(MoO_4)_3$ from Equation 3.7 and 3.8. Note that in order to visualize this behaviour in the $Nd_2(MoO_4)_3$ compound we multiplied $\cos\gamma_s$ by eighteen.

3.4. Bond Valence Model: Polyhedral Distortion and Ion Transport Pathways

The bond valence model (BVM), which is a generalization of Pauling's original concept of the electrostatic valence principle, has been reviewed extensively in the literature (Brown, 2016). There is a well-defined empirical relationship between bond valences and bond lengths in a given coordination polyhedron and it is this functional relationship which makes the BVM quantitatively useful, in particular for compounds related to the scheelite structures. For instance: 1) the evolution of the strain from the tetragonal phase to the monoclinic phase of $Bi_{1-x/3}V_{1-x}Mo_xO_4$ is correlated to the catalytic properties where the low BV sum over the Bi position for higher content of Mo is connected with the influence of vacancies (Terebilenko et al., 2016). 2) An increasing replacement of W^{6+} by Mo^{6+} switches the modulation from the (3+1)D to the (3+2)D regime in $CaEu_2(Mo_xW_{1-x}O_4)_4$; the tendency to optimize the bond valence balance for the A-cations and the oxygen atoms can be the main driving force for the compositional modulation (Abakumov et al., 2014). 3) The effects of the packing fraction and the BV affect the microwave dielectric properties of ABO_4 (A = Ca, Pb, Ba; B = Mo, W) ceramics (Kim et al., 2010). 4) The BV sum calculated for the cations in α-$Eu_2(MoO_4)_3$ explained the anomalies observed in the cell parameters, Raman modes and a non-linear red-shifting and broadening of the characteristic and luminescence peak for the Eu^{3+} (Guzmán-Afonso et al., 2015).

Regarding the studied compounds, from their atomic positions obtained by Rietveld refinements by using "amplimodes", we calculate bond lengths and angles, and polyhedral distortions in α-$Eu_2(MoO_4)_3$ and α-$Sm_2(MoO_4)_3$ (Guzmán-Afonso et al., 2011, 2013). In particular, because the positions of the oxygen atoms were rather well determined, the Bond Valence Model was used to study the temperature dependence of the stability of Mo tetrahedron and both the coordination polyhedron for the RE cation and the vacant in α-$Sm_2(MoO_4)_3$. The results were correlated to the anomalies observed in the thermal dependence of the Mo-O bonds and the a parameter at intermediate temperatures (Guzmán-Afonso et al., 2013). For this compound we plotted the contour maps of bond-valence sums (González-Platas et al. 1999), recently incorporated in FullProf suite. Bond-valence isosurfaces can help visualize the migration pathways of the mobile ions in solid electrolytes. The sum of bond valences for a particular atom (ion conductor) is calculated in any arbitrary point in the crystal; thus, by moving this atom through different positions, its valence-sum contour map can be displayed. At higher temperature, new possible sites for the oxygen atoms in α-$Sm_2(MoO_4)_3$ and α-$Eu_2(MoO_4)_3$ were detected which would favour their motion when an electric field is applied (in the regime of ionic transport) (Guzmán-Afonso et al., 2011 and 2013). For screening purposes, the analysis and prediction of ion transport in solids from static and dynamic structural models have become an interesting application for the bond valence approach. Specific adaptations of the bond valence approach, in particular, the bond valence site energy pathway model, derived from the static crystal structure for this application, is given by Adams and Rao (2015). Moreover, BVM as a description of a charge distribution has been considered consistent with both the small polaron and atomistic models which describe the electronic and ionic conductivity, respectively, in ceria phases (Shoko et al., 2009). A description of the charge distribution in stable structures may elucidate the microscopic mechanism of charge transport.

4. AC Conductivity in Scheelite and Related Compounds

The study of the electrical and optical properties in different types of materials has been the subject of extensive research with a view to gain a better understanding of the origin of these properties, and by this means improving and expanding their field of application. In this section we will first show the theoretical framework where the main concepts of electrical conductivity are developed and then we will review the study of ac conductivity in compounds with the scheelite or related structures.

4.1. AC Conduction in Disordered Solid and Macroscopic Models

The complex permittivity ε^* provides key information about the response of electric dipoles to changes in the electric field, frequency, and temperature. In this frame, the impedance spectroscopy is a suitable experimental technique for its measurement. However, in order to explain polaronic mechanisms of the electric transport one is more interested in the investigation of conductivity mechanisms than on dipolar relaxation. This information can be obtained from the relationship between ε^* and the real part of the complex conductivity $\sigma'(\nu)$

$$\varepsilon^*(\nu) = \varepsilon'(\nu) - i\frac{\sigma'(\nu)}{\nu} \tag{4.1}$$

where $\varepsilon'(\nu)$ is the real part of the complex permittivity. The spectral analysis of the conductivity is based on the so-called universal dielectric response (UDR) behaviour (Jonscher, 1983, Cuervo-Reyes, 2016), as an empirical power law (with a stochastic approach) which describes the frequency and temperature dependence of the dielectric response for a wide range of ionic and semiconductors materials, from glasses to crystals. Figure 4.1 shows the typical frequency dependence of the ac conductivity at three temperatures for $\alpha-Sm_2(MoO_4)_3$ (Guzmán-Afonso, 2013), where it is possible to fit the real part of the ac conductivity $\sigma'(\nu)$ to Jonscher's power law:

$$\sigma'(\nu) = \sigma_{dc} + A\nu^s \tag{4.2}$$

with σ_{dc} the dc conductivity (frequency-independent plateau in the low-frequency region). The prefactor A is a temperature dependent parameter which is proportional to the polarizability of a pair of sites (Singh et al., 2011), and s is the frequency exponent with typical values between 0 and 1, which represents the degree of interaction between the charge carriers and the lattice around them (Jonscher, 1983). As shown in Figure 4.1, the frequency that corresponds to the onset of the dispersive character of $\sigma'(\nu)$ is called the crossover frequency ν_p. Many authors (Punia et al., 2012, Shanmugapriya et al., 2016, Bhattacharya and Gosh, 2003, Dutta and Gosh, 2005) often express the Jonscher´s power law using the following equivalent equation:

$$\sigma'(\nu) = \sigma_{dc}\left[1 + \left(\frac{\nu}{\nu_p}\right)^s\right] \tag{4.3}$$

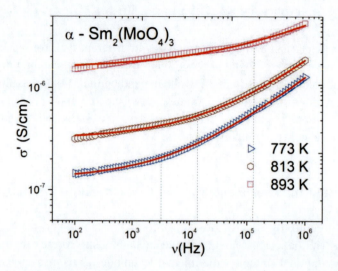

Figure. 4.1. Real part of the ac conductivity as a function of frequency at three different temperatures for α-Sm₂(MoO₄)₃. As the temperature is lowered, σ_dc decreases rapidly and, correspondingly, ν_p (dashed line) is displaced towards lower frequencies. The solid lines are the best fits to an equation of the type in Equation 4.3.

A plot of the logarithm of the dc conductivity σ_dc against 1/T (a so-called Arrhenius plot) then tells if the electrical conductivity is a thermally activated process, with the activation energy E_a given by the slope of the Arrhenius plot:

$$\sigma_{dc} = \sigma_o exp\left(-\frac{E_a}{2k_BT}\right) \qquad (4.4)$$

where k_B is the Boltzmann constant, and σ_0 is a pre-exponential factor. Different conduction mechanisms will be distinguished by different activation energies. Moreover, the thermal dependence of the crossover frequency (sometimes called hopping frequency) also follows an Arrhenius relation:

$$\nu_p = \nu_o exp\left(-\frac{E'_a}{k_BT}\right) \qquad (4.5)$$

when both activation energies coincide ($E_a \approx E'_a$), that is an evidence of the same mechanisms for relaxation and conduction (Murugavel and Upadhyay, 2011). As a general rule, for dielectric materials conductivity at low temperatures is due to impurity defects and mixed states, and the electronic or hole conduction (extrinsic) has a stronger dependence on the temperature than the ionic conduction (intrinsic), which usually takes place at higher temperatures. However, the frequency dependence of the ionic conduction is usually larger. Moreover, a strong dependence on temperature and frequency suggests a polaronic conduction and it is possible to model this dependence. Several theoretical approaches (Long, 1982, Elliott, 1987) have been proposed to correlate the conduction mechanism of ac conductivity with the behaviour of quantum-mechanical tunnelling (QMT) for the frequency exponent s. These theoretical models derive the temperature and frequency dependence of s from microscopic transport mechanisms, involving the classical hopping of a carrier across a

The Electrical and Structural Study of Compounds ... 329

potential barrier and the QMT through the barrier separating two equilibrium positions. In this way, for the correlated barrier hopping (CBH) model, s decreases gradually with increasing temperature. Within the processes associated with the QMT, the following models have been proposed:

- *quantum-mechanical tunnelling* (QMT), in which s is temperature independent but frequency dependent. This model assumes that there is no lattice distortion associated with the carrier whose motion gives rise to ac conductivity.
- *small polaron* (SP), in which the exponent s increases with increasing temperature. The lattice distortion created by the charge carrier is local, of the order of few interatomic distances.
- *overlapping large polaron* (OLP), the exponent s is both temperature and frequency dependent; s decreases with increasing temperature to a minimum value at a certain temperature and then increases with the rise of temperature. In this case, the lattice distortion created by the carrier extends over many interatomic distances (Austin and Mott, 1969, Mott, 1970).

Usually, small polarons can be easily distinguished from the large ones. Small polarons are expected when strong and short-range interactions are dominant, while large polarons represent the case of weaker and longer-range interactions. Figure 4.2 shows the thermal dependence of the s parameter for different models, which can be used to explain the transport behaviour in different scheelite-type materials.

From the fit of these models to the experimental data, the following parameters can be deduced: barrier height (W_M), hopping distance (R), concentration of pair sites (N) for the CBH model; and for QMT, the activation energy for polarons (W_H), the polaron radius (r_p), the tunnelling distance (R_w), the density of states [$N(E_F)$], and the spatial decay parameter for an exponential wave function which is assumed to be constant for all sites (α^{-1}). For both models, the characteristic relaxation time τ_0, often taken to be an inverse of the phonon frequency, can be obtained.

For each of these models, along with the dependence of the parameter s it is possible to give the corresponding expression for the real part of the ac conductivity, where any departure from linearity provides information on the particular type of loss mechanism involved. A nearly linear frequency dependence of $\sigma'(v)$ is predicted if the distribution of relaxation times $n(\tau)$ is inversely proportional to τ, which is $\tau_0\exp(\xi)$, where ξ is a random variable. This random variable can be expressed in different ways depending on the microscopic relaxation mechanisms, thus $\xi = W_M/k_BT$ for classical hopping and $\xi = 2\alpha R_w$ for quantum-mechanical tunnelling (Gosh, 1990).

The thermal dependence of the frequency exponent s and the prefactor A can be correlated with the atomic rearrangement when temperature increases, showing reverse behaviour. In Section 6 we will discuss the correlation between the evolution of the crystal structure and the mechanisms of the scheelite related materials taking into account these two parameters. On the other hand, the conductivity spectra, at different temperatures, can be scaled according to the following equation, which is more general than Jonscher's power law:

$$\frac{\sigma'}{\sigma_{dc}} = F\left(\frac{v}{v^*}\right) \tag{4.6}$$

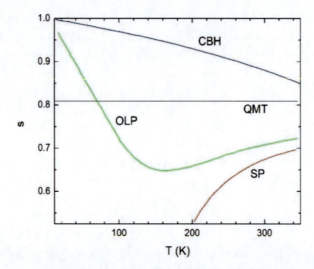

Figure 4.2. Temperature dependence of the frequency exponent s for different models of the ac conduction: correlated barrier hopping (CBH, blue line), quantum-mechanical tunnelling (QMT, black line), small polaron (SP, red line,) and overlapping large polaron (OLP, green line). For SP and OLP cases, the values $W_H = 0.2$ and 0.26 eV have been respectively used. For CBH, $W_H = 1.85$ eV. In all cases, $\omega\tau_0 = 10^{-9}$ has been assumed (Elliott, 1983).

where ν^* can be used as the scaling parameter with the advantage that it is a directly accessible quantity (Dyre et al., 2009). An indication that the relaxation dynamics of charge carriers are independent of the temperature in a certain region of temperatures comes from the fact that the curves collapse into a single master curve at such range. Typically, for ionic conduction, the concentration of charge carriers is weakly temperature dependent and the temperature dependence of the conductivity arises from the mobility of the ions.

Almond and West introduced the equation 4.3, widely applied for the description of conductivity spectra below 100 MHz, to describe defective crystals with an activated number of charge carriers. This formula has been used by many authors, which identify the crossover frequency (ν_p) with a "hopping rate". Thus, using the Nerst-Einstein relation, based on the net flux density of hopping charge carriers in a linear electric field being proportional to the random diffusion coefficient (Howard and Lidiard, 1964), we can determine the number density of mobile ions n_{AW} (assuming a typical jump length $a = 2\text{-}3$ Å):

$$n_{AW} = \frac{6 K_B T \sigma_{dc}}{q^2 a^2 \nu_p} \tag{4.7}$$

Then it is easy to obtain the mobility because for $n_{AW} \approx n_{mob}$ one has:

$$\sigma_{dc} = q n_{mob} \mu \tag{4.8}$$

4.2. Bismuth Vanadate and Related Glasses

Identifying the polaronic conduction and understanding its implications is essential for assessing materials and developing approaches for improving their efficiencies. Thus, for example, bismuth vanadate ($BiVO_4$) has been recognized as a promising visible-light active photocatalyst for water splitting (Kudo et al., 1999, Walter et al., 2010). However, the fact that the carriers are small polarons (with a broad absorption band about 1 eV, lying below the strong intrinsic interband at 2.5 eV) suggests exploiting its low dark conductivity by a sub-bandgap illumination (Rettie et al., 2015). Possible factors for the phase-dependent photocatalytic activity of $BiVO_4$ have been discussed by Kyoung and Gyeong (2013). Hybrid density functional theory (DFT) calculations show different hole localization behaviours in the bulk phases. An excess hole tends to spread widely over many lattice sites (large polaron) in bulk ms-$BiVO_4$ (monoclinic phase), whereas it localizes around a BiO_8 polyhedron with local lattice distortions (small polaron) in bulk ts-$BiVO_4$ (tetragonal phase). Cooper et al. (2014) have studied the implications of the fundamental electronic structure of ms-$BiVO_4$ on its photocatalytic behaviour, as well as making some considerations for improvements by substitutional incorporation of additional elements. In 2015, the same authors explained how the combination of strong visible light absorption and relatively long excited state lifetimes provides the basis for the high performance that can be achieved by $BiVO_4$ photoanodes during water splitting. The transition-metal oxide $BiVO_4$ is well-suited as a case study for elucidating the critical roles of charge localization, polarons, defects, and chemical theoretical models approaches to address inefficiencies and instabilities, and the prediction of new materials (Sharp et al., 2017). In spite of the interest on $BiVO_4$ and related doped materials we have not found a complete ac conductivity studies of its polaronic transport mechanisms. We highlight two studies in related glasses and some important parameters obtained very recently from the ac conductivity in pure $BiVO_4$.

The first measurements of the ac conductivity in the frequency range 10^2-10^5 Hz and in the temperature range 77-420 K, for various compositions of the bismuth-vanadate glassy semiconductors: $80(V_2O_5)$-$20(Bi_2O_3)$, were reported by Gosh in 1990. The observed dispersion behaviour of the ac conductivity obeys Jonscher's power law. The analysis of the experimental results shows that the OLP model can explain the temperature dependence of the frequency exponent s at low temperatures. However, this model predicts a temperature dependence of the ac conductivity much higher than what actual data shows. Instead, the CBH model is consistent with the temperature dependence of both the ac conductivity and its frequency exponent s, providing reasonable values of the maximum barrier heights (W_M), but higher values of the characteristic relaxation times (τ_0) and a worse fitting, at high temperature. These discrepancies are ascribed to the approximate nature of the CBH model.

In 2012 Punia et al., investigated the ac conductivity of bismuth zinc vanadate semiconducting glasses with formula $50V_2O_5$ xBi_2O_3 $(50-x)ZnO$ in the frequency range from 10^{-1} Hz to 2 MHz, and temperatures from 333.16 K to 533.16 K. They found that the ac conductivity increases with the Bi_2O_3 content (for $x > 5$, structural modifications can lead to a distorted scheelite structure). The fitting of the experimental data with Jonscher's power law yields a good agreement for all compositions, frequencies, and temperatures. The OLP model is the best candidate to explain the conduction mechanism, except in the intermediate frequency region at lower temperatures. The reason for this deviation may be due to either the limitation of considering the tunnelling distance R_w independent of frequency during the fitting, or the change of the local structure of the glass matrix with temperature. Another reason for this deviation may be that at low frequencies the contribution to the dielectric constant due to the hopping process levels off to a constant. The parameters r_p, α, $N(E_F)$, and W_H obtained from the fitting of the experimental data of the total ac conductivity agree well with those obtained from the fitting of the dc conductivity data within Mott's model for transition metals oxides (Mott, 1990).

Very recently, Oliveira et al. (2017) have identified pure and polycrystalline $BiVO_4$ as a giant dielectric ceramic with a very large dielectric permittivity ($\varepsilon' = 2.71 \times 10^6$ to 460°C and 1 Hz) upon a study using impedance spectroscopy techniques in the frequency range from 1Hz to 1MHz and temperatures ranging from 220°C to 460°C. The activation energy was calculated from the variation of the electric modulus' peak and the increase of conductivity at 1 Hz with temperature. For both measurements, the activation energy was found to be approximately, $E_a = 0.52$ eV which indicates that the charge carrier transport has to overcome the same energy barrier as for relaxation, so that the ion-hopping mechanism is more plausible as an explanation of the electric transport.

In order to completely understand the experimental discrepancies when describing transport mechanisms, it is necessary to perform a systematic study of the variation of the structure with the temperature, as suggested by Capaccioli et al. (1998) and van Staveren et al. (1991) twenty years ago.

4.3. Lead Tungstate, Doped by RE, Mo and Alkaline Earth Tungstates

Lead tungstate ($PbWO_4$) is a most attractive material for high-energy physics applications because of its high density, short decay time and high irradiation damage resistance. Annenkov et al. (2002) summarized the results of a research program on lead-tungstate (PWO) crystals performed by the CMS Collaboration at CERN, as well as by other groups who promoted the progress on PWO scintillating crystal. These authors concluded that only shallow polaronic and distorted regular centers, through relatively fast electron exchange with the conducting zone, contribute to the scintillation. The shallowness of them occurs in all crystals since it is an intrinsic defect: an additional electron autolocalized at an anionic WO_4^{2-} complex, via a Jahn-Teller distortion, creating a WO_4^{3-} polaronic center. The released electrons partly recombine radiatively, and partly are caught by deeper traps. Also, a systematic deficiency of lead in the crystal introduces additional oxygen vacancies creating irregular WO_3 groups where the luminescence is shifted to the green region while regular WO_4^{2-} groups show blue luminescence (Abraham et al., 2001). The theoretical calculations indicate that locally disordered Pb clusters are more favourable for the photoluminescence (PL) effect at generating localized states and inhomogeneous charge distribution in the cell, thus allowing the trapping of electrons and holes before excitation (Anicete-Santos et al., 2007). The radiative decay must occur through the same pathway, leading to a wide band emission. Anion vacancies, or their associations, contribute to slow components in scintillations, phosphorescence and additional transient or optical absorption induced by irradiation (Annenkov et al., 2002).

The effect of RE doping on the scintillating properties of lead tungstate single crystal has been intensively investigated and it was found to be dependent on the doping concentration (Huang et al., 2003). However, there has been much less research on their dielectric

The Electrical and Structural Study of Compounds ...

properties and conductivity. Huang et al., (2003a) studied a series of La-doped single crystals of $PbWO_4$, which were measured and analyzed by the impedance spectroscopic technique (frequency range of 20 Hz to 1MHz and temperature range from 100°C to 320°C). The experimental results show that the ac conductivities of PWO:La crystals follow the Jonscher's power law and that oxygen vacancies are responsible for the ionic conduction in all doped compounds. The similar activation energy of E_a and E'_a (equations 4.4 and 4.5) is evidence for an ionic-hopping mechanism for transport with an approximate activation energy of 0.78 eV. They interpret effects of La doping on the conductivity and dielectric relaxation: for low doping concentrations, positive charge caused by substitution of La ions on the Pb lattice are balanced by Pb vacancies, whereas it is balanced by interstitial O and La ions at W lattices for heavy doping.

The PL, as well as the electrical properties in nanoparticles of $PbMoO_4$, depend on the morphology, degree of crystallinity, and particle sizes. From ac conductivity measurements (in the frequency range from 1 Hz to 1MHz and the temperature range from 110°C to150°C) it was concluded that the conductivity is of polaronic type (Singh et al., 2013). Conduction mechanisms do not follow an Arrhenius behaviour completely and hence a variable activation energy can be expected. Above 10 KHz, the same authors explained that the dipole relaxation occurs mainly due to equilibrium stabilized in the backward and forward hopping of the dipoles. From the Arrhenius plot of $PbMoO_4$ at 10 KHz, the activation energy is about 0.57 eV.

The ac conductivity has been studied by Bouzidi et al. (2015) for ceramic $BaMo_{1-x}W_xO_4$ (x = 0.00, 0.05, 0.10, 0.15 and 0.20) in the frequency and temperature range of 40-10^6 Hz and 30-800°C, respectively. The ac conductivity versus frequency obeys Jonscher´s power law, where the exponent parameter s, calculated at higher frequencies, decreases with rising temperature, which is in good agreement with the predictions of the CBH model. The ac conductivity increases slightly with increasing temperature, which corresponds to transport by extrinsic or hopping conduction. The frequency dependence of the ac conductivity indicates that conduction is due to small polaron hopping. At low W concentration the hopping mechanisms between W sites is dominant. Compared to other works, the activation energy of $BaWO_4$ ceramics is lower than that of $BaMoO_4$, contrary to what happened in lead tungstates and molybdates.

Because of their important practical applications, there is a striking want of comparative ac studies on pure and different solid solutions including vanadates, tungstates and molybdates of lead and bismuth.

4.4. Double Tungstates

A complete study of the ac conductivity has been very recently performed for the first time in two double tungstates. In spite of numerous optical studies and the interest of this type of compounds for solid-state laser applications (Volkov et al., 2005, Cascales et al., 2006), we have not found similar studies in other double molybdates and tungstates.

The dc and ac conductivities of polycrystalline $LiGd(WO_4)_2$ (with scheelite structure, Bacha et al., 2017a) and $NaGd(WO_4)_2$ (with distorted scheelite structure, Bacha et al., 2017b), have been collected in the frequency range from 200 Hz to 5 MHz and the temperature range from 584 to 689 K and 613 to 711 K. Both compounds show similar ac conductivity spectra with two different regions as found in double phosphates (Miladi et al.,2016, Chakchouk et al., 2018), with both the high and low-frequency plateaus suggesting that two processes contribute to the conduction. The experimental results were fitted with Jonscher´s augmented equation (Dussouze, 2005). For the first compound, the dc conductivity exhibits two regions, which were attributed to the CBH model in the first region, and the SP model in the second region. The observed change must be due to the change in the conduction mechanism, because "there is no melting or phase transition as seen in the thermal analysis study reported earlier". However, modulated structures in similar compounds have been identified by the presence of additional satellite reflections in the lower angle part of the XRD pattern (Volkov et al., 2005 and Arakcheeva et al., 2012). These weak reflections are usually ignored because they can disappear at high temperatures while a phase transition takes place.

In the first region, from the thermal dependence of the frequency exponent s, the CBH model for polarons and bipolarons including Coulomb interactions led to fit values for the relaxation time τ_0, hopping distance R and the density of states $N(E_F)$. The two last parameters increase as the frequency increases, in agreement with this model. When the frequency increases, the height barrier W_M decreases and subsequently the charge carriers can jump easily between neighbours. In the second region, for the SP model, the fitting parameters besides the s exponent are the characteristic relaxation time τ_0, the polaron and the hopping energy W_H, the density of defect states and the tunneling distance R_w. The density of defects $N(E_F)$ increases with temperature whereas the hopping energy W_H and tunneling distance R_w decrease when the temperature increases. According to the model, it is clear that the tunneling distances must be decreasing as the frequency increases and it is in the order of the Li^+-Li^+ spacing. For the second compound, only one region was observed by the dc measurements with an activation energy more similar to that of the second region than that of the first region in the first compound. The s behaviour is well described by the SP model. Thus, the displacement of Na^+ ions can be the mechanism of conduction. Because the values obtained for the tunneling distance R_w vary in the range from 3.321 to 3.437 Å and are thus lower than the a lattice parameters of the structure, i.e., Na^+-Na^+ distance (5.245 Å), the conduction is ascribed to the movement of the small polaron Na^+ inserted into tunnels parallel to the a axis. From the scaling model a perfect overlap of the spectra at different temperatures is observed, which implies that the relaxation dynamics of charge carriers in the present compound is independent of temperature. For the first compound, it was found that the normalized modulus spectra superimposed perfectly at various temperatures indicating that the dynamic processes occur at different frequencies and independent temperatures, as in the second compound.

4.5. Transition Metal Tungstates and Molybdates

Transition-metal tungstate-type compounds (with wolframite or related structures) have attracted scientific interest due to their complex electronic structure and related electric, magnetic and optical properties. These materials have been found use as catalysts, humidity sensors, phosphors, and energy storage capacitors. However, there is a general lack of studies on their ac conductivity. In what follows we describe the ac conductivity in pure tungstates and molybdates where a complete study has been performed, although polaronic mechanisms based on the analysis of the thermal dependence of the frequency exponent is not always considered. The electrical transport of transition metal tungstates M^{2+} is typically characterized by a d-band near the Fermi level and formation of M^{3+} ions leads to electron hopping between adjacent metals. In general, the conductivity was observed to increase with temperature, from which a semiconducting behaviour for these materials is inferred.

The ac conductivity was measured in transitions metal tungstates and molybdates: polycrystalline $NiWO_4$, (Bhattacharya et al.,1997) nanoparticles of $CoWO_4$ (Shanmugapriya et al., 2016), nanocrystals of $CdWO_4$ (Vijayakumar et al., 2003) and spherical particles of α-$MnMoO_4$ (Sekar et al., 2012) between 0-10 Hz up to 1 MHz, except for cadmium molybdate (up to 1 KHz) and in the respective temperature ranges: 300-1000K, 380-640°C, 520-540°C, 400-540°C. The dispersion curves were fitted to Joncher's law following the formalism by Almond and West. The band gap calculated from the Arrhenius plot (twice the activation energy) appears to be too small (among 0.38 to 2.3 eV) for the excitation of electrons from the filled O^{2-}:2p band to the conduction band, as optical absorption measurements report this energy around 3.5 eV. Therefore, the conductivity mechanism would involve the excitation of electrons from a 3d valence band to an empty 3d conduction band. For $NiWO_4$ the authors suggested that surface defects might lead to the existence of Ni^{3+} ions. As $NiWO_4$ contains narrow 3d bands, small polarons may be formed when the local polarization produced by the carriers hopping between Ni^{2+} and Ni^{3+} is sufficient to trap them at one lattice site. For $CoWO_4$ the authors explain the increase of the charge mobility with temperature in terms of thermal lattice vibrations, which enhances the probability of charge transfer between close ions. The electron transfer from one Co^{2+} ion to a nearby Co^{2+} ion will lead to the formation of a Co^{3+} ion. This thermally activated hopping of the 3d electrons of the transition metal ions gives rise to a conduction behaviour as in $NiWO_4$. For $CdWO_4$, the authors do not give the activation energy and the concentration of carriers is two orders less than the one that we have described for other tungstates and molybdates. In general, an increase in temperature also increases the hopping frequency, the mobility of charge carriers and the dc conductivity but does not affect the carrier concentration, which indicates that all the ions involved in the conduction mechanism are in a "mobile state". On the other hand, in the high-frequency region the conductivity increases which indicates that the ions possess forward and backward hopping motions. The frequency exponent s > 1 shows that the hopping motion is localized because the backward hopping is faster than the site relaxation time (Cramer and Buscher, 1998).

4.6. Modulated Scheelites

The first electric characterization of the modulated scheelite-type structure was performed by Gaur et al. (1993) who calculated the conductivity (σ_{dc}) and the Seebeck coefficient (S) in the temperature range 450-1200 K for La, Ce, Pr, Nd, Sm and Eu trimolybdates. These molybdates were described as electrical insulators: materials whose internal electric charges do not flow freely and therefore makes it very hard to conduct an electric current under the influence of an electric field. These materials have a band gap which slowly increases when going down the series, from 2.30 eV for $La_2(MoO_4)_3$ to 3.20 eV for $Eu_2(MoO_4)_3$. In general, three linear region were observed, with two break temperatures, occurring due to changes in the conduction mechanism. At lower temperatures, the electrical conduction is mainly extrinsic, with impurities playing a dominant role. In addition, at higher temperatures, the charge carriers would become polarons. Calculated mobilities from the Seebeck effect are an order of magnitude less than the expected electrical mobility, wich must be ascribed to polaron formation. Moreover, the Seebeck coefficient due to large polarons, with intermediate coupling via band mechanisms, decreases with temperature (Sumi, 1972) as it was experimentally observed. We do not know of other studies on the ac conductivity until 2011 (Guzmán-Afonso et al., 2011).

In the next section we will complete this dc conductivity results with a detailed study of the ac conductivity in the three regions, in order to determine the different electrical transport mechanisms and shed further light on the behaviour of closely-related compounds.

5. AC CONDUCTIVITY OF MODULATED SCHEELITES $ND_2(MOO_4)_3$, $SM_2(MOO_4)_3$ AND $EU_2(MOO_4)_3$

This section is dedicated to the study, through measurements of the ac conductivity, of different conduction mechanisms as a function of frequency and temperature, for the compounds with modulated scheelite structure considered in section 3. The thermal dependence of the ac conductivity and the crystal structure have been correlated for the α-phase of $Sm_2(MoO_4)_3$ and $Eu_2(MoO_4)_3$ (Guzmán-Afonso et al., 2011 and Guzmán-Afonso et al., 2013). The study of $Nd_2(MoO_4)_3$ presented in this section, which has a different structural type, constitutes a new contribution. The analysis of the ac conductivity spectra, for each compound, is based on the Universal Dielectric Response behaviour and the scaling theories described in section 4.

This type of dielectric studies at high temperature has not usually been combined with the thermal dependence of the crystal structure in any family of compounds, including the family of SRCs, except when structural transitions or anomalous expansions are detected, as we will see in section 6. We have chosen these compounds because they exhibit different modulations related to the scheelite structure (sections 2 and 3). The dependence of their photoluminescence on the cation-vacancy ordering or RE cluster formation (Morozov et al., 2014) is also interesting; therefore, we expect differences in the ac conductivity between both structural types.

5.1. Impedance or Dielectric Spectroscopy: Experimental Conditions

The conductivity measurements were carried out using a computer-controlled Hewlett-Packard 4192 impedance analyzer. One pellet was selected (with the given dimensions) and their parallel surfaces were coated with a platinum paste. The pellet sample was then placed into the cell, which consists of a parallel plate capacitor located at the end of a coaxial line. This cell was enclosed in a resistance-heated furnace and the temperature was monitored using a thermocouple to an accuracy of 0.1 K. Measurements of the dielectric permittivity

The Electrical and Structural Study of Compounds ... 337

were recorded during a heating and cooling cycle, in the frequency range from 10^2 to 10^6 Hz with an applied voltage of 0.7 V_{pp}, through the temperature range from 500 to 1000 K, which was dynamically scanned (at a rate of 1.2 K min^{-1}) with measurements taken in steps of 5 K. The complex permittivity was obtained from the complex impedance $Z^*(\nu)$ using the relationship:

$$\varepsilon^*(\nu) = \frac{t}{i2\pi\nu\varepsilon_0 Z^*(\nu)a} \tag{5.1}$$

where ε_0 is the dielectric constant of vacuum, ν is the frequency and a and t represent the electrode area and sample thickness, respectively.

The imaginary part of $\varepsilon^*(\nu)$ was used to obtain the real part of the complex conductivity (σ'). The data for the dependence of σ' on the frequency (ν, in Hz) were scanned at each temperature.

5.2. AC Conductivity and Jonscher Universal Power Law

In a typical study of the ac conductivity, the dependence of the conductivity with the inverse of the temperature (log (σ') *vs.* 1/T) is analyzed at different frequencies. This type of representation can provide qualitative information about the existence of different conductivity mechanisms, which are manifested by their dependence or independence on frequency and/or temperature. As shown in Figure 5.1, σ' increases monotonically with the temperature and is strongly temperature and frequency dependent up to 800 K, at which point its dependence on frequency becomes less pronounced. Taking into account what was discussed in subsection 4.1, this behaviour could be associated with a polaronic conduction mechanism. These types of graphs have not been shown so far in any previous work on SRCs compounds, except for bismuth-vanadate glassy semiconductors where this behaviour is observed at a lower temperature (Ghosh, 1990).

All the results shown here, correspond to data obtained during the cooling cycles. As shown in Figure 5.1, the data collected in the heating cycle yield essentially the same results. However, for a more detailed analysis, it is convenient to measure the complete cycle.

To obtain more information about theses conduction mechanisms, the total ac conductivity measured as a function of the frequency is represented at different temperatures (Figure 5.2). The conductivity spectra observed for each of these compounds obeys Jonscher's power law (equation 4.2). For the fitting of the experimental data we have used in the present case a slightly different version of this law, also used by other authors (León et al., 1997), which is expressed as:

$$\sigma'(\nu) = \sigma_{dc}\left[1 + \cos\left(\frac{s\pi}{2}\right)\left(\frac{\nu}{\nu_p}\right)^s\right] \tag{5.2}$$

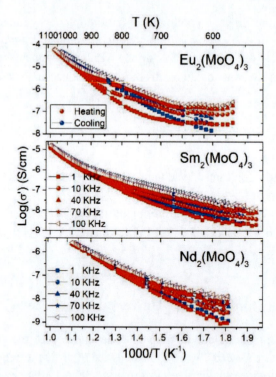

Figure 5.1. Temperature dependence of the real part of the complex conductivity (σ') at five frequencies for α-Eu$_2$(MoO$_4$)$_3$, α-Sm$_2$(MoO$_4$)$_3$ and Nd$_2$(MoO$_4$)$_3$ with the La$_2$(MoO$_4$)$_3$-type structure, obtained during a heating-cooling cycle.

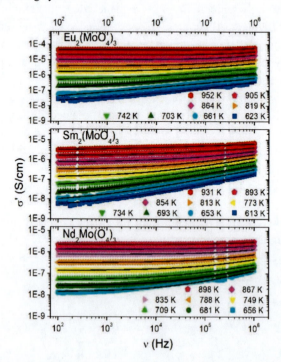

Figure 5.2. Frequency dependence of the real part of the complex conductivity (σ') at several temperatures for α-Eu$_2$(MoO$_4$)$_3$, α-Sm$_2$(MoO$_4$)$_3$ and Nd$_2$(MoO$_4$)$_3$ with the La$_2$(MoO$_4$)$_3$ type structure. The solid black lines are fits using equation 5.2.

We find that the fittings using this equation are better than those using equation 4.3. Comparing this expression with equation 4.2, it follows the relation $A = \sigma_{dc} \cos(s\pi/2)/v_p^s$ (pre-exponential factor). The inclusion of the term $\cos(s\pi/2)$ modulates the onset of the dispersive character of σ'. Also, the crossover frequency (v_p) is included in this equation and corresponds to the frequency where the σ' curves change slope. From these fittings we obtained the temperature dependence of σ_{dc}, v_p and s, for each compound.

5.3. Arrhenius Plots

From the study of the behaviour of these parameters with temperature, we obtain the dependence of σ_{dc} and v_p with the inverse of temperature (Arrhenius plot: $\ln \sigma_{dc}$ and $\ln v_p$ vs 1000/T, see equations 4.4 and 4.5). We can then calculate the activation energies E_a and E_a' associated to each possible conduction mechanism, as shown in Figures 5.3 and 5.4, respectively. For each compound, three thermally activated processes can be distinguished from Figure 5.3, and their corresponding activation energy E_a are calculated and listed in table 5.1.

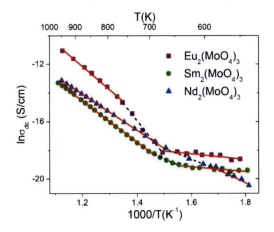

Figure 5.3. Arrhenius plot of σ_{dc}. The solid and dashed lines are linear fittings in the appropriate temperature intervals.

From approximately 550 K to around 650 K, in the lower temperature region, the conduction mechanism for the three studied compounds can be considered as of semiconducting type, with a real part of the complex conductivity depending strongly on both temperature and frequency, as shown in Figure 5.1. The low activation energy [0.34 and 0.29 eV for $Eu_2(MoO_4)_3$ and $Sm_2(MoO_4)_3$, respectively] suggests an extrinsic conduction due to impurity centres, donors or acceptors (Gaur et al., 1993). However, for $Nd_2(MoO_4)_3$ the activation energy obtained in this work, $E_a=1.61$ eV, differs from that obtained by Gaur et al. (1993) (0.21 eV). Our study is however not focused on this region of lower temperatures where microstructural effects are often typical.

In the intermediate temperature region the activation energies are higher for the three compounds. This indicates that there is a change in the conduction mechanism which is more evident in the compounds with α-phase. Moreover, as shown in Figure 5.1, σ' depends on frequency and temperature even more than at lower temperature. Such behaviour can be attributed to a conduction mechanism of polaronic type (Elliott, 1987, Gaur et al., 1993, Guzmán-Afonso et al., 2011 and Guzmán-Afonso et al., 2013).

Table 5.1. Activation energy values E_a, E_a' at different temperature regions

Samples	Temperature (K)	E_a (eV)	Temperature (K)	E_a' (eV)
$Eu_2(MoO_4)_3$	890-742	2.83	849-774	1.68
	742-669	4.11	766-710	4.38
	669-561	0.34	-	-
$Sm_2(MoO_4)_3$	899-686	2.50	876-798	1.88
	686-643	1.44	789-714	2.46
	643-553	0.29	-	-
$Nd_2(MoO_4)_3$	890-664	2.16	823-688	1.27
	664-598	1.35	656-598	0.16
	598-551	1.61	-	-

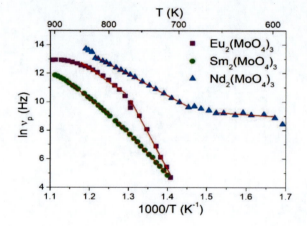

Figure 5.4. Arrhenius plot of v_p. The solid red lines are linear fittings in the appropriate temperature intervals.

In the final temperature region, the conductivity is less dependent on frequency than in the previous regions (see Figure 5.1). This behaviour is accompanied by a change in the activation energy, which could be due, again, to a change in the conduction mechanism. In this region, the activation energy is very similar for the three compounds (see Table 5.1).

The Arrhenius plot of v_p for the three compounds is shown in Figure 5.4. It can be seen from this plot that there are at least two regions with different activation energies E_a' in the same temperature interval in which the energies E_a were calculated (see Figure 5.3). The values of E_a' are also shown in Table 5.1 and it is observed that they are similar for the three compounds in the higher temperature region. This is an evidence of the same mechanism for relaxation and conduction operating for the three compounds in this region.

5.4. Frequency Exponent and Prefactor. Macroscopic Models for the AC Conductivity

As we mentioned in the previous section, the conduction mechanism for the intermediate temperature region can be of polaronic type. Several theoretical models explain the different behaviours, as briefly commented in section 4. In addition, the presence of some of these conduction mechanisms (small polaron and overlapping large polaron) in the materials is related to a distortion of the crystal lattice. It is thus convenient to compare the ac conductivity measurements with the temperature evolution of the crystal structure. This latter analysis will be discussed in the next section. Figure 5.5 shows the temperature dependence of the frequency exponent s, obtained from the fitting of the experimental data shown in Figure 5.2 using equation 5.2, for each temperature.

As shown in Figure 5.5, the behaviour of the exponent s is similar for the three compounds $Eu_2(MoO_4)_3$, $Sm_2(MoO_4)_3$ and $Nd_2(MoO_4)_3$ in the intermediate range of temperatures; s initially decreases with increasing temperature to a minimum value at a certain temperature and then it increases with the rise of temperature. Such behaviour is predicted by the overlapping large polaron model (OLP) (Elliott, 1987):

$$s = 1 - \frac{8\alpha R_\omega + 6W_{HO} r_p / R_\omega k_B T}{[2\alpha R_\omega + W_{HO} r_p / R_\omega k_B T]^2} \tag{5.3}$$

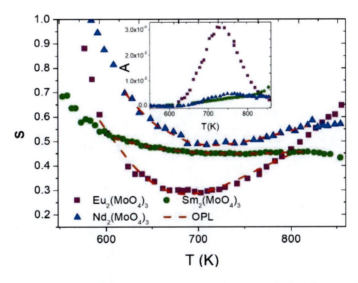

Figure 5.5. Temperature dependence of the frequency exponent s. The dashed lines are the fits to the overlapping large polaron model (equations 5.3 and 5.4). The inset shows the temperature dependence of the pre-exponential factor A. The data for $Nd_2(MoO_4)_3$ in the inset are multiplied by 10.

where W_{HO} and r_p are the energy barrier and the polaron radius, respectively, and R_ω is the tunneling distance at a frequency ($\omega = 2\pi\nu$), given by:

$$R_\omega = \frac{1}{4\alpha}\left[\ln\left(\frac{1}{\omega\tau_0}\right) - \frac{W_{HO}}{k_B T}\right] + \left\{\left[\ln\left(\frac{1}{\omega\tau_0}\right) - \frac{W_{HO}}{k_B T}\right]^2 + \frac{8\alpha r_p W_{HO}}{k_B T}\right\}^{1/2} \tag{5.4}$$

Although this model was developed for the OLP case, it has been later reported that it is also applicable to small polarons (van Staveren et al., 1991). The fitting of this model, for each compound, in their respective temperature ranges, is depicted by a dashed line in Figure 5.5 and the calculated fitting parameters are shown in Table 5.2.

The value of $\ln(1/\omega\tau_0)$ is within the typical order of magnitude for the overlapping large polaron model (Long, 1982). The value obtained for the polaron radius r_p for $Sm_2(MoO_4)_3$ is adequate within the OLPT model (Elliott, 1987, Punia et al., 2012). However, the radii of the polarons for $Eu_2(MoO_4)_3$ and $Nd_2(MoO_4)_3$ are more typical of small polarons. The values of the barrier energy W_{HO} for $Eu_2(MoO_4)_3$ and $Nd_2(MoO_4)_3$ are large compared with other studies (Punia et al., 2012, Bhattacharya and Ghosh, 2006). However, we must bear in mind that these mechanisms occur in a region of high temperatures and, it is known that when the temperature increases the thermal motion is enhanced and the ability to polarize the ions locally in the network is reduced, so that r_p decreases and W_{HO} can indeed increase (Lal et al., 1974, van Staveren et al., 1991).

The inset in Figure 5.5 shows the temperature dependence of the pre-exponential factor A. It can be seen that this A parameter has a maximum in the temperature range where s displays a minimum and it decreases with further increase in temperature. The high value of the pre-exponential factor A is associated with an increase of the polarizability, which corresponds to an increase in the strong interaction between the charge carriers and the surrounding crystal lattice. As we will see in section 6, some structural differences between the two modulated scheelite-types must be detected. The values obtained for A are within the typical order of magnitude for compounds with scheelite-type or related structures (Bouzidi et al., 2015, Parida et al., 2013, Parida et al., 2012).

Table 5.2. Parameters obtained from the fitting using the overlapping large polaron tunnelling (OLPT) model for the α-phase in $Eu_2(MoO_4)_3$, $Sm_2(MoO_4)_3$ and $Nd_2(MoO_4)_3$, in the intermediate temperature region

	Temperature Interval (K)	$\ln(1/\omega\tau_0)$	W_{HO} [eV]	r_p [Å]	α [Å^{-1}]
$Eu_2(MoO_4)_3$	630-800	42.58	2.62	0.57	0.25
$Sm_2(MoO_4)_3$	608-818	18.9	1.01	1.6	0.39
$Nd_2(MoO_4)_3$	620-818	43.73	2.53	0.71	0.42

5.5. Master Curve, Mobility and Concentration of Carriers

We discuss now the main conduction mechanism in the highest temperature region. In the study of the conductivity spectra of many materials at different temperatures, it is frequent to use a scaling law (Dyre et al., 2009, Singh et al., 2011), see equation 4.6. According to this law, a collapse within a temperature interval of all the curves to a single master curve indicates that the conduction mechanism is independent of the temperature. In our study, the scaling parameter used was $\sigma_{dc}T$ (Kumar and Ye, 2005, Dyre et al., 2009). Figure 5.6 shows the scaling plot of the conductivity at high temperatures for our three compounds. We see that the conductivity collapses on a single curve above 809 K for Eu$_2$(MoO$_4$)$_3$, and, although less clearly, at 808 K for Sm$_2$(MoO$_4$)$_3$ and above 668 K for Nd$_2$(MoO$_4$)$_3$. As an example, it can also be seen from this figure that the curves at 753 K for Eu$_2$(MoO$_4$)$_3$ and at 751 K for Sm$_2$(MoO$_4$)$_3$ do not collapse properly into the single master curve. This behaviour, which is more clearly observed for Eu$_2$(MoO$_4$)$_3$ than for Sm$_2$(MoO$_4$)$_3$, is attributed to a more abrupt change in the conductivity through the different regions (see Figure 5.3). Therefore, it is less evident in Nd$_2$(MoO$_4$)$_3$.

This behaviour is typical of materials exhibiting ionic conduction, for which the concentration of charge carriers is weakly temperature dependent and the temperature dependence of the conductivity arises from the mobility of the ions. Moreover, the interval of temperatures where the scaling curves collapse and the region (at higher temperatures) where the activation energies are similar in the three compounds, approximately coincide. In this case, it is possible to estimate the concentration of the charge carriers (n$_{AW}$) from the Nerst-Einstein relation (equation 4.7), taking a = 2.8 x 10^{-8} cm, for each temperature. The results for n$_{AW}$ in this region were almost constants and with mean values of 7.8x10^{22}, 5.1 x10^{22} and 3.7 x10^{21} cm^{-3} for Eu$_2$(MoO$_4$)$_3$, Sm$_2$(MoO$_4$)$_3$ and Nd$_2$(MoO$_4$)$_3$, respectively. Similar results have been found in other compounds with similar structures (Esaka, 2000). The mobility can be calculated using Equation 4.8, resulting in values which are four orders of magnitude lower than those obtained by Gaur et al. (1993), which vary from 10^{-6} to 10^{-4} m^{2}V^{-1}s^{-1}. Note that our calculation has been performed at high temperature and ionic mobilities are much smaller than electron or hole mobilities.

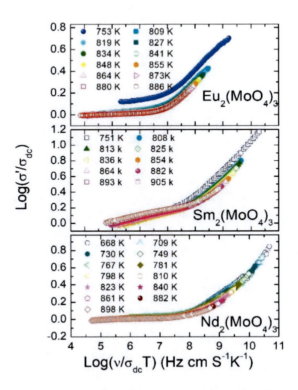

Figure 5.6. Scaled conductivity spectra of the α-phase in Eu$_2$(MoO$_4$)$_3$, Sm$_2$(MoO$_4$)$_3$ and Nd$_2$(MoO$_4$)$_3$ in the temperature range from 809-886 K, 808-905 K and 668-898 K, respectively.

6. Correlations between the Crystal Structure and the Electrical Properties. Conclusions

6.1. Previous Studies: Correlations Crystal Structure - AC Conductivity in the α-Phases of Samarium and Europium Molybdates

We have analysed the thermal dependence of the crystal structure and the dielectric properties of the α-phase of $Sm_2(MoO_4)_3$ and $Eu_2(MoO_4)_3$ (Guzmán-Afonso et al., 2011 and Guzmán-Afonso et al., 2013). From these analyses, we have correlated, for the first time, the thermal dependence of the electric properties in these compounds with the evolution of their crystalline structure with temperature. Figure 6.1 shows three coincident regions for both the structural and electrical behaviours. Although the volume of the unit cell expands monotonously in the whole temperature range, the lattice parameter a displays an anomalous temperature dependence according to the possible three conduction regimens. In what follows, we will explain the different types of mechanisms that were established in each region.

1) At the lower temperature region [below 643 K for $Sm_2(MoO_4)_3$ and 669 K for $Eu_2(MoO_4)_3$] the conduction mechanism for the two studied compounds is considered as semiconductor-type because the logarithm of the ac conductivity against 1000/T depends strongly on temperature and frequency showing in both cases a linear increase with T. The low activation energy [0.34 and 0.27 eV for $Eu_2(MoO_4)_3$ and $Sm_2(MoO_4)_3$] which is also obtained by Gaur et al. (1993) from dc conductivity measurements, suggests an extrinsic conduction due to the impurity centres, donors or acceptors. The anomalous contraction of the parameter a does not favour the volume expansion.

2) From the results shown in subsection 5.3, at intermediate temperatures the main conduction mechanism changes, displaying now features of polaronic-type conduction. The frequency exponent s shows a dependence with the temperature consistent with the overlapping large polaron model (Elliott, 1987, Ghosh et al., 2000). The first evidence of this possible polaronic behaviour can be observed in Figure 6.1, where the conductivity is strongly dependent on temperature and frequency. This conduction by polarons occurs approximately when the lattice parameter a reaches a minimum [at 620 K for $Sm_2(MoO_4)_3$ and 753 K for $Eu_2(MoO_4)_3$].

3) In the higher temperature region [818-905 K for $Sm_2(MoO_4)_3$ and 809-910 K for $Eu_2(MoO_4)_3$], the conductivity is less dependent on frequency than in the other regions (see Figure 6.1) and this coincides with the temperature interval in which we found a master curve (see subsection 5.5). This implies that the relaxation dynamics of the charge carriers is independent of temperature in this region. Also, it was possible to calculate the charge carrier concentration which is almost constant for each temperature (Howard et al., 1964). This means that the temperature dependence of the conductivity arises from the mobility of the ions rather than from the concentration of the charge carriers. Finally, the lattice parameter a increases at 800

K, therefore, the coefficient of the volume expansion increases with respect to the previous stage. All these behaviours suggest that the ionic transport must be predominant.

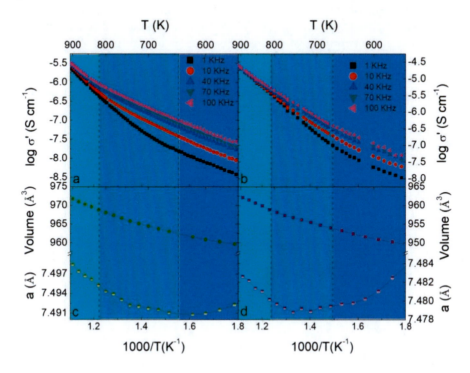

Figure 6.1. Temperature dependence of the real part of the complex conductivity (σ′) at five frequencies and temperature dependence of the lattice parameter a and the unit-cell volume for the α-phase of $Sm_2(MoO_4)_3$ (a, c) and $Eu_2(MoO_4)_3$ (b, d). Colours show the temperature regions with different behaviours.

6.2. New Contributions and Comparisons between Two Types of Modulated Scheelites with Respect to the Structure-Property Correlations

Slight differences in the thermal behaviour of the crystal structure and the ac conductivity have been found for the two modulated scheelite structures. In Figure 6.2 we have represented the logarithm of the polarizability ($α_p$) and the supercell parameter a and $β$ angle for $Eu_2(MoO_4)_3$ and $Nd_2(MoO_4)_3$, respectively, normalized in the temperature range 550-950 K, in order to show the similar behaviour of the parameters obtained with two completely different techniques (X-ray diffraction and impedance spectroscopy). The polarizability is proportional to the dipolar moment, which can increase when the interaction with the charges and its crystal field is greater. Lattice contraction, given by the shortening of the cell parameter, favours an increment in the crystal field and thus in the interaction with mobile charges. There is a complete correlation between the thermal dependence of the crystal structure and this electrical feature. On the other hand, from Figure 3.6 the contraction in $Eu_2(MoO_4)_3$ is greater than in $Sm_2(MoO_4)_3$, hence the polarizability must be greater too, as we can see in the inset of Figure 5.5. Moreover, the polarizability in Nd$_2$(MoO$_4$)$_3$ is one order of magnitude less than in Eu$_2$(MoO$_4$)$_3$ or Sm$_2$(MoO$_4$)$_3$. The lattice contraction in this compound cannot be directly observed from the corresponding parent cell as was done for the compounds with α-phase. For that, it is necessary to compare the projection of the supercell parameter a (as explained in section 3.3). We have thus definitely differentiated the electric properties between the two types of modulated structures with different orderings of the rare earth and vacancies.

Finally, if we refer to the mechanisms of the conductivity in the intermediate region, the overlapped large polaron model intermediates between an electronic conduction in all compounds, which could be explained by a correlated barrier hopping model as in other similar materials, to an ionic model typical of other metal oxides. When the exponent s is increased, the best-fit model would be the non-overlapped small polaron, but in our samples at high temperatures ionic conductivity is expected, as we have explained in section 5. We cannot differentiate between both structural types based solely on the fitting parameters that describe the OLP model.

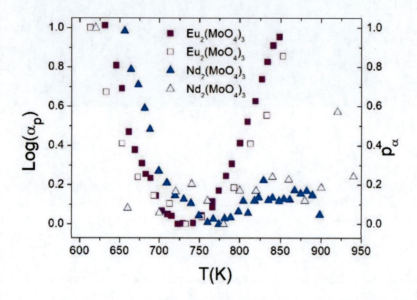

Figure 6.2. The open symbols represent normalized cell parameters [a for the α-phase and angle β for the La$_2$(MoO$_4$)$_3$-phase] and the filled symbols correspond to the normalized logarithm of the polarizability (proportional to the pre-exponential factor A calculated in section 5).

6.3. Thermal Anomalies and/or Structural Phase Transitions

As we have commented in previous sections, to our knowledge there have not been studies in which the thermal dependence of structural parameters such as bond distances and cell parameters were correlated to the thermal dependence of the ac conductivity or parameters such as the dc conductivity, the frequency pre-exponential factor A and the exponent s. However, there are particular cases in which in order to understand structural phase transition mechanisms, the knowledge of the dielectric behaviour close to the transition temperature is required. This is for example the case with ferroelectrics, where the study of the electrical conductivity is very important, since it is associated to physical properties, such as piezoelectricity and/or pyroelectricity local polar characteristics, dielectric/tunability properties for microwave applications etc. (Ciomaga et al., 2007).

In section 5 we have seen how, for the three compounds studied here, the value of the exponent s decreased with the rise in temperature and reached a minimum, afterwards increasing with further increase in temperature, whereas the pre-exponential factor A displays the opposite trend. A similar behaviour is found in compounds with ferroelectric-paraelectric phase transition: perovskites (Barick et al., 2011, Padhee et al., 2013), tungsten bronzes (Venkataraman and Varma, 2004, Rao et al., 2008, Parida et al., 2012, Parida et al., 2013), and even in compounds consisting of arrangements of tetragonal clusters linked to other larger polyhedrons as in SCRc (Pati et al., 2013 and Pati et al., 2015). Therefore, when the phase transition occurs this behaviour can be explained. If the exponent s represents the interaction between mobile charges surrounded by the lattice ions and the pre-exponential factor A determines the strength of the polarizability, then, the observed minima for s (maxima for A) close to T_c imply a strong interaction between the lattice and the mobile charges leading to higher polarizability (maximum dielectric constant). According to the dynamical theory, transverse optical soft modes may be weakened and the restoring forces tend to zero at the transition temperature T_c. If the charge carriers couple with a soft mode then, the charge carriers become very mobile at T_c, and hence the conductivity increases (Yamada and Shirane, 1969; Jonscher, 1983).

The evolution of the crystal structure with increasing the temperature, in the studied compounds, presents many similarities to the phase transitions described above. In our case, it is manifested by an anomalous contraction of the lattice parameter up to a certain temperature, to which follows a normal thermal expansion. This fact makes it possible an increase of the electric polarizability until reaching a maximum, and then decreasing again. Also, the fact that longer and shorter Mo ... Mo contacts tend to be the same length can be interpreted as an increase in the structural disorder manifested by a decrease in the exponent s. The higher activation energy for the dc conductivity in $Eu_2(MoO_4)_3$ at the intermediate region was not explained in the subsection 5.3. Now, this could be understood if the mobile charges were more strongly coupled to the lattice in this compound (major contraction) and a possible soft mode would diminish the restoring forces, therefore more free charges would help increase the conductivity.

We plan to do a more exhaustive experiment using neutron thermodiffraction in molybdates and tungstates with modulated scheelite structures in order to monitor the thermal dependence of the atomic coordinates from which we will be able to correlate the crystal structure with the thermal dependence of the polarizability and other parameters calculated from the ac conductivity models, and in particular from the overlapping large polaron model. Furthermore, possible soft modes could be evidenced by infrared or Raman spectroscopy.

Conclusion

The implications of the polaronic conductivity in the properties of two well-known compounds with slightly distorted scheelite phase (the scintillator $PbWO_4$ and the photocatalyst $BiVO_4$) have been investigated. Theoretical calculations have been compared with experimental data obtained using different techniques. However, a complete ac conductivity study of these compounds has not been performed yet. In general, the dielectric properties of compounds with scheelite related structure have not been enough studied, in spite of their outstanding optical properties. The few studies on double tungstates with scheelite structure and metal tungstates with wolframite structure show well-differentiated behaviours, depending on their crystal structures. The ac conductivity of the former compounds is due to small polarons at temperatures beyond 625 K. However, the CBH model gives a better fitting at lower temperatures. Nickel, cobalt, cadmium and manganese tungstates show a similar semiconductor behaviour: by increasing the temperature, the exponent s (greater than one) in Jonscher's Universal Power Law, hopping frequency v_p, mobility of charge carriers and dc conductivity σ_{dc}, all increase, while the carrier concentration remains constant.

The interest in modulated scheelites has arisen quite recently and much research is being dedicated to their optical properties. Different orderings for stoichiometric vacancies (or different sizes of rare earth clusters) affect these properties. We have transferred these conclusive investigations to the field of the electrical properties. In particular we have studied the thermal dependence of the ac conductivity and the crystal structure in two different modulated scheelites phases: the α-phase and the $La_2(MoO_4)_3$-phase.

Regarding the crystal structure, we have reviewed the thermal dependence of the supercell of the samarium and europium molybdates in the α-phase, for which an anomalous behaviour is observed in the parameter a. The thermal dependence of the supercell with the $La_2(MoO_4)_3$ structure for neodymium molybdate has been reported for the first time, for which an anomalous dependence of the monoclinic β angle is observed. We have succeeded in distinguishing between the evolution of the crystal structure in compounds with the α-phase and compounds with the $La_2(MoO_4)_3$-phase. To do this, we have calculated a parameter p for both structures by projecting the parameter a_s (or the parameter b_s) onto the ab plane, with γ_s being the angle between them (here, the sub-index "s" refers to the parent or scheelite cell). This parameter displays a similar temperature dependence in both phases, if we multiply $\cos(\gamma_s)$ by 18, in the case of $Nd_2(MoO_4)_3$. That is, this projection will have a longer length in a longer period, considering that γ_s is almost 90° in the second phase and 91.2° in the first phase. (Remember that the supercell of the $La_2(MoO_4)_3$-phase is nine times the cell of the scheelite structure.)

Additionally, we have given the results of a systematic analysis of the measurements of the ac conductivity by comparing the two structural types. In the three samples, a greater dependence of the conductivity with the frequency and the temperature is detected, in an intermediate region between about 600 K and 800 K. Three regions with different activation energies from the analysis of Arrhenius plot of σ_{dc} were observed. This feature is less clear for $Nd_2(MoO_4)_3$. In the region of higher temperature, the activation energies obtained from the Arrhenius plot of the crossover frequency (v_p) are similar to those calculated for σ_{dc}, indicating that the conduction and relaxation mechanisms are similar. For $Sm_2(MoO_4)_3$ the Arrhenius type behaviour for v_p at high temperature is not so clear, the relaxation mechanisms are more sensitive to the proximity of a structural phase transition. We have focused our study of conductivity mechanisms in the central region where the exponent s fits to the overlapped large polaron model in the three compounds studied. The thermal dependence of the pre-exponential factor A has a behaviour opposite to that of s. In the low temperature region, we consider an extrinsic electronic conduction and in the high temperature region, an ionic conductivity for the three compounds that is confirmed when the master curve is performed. At very high temperatures, the transition to the β-phase is expected, which in the case of $Sm_2(MoO_4)_3$ is observed in the behaviour of the parameters s and the parameter v_p.

Once the structural and electric mechanisms were explained, we have correlated the slight differences between the cell parameters (i.e., crystal structure) of both modulated phases to differences in the frequency prefactor that is related to the polarizability (i.e., an ac conductivity parameter). We have observed that there must be a contraction in both phases around the RE^{3+} ions and the vacancies in the ab plane, which is clearer in the α-phase (contraction of one of the diagonals of all the parent cells that form the supercell). In the $La_2(MoO_4)_3$-phase, this contraction does not occur in all the parent cells that form the supercell, but finally, it is transmitted in a contraction of the β-angle of the corresponding supercell (which is 9 times the parent cell). This explains why the polarizability is an order of magnitude smaller in $Nd_2(MoO_4)_3$. However, we cannot observe these differences in the exponent s, which is a statistical parameter not directly related to the crystal structure. This parameter indicates that, in all the compounds, there is a greater degree of disorder at the temperature for which the contraction is maximum. This behaviour is very similar to that observed in ferroelectric-paraelectric or other second-order transitions where the structural change is very small. We can interpret that a transition occurs between two structures [α or $La_2(MoO_4)_3$] that have the same space group but are nonetheless slightly different when the conduction mechanisms are different too. The overlapped large polaron behaviour occurs during the transition.

In this chapter, we have also presented a protocol for the analysis of the ac conductivity. The largest possible number of measurements should be taken in the widest range of temperatures and frequencies (v) in a complete temperature cycle. Then, the plots "log σ' versus $1/T$" and "σ' versus v at several temperatures" must be performed. Next, one has to fit using Jonscher Universal Power Law, possibly distinguishing different intervals where it is performed. From this fitting, σ_{dc}, v_p, A and s parameters will be obtained. The σ_{dc} and v_p parameters must be adjusted to a possible Arrhenius behaviour. The thermal dependence of the pre-exponential factor A and the frequency exponent must also be shown within the complete interval where the fitting has been made. In the case of adjusting to different conduction models, it is essential to check the values of the parameters obtained [τ_0, α, r_p, W_M, R, N, W_{H0}, R_w, $N(E_F)$]. Finally, one can draw the master curve and estimate the concentration of carriers and mobility; this study is usually employed when the concentration of carriers is temperature independent (more typical in ionic conduction). We take this opportunity to recommend the possibility of organizing an accessible data repository of impedance spectroscopy, so that different conduction models can be checked, in particular the polaronic conduction where the dependence on temperature and frequency is more complex. Note that we have not interpreted the results of the measurements of electrical permittivity that are also obtained from impedance spectroscopy, nor have we taken into account how the micro/nano-structure (grain size, microstrains, etc.) affects the conductivity mechanisms. We are interested in correlating differences in crystal structures with small changes in the parameters that feature the ac conductivity.

Finally, we want to emphasize the importance of knowing the thermal dependence of the crystalline structure or the ionic environment when there is no long-range ordering. Simple experiments of thermo-diffraction or infrared and raman spectroscopy as a function of temperature will be key to the interpretation of the mechanisms of the electric charge transport; especially, when important changes are observed in the activation energies of the dc conductivity and the crossover frequency. One must also pay attention to the changes in the curvature (maximum and minimum) in the thermal dependence of the pre-exponential factor A and exponent s of the frequency obtained from the Jonscher's power law, since they can be associated with structural anomalies and even phase transitions.

REFERENCES

Abakumov, A. M., V. A. Morozov, A. A. Tsirlin, J. Verbeeck, J. Hadermann. 2014. "Cation Ordering and Flexibility of the BO_4^{2-} Tetrahedra in Incommensurately Modulated $CaEu_2(BO_4)_4$ (B = Mo, W) Scheelites." *Inorganic Chemistry* 53 (17): 9407. doi:10.1021/ic5015412.

Abraham, Y.B., Naw Holzwarth, R. Williams, Ge Matthews A. Tackett. 2001. "Electronic Structure of Oxygen-Related Defects in $PbWO_4$ and $CaMoO_4$ Crystals." *Physical Review B* 64 (24) 245109. doi:10.1103/PhysRevB.64.245109.

Abrahams, S. C. and J.L. Bernstein. 1966. "Crystal Structure of the Transition-Metal Molybdates and Tungstates. II. Diamagnetic $Sc_2(WO_4)_3$. " *The Journal of Chemical Physics* 45 (8): 2745-2752. doi: 10.1063/1.1728021.

Adams, S., and Rao, R. P. 2014. Understanding ionic conduction and energy storage materials with bond-valence-based methods. In *Bond Valences*, vol. (158). Springer, Berlin, Heidelberg. doi:10.1007/430_2013_137.

Alexandrov, A. S. and Kornilovitch, P. E. 1999. "Mobile small polaron." *Physical Review Letters*, 82 (4): 807. doi:10.1103/PhysRevLett.82.807.

Anicete-Santos, M., E. Orhan, De Maurera, M. A. M. A., L. G. P. Simões, A. G. Souza, P. S. Pizani, E. R. Leite, et al. 2007. "Contribution of Structural Order-Disorder to the Green Photoluminescence of $PbWO_4$." *Physical Review B – Condensed Matter and Materials Physics* 75 (16): 165105, doi:10.1103/PhysRevB.75.165105.

Anisimov, V. I., F. Aryasetiawan, and A. I. Lichtenstein. 1997. "First-Principles Calculations of the Electronic Structure and Spectra of Strongly Correlated Systems: The LDA + U Method." *Journal of Physics Condensed Matter* 9 (4): 767-808. doi:10.1088/0953-8984/9/4/002.

Annenkov, A. A., M.V. Korzhik, P. Lecoqc. 2002. "Lead tungstate scintillation material." *Nuclear Instruments and Methods in Physics Research A*, 490:30–50. doi:10.1016/S0168-9002(02)00916-6.

Arakcheeva, A. and G. Chapuis. 2008. "Capabilities and Limitations of a (3+D)-dimensional Incommensurately Modulated Structure as a Model for the Derivation of an Extended Family of Compounds: Example of the Scheelite-like Structures." *Acta Crystallographica. B* 64 (1): 12-25. doi:10.1107/S010876810705923X.

Arakcheeva, A., D., Logvinovich, G., Chapuis, V. Morozov, S.V. Eliseeva, J. C., G. Bnzli, and P. Pattison. 2012. "The Luminescence of $Na_x Eu_{3+(2X)/3}MoO_4$ Scheelites Depends on the Number of Eu-Clusters Occurring in their Incommensurately Modulated Structure." *Chemical Science* 3 (2): 384-390. doi:10.1039/c1sc00289a.

Austin I. and Mott N.F. 1969. "Polarons in Crystalline and Non-Crystalline Materials." *Advances in Physics* 18 (71): 41-102. doi:10.1080/00018736900101267.

Bacha, F. B., K. Guidara, M. Dammak, and M. Megdiche. 2017a. "AC and DC Conductivity Study of Ceramic Compound NaGd(WO$_4$)$_2$ using Impedance Spectroscopy." *Journal of Materials Science: Materials in Electronics* 28 (14): 10630-10639. doi:10.1007/s10854-017-6838-1.

Bacha, F. B., S. Megdiche Borchani, M. Dammak, and M. Megdiche. 2017b. "Optical and Complex Impedance Analysis of Double Tungstates of Mono- and Trivalent Metals for LiGd(WO$_4$)$_2$ Compound." *Journal of Alloys and Compounds* 712: 657-665. doi:10.1016/j.jallcom.2017.04.107.

Barick, B. K., K. K. Mishra, A. K. Arora, R. N. P. Choudhary, and Dillip K. Pradhan. 2011. "Impedance and Raman Spectroscopic Studies of (Na$_{0.5}$Bi$_{0.5}$)TiO$_3$." *Journal of Physics D: Applied Physics* 44 (35): 355402. doi:10.1088/0022-3727/44/35/355402.

Basiev, T. T., Sobol, A. A., Voronko, Y. K., Zverev, P. G., 2000. "Spontaneous Raman spectroscopy of tungstate and molybdate crystals for Raman lasers." *Optical Materials*, 15 (3): 205-216. doi: 10.1016/S0925-3467(00)00037-9.

Batuk, D., Batuk, M., Abakumov, A. M., Hadermann, J. 2015. "Synergy between transmission electron microscopy and powder diffraction: application to modulated structures." *Acta Crystallographica.* B: Structural Science, Crystal Engineering and Materials, 71 (2): 127-143. doi: 10.1107/S2052520615005466.

Bhattacharya, S. and A. Ghosh. 2003. "Ac Relaxation in Silver Vanadate Glasses." *Physical Review B* 68 (22). doi:10.1103/PhysRevB.68.224202.

Bhattacharya, S. and A. Ghosh. 2006. "Relaxation dynamics of Ag$^+$ ions in superionic glass nanocomposites embedded with ZnO nanoparticle." *Physical review. B, Condensed matter 74(18)* 184308 doi: 10.1103/PhysRevB.74.184308.

Bhattacharya, A., R. Biswas, and A. Hartridge. 1997. "Environment Sensitive Impedance Spectroscopy and Dc Conductivity Measurements on NiWO$_4$." *Journal of Materials Science* 32 (2): 353-356. doi:1018545131216.

BIBAO CRYSTALLOGRAPHIC SERVER: http://www.cryst.ehu.es/

Borchardt, H. J. and P. E. Bierstedt. 1966. "Gd$_2$(MoO$_4$)$_3$: A Ferroelectric Laser Host." *Applied Physics Letters* 8 (2): 50-52. doi:10.1063/1.1754477.

Boulahya, K., M. Parras, J. M. González-Calbet. 2005. "Synthesis, Structural and Magnetic Characterization of a New Scheelite Related Compound: Eu$_2$Mo$_3$O$_{12}$." *European Journal of Inorganic Chemistry* 5, 967-970. doi:10.1002/ejic.200400662.

Bouzidi, C., N. Sdiri, A. Boukhachem, H. Elhouichet, M. Férid. 2015. "Impedance Analysis of BaMo$_{1-x}$W$_x$O$_4$ Ceramics." *Superlattices and Microstructures* 82: 559-573. doi:10.1016/j.spmi.2015.03.019.

Brixner, L. H., J. R. Barkley, and W. Jeitschko. 1979. *Chapter 30 Rare Earth Molybdates (VI). Handbook on the Physics and Chemistry of Rare Earths*, V3. North Holland: Elsevier doi:10.1016/S0168-1273(79)03013-0.

Brixner, L. H. and A. W. Sleight. 1973. "Crystal Growth and Precision Lattice Constants of some Ln$_2$(WO$_4$)$_3$-Type Rare Earth Tungstates." *Materials Research Bulletin* 8 (10): 1269-1273. doi:10.1016/0025-5408(73)90165-7.

Brown, I. D. 2006. *The chemical bond in inorganic chemistry: the bond valence model* (Vol. 27). Oxford University Press.

Bubb, D. M., D. Cohen, and S. B. Qadri. 2005. "Infrared-to-Visible Upconversion in Thin Films of LaEr(MoO$_4$)$_3$." *Applied Physics Letters* 87 (13): 131909. doi:10.1063/1.2067712.

Capaccioli, S., M. Lucchesi, P. A. Rolla, G. Ruggeri. 1998. "Dielectric Response Analysis of a Conducting Polymer Dominated by the Hopping Charge Transport." *Journal of Physics Condensed Matter* 10 (25): 5595-5617. doi:10.1088/0953-8984/10/25/011.

Cascales, C., M. D. Serrano, F. Esteban-Betegon, C. Zaldo, R. Peters, K. Petermann, G. Huber, et al. 2006. "Structural, Spectroscopic, and Tunable Laser Properties of Yb^{3+}-Doped NaGd(WO$_4$)$_2$." *Physical Review B* 74 (17): 4114. doi:10.1103/PhysRevB.74.174114.

Chakchouk, N., B. Louati, K. Guidara. 2018. "Electrical Properties and Conduction Mechanism Study by OLPT Model of NaZnPO$_4$ Compound." *Materials Research Bulletin* 99: 52-60. doi:10.1016/j.materresbull.2017.10.046.

Chen, H., Fu, X., An, Q., Tang, B., Zhang, S., Yang, H., Li, Y. 2017. "Determining the Quality Factor of Dielectric Ceramic Mixtures with Dielectric Constants in the Microwave Frequency Range." *Scientific reports*, 7(1): 14120-1-11. doi:10.1038/s41598-017-14333-9.

Ciomaga, C., M. Viviani, M. Buscaglia, V. Buscaglia, L. Mitoseriu, A. Stancu, and P. Nanni. 2007. "Preparation and Characterisation of the Ba(Zr,Ti)O$_3$ Ceramics with Relaxor Properties." *Journal of the European Ceramic Society* 27 (13-15): 4061-4064. doi:10.1016/j.jeurceramsoc.2007.02.095.

Cooper, J., S. Gul, F. Toma, L. Chen, P. A. Glans, J. Guo, J. W. Ager, J. Yano, and I. Sharp. 2014. "Electronic Structure of Monoclinic BiVO$_4$." *Chemistry of Materials* 26 (18): 5365-5373. doi:10.1021/cm5025074.

Cooper, J., S. Gul, F. Toma, L. Chen, Ys Liu, J. Guo, J. W. Ager, J. Yano, and Id Sharp. 2015. "Indirect Bandgap and Optical Properties of Monoclinic Bismuth Vanadate." *Journal of Physical Chemistry C* 119 (6): 2969-2974. doi:10.1021/jp512169w.

Cramer, C. and M. Buscher. 1998. "Complete Conductivity Spectra of Fast Ion Conducting Silver Iodide/Silver Selenate Glasses." *Solid State Ionics* 105 (1): 109-120. doi:10.1016/S0167-2738(97)00456-6.

Cuervo-Reyes, E. 2016. "Why the Dipolar Response in Dielectrics and Spin-Glasses is Unavoidably Universal." *Scientific Reports* 6 (1). doi:10.1038/srep29021.

Dissado, L. A., and Hill, R. M. 1984. "Anomalous low-frequency dispersion. Near direct current conductivity in disordered low-dimensional materials." *Journal of the Chemical Society*, Faraday Transactions 2: *Molecular and Chemical Physics*, 80 (3): 291-319. doi: 10.1039/F29848000291.

Dutta, A. and A. Ghosh. 2005. "Effect of Alkaline-Earth Ions on the Dynamics of Alkali Ions in Bismuthate Glasses." *Physical Review B - Condensed Matter and Materials Physics* 72 (22). doi:10.1103/PhysRevB.72.224203.

Dussouze, M. 2005. *Second Harmonic Generation in Glasses Borophosphate Sodium and Niobium Thermal Polarization*. PhD diss., University of Bordeaux

Dyre, J. C., P.Maass, B. Roling, and D. L. Sidebottom. 2009. "Fundamental Questions Relating to Ion Conduction in Disordered Solids." *Reports on Progress in Physics* 72. doi:10.1088/0034-4885/72/4/046501.

Elliott, S. R. 1987. "A.C. Conduction in Amorphous Chalcogenide and Pnictide Semiconductors." *Advances in Physics* 36 (2): 135-217. doi:10.1080/00018738700101971.

Emin. D. 1993 "Optical properties of large and small polarons and bipolarons." *Physical Review B*, 48, (18): 13691-13702. doi: 10.1103/PhysRevB.48.13691.

Errandonea, D., Kumar, R. S., Achary, S. N., Tyagi, A. K. 2011. "In Situ High-Pressure Synchrotron X-Ray Diffraction Study of $CeVO_4$ and $TbVO_4$ Up to 50 GPa." *Physical Review.B, Condensed Matter and Materials Physics* 84 (22): 224121. doi:10.1103/PhysRevB.84.224121.

Errandonea, D. and F. J. Manjón. 2008. "Pressure Effects on the Structural and Electronic Properties of ABX_4 Scintillating Crystals." *Progress in Materials Science* 53 (4): 711-773. doi:10.1016/j.pmatsci.2008.02.001.

Esaka, T. 2000. "Ionic conduction in substituted scheelite-type oxides." *Solid State Ionics* 136–137:1–9. doi: 10.1016/S0167-2738(00)00377-5.

Evans, J. S. O., T. A. Mary, and A. W. Sleight. 1998. "Negative Thermal Expansion in $Sc_2(WO_4)_3$." *Journal of Solid State Chemistry* 137 (1): 148-160. doi:10.1006/jssc.1998.7744.

Fabbri, E., Lei Bi, D. Pergolesi, E. Traversa. 2012. "Towards the Next Generation of Solid Oxide Fuel Cells Operating Below 600 °C with Chemically Stable Proton-Conducting Electrolytes." *Advanced Matererials*, 24, 195–208. doi: 10.1002/adma.201103102.

Friedman, L., and Holstein, T. 1963. "Studies of polaron motion: Part III: The Hall mobility of the small polaron. "*Annals of Physics*, 21 (3):494-549. doi: 10.1016/0003-4916(63)90130-1.

FullProf: https://www.ill.eu/sites/fullprof/

Gai, S., Li, C., Yang, P., Lin, J., 2014. "Recent progress in rare earth micro/nanocrystals: soft chemical synthesis, luminescent properties, and biomedical applications." *Chemical reviews*, 114 (4):2343-2389. doi: 10.1021/cr4001594.

Ganesamoorthy, S., I. Bhaumik, A. K. Karnal, and V. K. Wadhawan. 2004. "Optical, Thermal and Defect Studies on $PbWO_4$ Single Crystals Grown by the Czochralski Method." *Journal of Crystal Growth* 264 (1): 320-326. doi:10.1016/j.jcrysgro.2004.01.004.

Gaur, K., M. Singh, and H. B. Lal. 1993. "Electrical Transport in Light Rare-Earth Molybdates." *Journal of Materials Science* 28 (14): 3816-3822. doi:10.1007/BF00353184.

Ghosh, A. 1990. "Frequency-Dependent Conductivity in Bismuth-Vanadate Glassy Semiconductors." *Physical Review* B 41 (3): 1479-1488. doi:10.1103/PhysRevB.41.1479.

Ghosh, A., A. Pan. 2000. "Scaling of the Conductivity Spectra in Ionic Glasses: Dependence on the Structure." *Physical Review* Letters. 84 (10): 2188-2190. doi: 10.1103/PhysRevLett.84.2188.

Gong, C., Yu-Jun Bai, Jun Feng, Rui Tang, Yong-Xin Qi, Ning Lun, Run-Hua Fan. 2013. "Enhanced Electrochemical Performance of $FeWO_4$ by Coating Nitrogen-Doped." *ACS Applied Materials and Interfaces*, 5 (10):4209–4215. doi: 10.1021/am400392t.

González-Platas, J., C. González-Silgo, and C. Ruiz-Pérez. 1999. "VALMAP 2.0: Contour Maps using the Bond-valence-sum Method." *Journal of Applied Crystallography* 32 (2): 341-344. doi:10.1107/S0021889898010279.

González-Silgo, C., C. Guzmán-Afonso, V. Sanchez-Fajardo, S. Acosta-Gutierrez, A. Sanchez-Soares, M. E. Torres, N. Sabalisck, E. Matesanz, and J. Rodríguez-Carvajal.

2013. "Polymorphism in $Ho_2(MoO_4)_3$." *Powder Diffraction* 28: S40. doi:10.1017/S0885715613001176.

Gracia, L., V. M. Longo, L.S. Cavalcante, A. Beltrán, W. Avansi, Má Li, V. R. M., J. A. Varela, E. Longo, and J. Andrés. 2011. "Presence of Excited Electronic State in $CaWO_4$ Crystals Provoked by a Tetrahedral Distortion: An Experimental and Theoretical Investigation." *Journal of Applied Physics* 110 (4): 3501. doi:10.1063/1.3615948.

Guzmán-Afonso, C., M. E. Torres, C. González-Silgo, N. Sabalisck, J. González-Platas, E. Matesanz, and A. Mujica. 2011. "Electrical Transport and Anomalous Structural Behavior of at High Temperature." *Solid State Communications* 151 (22): 1654-1658. doi:10.1016/j.ssc.2011.08.009.

Guzmán-Afonso, C., C. González-Silgo, J. González-Platas, M. E. Torres, A. D. Lozano-Gorrón, N. Sabalisck, V. Sánchez-Fajardo, J. Campo, and J. Rodríguez-Carvajal. 2011. "Structural Investigation of the Negative Thermal Expansion in Yttrium and Rare Earth Molybdates." *Journal of Physics Condensed Matter* 23 (32): 325402. doi:10.1088/0953-8984/23/32/325402.

Guzmán-Afonso, C., C. González-Silgo, M. E. Torres, E. Matesanz, and A. Mujica. 2013. "Structural Anomalies Related to Changes in the Conduction Mechanisms of α-$Sm_2(MoO_4)_3$." *Journal of Physics Condensed Matter* 25 (3): 035902. doi:10.1088/0953-8984/25/3/035902.

Guzmán-Afonso, C., J. López-Solano, C. González-Silgo, S. F. León-Luis, E. Matesanz, and A. Mujica. 2014. "Pressure Evolution of Two Polymorphs of $Tb_2(MoO_4)_3$." *High Pressure Research* 34 (2): 184-190. doi:10.1080/08957959.2014.895342.

Guzmán-Afonso C., León-Luis S. F., Sans J. A., González-Silgo C., Rodríguez-Hernández P., Radescu S., Muñoz A., López-Solano J., Errandonea D., Manjón F. J., Rodríguez-Mendoza U. R., Lavín V. 2015. "Experimental and Theoretical Study of α-Eu2(MoO4)3 Under Compression." *Journal of Physics Condensed Matter* 27 (46): 465401. doi:10.1088/0953-8984/27/46/465401.

Hamdi H., E. K. H. Salje, P. Ghosez, E. Bousquet. 2016. "First-Principles Re-Investigation of Bulk WO_3." *Physical Review B* 94: 245124. doi:10.1103/PhysRevB.94.245124.

Han, J., McBean, C., Wang, L., Jaye, C., Liu, H., Fischer, D. A., Wong, S. S. 2015. "Synthesis of Compositionally Defined Single-Crystalline Eu3+-Activated Molybdate–Tungstate Solid-Solution Composite Nanowires and Observation of Charge Transfer in a Novel Class of 1D $CaMoO_4$–$CaWO_4$: Eu3+–0D CdS/CdSe QD Nanoscale Heterostructures." *The Journal of Physical Chemistry C*, 119 (7): 3826-3842. **doi** 10.1021/jp512490d.

Hara, K., Ishii, M., Nikl, M., Takano, H., Tanaka, M., Tanji, K., Usuki, Y. 1998. "La-doped $PbWO_4$ scintillating crystals grown in large ingots." *Nuclear Instruments and Methods in Physics Research* A, 414: 325. doi 10.1016/S0168-9002(98)00634-2.

Hitchcock, Fanny R. M. 1895. "The Tungstates and Molybdates of the Rare Earths." *Journal of the American Chemical Society* 17 (7): 520-537. doi:10.1021/ja02162a006.

Hoang, K. 2017. "Polaron formation, native defects, and electronic conduction in metal tungstates." *Physical Review Materials*, 1, 024603. doi: 10.1103/PhysRev Materials.1.024603.

Hoffart, L., U. Heider, L. Jrissen, R. A. Huggins, W. Witschel. 1994. "Transport Properties of Materials with the Scheelite Structure." *Solid State Ionics* 72 (PART 2): 195-198. doi:10.1016/0167-2738(94)90146-5.

Holstein, T. 1959a. "Studies of polaron motion: Part I. The molecular-crystal model." *Annals of Physics*, 8 (3): 325-342. doi:10.1016/0003-4916(59)90002-8.

Holstein, T. 1959b. "Studies of polaron motion: Part II. The small polaron." *Annals of Physics*, 8 (3):343-389. doi:10.1016/0003-4916(59)90003-X.

Howard, R. E. and A. B. Lidiard. 1964. "Matter Transport in Solids." *Reports on Progress in Physics* 27 (1): 161-240. doi:10.1088/0034-4885/27/1/305.

Huang, H., X. Feng, Z. Man, T. B. Tang, M. Dong, and Z. G Ye. 2003a. "Impedance Spectroscopy Analysis of La-Doped $PbWO_4$ Single Crystals." *Journal of Applied Physics* 93 (1): 421-425. doi:10.1063/1.1519956.

Huang, Y. L., W. L. Zhu, X. Q. Feng, and Z. Y. Man. 2003b. "The Effects of La^{3+} Doping on Luminescence Properties of $PbWO_4$ Single Crystal." *Journal of Solid State Chemistry* 172 (1): 188-193. doi:10.1016/S0022-4596(03)00013-6.

ICSD https://icsd.fiz-karlsruhe.de

Imanaka, N., Y. Kobayashi, S. Tamura, and G. Adachi. 2000. "Trivalent Ion Conducting Solid Electrolytes." *Solid State Ionics* 136: 319-324. doi:10.1016/S0167-2738(00)00464-1.

Imanaka, N., Y. Kobayashi, K. Fujiwara, T. Asano, Y. Okazaki, and G. Adachi. 1998. "Trivalent Rare Earth Ion Conduction in the Rare Earth Tungstates with the $Sc_2(WO_4)_3$-Type Structure." *Chemistry of Materials* 10 (7): 2006-2012. doi:10.1021/cm980157e.

Ivanova, M., Ricote, S., Baumann, S., Meulenberg, W. A., Tietz, F., Serra, J. M., and H., Richter. 2013. "Ceramic materials for energy and environmental applications: Functionalizing of properties by tailored compositions." In *Doping: Properties, Mechanisms and Applications*. Nova Science Publishers, Incorporated.

JANA: http://jana.fzu.cz/

Janssen, T., Chapuis, G.and M. De Boissie. 2007. *Aperiodic Crystals: From Modulated Phases to Quasicrystals. IUCr Monographs on Crystallography, no. 20.* Oxford: IUCr/Oxford University Press, ISBN 978-0-19-856777-6. doi:10.1107/S010876730 8009343.

Jayaraman, A., S. K. Sharma, Z. Wang, and S. Y. Wang. 1997. "Pressure-Induced Amorphization in the A-Phase of $Nd_2(MoO_4)_3$ and $Tb_2(MoO_4)_3$." *Solid State Communications* 101 (4): 237-241. doi:10.1016/S0038-1098(96)00587-X.

Jeitschko, W. 1972. "A Comprehensive X-Ray Study of the Ferroelectric–ferroelastic and Paraelectric–paraelastic Phases of $Gd_2(MoO_4)_3$." *Acta Crystallographica. B Structural Cryst. and Crystal Chemistry* 28 (1): 60-76. doi:10.1107/S0567740872001876.

Jeitschko, W. 1973. " Crystal structure of $La_2(MoO_4)_3$, a new ordered defect Scheelite type." *Acta Crystallographica. B: Structural Crystallography and Crystal Chemistry* 29 (10): 2433-2436. doi: 10.1107/S0567740873006138.

Jonscher, A. K. 1977. "The 'Universal' Dielectric Response." *Nature* 267 (5613): 673-679. doi:10.1038/267673a0.

Jonscher, K.A. 1983. "Dielectric Relaxation in Solids" London: *Chelsea Dielectrics Press.* ISBN: 0950871109, 9780950871103.

Kaminskii, A.A., Butashin, A., Eichler, Hans Joachim and Grebe, D., Macdonald, R., Ueda, K., Nishioka, H., Odajima, W., Tateno, M., Song, J., Musha, M., Bagaev, S.N., Pavlyuk,

A.A. 1997. "Orthorhombic Ferroelectric and Ferroelastic $Gd_2(MoO_4)_3$ Crystal — a New Many-Purposed Nonlinear and Optical Material: Efficient Multiple Stimulated Raman Scattering and CW and Tunable Second Harmonic Generation." *Optical Materials* 7 (3): 59-73. doi:10.1016/S0925-3467(97)00006-2.

Keve, E. T., Abrahams, S. C., J. L. Bernstein. 1971. " Ferroelectric Ferroelastic Paramagnetic Beta-$Gd_2(MoO_4)_3$ Crystal Structure of the Transition-Metal Molybdates and Tungstates. VI." *J. Chemical Physics* 54 (7): 3185-3194. doi: 10.1063/1.1675308.

Kim, E. S., B. S. Chun, R. Freer, and R. J. Cernik. 2010. "Effects of Packing Fraction and Bond Valence on Microwave Dielectric Properties of A 2+B 6+O 4 (A2+: Ca, Pb, Ba; B6+: Mo, W) Ceramics." *Journal of the European Ceramic Society* 30 (7): 1731-1736. doi:10.1016/j.jeurceramsoc.2009.12.018.

Kobayashi, M.; Ishii M., Usuki, Y. 1998. "Comparison of Comparison of radiation damage in different $PbWO_4$ scintillating crystal" *Nuclear Instruments and Methods in Physics Research* A, 406 (3): 442-450. doi: 10.1016/S0168-9002(98)00015-1.

Kudo, A., K. Omori, and H. Kato. 1999. "A Novel Aqueous Process for Preparation of Crystal Form-Controlled and Highly Crystalline $BiVO_4$ Powder from Layered Vanadates at Room Temperature and its Photocatalytic and Photophysical Properties." *Journal of the American Chemical Society* 121 (49): 11459-11467. doi:10.1021/ ja992541y.

Kumar, M. M. and Ye Z. G., 2005. " Scaling of conductivity spectra in the acceptor-doped ferroelectric $SrBi_2Ta_2O_9$." *Physical Review B* 72, 024104, doi:10.1103/ PhysRevB.72.024104.

Kweon, Ke and Gs Hwang. 2013. "Structural Phase-Dependent Hole Localization and Transport in Bismuth Vanadate." *Physical Review B* 87 (20): 205202. doi:10.1103/PhysRevB.87.205202.

Kweon, K. E.; Hwang, G. S.; Kim, J.; Kim, S.; Kim, S. 2015. "Electron Small Polarons and Their Transport in Bismuth Vanadate: A First Principles Study." *Physical Chemistry Chemical Physics.* 2015, 17, 256−260. doi:10.1039/C4CP03666B.

Kyoung E. K. and Gyeong S. H. 2013. "Surface Structure and Hole Localization in Bismuth Vanadate: A First Principles Study." *Applied Physics Letters* 103 (13). doi:10.1063/1.4822270.

Lacomba-Perales, R., D. Errandonea, D. Martinez-Garcia, P. Rodriguez-Hernandez, S. Radescu, A. Mujica, A. Munoz, J.C. Chervin, and A. Polian. 2009. "Phase Transitions in Wolframite-Type CdWO4 at High Pressure Studied by Raman Spectroscopy and Density-Functional Theory." *Physical Review B* 79 (9): 094105. doi:10.1103/ PhysRevB.79.094105.

Lacomba-Perales, R., D. Errandonea, A. Segura, J. Ruiz-Fuertes, P. Rodríguez-Hernández, S. Radescu, J. López-Solano, A. Mujica, A. Muñoz. 2011. "A Combined High-Pressure Experimental and Theoretical Study of the Electronic Band-Structure of Scheelite-Type AWO_4 (A = Ca, Sr, Ba, Pb) Compounds." *Journal of Applied Physics* 110 (4). doi:10.1063/1.3622322.

Lahoz, F., N. P. Sabalisck, E. Cerdeiras, L. Mestres. 2015. "Nano-to Millisecond Lifetime Luminescence Properties in $Ln_2(WO_4)_3$ (Ln = La, Ho, Tm and Eu) Microcrystalline Powders with Different Crystal Structures." *Journal of Alloys and Compounds* 649: 1253-1259. doi:10.1016/j.jallcom.2015.07.155.

Lal, H. B., N. Dar, and K. Kumar. 1974. "On the electrical conductivity, dielectric constant and magnetic susceptibility of $EuWO_4$." *Journal of Physics C: Solid State Physics* 7(23): 4335.

Lavín, V., Th. Tröster, U. R. Rodríguez-Mendoza, I. R. Martín, and V. D. Rodríguez. 2002. "Spectroscopic monitoring of the Eu^{3+} ion local structure in the pressure induced amorphization of $EuZrF_7$ polycrystal." *High Pressure Research* 22 (1): 111-114. doi:10.1080/08957950211348.

Le Bacq, O., D. Machon, D. Testemale, and A. Pasturel. 2011. "Pressure-Induced Amorphization Mechanism in $Eu_2(MoO_4)_3$." *Physical Review B* 83 (21): 214101. doi:10.1103/PhysRevB.83.214101.

Le Bail. A. 2005. "Whole powder pattern decomposition methods and applications: A retrospection" *Powder Diffraction* 20 (4): 316-326. doi: 10.1154/1.2135315.

León, C., Rivera, A., Várez, A., Sanz, J., Santamaria, J., Ngai, K. L. 2001. "Origin of constant loss in ionic conductors." *Physical review let*ters, 86 (7): 1279. doi: 10.1103/PhysRevLett.86.1279.

León, C., M. L. Lucía, J. Santamaría. 1997. "Correlated ion hopping in single-crystal yttria-stabilized zirconia." *Physical review B*. 55 (2): 882-887. doi:10.1103/PhysRevB.55.882.

Li, Guo-Ling. 2017. "First-Principles Investigation of the Surface Properties of Fergusonite-Type Monoclinic $BiVO_4$ Photocatalyst." *RSC Advances* 7 (15): 9130-9140. doi:10.1039/c6ra28006d.

Liu, C., Li, F., Ma, L. P., Cheng, H. M., 2010. "Advanced materials for energy storage." *Advanced materials*, 22(8): E28. doi:10.1002/adma.200903328.

Logvinovich, D., A. Arakcheeva, P. Pattison, S. Eliseeva, P. Tomes, I. Marozau, and G. Chapuis. 2010. "Crystal Structure and Optical and Magnetic Properties of $Pr_2(MoO_4)_3$." *Inorganic Chemistry* 49 (4): 1587. doi:10.1021/ic9019876.

Long, A. R. 1982. "Frequency-Dependent Loss in Amorphous Semiconductors." *Advances in Physics* 31 (5): 553-637. doi:10.1080/00018738200101418.

López-Solano, J., P. Rodríguez-Hernández, A. Muñoz, F. J. Manjón. 2006. "Theoretical study of the scheelite-to-fergusonite phase transition in $YLiF_4$ under pressure" *Journal of Physics and Chemistry of Solids* 67 (9–10): 2077-2082. doi: 10.1016/j.jpcs.2006.05.026.

López-Solano, J., P. Rodríguez-Hernández, S. Radescu, A. Mujica, A. Munoz, D. Errandonea, F. Manjon, J. Pellicer-Porres, N. Garro, J. Pellicer-Porres, A. Segura, Ch. Ferrer-Roca, R. S. Kumar, O. Tschauner and G. Aquilanti. 2007. "Crystal Stability and Pressure-Induced Phase Transitions in Scheelite AWO_4 (A=Ca, Sr, Ba, Pb, Eu) Binary Oxides. I: A Review of Recent Ab Initio Calculations, ADXRD, XANES, and Raman Studies." *Physica Status Solidi B* 244 (1):295-302. doi:10.1002/pssb.200672559.

Macdonald, J. R. and W.B. Johnson. 2005. "Fundamentals of Impedance Spectroscopy." *Impedance Spectroscopy*. John Wiley & Sons, Inc., pp: 1-26. doi:10.1002/047171 6243.ch1.

Maczka, M., A. G. Souza Filho, W. Paraguassu, P. T. C. Freire, J. Mendes Filho, and J. Hanuza. 2012. "Pressure-Induced Structural Phase Transitions and Amorphization in Selected Molybdates and Tungstates." *Progress in Materials Science* 57 (7): 1335-1381. doi:10.1016/j.pmatsci.2012.01.001.

Marinkovic, B. A., M. Ari, Rr de Avillez, F. Rizzo, Ff Ferreira, Kj Miller, M. B. Johnson, and M. A. White. 2009. "Correlation between AO(6) Polyhedral Distortion and Negative

Thermal Expansion in Orthorhombic $Y_2Mo_3O_{12}$ and Related Materials." *Chemistry of Materials* 21 (13): 2886-2894. doi:10.1021/cm900650c.

Martinez-Garcia, J., A. Arakcheeva, P. Pattison, V. Morozov, and G. Chapuis. 2009. "Validating the Model of a (3 + 1)-Dimensional Incommensurately Modulated Structure as Generator of a Family of Compounds for the $Eu_2(MoO_4)_3$ Scheelite Structure." *Philosophical Magazine Letters* 89 (4): 257-266. doi:10.1080/09500830902802657.

Mather, G. C., C. A. J. Fisher, M. Saiful Islam. 2010. "Defects, Dopants, and Protons in $LaNbO_4$." *Chemistry of Materials* 2010, 22 (21): 5912–5917. doi: 10.1021/cm1018822.

Méndez-Blas, A., Rico, M., Volkov, V., Zaldo, C., Cascales, C., 2007. "Crystal field analysis and emission cross sections of Ho3+ in the locally disordered single-crystal laser hosts $M^+Bi(XO_4)_2$ (M+ = Li, Na; X = W, Mo)." *Physical Review B*, 75 (17), 174208. doi: 10.1103/PhysRevB.75.174208.

Miladi, L., A. Oueslati, K. Guidara. 2016. "Phase Transition, Conduction Mechanism and Modulus Study of $KMgPO_4$ Compound." *RSC Advances* 6 (86): 83280-83287. doi:10.1039/c6ra18560f.

Moreau, J. M., R. E. Gladyshevskii, P. Galez, J. P. Peigneux, M. V. Korzhik. 1999. "A New Structural Model for Pb-Deficient $PbWO_4$." *Journal of Alloys and Compounds* 284 (1): 104-107. doi:10.1016/S0925-8388(98)00750-6.

Mormann, T. J. and W. Jeitschko. 2000. "Mercury(I) Molybdates and Tungstates: Hg_2WO_4 and Two Modifications of Hg_2MoO_4." *Inorganic Chemistry* 39 (19): 4219-4223. doi:10.1021/ic000331c.

Morozov, V. A., Arakcheeva, A. V., Chapuis, G., Guiblin, N., Rossell, M. D., Van Tendeloo, G., 2006. "$KNd(MoO_4)_2$: A new incommensurate modulated structure in the scheelite family." *Chemistry of materials* 18(17), 4075-4082. doi: 10.1021/cm0605668.

Morozov, V., Arakcheeva, A., Redkin, B., Sinitsyn, V., Khasanov, S., Kudrenko, E., Van Tendeloo, G. 2012. "$Na_{2/7}Gd_{4/7}MoO_4$: αModulated Scheelite-Type Structure and Conductivity Properties." *Inorganic chemistry,* 5 (9): 5313-5324.d: 10.1021/ic300221m.

Morozov, V. A., Bertha, A., Meert, K. W., Van Rompaey, S., Batuk, D., Martinez, G. T., Martinez, S. Van Aert, P. F. Smet, M. V. Raskina, D. Poelman, A.M. Abakumov, J. Hadermann. 2013. "Incommensurate Modulation and Luminescence in the $CaGd_{2(1-x)}Eu_{2x}(MoO_4)_{4(1-y)}(WO_4)_4$ y $(0 \leq x \leq 1, 0\leq y \leq 1)$ Red Phosphors." *Chemistry of Materials,* 25 (21), 4387-4395. doi:10.1021/cm402729r.

Morozov, V. A., Bi Lazoryak, Sz Shmurak, A. P. Kiselev, O. I. Lebedev, N. Gauquelin, J. Verbeeck, J. Hadermann, G. Van Tendeloo. 2014. "Influence of the Structure on the Properties of $Na_xEuy(MoO_4)(Z)$ Red Phosphors." *Chemistry of Materials* 26 (10): 3238-3248. doi:10.1021/cm500966g.

Murugavel, S., and M. Upadhyay. 2011. "A. C. Conduction in Amorphous Semiconductors." *Journal of the Indian Institute of Science* 91 (2): 303-317. ISSN: 0970-4140.

Mott, N. F. 1970. "Conduction in Non-Crystalline Systems IV. Anderson Localization in a Disordered Lattice." *Philosophical Magazine* 22 (175): 7-29. doi:10.1080/14786437008228147.

Mott, N. 1990 "On metal-insulator transitions." *Journal of Solid State Chemistry* 88 (1): 5-7.

Mott, N. F. 2001. "Electrons in disordered structures." *Advances in Physics*, 50 (7):865-945. doi: 10.1080/00018730110102727.

Nassau, K., J. W. Shiever, and E. T. Keve. 1971. "Structural and Phase Relationships among Trivalent Tungstates and Molybdates." *Journal of Solid State Chemistry* 3 (3): 411-419. doi:10.1016/0022-4596(71)90078-8.

Nayak, Priyambada, Tanmaya Badapanda, Anil Kumar Singh, and Simanchalo Panigrahi. 2017. "An Approach for Correlating the Structural and Electrical Properties of Zr^{4+} - Modified $SrBi_4Ti_4O_{15}$/SBT Ceramic." *RSC Advances* 7 (27): 16319-16331. doi:10.1039/c7ra00366h.

Nithiyanantham, U., Sr Ede, S. Anantharaj, S. Kundu. 2015. "Self-Assembled $NiWO_4$ Nanoparticles into Chain-Like Aggregates on DNA Scaffold with Pronounced Catalytic and Supercapacitor Activities." *Crystal Growth & Design* 15 (2): 673-686. doi:10.1021/cg501366d.

Ofelt, G. S., 1963. "Structure of the f6 Configuration with Application to Rare-Earth Ions." *The Journal of Chemical Physics* 38(9): 2171-2180. doi: 10.1063/1.1733947.

Oliveira, R. G. M., J. W. O. Bezerra, de Morais, J E V, M. A. S. Silva, J. C. Goes, M. M. Costa, and A. S. B. Sombra. 2017. "Identification of Giant Dielectric Permittivity in the $BiVO_4$." *Materials Letters* 205: 67-69. doi:10.1016/j.matlet.2017.05.105.

Orera, A. and Pr Slater. 2010. *New Chemical Systems for Solid Oxide Fuel Cells.* Vol. 22. doi:10.1021/cm902687z.

Orobengoa, D., C. Capillas, M. I. Aroyo, and J. M. Perez-Mato. 2009. "AMPLIMODES: Symmetry-mode Analysis on the Bilbao Crystallographic Server." *Journal of Applied Crystallography* 42 (5): 820-833. doi:10.1107/S0021889809028064.

Overhof, H. 1998. "Fundamental concepts in the physics of amorphous semiconductors." *Journal of non-crystalline solids*, 227, 15-22. doi.org/10.1016/S0022-3093(98)00020-9.

Padhee, R., Piyush R. Das, B. N. Parida, R. N. P. Choudhary. 2013. "Electrical and Pyroelectric Properties of Lanthanum Based Niobate." *Journal of Physics and Chemistry of Solids* 74 (2): 377-385. doi:10.1016/j.jpcs.2012.10.017.

Pang, L. X., Sun, G. B., Zhou, D. 2011. "$Ln_2Mo_3O_2$ (Ln= La, Nd): A novel group of low loss microwave dielectric ceramics with low sintering temperature." *Materials Letters*, 65 (2), 164-166. doi: 10.1016/j.matlet.2010.09.064.

Pang, L. X., Zhou, D., Qi, Z. M., & Yue, Z. X. 2017. "Influence of W substitution on crystal structure, phase evolution and microwave dielectric properties of $(Na_{0.5}Bi_{0.5})MoO_4$ ceramics with low sintering temperature." *Scientific Reports*, 7: 3201-1-6. doi:10.1038/s41598-017-03620-0.

Parida, B. N., P. R. Das, R. Padhee, R. N. P. Choudhary. 2012. "Phase Transition and Conduction Mechanism of Rare Earth Based Tungsten-Bronze Compounds." *Journal of Alloys and Compounds* 540: 267-274. doi:10.1016/j.jallcom.2012.06.077.

Parida, B. N., P. R. Das, R. Padhee, R. N. P. Choudhary. 2013. "Ferroelectric, Pyroelectric and Electrical Properties of New Tungsten–bronze Tantalate." *Current Applied Physics* 13 (9): 1880-1888. doi:10.1016/j.cap.2013.07.018.

Park, P., K. J. McDonald, K. S. Choi., 2013. "Progress in bismuth vanadate photoanodes for use in solar water oxidation." *Chemical Society Reviews*, 2013, 42(6): 2321. doi:10.1039/c2cs35260e.

Pati, B., R. N. P. Choudhary, and P. R. Das. 2013. "Phase Transition and Electrical Properties of Strontium Orthovanadate." *Journal of Alloys and Compounds* 579: 218-226. doi:10.1016/j.jallcom.2013.06.050.

Pati, B., R. N. P. Choudhary, and P. R. Das. 2015. "Pyroelectric Response and Conduction Mechanism in Highly Crystallized Ferroelectric $Sr_3(VO_4)_2$ Ceramic." *Journal of Electronic Materials* 44 (1): 313-319. doi:10.1007/s11664-014-3476-8.

Peng, H.; Lany, S. 2012. "Semiconducting Transition-Metal Oxides Based on d5 Cations: Theory for MnO and Fe_2O_3." *Physical Review B: Covering Condensed Matter and Materials Physics* 85, 201202. doi: 10.1103/PhysRevB.85.201202.

Petrov, A. and P. Kofstad. 1979. "Electrical Conductivity of $CaMoO_4$." *Journal of Solid State Chemistry* 30 (1): 83-88. doi:10.1016/0022-4596(79)90133-6.

Punia, R., R. S. Kundu, M. Dult, S. Murugavel, and N. Kishore. 2012. "Temperature and Frequency Dependent Conductivity of Bismuth Zinc Vanadate Semiconducting Glassy System." *Journal of Applied Physics* 112 (8). doi:10.1063/1.4759356.

Rao, K. S., D. M. Prasad, P. M. Krishna, and J. H. Lee. 2008. "Synthesis, Electrical and Electromechanical Properties of a Tungsten-Bronze Ceramic Oxide: $Pb_{0.68}K_{0.64}Nb_2O_6$." *Physica B: Physics of Condensed Matter* 403 (12): 2079-2087. doi:10.1016/j.physb.2007.11.031.

Rettie, Aje, W. Chemelewski, D. Emin, C. Mullins. 2016. "Unravelling Small-Polaron Transport in Metal Oxide Photoelectrodes." *Journal of Physical Chemistry Letters* 7 (3): 471-479. doi:10.1021/acs.jpclett.5b02143.

Rettie, A. J. E., W. D. Chemelewski, J. Lindemuth, J. S. Mccloy, L. G. Marshall, J. Zhou, D. Emin, and C. B. Mullins. 2015. "Anisotropic Small-Polaron Hopping in W:BiVO4 Single Crystals." *Applied Physics Letters* 106 (2): 022106. doi:10.1063/1.4905786.

Röntgen, W. C. 1896. "On a new kind of rays." *Science*, 3 (59): 227-231. doi: 10.1126/science.3.59.227.

Sabalisck, N. P., J. López Solano, C. Guzmán-Afonso, D. Santamaría Pérez, C. González-Silgo, A. Mújica, A. Muñoz, P. Rodríguez-Hernández, S. Radescu, X. Vendrell, L. Mestres, J. A. Sans, and F. J. Manjón. 2014. "Effect of Pressure on $La_2(WO_4)_3$ with α Modulated Scheelite-Type Structure." *Physical Review B* 89: 174112. doi:10.1103/PhysRevB.89.174112.

Sabalisck, N.P., Lahoz, F., González-Silgo, C., Padilla, J.D., Cerdeiras Montero, E., Mestres, L. 2017a. "Control of the luminescent properties of $Eu_{2-x}Dy_x(WO_4)_3$ solid solutions for scintillator applications." *Journal of Alloys and Compounds*. 726. 796-802. doi: 10.1016/j.jallcom.2017.07.283.

Sabalisck, N. P., Guzmán-Afonso, C., González-Silgo, C., Torres, M. E., Pasán, J., del-Castillo, J., Ramos-Hernández, D., Hernaández-Suárez, A. and Mestres, L., 2017b. "Structures and thermal stability of the α-$LiNH_4SO_4$ polytypes doped with Er^{3+} and Yb^{3+}." *Acta Crystallographica. B: Structural Science, Crystal Engineering and Materials*, 73 (1), 122-133. doi: 10.1107/S2052520616019028.

Schirmer, O. F. 2011. "Holes Bound as Small Polarons to Acceptor Defects in Oxide Materials: Why are their Thermal Ionization Energies so High?" *Journal of Physics Condensed Matter* 23 (33): 334218. doi:10.1088/0953-8984/23/33/334218.

Schreiber, M. 2008. Aperiodic Crystals: From Modulated Phases to Quasicrystals. by T. Janssen, G. Chapuis, and M. De Boissieu. *IUCr Monographs on Crystallography*, no. 20. Oxford: IUCr/Oxford University Press, 2007. Pp. 466. Price GBP 75.00. ISBN 978-0-19-856777-6. (Book Review). Vol. 64. doi:10.1107/S0108767308009343.

Sekar, C., R. K. Selvan, S. T. Senthilkumar, B. Senthilkumar, and C. Sanjeeviraja. 2012. "Combustion Synthesis and Characterization of Spherical α-MnMoO$_4$ Nanoparticles." *Powder Technology* 215–216: 98-103. doi:10.1016/j.powtec.2011.09.016.

Shanmugapriya, S., S. Surendran, V. D. Nithya, P. Saravanan, and R. Kalai Selvan. 2016. "Temperature Dependent Electrical and Magnetic Properties of CoWO$_4$ Nanoparticles Synthesized by Sonochemical Method." *Materials Science & Engineering B* 214: 57-67. doi:10.1016/j.mseb.2016.09.002.

Sharp, I. D., J. K. Cooper, F. M. Toma, and R. Buonsanti. 2017. "Bismuth Vanadate as a Platform for Accelerating Discovery and Development of Complex Transition-Metal Oxide Photoanodes." *ACS Energy Letters* 2 (1): 139-150. doi:10.1021/acsenergylett. 6b00586.

Shoko, E., M. Smith, and R. H. Mckenzie. 2009. "Mixed Valency in Cerium Oxide Crystallographic Phases: Valence of Different Cerium Sites by the Bond Valence Method." *Physical Review B* 79 (13): 4108. doi:10.1103/PhysRevB.79.134108.

Shim, H. W., In-Sun Cho, Kug Sun Hong, Ah-Hyeon Lim, Dong-Wan Kim. 2011. "Wolframite-type ZnWO4 Nanorods as New Anodes for Li-Ion Batteries." *The Journal of Physical Chemistry* 115, 16228–16233. doi:10.1021/jp204656v.

Singh, B. P., A. K. Parchur, S. B. Rai, P. Singh. 2013. *Luminescence and Electrical Behavior of Lead Molybdate Nanoparticles. AIP Conference Proceedings* 1512, 248. doi: 10.1063/1.4791004.

Singh, P., Raghvendra, O. Parkash, and D. Kumar. 2011. "Scaling of Low-Temperature Conductivity Spectra of BaSn$_{1-x}$Nb$_x$O$_3$ (X <= 0.100): Temperature and Compositional-Independent Conductivity." *Physical Review B* 84 (17): 4306. doi: 10.1103/ PhysRevB.84.174306.

Sleight, A. W., K. Aykan, and D. B. Rogers. 1975. "New Nonstoichiometric Molybdate, Tungstate, and Vanadate Catalysts with the Scheelite-Type Structure." *Journal of Solid State Chemistry* 13 (3): 231-236. doi:10.1016/0022-4596(75)90124-3.

Sleight, A. W., H. Chen, A. Ferretti, D. E. Cox. 1979. "Crystal Growth and Structure of BiVO4." *Materials Research Bulletin* 14 (12):1571-1581. doi:10.1016/0025-5408(72)90227-9.

Stoltzfus, M., P. Woodward, R. Seshadri, J. Klepeis, B. Bursten. 2007. "Structure and Bonding in SnWO$_4$, PbWO$_4$, and BiVO$_4$: Lone Pairs Vs Inert Pairs." *Inorganic Chemistry* 46 (10): 3839-3850. doi:10.1021/ic061157g.

Sumi, H. 1972. "Polaron Conductions from Band to Hopping Types." *Journal of the Physical Society of Japan* 33 (2): 327-342. doi:10.1143/JPSJ.33.327.

Sumithra, S. and A. M. Umarji. 2006. "Negative Thermal Expansion in Rare Earth Molybdates." *Solid State Sciences* 8 (12): 1453-1458. doi:10.1016/j. solidstatesciences.2006.03.010.

Sumithra, S. and A. M. Umarji. 2004. "Role of Crystal Structure on the Thermal Expansion of Ln$_2$W$_3$O$_{12}$ (Ln = La, Nd, Dy, Y, Er and Yb)." *Solid State Sciences* 6 (12): 1313-1319. doi:10.1016/j.solidstatesciences.2004.07.023.

Templeton, D. H. and A. Zalkin. 1963. "Crystal Structure of Europium Tungstate." *Acta Crystallographica.* 16 (8): 762-766. doi:10.1107/S0365110X63001985.

Terebilenko, K. V., K. L. Bychkov, V. N. Baumer, N. S. Slobodyanik, M. V. Pavliuk, A. Thapper, I. I. Tokmenko, I. M. Nasieka, and V. V. Strelchuk. 2016. "Structural Transformation of Bi$_{1-x/3}$ V$_{1-x}$ Mo$_x$O$_4$ Solid Solutions for Light-Driven Water Oxidation."

Journal of the Chemical Society, Dalton Transactions 45 (9): 3895-3904. doi:10.1039/c5dt04829j.

Tripathi, A. K. and H. B. Lal. 1980. "Electrical Transport in Rare-Earth Molybdates: $Gd_2(MoO_4)_3$ and $Tb_2(MoO_4)_3$." *Journal of the Physical Society of Japan* 49 (5): 1896-1901. doi:10.1143/JPSJ.49.1896.

van den Elzen, A. F. and G. D. Rieck. 1973. " The crystal structure of $Bi_2(MoO_4)_3$" *Acta Crystallographica. B: Structural Crystallography and Crystal Chemistry* 29 (11): 2074-2081. doi: 10.1107/S0567740873006795.

van Staveren, M. P. J., H. B. Brom, L. J. de Jongh. 1991. "Metal-Cluster Compounds and Universal Features of the Hopping Conductivity of Solids." *Physics Reports* 208 (1): 1-96. doi:10.1016/0370-1573(91)90013-C.

van Vliet, J. P. M., G. Blasse, and L. H. Brixner. 1988. "Luminescence Properties of Alkali Europium Double Tungstates and Molybdates $AEuM_2O_8$." *Journal of Solid State Chemistry* 76 (1): 160-166. doi:10.1016/0022-4596(88)90203-4.

Venkataraman, B. H. and K. Varma. 2004. "Frequency-Dependent Dielectric Characteristics of Ferroelectric $SrBi_2Nb_2O_9$ Ceramics." *Solid State Ionics* 167 (1-2): 197-202. doi:10.1016/j.ssi.2003.12.020.

Vijayakumar, M., G. Hirankumar, M. S. Bhuvaneswari, and S. Selvasekarapandian. 2003. "Influence of B_2O_3 Doping on Conductivity of $LiTiO_2$ Electrode Material." *Journal of Power Sources* 117 (1): 143-147. doi:10.1016/S0378-7753(03)00110-1.

Volkov, V., C. Cascales, A. Kling, C. Zaldo. 2005. "Growth, Structure, and Evaluation of Laser Properties of $LiYb(MoO_4)_2$ Single Crystal." *Chemistry of Materials* 17 (2): 291-300. doi:10.1021/cm049095k.

Walter, M. G., E. L. Warren, J. R. McKone, S. W. Boettcher, Q. Mi, E. A. Santori, and N. S. Lewis. 2010. "Solar Water Splitting Cells." *Chemical Reviews* 110 (11): 6446-6473. doi:10.1021/cr1002326.

Yamada, Y. and G. Shirane. 1969. "Neutron Scattering and Nature of the Soft Optical Phonon in $SrTiO_3$." *Journal of The Physical Society of Japan* 26 (2): 396-403. doi:10.1143/JPSJ.26.396.

Yi, G., Sun, B., Yang, F., Chen, D., Zhou, Y., Cheng, J., 2002. "Synthesis and characterization of high-efficiency nanocrystal up-conversion phosphors: ytterbium and erbium codoped lanthanum molybdate." *Chemistry of materials*, 14 (7): 2910-2914. doi:10.1021/cm0115416.

Young, R.A. 1993. *The Rietveld method* (Vol. 5). International union of crystallography.

Zaafouri, A., M. Megdiche, M. Gargouri. 2015. "Studies of Electric, Dielectric, and Conduction Mechanism by OLPT Model of $Li_4P_2O_7$." *Ionics* 21 (7): 1867-1879. doi:10.1007/s11581-015-1365-7.

Zalkin A., and D. H. Templeton. 1964. "X-Ray Diffraction Refinement of the Calcium Tungstate Structure." *The Journal of Chemical Physics* 40 (2): 501-504. doi: 10.1063/1.1725143.

Zhang, Y., Li, L., Su, H., Huang, W., Dong, X. 2015. "Binary metal oxide: advanced energy storage materials in supercapacitors." *Journal of Materials Chemistry A*, 3(1), 43-59. doi: 10.1039/C4TA04996A.

Zhou S., G. Barim, B. J. Morgan, B. C. Melot, and R. L. Brutchey. 2016. "Influence of Rotational Distortions on Li+- and Na+- Intercalation in Anti-NASICON Fe2(MoO4)3." *Chemistry of Materials* 28 (12): 4492–4500 doi: 10.1021/acs.chemmater.6b01806.

Zverev, P.G. 2012 "Tungstate and Molybdate Crystals for Raman Lasers, Paper presented at *the International Photonics and Optoelectronics Meetings, OSA Technical Digest*, 2012, Wuhan China, November 1–2. doi.org/10.1364/LTST.2012.MTh5A.4.

In: Polarons: Recent Progress and Perspectives
Editor: Amel Laref

ISBN: 978-1-53613-935-8
© 2018 Nova Science Publishers, Inc.

Chapter 9

UNRAVELLING THE EFFECTS OF POLARON CONDUCTION ON MIXED CONDUCTIVITY GLASSES

Marisa A. Frechero, Evangelina C. Cardillo, Pablo di Prátula, Soledad Terny, Luis A. Hernandez García, Mariela E. Sola and Magalí C. Molina*

INQUISUR-Departamento de Química, Universidad Nacional del Sur
(UNS), Bahía Blanca, Argentina

ABSTRACT

Oxide glasses are the oldest ever-known glasses. Probably, silicate glasses are the most ancient material of industrial interest. However, glasses formed by another kind of oxides are of uppermost importance because of the special properties which they are able to develop for many applications in electronics, optics, biology, etc. Every oxide that originates a three-dimensional network built by corner connected oxygen polyhedral - with a particular coordination number- conforms a glassy matrix of singular properties. Such properties can be modified by the incorporation of other oxides. In the present chapter, it is analyzed *polaron* conductivity in the presence of large ion concentrations where the ionic conductivity is not negligible. The incorporation of alkaline metal oxides in non-conventional oxide glasses enlarges the number of mobile cations, which affects the transportation of the polaron (given by the presence of transition metal oxides), through the glassy matrix.

INTRODUCTION

Oxide glasses are the oldest ever-known glasses. Probably, silicate glasses are the most ancient material of industrial interest. However, glasses formed by another kind of oxides are of uppermost importance because of the special properties which they are able to develop for many applications in electronics, optics, biology, etc. Every oxide that originates a three-

---

*Corresponding author: frechero@uns.edu.ar.

dimensional network built by corner connected oxygen polyhedral, with a particular coordination number, conforms a glassy matrix of singular properties. Such properties can be modified by the incorporation of other oxides. In general, each oxide has a specific function, i.e., it could behave as a glass former, a former and a modifier or just a modifier. A huge number of glass compositions can be prepared by different methods which also have an influence on the glass properties.

As we are particularly interested in the electrical properties of glasses, in the present chapter, we analyze polaron conductivity in the presence of large ion concentrations, i.e., when the ionic conductivity is not negligible. We analyze the electrical response of uncommon oxide glasses like TeO_2, P_2O_5, B_2O_3 and Bi_2O_3 into which several transition metal oxides like V_2O_5, Cu_2O, Ag_2O, Nb_2O_5 and MoO_3 have been incorporated in a single or mixed in presence of some modifier glasses (Li_2O, Na_2O, BaO, MgO, SrO); the most important difference among them is that they are univalent or divalent cations.

Towards the end of the second part of the last century, the appearance of materials with low charge carrier mobility (<0.1 cm/V.s) represented a challenging problem because the classical transport theory was not able to explain such behavior. In 1959, Holstein proposed that an electron could be trapped and it would not be capable of moving unless the lattice -in where the electron is- moves along with it; thus, introducing the polaron concept. That charge carrier (the electron, for example) induces a dipole moment on its surrounding and as a consequence, a new entity, the polaron, is born. Then, polaron conductivity results from the displacement of polarons in a material. During its transport, the polaron has to displace through the material. If the material is a glass, it has to displace through the glassy matrix, i.e., the charge carrier and the distortion in its surrounding. Therefore, we also analyze here, how the presence of large number of mobile cations affects the polaron transportation originated by the existence of transition metal oxides incorporated in non conventional oxide glasses.

Theory

When an electron (or a hole) is associated to the polarization that it induces to its surrounding a quasi-particle called *polaron* appears. Such concept is of great interest because it describes the physical phenomenon behind the charge carriers in solid polarized materials. Polaron concept was proposed by Lev Landau in 1933 [1] to describe an electron moving in a dielectric crystalwhere the atoms move from their equilibrium positions to effectively screen the charge of an electron, known as a phonon cloud. The polaron is characterized by its binding energy, effective mass and by its response to external electric and magnetic fields (e.g.,mobility and impedance). In 1951 Landau and Pekar studied the self-energy and the effective mass of the polaron and in 1954 Fröhlich showed that it corresponds to the adiabatic or strong-coupling regime [2].

Many oxide glasses exhibit electronic conduction through a polaronic conduction mechanism [3]. The conductivity behavior results from the metal ions being in two different oxidation states and the conduction mechanism is due to the thermally activated polaron hopping from $M^{reduced+}$ to $M^{oxidated+}$ sites. As a result, the polaron transport depends on the metal oxide content and the total number of metal ions in different oxidation states.

Transport properties of polar and ionic solids are influenced by the polaron coupling. By intuition, it is expected that the mobility of polarons will be inversely proportional to the number of real phonons present. A polaron can be bound to a charged vacancy or to a charged interstitial.

In 1996 Bazan et al., [4] showed a confirmation of the Kraevsky's earlier findings [5], and additionally, advanced on the hypothesis of the interactions between polarons and mobile cations as the main origin of anomaly of the conductivity behavior considering that the mobile electrons or polarons are attracted to the oppositely charged mobile cations. Therefore, such cation-polaron pairs move together as neutral entities and they do not transport electric charge. Consequently, those entities do not contribute to the electrical conductivity. Eventually, when the concentration of mobile cations was high enough, the ionic conductivity could be dominating the electrical response. The authors in that work explained a comparison between the ion-polaron effect and the better known mixed alkali effect, where the ion-polaron effect involves a conductivity minimum but the maxima in both activation energies (E_a) and the pre-exponential factors of the conductivity Arrhenius expression (σ_0) are much less in evidence. The previous explanation was made under the consideration that polarons and cations stay in different positions within the structure and there is not a mismatch effect between their sites. The ion–polaron interaction is responsible for decreasing the electrical conductivity in mixed ionic–electronic conductors because slowing down both electronic and ionic displacement due to Coulomb interactions between them. A gradual change in the electrical conductivity response -when a glass composition is modified- allows understanding the relationship between polaronic and ionic contribution to the whole conductivity in glassy mixed conductors [6].

The interaction of a localized electron (hole) with glassy matrix vibrations incites a charge carrier to jump from one position to an appropriate neighboring one. This mechanism known as *hopping* was explained in 1962 [7] as a sequence in which a polaron appears and then disappears and at sufficiently high temperature. The time between consecutive jumps Δt satisfies: $t_0 \ll \Delta t \ll t_p$, where $t_0 \sim \hbar/[(W_H k_B T)^{1/2}]$ (with W_H, the thermal activation energy for hopping) denotes the jump-over time and t_p is the tunneling time. Hence, an electron remains most of the time at a site, suffering a hopping transition from site to site rather rarely, but on average earlier than a tunneling occurrence. Therefore, polarons -at sufficiently high temperature- are characterized by diffusive motion.

Holstein [8], in1959, gave a quantitative treatment of small polarons through the so-called molecular crystal model. The main point in his model is: a unidimensional chain is considered with N diatomic molecules in which an excess electron is moving. With this model, the occurrence of two regimes, is established theoretically: a) hopping induced by phonons and b) Bloch-type band motion. An important role in *small-polaron* theory belongs to the distinction between adiabatic and non-adiabatic hopping transitions [roughly speaking, the adiabatic regime is characterized by the fact that the electron follows the ionic motion instantaneously]. The formation of small bipolarons by coupling of electrons to acoustic phonons in disordered media was examined by Cohen [9]. The study of small-polaron properties has been extended and carried out in depth by Mott [10] who identified and analyzed many instances of small-polaron transport including variable-range hopping, in which electrons hop over a range of distances and not only between the nearest neighbouring sites. The coherence and dynamics of small polarons in presence of disorder were represented in terms of two characteristic energies: the polaron bandwidth and the bare electron bandwidth. The first one specifies the energy scale of disorder at which the polarons become localized as composite particles, while the second one defines the energy scale at which the polaron ceases to be a composite particle [2].

Therefore, it is necessary to understand that in the polaronic transport theory there are two relevant length scales: one, is the average distance between polaron probable sites, **R**, and the other is the polaron radius itself, $\mathbf{r_p}$, i.e., the radius of the polaron entity which is associated to the range of the polarization field experienced by the electron as it jumps from $M^{reduced+}$ to $M^{oxidezed+}$. Experimental determination of the polaron radius in glasses is quite complicated and it is usually estimated through the Bogomolov equation and the Mirilin equation [11]. It is worth noticing that both of the length scales, R and r_p, are calculated based on the glass composition and density. Hence, their values are based on an average structure. In literature there are other different methods which describe the estimation of mean-squared localized displacement of ions either from the conductivity spectra or from the permittivity spectra. The polaron radius, r_p must be greater than the radius of the atom, on which the electron is localized, but less than the distance, R, separating these sites [12] and ref. therein.

RESULTS AND ANALYSIS

In the present chapter we make a comparative analysis relative to a set of glasses with polaronic conductivity due to the presence of a metal transition modifier oxide. Some of those glasses have also incorporated alkaline or alkaline earth oxides and when the concentration of them increases, the studied glasses become mixed conductors (polaronic + ionic). Particularly, we compare only tellurite glasses in order to avoid other possible contributions from the matrix (glass making oxide).

The analyzed systems in the present chapter are:

Short denomination	Glass Formula
Mg	xMgO (1-x)[0.5V$_2$O$_5$ 0.5MoO$_3$] 2TeO$_2$
Cu	xCu$_2$O (1-x)[0.5V$_2$O$_5$ 0.5MoO$_3$] 2TeO$_2$
Na	xNa$_2$O (1-x)[0.5V$_2$O$_5$ 0.5MoO$_3$] 2TeO$_2$
Ba-Mg	0.8[xBaO (1-x)MgO] 0.2Nb$_2$O$_5$ 2TeO$_2$

As it is mentioned above, polaron hopping theory [8] was originally used to explain transport in doped or undoped semiconductors where electrons occupying hydrogenic orbitals with wave function: $\varphi = \varphi_0 \exp(-2\alpha R)$ are localized by a landscape potential generated by the presence of a modifier incorporated in a glassy matrix. There is a competition between the potential energy difference and the distance that electrons are allowed to hop. This is reflected in the expression of the hopping rate (η) to a site at a distance R where the energy of the carrier is ΔW higher than that at the origin: $\eta = \eta_0.exp(-2\alpha R).exp(-\Delta W/kT)$. H is proportional to the electrical conductivity (σ) [13].

The number of transition metal ions per unit volume present in the glassy matrix allows estimating the ion–ion separation which influences directly the conduction mechanism. Taking into account that the factor $exp(-2R\alpha)$ in the equation of the hopping rate previously presented is the overlap integral, when $\alpha \sim 0$ it is possible to assume that the hopping mechanism is adiabatic and the semiconducting behaviour is mostly controlled by the activation energy, otherwise the hopping conduction mechanism is non-adiabatic ($\alpha \neq 0$) and $\sigma_{dc} \sim \sigma_0 exp(-W/kT)$.

Therefore, we have to consider that in the theory of polaronic transport there are those two significant characteristic lengths: one, the average distance between potential polaron sites, given by the metal transition ions in glass R, and the second, the polaron radius, r_p, the radius of the polaron region which is an estimation of the localized length of the polarization field experienced by the electron as it hops from M^{red+} to M^{oxid+} ion. In the following chapter we revise a way to estimate those and we analyze the conductivity response through their behaviour as function of the vanadium ion content in our tellurite glasses. Both lengths, R and r_p, are calculated based on the glass composition and density applying the following equations:

N=number of metal transition ions/ molar volume $\hspace{2cm}$ (1)

metal transition ions = vanadium or niobium ions in the present systems (explained below in the text).

$$R= N^{-1/3} \hspace{6cm} (2)$$

$$r_p= 0.5 \ (\pi/6)^{1/3} R \hspace{5cm} (3)$$

Table I. Molar Volume [cm3]

x	Mg	Cu	Na	Ba-Mg
0.1	22.90	33.86	33.79	26.29
0.2	22.87	33.03	32.60	27.01
0.3	22.80	32.06	32.88	27.61
0.4	23.43	31.46	31.27	28.19
0.5	23.62	30.96	30.77	29.22

Table II. R [nm]

x	Mg	Cu	Na	Ba-Mg
0.1	0.50	0.57	0.57	0.70
0.2	0.52	0.59	0.59	0.71
0.3	0.55	0.61	0.61	0.72
0.4	0.58	0.64	0.64	0.72
0.5	0.62	0.68	0.67	0.73

Table III. W [eV], values of W estimated from experimental conductivity data, i.e., the slope of Arrhenius plot presented in figure 2

x	Mg	Cu	Na	Ba-Mg
0.1	0.52	0.47	0.49	1.35
0.2	0.54	0.48	0.49	1.34
0.3	0.55	0.53	0.64	1.30
0.4	0.61	0.57	0.72	1.27
0.5	0.64	0.62	1.03	1.24

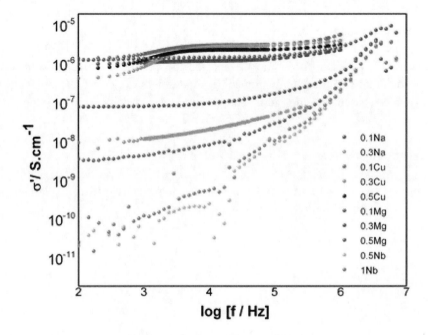

Figure 1. Real part conductivity spectra of the set of glasses presented in this chapter, at around 180°C.

The tellurite glasses that are presented here have two possible electric carriers: electrons (polarons) and/or ions according to the type and concentration of the modifier oxide incorporated. Taking into account that vanadium oxide is present in three of the four glassy matrices analyzed here and, that the vanadium cations usually appear in different (mixed) oxidation states, $V^{reduced}$ and $V^{oxidized}$, the conduction phenomenon is considered through an electron transfer between those available centers [14, 15, 16, 17]. As the electron travel inside the matrix the distortion emerges, and it moves joint to the electron as a quasi particle that we call polaron and the dc conductivity in the system is expressed as [18]:

$$\sigma = c.(1-c).N.\frac{e^2 R^2 v_0}{6kT}.\exp[-2\alpha R].\exp\left[-\frac{W}{k.T}\right] \quad (4)$$

where c and (1-c) are the occupied and available sites ($V^{reduced}$ and $V^{oxidized}$) of the total concentration N; R is the average distance between two adjacent V with different oxidation state; v_0 is the phonon frequency; α is the tunneling factor; W is the activation energy; k_B and T are the Boltzmann constant and the absolute temperature.

Temperature dependence of the conductivity is expressed as:

$$\sigma.T = \sigma_0 .\exp\left[-\frac{W}{k.T}\right] \quad (5)$$

Figures 2 *a-d* show the variation of log (σ.T) vs. T^{-1} of every system that we are analyzing.

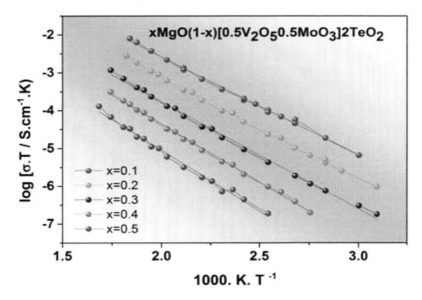

Figure 2a. Temperature dependence of dc conductivity of systems: xMgO (1- x) (0.5V$_2$O$_5$0.5MoO$_3$) 2TeO$_2$.

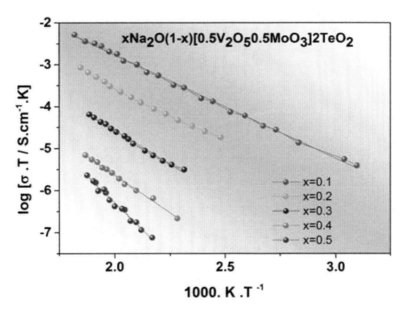

Figure 2b. Temperature dependence of dc conductivity of systems: xNa$_2$O (1- x) (0.5V$_2$O$_5$0.5MoO$_3$) 2TeO$_2$.

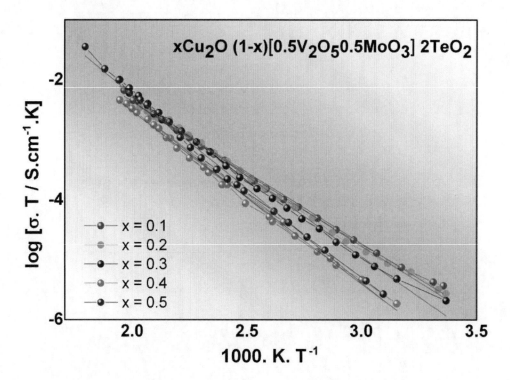

Figure 2c. Temperature dependence of dc conductivity of systems: $x\text{Cu}_2\text{O}$ $(1-x)$ $(0.5\text{V}_2\text{O}_5 0.5\text{MoO}_3)$ 2TeO_2.

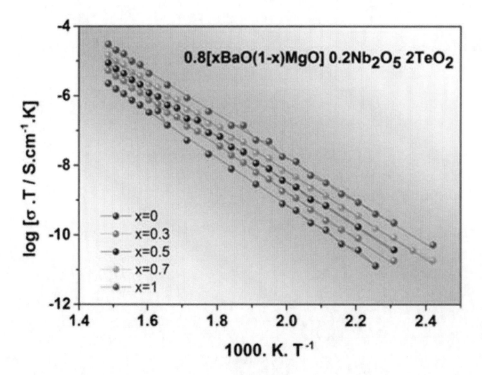

Figure 2d. Temperature dependence of dc conductivity of systems: $0.8[x\text{BaO} (1-x)\text{MgO}]$ 0.2 $(0.5\text{V}_2\text{O}_5 0.5\text{MoO}_3)$ 2TeO_2.

The W in Eq. 4 can be considered as the potential barrier that the electron has to overwhelm because of the phonon vibrations in order to move from the V^{red} to V^{oxid} site [19]. We see that as the vanadium oxide concentration decreases, a rise of such barrier appears because of the extra difficulty for the charge carrier due to the increase in R. The electrical conductivity of those systems has been studied previously [17, 20, 21]. That works show that the conductivity continuously decreases and the activation energy increases when the V_2O_5 content decreases (R increases) and similarly, if vanadium oxide is replaced by Nb_2O_5 and the modifier oxides are MgO and/or BaO these are responsible for the variation of the R value.

It is possible to consider that only the vanadium ions are involved in the process of conduction because the incorporation of molybdenum ions do not involve a higher energy barrier as Frechero et al., have shown through FTIR. Such work gives experimental evidenceof that molybdenum ions are homogeneously mixed in the tellurite glassy matrix meanwhile vanadium ionsare clearly distinguished in the spectra through its own bands [22]. As the V_2O_5 concentration decreases, the average distance between two vanadium atoms increases and therefore the electrical conductivity diminishes. Additionally, the ratio V^{red}/V^{oxid} is something that is conditioned by its composition because it depends strongly on its redox environment [23, 24].

Figure 3 shows a significant reduction in the conductivity when x of the V_2O_5 is replaced by MgO or Na_2O; but, this behavior is almost imperceptible when V_2O_5 is replaced by Cu_2O. In the three systems compared in figure 3, the diminution of the content of V_2O_5 is the same. Nevertheless, such reduction in the content of vanadium oxide provokes a similar change on the V^{red}- V^{oxid} average distance as we can see in figure 4. However, the replacement of V_2O_5 by Cu_2O does not involve a change on the conductivity values (the conductivity remains almost constant) as if the polaron hopping would not be affected by the mix type, weather it was V_2O_5 or Cu_2O, just a path of reduced/oxidized sites is available for the polaron to hop.

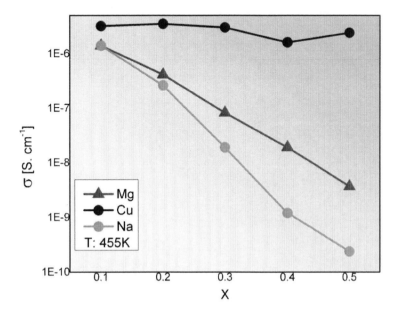

Figure 3. Isotherm of conductivity at around 180°C as afunction of modifier oxide content (x) of systems: xMgO (1- x) 0.5$V_2O_5$0.5MoO_3) 2TeO_2; xCu_2O (1- x) (0.5V_2O_5 0.5MoO_3) 2TeO_2 and xNa_2O (1- x) (0.5V_2O_5 0.5MoO_3) 2TeO_2.

On the other hand, when MgO replaces V_2O_5, the conductivity value diminishes due to the diminution of polaron concentration (the V_2O_5 is getting lower) but also, the polaron feels a very different environment in presence of the MgO incorporated in the glassy matrix. Furthermore, such effect is even stronger when Na_2O replaces V_2O_5. The $V^{red} \rightarrow V^{oxid}$ path seems to be harshly interrupted, in addition, the ionic conductivity due to the presence of free mobile Na cations causes a stronger perturbation on the polaron. Therefore, the mixed conductivity (ionic + polaronic) explains the changes of the W values showed in figure 4 and how those conductivity values are hardly affected.

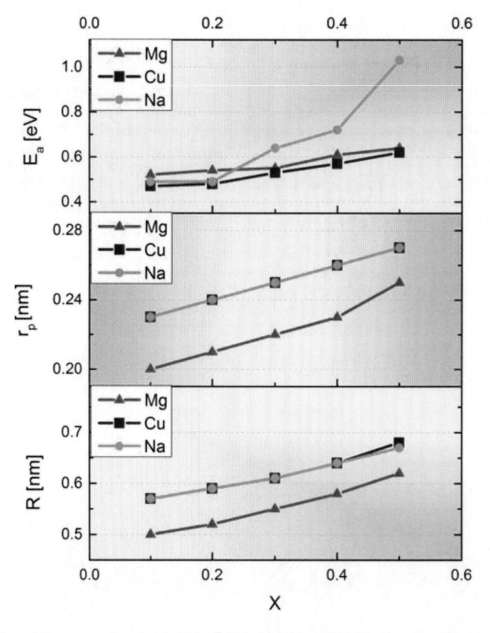

Figure 4. Temperature dependence of dc conductivity of systems: xMgO (1- x) (0.5V_2O_5 0.5MoO_3) 2TeO_2.

Unravelling the Effects of Polaron Conduction on Mixed Conductivity Glasses 375

The interpretation of the electrical conduction phenomenon plotted in figure 4 is explained through the analysis of the pre-exponential factor in the Eq. 3 in a comparative way, as follows:

In the Eq. 3 the displacement of one electron from one site to an adjacent one when the electron is between two sites with similar energy level is considered; the phonon scattering gives the temporary equalization of two adjacent positions, additionally, the overlapping of the wave function is described by the exp(-2αR) factor. This expression can be analyzed as composed by three fundamental factors [25]:

$$\sigma_0 = c.(1-c).N.\frac{e^2 R^2 v_0}{6kT}.\exp[-2\alpha R]$$

(6)

(a) probability of having a donor and an acceptor sites in adjacent position: c. (1-c)
(b) probability that two adjacent sites have the equivalent energy: $v_0.exp$ [$W_e + W_D/2kT$]
(c) probability for an electron to hop from site to site: $exp(-2\alpha R)$, i.e., the tunneling factor

Therefore, if we compare the conductivity results for each x value presented in figure 4, while taking into account the results from the figures 2 and the parameters presented in the tables above, is easy to understand that:

$$\frac{\sigma_{0Mg}}{\sigma_{0Na}} \rangle 1 \Rightarrow \left\{ \left(\ln N_{Mg} + 2\ln R_{Mg} - 2\alpha_{Mg} R_{Mg} \right) - \left(\ln N_{Na} + 2\ln R_{Na} - 2\alpha_{Na} R_{Na} \right) \right\}$$

And considering that:

$$\left[-2\alpha_{Mg} R_{Mg} + 2\alpha_{Na} R_{Na} \right] \cong 0$$
$$\left[2\ln R_{Mg} - 2\ln R_{Na} \right] \langle 0$$
$$\left[\ln N_{Mg} - \ln N_{Na} \right] \Rightarrow \rangle\rangle 1$$

It results that is possible to assume that: $N_{Mg} \gg N_{Na}$
And, in the same way: $N_{Cu} \gg N_{Mg}$
This is not a minor result.

Before finishing the analysis presented here, it is worth to mention that the original theory developed by Anderson in 1958 considered that a random small non-periodic potential affected the migration of the electron inside the system [x]. Such model condition is not strictly satisfied as x in our systems is increased (i.e., when the modifier oxide concentration is large in comparison to the concentration of polarons). Nevertheless, it is possible to analyze with enough success the results presented here, except for the system $0.8[xBaO(1-x)MgO]$ $0.2Nb_2O_5$ $2TeO_2$ and $0.5Na_2O$ $0.5[0.5V_2O_5$ $0.5MoO_3]$ $2TeO_2$, which clearly does not satisfy such condition.

CONCLUSION

The analysis that we have made here reveals that the entity originated from the interaction between a charge carrier (electron) and the dipole moment induced on surrounding, called polaron, involves a conductivity phenomenon in a material. During the charge transport process, the polaron displaces through the material. When the material is a tellurite glass, the charge carrier and its distortion in its surrounding are strongly affected by the presence of free mobile ions in the glassy matrix. Therefore, the presence of a large number of mobile cations affects the polaron migration. The model condition of the original theory developed by Anderson is not strictly satisfied. The presence of a not small perturbation given by a free cation -contrary to what the model assumes- provokes a large effect on the non-periodic potential. Additionally, it was possible to evidence the existence of ion-polaron entity in modified tellurite glassy matrices.

REFERENCES

[1]   Landau L. D. Motion of Electrons in Crystal Lattice. *Phys. Z. Sowjetunion* 3, 1933, 664.

[2]   Devreese, J. T.. *arXiv*: cond-mat/0004497v2 [cond-mat.str-el] 2000.

[3]   Sayer, M. and Mansingh A., *Phys. Rev. B*: Solid State, 1972, 6, 4629.L. Murawski, C. H. Chung and J. D. Mackenzie, *J. Non-Cryst. Solids*, 1979, 32, 91.N. F. Mott, *J. Non-Cryst. Solids*, 1968, 1, 1. A. Mogus-Milanković, D. E. Day and B. Santic´, *Phys. Chem. Glasses*, 1999, 40, 69.A.S˘antic´ and A. Mogus-Milanković, *Croat. Chem.* Acta, 2012, 85, 245.

[4]   Bazan, J. C. Duffy, J. A. Ingramb, M. D., M. R. Mallace. *Solid State Ionics* 86-88, 1996, 497-501.

[5]   Kraevski, S. L. Evdokimov, T. F., Solonov, U. F. and E. Shishmentseva, *Fiz. Khim. Stekla* 4, 1978, 366.

[6]   Mogus-Milankovic, A. A. Santic, S. T. Reis, K. Furic, D. E. Day. *Journal of Non-Crystalline Solids* 342, 2004, 97–109.

[7]   Lang, I. G., Firsov, Y. A.: Zh. Eksp. *Teor. Fiz.* 43, 1962, 1843.

[8]   Holstein, T., *Ann. Phys.* (USA), 8, 1959, 343 - 389.

[9]   Cohen, M. H., Economou, E. N., Soukoulis C. M. *Phys. Rev.* B29, 4496 – 4499, 1984, 4500 - 4504.

[10] Mott, N. F. Davis. E. A., Electronic Processes in *Non-Crystalline Materials Clarendon- Press*, Oxford 1979. ISSN 978 019 964533-6.

[11] Bogomolov, V. N. Mirilin, D. N., *Phys. Status Solid B*, 1968, 27, 443.

[12] Santić, A. Banhatti, R. D. L. Pavić, H. Ertap, M. Yüksek, M. Karabulut, A. Mogus-Milanković. *Phys. Chemistry Chemical Phys.*, 2017, 19, 3999.

[13] Banerjee, S. Pal, Rozenberg, E., Chaudhuri, B. K. *J. Phys. Condens. Matter* 13,2001 9489–9504

[14] Austin, I. G., Mott, N. F., *Adv. Phys.* 18, 1969, 41e102.

[15] Vijaya Prakash, G. Narayana Rao, D., A. K. Bhatnagar, *Solid State Commun.* 1192001, 39e44.

[16] Szu, S., Shing Gwo Lu. *Phys.* B 391, 2007, 231e237.

[17] di Prátula, P. E. Terny, S., Cardillo, E. C., M. A. Frechero, *Solid State Sciences* 49, 2015, 83e89.

[18] Ghosh, A. *J. Appl. Phys.* 74, 1993, 3961e3965.

[19] L. Murawski, C. Chung, J. Mackenzie, *J. Non Cryst. Solids* 32, 1979, 91e96.

[20] Terny, S. di Prátula, P. E., J. De Frutos, M. A. Frechero. *Journal of Non-Crystalline Solids* 444, 2016, 49-54.

[21] Terny, S. De la Rubia, M. A., J. De Frutos, M. A. Frechero. *Journal of Non-Crystalline Solids* 433, 2016, 68-74.

[22] Frechero, M. A., Quinzani, O., R. Pettigrosso, M. Villar, R. Montani, *Journal of Non-Crystalline Solids* 353, 2007, 2919e2925.

[23] Das, B. B., Mohanty D., *Indian J. Chem.* 45, 2006, 2400e2405.

[24] Sen, S. Ghosh, A. *J. Appl. Phys.* 87 (7), 2000, 3355e3359.

[25] Lebrun, N. Levy, M., J.L. Souquet, *Solid State Ionics* 40/41, 1990, 718e722.

In: Polarons: Recent Progress and Perspectives
Editor: Amel Laref

ISBN: 978-1-53613-935-8
© 2018 Nova Science Publishers, Inc.

Chapter 10

SMALL POLARON HOPPING CONDUCTION MECHANISM IN V_2O_5-BASED GLASS-CERAMIC NANOCOMPOSITES

M. M. El-Desoky[1,], M. S. Ayoub[1], A. E. Harby[1] and A. M. Al-Syadi[1,2]*

[1]Physics Department, Faculty of Science, Suez University, Suez, Egypt
[2]Physics Department, Faculty of Education, Ibb University, Ibb, Yemen

ABSTRACT

Three systems of V_2O_5-based glasses, $BaTiO_3$-V_2O_5, $BaTiO_3$-V_2O_5-Bi_2O_3 and $SrTiO_3$-PbO_2-V_2O_5, were studied before and after nanocrystallization. Nanostructural behavior and electrical properties of these glass systems and their corresponding glass–ceramic nanocomposites were studied. Differential scanning calorimeter (DSC) and the overall features of X-ray diffraction (XRD) of quenched glasses confirmed the amorphous nature of the present glass systems. Transmission electron micrograph (TEM) and XRD of the present glass-ceramic nanocomposite samples indicate nanocrystals with an average particle size between 20–35 nm. It was found that density (d) increased gradually with the increase of the $BaTiO_3$ and $SrTiO_3$ content in the present glass systems and corresponding glass–ceramic nanocomposites. The glass–ceramic nanocomposites exhibited high conductivity compared to the samples in glassy phase. The enhancement of the electrical conductivity after nanocrystallization was attributed to two interdependent factors: (i) an increase of concentration of V^{4+}–V^{5+} pairs; and (ii) formation of defective, well-conducting regions along the glass–crystallites interfaces. On the other hand, the high conductivity of these glass–ceramic nanocomposites is considered to be due to the presence of nanocrystals. This is attributed to the formation of extensive and dense network of electronic conduction paths which are situated between V_2O_5 nanocrystals and on their surface. The electrical conductivity of these systems can be fitted with Mott's model of nearest neighbor hopping at high temperature. From the conductivity temperature relation, it was found that small polaron hopping (SPH) model

---

* Corresponding Author (M.M. El-Desoky): Physics Department, Faculty of Science, Suez University, Suez, Egypt. Email: mmdesoky@suezuniv.edu.eg; mmdesoky@gmail.com.

was applicable at temperature above $\theta_D/2$ (θ_D, the Debye temperature). The electrical conduction at $T > \theta_D/2$ was due to non-adiabatic SPH of electrons between vanadium ions. The various parameters such as optical phonon frequency (v_0), polaron radius (r_p), density of states ($N(E_F)$) and small polaron coupling constant (γ_p), obtained from the best fits were found to be consistent with the glasses and glass–ceramics nanocomposites.

Keywords: V_2O_5, glass–ceramic nanocomposites, dc conductivity, small polaron hopping

INTRODUCTION

Glass-ceramic nanocomposites are polycrystalline materials containing nano sized crystallites dispersed in the glass matrix formed during heat treatment process at the early stage of crystallization. The controlled crystallization process leads to the separation of a crystalline phase from the glassy parent phase in the form of tiny crystals, where the number of crystals, their growth rate, micro structure and their final size are controlled by suitable heat treatment [1, 2]. Glass ceramic nanocomposites are important because of their physical properties which are not obtainable in other classes of materials. Although there are numerous reports on the properties of nanomaterials, much smaller attention has been paid to understand the properties of nanomaterials embedded within a glass matrix. Such nanocomposite materials can be obtained by crystallization of glass. Since the glass-ceramic nanocomposites could be obtained in the early stage of crystallization of glass, we must find an alternative method that could identify a very small quantity of crystals with sizes in the nanometer range [3]. Since the glass–ceramic nanocomposites could be obtained in the early stage of crystallization of glass. A number of methods that are known to be sensitive to the local change of atomic structure was attempted to identify the nanocrystallization, like X-ray diffraction (XRD), scanning electron microscopy (SEM), transmission electron micrograph (TEM), Möessbauer, infrared and Raman spectra, electrical conductivity, and dielectric properties [4-7]. Although these methods successfully detected some changes upon the nanocrystallization, they measured specimens that were quenched after the specified heat treatment. A variety of methods have been explored for preparing these material, such as the sol–gel method [8], the solid reaction method [9], the aerosol method [10] and the hydrothermal method [11]. However, each method has its limitations and disadvantages. For example, a solid reaction method requires a high temperature and a long reaction time. Further, only submicron particles or micron-sized particles can be obtained, and impurities may be introduced during their post-treatments, such as in the milling process [9].

Glasses and glass-ceramics nanocomposites in system containing V_2O_5 belong to the best electronic conductors [12-15]. They inherit some structural features and transport properties from the nanocrystalline forms of V_2O_5 and in particular of its α-phase, being often cited as a model of an electronic conductor [15]. On the other hand, in recent years it has been established that the presence of fine grains of foreign phases distributed in moderately conducting matrices (glasses, but also polymer electrolytes or polycrystalline materials) can significantly improve their conductivity [12-14, 16].

Owing to their semiconducting properties, oxide glasses and corresponding glass-ceramics nanocomposites containing large amounts of transition metal oxides (TMO) show interesting electrical properties [17-22]. This behavior is strongly influenced by the simultaneous presence in the glass network of transition metal ion in two different valance states, due to redox processes accruing in the melt at high temperatures in course of preparation. In the conduction of V_2O_5 containing glasses changes, $V^{4+} \rightarrow V^{5+}+e$ takes place between two vanadium ions in glass. The charge transfer is usually termed small polaron hopping (SPH) [23, 24]. The electrical conductivity for such glasses depends strongly upon the local interaction of an electron with its surroundings and distance between vanadium ions [12-15]. In several works [25-29], nanostructural and electrical properties of V_2O_5-based glasses and their corresponding glass-ceramic nanocomposites were studied. TEM and XRD of glass-ceramic nanocomposites of these samples were indicated nanocrystals with a particle size of 20-50 nm. Also, it was observed that the conductivity of the glass-ceramic nanocomposites is higher than that of the corresponding glassy phase. The high conductivity of these glass-ceramic nanocomposites was considered to be due to the presence of nanocrystals. This is attributed to formation of extensive and dense network of electronic conduction paths which were situated between nanocrystals and on their surface.

Strontium and barium titanate are ferroelectric materials that have a remnant polarization and piezoelectric properties. Due to these properties, they have wide range of applications in nonvolatile memory devices, sensors and actuators [27-29]. It is necessary to investigate these properties in order to understand their physical phenomenon and to have better device performance. Furthermore, glass-ceramic nanocomposites containing nanocrystalline ferroelectric phases have been of much interest because of their promising non-linear optical properties [30, 31].

In this chapter, three systems of V_2O_5-based glass and corresponding glass-ceramic nanocomposites are investigated. The first system is $BaTiO_3$-V_2O_5 [27], the second system is $BaTiO_3$–V_2O_5–Bi_2O_3 [28] and the third system is $SrTiO_3$-PbO_2-V_2O_5 [29], which will be denoted by BTV, BTVB and STPV, respectively. These glasses were prepared by the press-quenching method. The glass-ceramic nanocomposites were prepared by the controlled crystallization of glass. The first objective is to investigate the compositional dependence of the nanostructural and transport properties of V_2O_5-based glasses and corresponding glass-ceramic nanocomposites in view of Mott's SPH model. The second objective is to clarify the mechanism of electrical conduction of these glasses and the corresponding glass-ceramic nanocomposites.

STRUCTURAL BEHAVIOR

Differential Scanning Calorimeter (DSC)

Figure 1 shows DSC thermogram for BTV, BTVB and STPV glasses and the composition dependence of glass transition temperature (T_g), crystallization temperature (T_c) and temperature difference $\Delta T = T_c - T_g$ respectively, for the present glass systems. The shapes of all the curves are similar and confirm the glassy nature of all the as-received samples. Each curve exhibits an endothermic dip due to T_g and exothermic peak corresponding to the glass crystallization process (T_c).

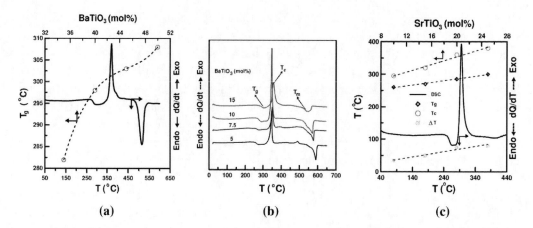

Figure 1. DSC thermogram, T_g, T_c and ΔT as a function of BaTiO$_3$ and SrTiO$_3$ content for (a) BTV (b) BTVB (c) STPV glass systems [27, 28, 29].

Figure 1 (a) shows the composition dependence of glass transition temperature (T_g) for BTV glass system and DSC thermogram. It is clear that T_g increases with increasing BaTiO$_3$ content and lies between 282°C and 308°C. Moreover, T_c, lies between 370°C and 385°C for BTV glass. It is known that the thermal stability and viscosity of a glass depends on the temperature difference ΔT of the glass [32]. The difference between T_g and T_c is about 82°C and it decreases with the increase of BaTiO$_3$ content in the glasses indicating that the stability of the glasses decreases with BaTiO$_3$ content.

Figure 1(b) shows the DSC curves of BTVB glass system. The exothermic peak in the temperature range 347-357°C correspond to the crystallization, and the peak positions are defined as crystallization peak temperature of T_c. The endothermic dip corresponding to the glass transition (T_g) are observed at 270-305°C. Also, the endothermic dip due melting temperature (T_m) are observed at 548-595°C. The difference between T_g and T_c decreases with the increase of BaTiO$_3$ content in the glasses indicating that the thermal stability of the glasses decreases with BaTiO$_3$ content. The stability of the glasses could also be predicted by the factor T_g/T_m, i.e., the ratio of the glass transition and melting temperature. For very stable glass the ideal value of T_g/T_m is 0.67 [33]. The T_g/T_m values all the compositions studied in present system fall in the range 0.49-0.51, being near to that of the ideal value.

Also from Figure 1(c), it is clear that T_g increases with increasing SrTiO$_3$ content and lies between 260°C and 300°C. Moreover, T_c lies between 295°C and 380°C for the present glass system. It is known that the thermal stability and viscosity of a glass depends on the temperature difference ΔT of the glasses [21, 22, 34-37]. The difference between T_g and T_c is about 60°C and it increases with the increase of SrTiO$_3$ content in the glasses indicating that the thermal stability of the glasses increases with SrTiO$_3$ content. Also, since these glasses are highly viscous and the difference ΔT is quite large, many shapes such as wires, tapes or thick films can be prepared from viscous metals.

Also, since these glasses are highly viscous and the difference ΔT is quite large, many shapes such as wires, tapes or thick films can be prepared from viscous metals. On the other hand, DSC studies [18] on the structure of many glasses have already showed that T_g shows an obvious correlation with the change in the coordination number of the network former and the construction of non-bridging oxygen (NBO) atoms, which means destruction of the network structure [18]. Generally, T_g shows a distinct increase when the coordination number of the network former increases. Converse to this, a construction of NBO causes a decrease into the T_g. The continuous increase in the T_g in the present system, therefore, seems to suggest continuing decrease in the coordination number of V^{5+} and V^{4+} ions and destruction of NBO atoms [18].These results are consistent with the density and oxygen molar volume data. This is ascribed to the decreased interatomic distances between metal and oxygen ions leading to increase in the density and decrease in the oxygen molar volume in the present glass system [18, 38].

X-Ray Diffraction (XRD)

XRD patterns that were obtained for the as-received as well as heat treated samples of the compositions under study are depicted in Figure 2(a, b and c), respectively. In the case of as-received samples, there is only a wide halo observed with no indication of diffraction peaks. This confirms the amorphous state initial glasses. In the case of the samples heat-treated at crystallization temperature. It contains a number of peaks corresponding to a nanocrystalline phases superimposed on a wide halo indicating that there is still substantial amount of amorphous phase. The XRD patterns of the heat-treated samples are a clear indication that annealing at temperature T_c only starts to produce small nanocrystallites in the glass matrix [13-15]. This is a valuable hint, since it means that by adjusting the temperature of annealing, one can control the amount of nanocrystalline grains formed in the material and thus enhance its electrical conductivity.

Figure 2a shows representative XRD patterns of the BTV glass and corresponding glass-ceramic nanocomposite. The overall features of Figure 2a(*a*) confirm again the amorphous nature of the present glasses. The pattern of the corresponding glass-ceramic nanocomposite (Figure 2a(*b*)) showed peaks identified to be nanocrystal $BaTiO_3$, V_2O_5, TiO_2 etc. phases. Similar results were exhibited for the other samples. The average particle sizes, D, of the precipitated nanocrystals were calculated according to Sherrer formula [39];

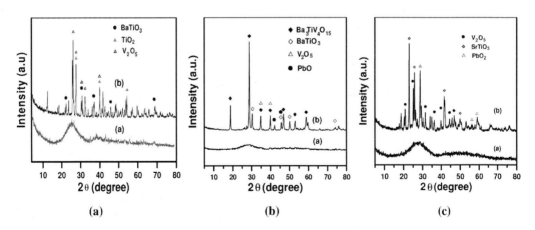

Figure 2. X-ray diffraction for representative sample of (a) BTV, (b) BTVB (c) STPV glass and corresponding glass-ceramic nanocomposite systems [27, 28, 29].

$$D = \frac{K\lambda}{\beta \cos\theta} \qquad (1)$$

where $K \sim 1$, λ (nm) represents the wavelength of CuK_α radiation, θ is the Bragg angle of the X-ray diffraction peak and β represents the corrected full width at half maximum of the diffraction peak in radians. D was found to be about 34 nm.

Moreover, Figure 2b(*a*) shows a representative XRD of the BTVB glass that confirms the amorphous nature of the present glasses. The nanocrystalline phases were recognized as $Ba_3TiV_4O_{15}$, $BaTiO_3$, V_2O_5 and Bi_2O_3 as shown in Figure 2b(*b*) for the BTVB glass. Additionally, there are some new peaks corresponding to the phase which we have not identified yet. Average size of nanocrystallites was predictable from the widths of diffraction peaks, using the Scherrer formula, to ca. 25 nm [13-15]. The average grain size of the particles of $BaTiO_3$ and V_2O_5 obtained from the X-ray line-broadening studies is found to be 20-30 nm.

Figure 2c shows representative XRD patterns of STPV glass and corresponding glass-ceramic nanocomposite. In the case of glass sample (Figure 2c(*a*)), this confirms the amorphous state of initial glasses. A pattern shown in Figure 2c(*b*) was collected for sample after its annealing at crystallization temperature T_c, determined from DSC studies. It contains a number of peaks corresponding to a nanocrystalline $SrTiO_3$, PbO_2, V_2O_5 phase. XRD pattern in Figure 2c (*b*) is a clear indication that annealing at temperature T_c only starts to produce small nanocrystallites embedded in the glass matrix [13, 14, 40]. Additionally, there are some new peaks corresponding to the phase which we have not identified yet. The average particle sizes, D, of the precipitated nanocrystals were calculated according to Sherrer formula [39], D was found to be about 32 ±3 nm for all studied samples.

High-Resolution Transmission Electron Micrograph (HR-TEM)

The HR-TEM along with the selective area electron diffraction (SAED) of BTV glass-ceramic nanocomposite is shown in Figure 3a. The particle size estimated based on these studies lies in the range 20-35 nm, dispersed in the glass matrix. The SAED pattern (inset of Figure 3a) and the lattice spacings, obtained therein confirm the presence of barium titanate crystallites.

Figure 3. TEM of representative sample for (a) BTV, (b) BTVB (c) STPV glass-ceramic nanocomposite systems [27, 28, 29].

Figure 3b shows TEM along with the SAED of the BTVB glass-ceramic nanocomposite sample. The SAED pattern (inset of Figure 3b) and the lattice spacings obtained therein confirm the presence of $Ba_3TiV_4O_{15}$ nanocrystallites. The SAED pattern recorded from this sample show a glassy halo but this may be attributed to the strong diffraction of glassy matrix masking the diffraction from crystalline nuclei [27]. Similar results were obtained on the other samples

TEM of STPV glass-ceramic nanocomposite is shown in Figure 3c. The nearly spherical crystallites of fairly uniform size laying in the range 30-35 nm, evenly dispersed in a glass matrix. These results obtained from TEM are agreed with the results obtained by XRD results.

Density and Molar Volume

The density, d, and molar volume, V_m, for the glass-ceramic nanocomposites systems are shown in Figure 4(a, b and c). These properties changed linearly as a function of glass ceramic nanocomposites. It is observed that the density increases gradually with the increase of the $BaTiO_3$ and $SrTiO_3$ content in the glass-ceramic nanocomposites for all systems. The density results show that as the barium, strontium or titanium cation concentration increases the glass ceramic structure becomes less open, allowing for the probable formation of decreasing number of non-bridging oxygen (NBO) [41]. In the BTV glass-ceramic nanocomposites system the densities vary from 3.47 up to 3.81 g/cm³. Also from Figure 4a, it is found that the molar volume decreases with increasing $BaTiO_3$ content. Here the composition dependence of molar volume gives information about the coordination state of the $BaTiO_3$ cations. The observed decrease of molar volume suggests that most vanadium cations have on the average a large coordination number of oxygen atoms [20, 42]. Moreover, in the BTVB system as shown in Figure 4b, the densities vary from 4.24 up to 4.36 g/cm³ and 4.29 up to 4.48 g/cm³ for glasses and corresponding glass-ceramic nanocomposites, respectively. Additionally, in the STPV system the densities vary from 3.87 up to 4.11 g/cm³ and 3.91 up to 4.14 g/cm³ for glasses and corresponding glass-ceramic nanocomposites, respectively. Also from Figure 4c, it is found that the molar volume decreases with increasing $SrTiO_3$ content. Here the composition dependence of molar volume gives information about the coordination state of the Ti cations.

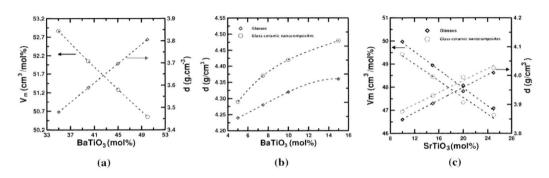

Figure 4. Composition dependence of density and molar volume (V_m) as a function of $BaTiO_3$ and $SrTiO_3$ content for (a) BTV (b) BTVB (c) STPV systems [27, 28, 29].

Dc Conductivity

Conductivity and Activation Energy

The logarithmic dc conductivity, σ, as a function of reciprocal temperature of the present glasses and corresponding glass-ceramic nanocomposites are shown in Figure 5(a, b and c) and Figure 6(a, b and c), respectively. It is observed from the figures that a linear temperature dependence up to a certain temperature ($\theta_D/2$). Such behavior arises from the hopping of electrons or polarons between mixed valance states [21, 22, 34-37]. So the experimental conductivity data above $\theta_D/2$ were fitted with the small polaron hopping (SPH) model proposed by Mott, [23, 24]. The high temperature activation energy was computed from the slope of each curve in the highest range of the temperature measured. The experimental conductivity data in such a situation is well described by activation energy for conduction given by Mott formula [23, 24].

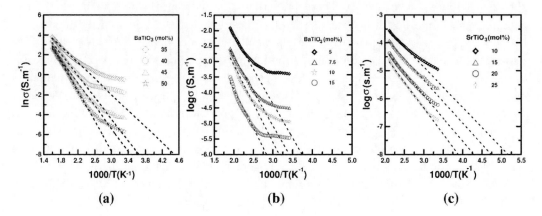

Figure 5. Temperature dependence of dc conductivity(σ) as a function of BaTiO$_3$ and SrTiO$_3$ content for (a) BTV (b) BTVB (c) STPV glass systems [27, 28, 29].

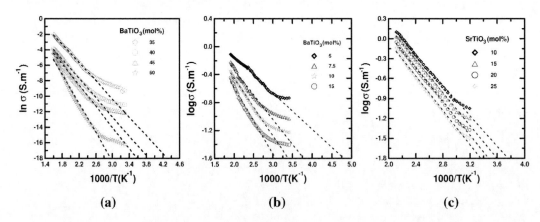

Figure 6. Temperature dependence of dc conductivity (σ) as a function of BaTiO$_3$ and SrTiO$_3$ content for (a) BTV (b) BTVB (c) STPV glass-ceramic nanocomposite systems [27, 28, 29].

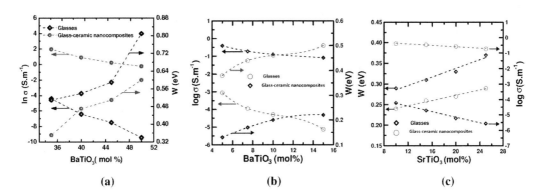

Figure 7. The relation between conductivity and activation energy as a function of BaTiO$_3$ and SrTiO$_3$ content for (a) BTV (b) BTVB (c) STPV systems [27, 28, 29].

$$\sigma = \sigma_0 \exp(-W/kT) \qquad (2)$$

where σ_0 is a pre-exponential factor, W is the activation energy and k is the Boltzmann constant. The variation of the high temperature conductivity and the high temperature activation energy of glasses and corresponding glass-ceramic nanocomposites with composition are shown in Figure 7(a, b and c). It is clear from the figure that the conductivity increases while the activation energy decreases with the increase of the BaTiO$_3$ or SrTiO$_3$ content. Such a behavior is a feature of SPH [20, 23]. This change in conductivity and activation energy may help to detect the structural changes as a consequence of increasing BaTiO$_3$ or SrTiO$_3$ and decreasing V$_2$O$_5$ content.

Table 1. Some physical and SPH parameters of BTV glasses [27]

Nominal composition (mol%) BaTiO$_3$	V$_2$O$_5$	d ± 0.02 (g.cm^{-3})	W_{glass} ± 0.01 (eV)	log σ ± 0.02 (at 500 K) (S.cm^{-1})	R ± 0.001 (nm)	N ± 0.01 (× 10^{22} cm^{-3})	r_P ± 0.001 (nm)	θ_D ± 1 (K)	ν_0 ±0.01 (× 10^{13} Hz)	$N(E_F)$ (×10^{21}) (eV^{-1}.cm^{-3})
35	65	3.64	0.51	-4.39	0.450	1.10	0.182	784	10.27	4.59
40	60	3.76	0.54	-6.44	0.447	1.12	0.180	754	9.87	4.61
45	55	3.87	0.59	-7.50	0.444	1.14	0.179	740	9.69	4.54
50	50	3.98	0.81	-9.46	0.442	1.16	0.179	715	9.41	4.45

Table 2. Some physical and SPH parameters of BTV glass-ceramic nanocomposites [27]

Nominal composition (mol%) BaTiO$_3$	V$_2$O$_5$	d ± 0.02 (g.cm^{-3})	W_{nano} ± 0.01 (eV)	log σ ± 0.02 (at 500 K) (S.cm^{-1})	R ± 0.001 (nm)	N ± 0.01 (× 10^{22} cm^{-3})	r_P ± 0.001 (nm)	θ_D ± 1 (K)	ν_0 ± 0.01 (× 10^{13} Hz)	$N(E_F)$ (×10^{21}) (eV^{-1}.cm^{-3})
35	65	3.77	0.35	1.95	0.444	1.13	0.179	850	11.14	7.79
40	60	3.88	0.47	0.9	0.443	1.15	0.178	830	10.87	5.84
45	55	3.99	0.51	0.22	0.440	1.72	0.177	740	9.69	4.79
50	50	4.11	0.60	-0.235	0.437	1.92	0.176	730	9.56	4.77

Table 3. Some physical properties and SPH parameters of BTVB glasses [28]

Nominal composition (mol%)			d ± 0.02 (g.cm^{-3})	W_{glass} ± 0.01 (eV)	$\log \sigma$ ± 0.02 (at 450 K) (S.cm^{-1})	R ± 0.001 (nm)	N ± 0.001 (× 10^{22} cm^{-3})	r_P ±0.001 (nm)	θ_D ±1 (K)	ν_0 ±0.01 (× 10^{13} Hz)	$N(E_F)$ (×10^{21}) (eV^{-1}. cm^{-3})
BaTiO$_3$	V$_2$O$_5$	Bi$_2$O$_3$									
5	75	20	4.42	0.38	-3.05	0.454	1.064	0.183	833	1.74	6.72
7.5	72.5	20	4.28	0.44	-3.95	0.452	1.079	0.182	810	1.70	5.88
10	70	20	4.32	0.46	-4.30	0.451	1.089	0.181	775	1.55	5.66
15	65	20	4.36	0.50	-5.13	0.449	1.098	0.180	743	1.48	5.28

Table 4. Some physical and SPH parameters of BTVB glass-ceramic nanocomposites [28]

Nominal composition (mol%)			d ± 0.02 (g.cm^{-3})	W_{nano} ± 0.01 (eV)	$\log \sigma$ ± 0.002 (at 450 K) (S.cm^{-1})	R ± 0.001 (nm)	N ± 0.01 (× 10^{22} cm^{-3})	r_P ± 0.001 (nm)	θ_D ± 1 (K)	ν_0 ± 0.01 (× 10^{13} Hz)	$N(E_F)$ (×10^{21}) (eV^{-1}.cm^{-3})
BaTiO$_3$	V$_2$O$_5$	Bi$_2$O$_3$									
5	75	20	4.29	0.13	-0.050	0.452	1.08	0.182	645	1.29	19.89
7.5	72.5	20	4.37	0.17	-0.085	0.450	1.10	0.180	666	1.33	15.41
10	70	20	4.42	0.20	-0.150	0.447	1.12	0.179	689	1.37	13.37
15	65	20	4.48	0.22	-0.160	0.445	1.13	0.178	714	1.43	12.32

Table 5. Some physical and SPH parameters of STPV glasses [29]

Nominal composition (mol%)			d ± 0.02 (g.cm^{-3})	W_{glass} ± 0.01 (eV)	$\log \sigma$ ± 0.02 (at 400 K) (S.cm^{-1})	R ± 0.001 (nm)	N ± 0.01 (× 10^{22} cm^{-3})	r_P ± 0.001 (nm)	θ_D ± 1 (K)	ν_0 ± 0.01 (× 10^{13} Hz)	$N(E_F)$ (×10^{21}) (eV^{-1}.cm^{-3})
SrTiO$_3$	PbO$_2$	V$_2$O$_5$									
10	20	70	3.87	0.29	-4.22	0.436	1.20	0.175	740	1.53	9.901
15	20	65	3.95	0.31	-4.72	0.433	1.23	0.174	690	1.43	9.377
20	20	60	4.04	0.33	-5.22	0.429	1.26	0.173	645	1.34	9.057
25	20	55	4.11	0.37	-5.56	0.428	1.28	0.172	625	1.30	8.135

Table 6. Some physical and SPH parameters of STPV glass-ceramic nanocomposites [29]

Nominal composition (mol%)			d ± 0.02 (g.cm^{-3})	W_{nano} ± 0.01 (eV)	$\log \sigma$ ± 0.02 (at 450 K) (S.cm^{-1})	R ± 0.001 (nm)	N ± 0.01 (× 10^{22} cm^{-3})	r_P ± 0.001 (nm)	θ_D ± 1 (K)	ν_0 ±0. 01 (× 10^{13} Hz)	$N(E_F)$ (×10^{21}) (eV^{-1}. cm^{-3})
SrTiO$_3$	PbO$_2$	V$_2$O$_5$									
10	20	70	3.91	0.24	-0.39	0.434	1.22	0.175	666	1.38	0.12
15	20	65	3.99	0.26	-0.47	0.432	1.24	0.174	655	1.36	0.11
20	20	60	4.09	0.27	-0.59	0.428	1.27	0.173	646	1.33	2.39
25	20	55	4.14	0.29	-0.73	0.427	1.29	0.172	635	1.32	2.36

It is interesting to note that the corresponding glass-ceramic nanocomposites show high conductivity compared to the samples in glassy phase [13, 14, 40]. The enhancement of conductivity of these corresponding glass-ceramic nanocomposites is considered to be due to the presence of nanocrystals. With an increase in conductivity by nanocrystallization, the activation energies for conduction were found to be W_{nano} = 0.35 – 0.60 eV at high temperatures which are much lower than those for the as-received glasses W_{glass}= 0.57- 0.81 eV at high temperatures (Tables 1and 2) for BTV system and W_{nano} = 0.13 – 0.22 eV, W_{glass}= 0.50 – 0.38 eV at high temperatures (Tables 3and 4) for BTVB system. And for STPV system were found to be W_{nano} = 0.24 – 0.29 eV at high temperatures which are much lower than those for the as-received glasses W_{glass} = 0.29 – 0.33 eV at high temperatures (Tables 5and 6). This means an increase of V^{4+} ion ratio causing the activation energy, W, to decrease and σ to increase. In the present glass systems of our investigation, the $BaTiO_3$ or the $SrTiO_3$ addition decreased the conductivity. Generally, it is known that the addition of $BaTiO_3$ or $SrTiO_3$ to glass decreases the conductivity as a result of decreasing non-bridging oxygen (NBO) cations [39]. This may decrease the open structure, through which the charge carriers can move with lower mobility.

On the other hand, the improvement of electrical conductivity of glass-ceramic nanocomposites under study can be explained in the following way. The most important for electronic conduction in the glass systems with high amount of V_2O_5 is the spatial distribution of V^{4+} and V^{5+} ions which are centers of hopping for electrons [13, 14, 20, 21, 23, 24, 36, 37, 40]. In the initial glasses, there is a random distribution of such centers. The annealing at temperatures close to crystallization temperature leads to formation of nanocrystallites of V_2O_5 embedded in the glass matrix. Since the average size of these grains is small, the interface between crystalline and amorphous phases is very extensively ramified and strongly influences overall electrical properties of the nanomaterial as reported in XRD. In particular, it may contain the improved concentration of V^{4+} and V^{5+} centers dispersed on the surface of V_2O_5 crystallites [13, 14, 40]. However, This enhancement of electrical conductivity can be attributed to (i) an increasing of concentration of V^{4+}-V^{5+} pairs (a possible reason of this increase between surfaces of nanocrystallites and glassy phase) and (ii) formation of defective, well-conducting regions along the glass-crystallites interfaces [13, 14, 40].

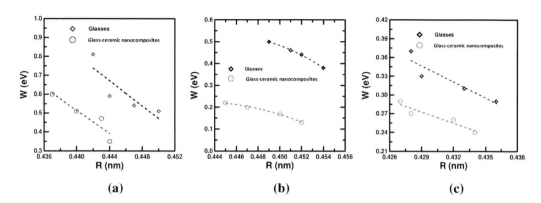

Figure 8. Effect of average distance (R) on activation energy (W) for (a) BTV (b) BTVB (c) STPV systems [27, 28, 29].

In addition, the decrease in dc conductivity and the increase in activation energy for the present systems suggest some changes in conduction mechanisms. It has been previously reported [20-22, 34-37, 41] that in glasses containing vanadium oxide the dc conductivity is electronic and depends strongly upon the average distance, R, between the vanadium ions. The average distance, R, was calculated for the present systems (see Tables 1 → 6) from the relation $R = (1/N)^{1/3}$, where N is the concentration of vanadium ions per unit volume, calculated from batch composition and the measured density. The concentration of vanadium ions per unit volume, N, and average distance, R, are given in Tables 1 → 6 for the glasses and corresponding glass-ceramic nanocomposites. The relation between the average distance, R, and activation energy, W, for glasses and corresponding glass-ceramic nanocomposites for all glass systems are illustrated in Figure 8 (a, b and c). These results agree with the results suggested by El-Desoky [20, 37, 41] described the dependence of W on the V-O-V site distance.

On the other hand, the theoretical expression for that energy includes a term $W = W_0$ $(1-r_p/R)$, where W_0 is constant and r_p denotes a radius of small polaron [23, 24]. All above formulas indicate that electronic conductivity increases and activation energy decreases when the distance R between hopping centers decreases. In our case one can expect that due to nanocrystallization process the concentration of V^{4+}-V^{5+} pairs is higher near the surfaces of the newly formed crystallites than in the remaining glassy phase and inside the crystallites. It is widely known that many important characteristics of nanomaterials originate from the defective nature of the interfaces between crystalline and amorphous phases [40]. Higher concentration of V^{4+}-V^{5+} pairs leads to smaller average distance between the hopping centers, and according to Mott model [23, 24] it causes an increase in conductivity. The interface regions of higher than average conductivity form a kind of "easy conduction paths" for electrons [40].

Conduction Mechanism

Mott [23] proposed a model for conduction processes in the transition metal oxides (TMO) glasses. In this model, the conduction process is considered in terms of phonon assisted hopping of small polarons between localized states. The dc conductivity in the Mott model for the nearest neighbours hopping in non-adiabatic regime at high temperatures $T > \theta_D/2$ is given by [23, 24]:

$$\sigma = \frac{v_0 N e^2 R^2}{kT} C(1 - C) \, exp(-2\alpha R) \, exp(-W/kT) = \sigma_0 exp(-W/kT) \tag{3}$$

The pre-exponential factor (σ_0) in Eq. (2) is given by

$$\sigma_0 = \frac{v_0 N e^2 R^2}{kT} C(1 - C) \, exp(-2\alpha R) \tag{4}$$

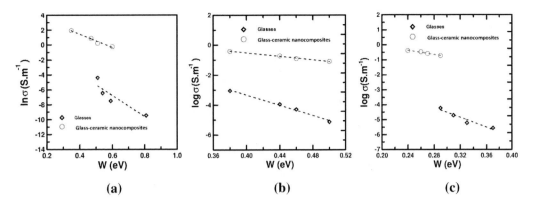

Figure 9. Effect of activation energy (W) on dc conductivity (σ) at fixed temperature for (a) BTV (b) BTVB (c) STPV systems [27, 28, 29].

where ν_o is the optical phonon frequency, α is the tunneling factor (the ratio of wave function decay), N the transition metal density, C the fraction of reduced transition metal ion and W is the activation energy for hopping conduction. Assuming a strong electron–phonon interaction, Austin and Mott [24] have shown that

$$W = W_H + W_D/2 \text{ (for } T > \theta_D/2) \tag{5a}$$

$$W = W_D \text{ (for } T < \theta_D/4) \tag{5b}$$

where W_H is the polaron hopping energy, and an energy difference W_D which might exist between the initial and final sites due to variation in the local arrangements of ions.

The nature of polaron hopping mechanism (adiabatic or non-adiabatic), of all these glasses and corresponding glass-ceramic nanocomposites can be estimated from a plot of logarithm of the conductivity against activation energy (Figure 9(a, b and c)) at fixed experimental temperature T [20, 43]. It is expected that the hopping will be in the adiabatic regime if the temperature estimated T_e, from the slope of such a plot, is close to the experimental temperature T. Otherwise the hopping will be in the non-adiabatic regime. From the plot of $\log\sigma$ against W at $T = 400$ K as shown in Figure 9 (a, b and c), the experimental slopes were not equal to the theoretical slopes. Figure 10(a, b and c) presents the effect of $BaTiO_3$ and $SrTiO_3$ content on the pre-exponential factors (σ_o) obtained from the least squares straight line fits of the data indicating an increase in σ_o with $BaTiO_3$ and $SrTiO_3$ content. For both results, it can be concluded that the conduction mechanism in the present glasses and corresponding glass-ceramic nanocomposites is due to the non-adiabatic hopping of the polarons [43].

Polaron Hopping Parameters

Holstein [43] suggests a method for calculating the polaron hopping energy W_H as:

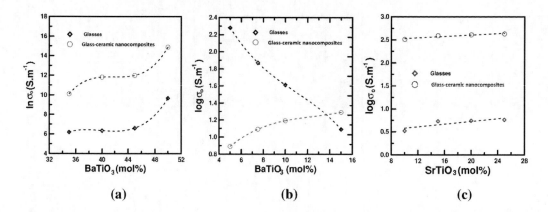

Figure 10. Effect of BaTiO$_3$ and SrTiO$_3$ content on pre-exponential factor (σ_o) for (a) BTV (b) BTVB (c) STPV systems [27, 28, 29].

$$W_H = (1/4N)\Sigma_p[\gamma_p]^2\hbar\omega_p \qquad (6)$$

where $[\gamma_p]^2$ is the electron–phonon coupling constant and ω_p is the optical phonons frequency. On the other hand, Bogomolov et al. [44] have calculated the polaron radius, r_p, for a non-dispersive system of frequency ν_o for Eq. (6) as:

$$r_p = \left(\frac{\pi}{6}\right)^{\frac{1}{3}} \frac{R}{2} \qquad (7)$$

The values of the polaron radii calculated from Eq. (7), using R from Tables 1 → 6 are shown in the same tables for the present systems. Although the possible effect of disorder has been neglected in the above calculation, the small values of polaron radii suggest that the polarons are highly localized.

The polaron hopping energy W_H is given by [45]

$$W_H = \frac{e^2}{4\varepsilon_p}\left(\frac{1}{r_p} - \frac{1}{R}\right) \qquad (8)$$

where

$$\frac{1}{\varepsilon_p} = \frac{1}{\varepsilon_\infty} - \frac{1}{\varepsilon_S} \qquad (9)$$

ε_S and ε_∞ are the static and high frequencies dielectric constants of the glasses. An estimate of W_H can be made from Eq. (8) from the known values of R and r_p, while ε_P were estimated from Cole–Cole plot [45]. The small polaron coupling constant (γ_p) a measure of the electron–phonon interaction is given by $\gamma_p = 2W_H/h\nu_o$ [23]. The estimated values of γ_p are 8.49–12.6 and 8.1–10.26 for STPV glasses and corresponding glass–ceramic nanocomposites, respectively. Austin and Mott [24] suggested that a value of $\gamma_p > 4$ usually indicates a strong electron–phonon interaction in glasses. Thus, we conclude that there is a strong electron–phonon interaction of the present glass systems.

The density of states for thermally activated electron hopping near the Fermi level from basic principles is given as [45, 46].

$$N(E_F) = 3/(4\pi R^3 W) \tag{10}$$

The results for the glasses and corresponding glass-ceramic nanocomposites are listed in Tables $1 \to 6$. The values of $N(E_F)$ are reasonable for localized states [20-22, 34-37, 41].

The phonon–assisted hopping model of Mott [23, 24] is consistent with dc conductivity data presented in Figure 5(a, b and c) and Figure 6(a, b and c) in the high temperature region. Eq. (3) predicted by Mott model is fitted in Figure 5(a, b and c) and Figure 6(a, b and c) with the experimental data at high temperatures, using v_o, α and W as variable parameters. The best fits are observed above $\theta_D/2$ (Tables $1 \to 6$) for the values of the parameters shown in Tables $1 \to 6$.

We estimated the optical phonon frequency, (v_o) in Eq. (3) using the experimental data from Tables $1 \to 6$, according to $k\theta_D = h v_o$ (h is the Planck's constant). To determine v_o for the different compositions, the Debye temperature θ_D was estimated by $T > \theta_D/2$ using the θ_D values given in Tables $1 \to 6$ at the point of significant change of the slope of the curves shown in Figure 5(a, b and c) and Figure 6 (a, b and c). Thus, the estimated θ_D values are considered to be physically reasonable. Then, using θ_D value, v_o was calculated. The values of θ_D and v_o are summarized in Tables $1 \to 6$.

CONCLUSION

V_2O_5-based glasses in the systems $BaTiO_3$-V_2O_5, $BaTiO_3$-V_2O_5-Bi_2O_3 and $SrTiO_3$-PbO_2-V_2O_5 have been transformed into glass–ceramic nanocomposites via annealing at crystallization temperature (T_c) determined from DSC thermograms. The nanostructural and electrical properties were investigated by DSC, TEM, XRD, density, and dc conductivity. The overall features of the XRD patterns confirm the amorphous nature of the present glass systems. Both TEM and XRD studies indicate the presence of nanocrystalline phases with average particle size of 20–35 nm in the glass matrix. The density increases with increasing $BaTiO_3$ or $SrTiO_3$ content. In these systems, the glass–ceramic nanocomposites exhibit much higher electrical conductivity than the initial glasses. It was postulated that the major role in the conductivity enhancement of these nanomaterials is played by the developed interfacial regions between crystalline and amorphous phases, in which the concentration of V^{4+}–V^{5+} pairs responsible for electron hopping has values higher than that inside the glassy matrix. From the best fits, reasonable values of various SPH parameters are obtained. The conduction is attributed to non-adiabatic hopping of small polaron.

REFERENCES

[1] Bahgat, A. A.; Moustafa, M. G.; Shaisha, E. E. *J Mater Sci Technol* 2013, *29*, 1166-1176.
[2] Shankar, J.; Deshpande, V. K. *Phys B: Condens Matter* 2011, *406*, 588-592.
[3] Ohta, Y.; Kitayama, M.; Kaneko, K.; Toh, S.; Shimiz, F.; Morinaga, K. *J Am Ceram Soc* 2005, *88*, 1634-1636.

[4] Nishida, T.; Kubuki, S.; Shibata, M.; Maeda, Y.; Tamaki, T. *J Mater Chem* 1997, *7,* 180-1806.

[5] Prasad, N. S.; Varma, K. B. R. *Mater Sci Eng B* 2002, *90,* 246-253.

[6] Barquin, L. F.; Barandiaran, J. M.; Sal, J. C. G.; Telleria, I. *Phys Status Solidi A* 1996, *55,* 439-450.

[7] Graca, M. P. F.; Valente, M. A.; Da Silva, M. G. F. *J Non-Cryst Solid* 2003, *325,* 267-274.

[8] Beck, C.; Hartl, W.; Hempelmann, R. *J Mater Res* 1998, *13,* 3174-3180.

[9] Berbenni, V.; Marini, A.; Bruni, G. *Thermochim Acta* 2001, *374,* 151-158.

[10] Funakubo, H.; Nagano, D.; Saiki, A.; Inagaki, Y.; Shinozaki, K.; Mizutani, N. *Jpn J Appl Phys* 1997, *36,* 5879-5884.

[11] Dutta, P. K.; Asiaie, R.; Akbar, S. A.; Zhu, W. D. *Chem Mater* 1994, *6,* 15421548.

[12] El-Desoky, M. M. *Mater Chem Phys* 2010, *119,* 389-394.

[13] Garbarczyk, J. E.; Jozwiak, P.; Wasiucione, M.; Nowinski, J. L. *Solid State Ionics* 2006, *177,* 2585-2588.

[14] Garbarczyk, J. E.; Jozwiak, P.; Wasiucionek, M.; Nowinski, J. L. *J Power Sources* 2007, *173,* 743-747.

[15] El-Desoky, M. M.; Ibrahim, F. A.; Mostafa, A. G.; Hassaan M. Y. *Materials Research Bulletin* 2010, *45,* 1122-1126.

[16] Hirao, K. *Bull Ceram Soc Japan* 2001, *36,* 652-656.

[17] El-Desoky, M. M.; Al.-Shahrani, A. *J Mater Sci: Mater Electronics* 2005, *16,* 221-224.

[18] El-Desoky, M. M.; Al-Assiri, M. S. *Mat Sci Eng B* 2007, *137,* 237-246.

[19] Al-Hajry, A.; Al-Shahrani, A.; El-Desoky, M. M. *Mater ChemPhysics* 2006, *95,* 300-306.

[20] El-Desoky, M. M.; Al-Shahrani A. *Physica B* 2006, *371,* 95-99.

[21] Al-Shahrani, A.; Al- Hajry, A.; El- Desoky, M. M. *Phys Stat Sol (a)* 2003, *200,* 378-387.

[22] Al-Assiri, M. S.; Salem, S. A.; El-Desoky, M. M. *J Phys Chem Solids* 2006, *67,* 1873-1881.

[23] Mott, N. F. *J Non-Cryst Solids* 1968, *1,* 1-17.

[24] Austin, I. G.; Mott N. F. *Adv Phys* 1969, *18,* 41-102.

[25] Al-Syadi, A. M.; Al-Assiri, M. S.; Hassan, H. M. A.; El-Desoky, M. M. *J Mater Sci: Mater Electron* 2016, *27,* 4074-4083.

[26] Harby, A. H.; Hannora, A. E.; Al-Assiri, M. S.; El-Desoky, M. M. *J Mater Sci: Mater Electron* 2016, *27,* 8446-8454.

[27] Al-Assiri, M. S.; El-Desoky, M. M.; Al-Hajry, A.; Al-Shahrani, A.; Al-Mogeeth, A. M.; Bahgat A. A. *Physica B* 2009, *404,* 1437-1445.

[28] Al-Assiri, M. S.; El-Desoky, M. M. *J Alloys Compounds* 2011, *509,* 8937-8943.

[29] El-Desoky, M. M.; Zayed, H. S. S.; Ibrahim, F. A.; Ragab H. S. *Physica B* 2009, *404,* 4125-4131.

[30] Borelli, N. F.; Layton, M. M. *IEEE Trans Electron Devices* 1969, *ED-16,* 511-514.

[31] Tanaka, K.; Kashima, K.; Soga, N.; Mito, A.; Nasu, H. *J Non-Cryst Solids* 1995, *185,* 123-126.

[32] El-Desoky, M. M. *J Non-Cryst Solids* 2005, *351,* 3139-3146.

[33] Sakka, S.; Mackenzie, J. D. *J Non-Cryst Solids* 1971, *6,* 145-162.

[34] El-Desoky, M. M.; Kashif, I. *Phys Stat Sol (a)* 2002, *194,* 89-105.

[35] El-Desoky, M. M. *J Mater Sci: Mater Electronics* 2003, *14,* 215-221.

[36] Mori, H.; Kitami, T.; Sakata, H. *J Non-Crst Solids* 1994, *168,* 157-166.

[37] El-Desoky, M. M.; Tashtoush, N. M.; Habib, M. H. *J Mater Sci: Mater Electronics* 2005, *16,* 533-539.

[38] Yoneda, Y.; Mizuki, J.; Kohara, S.; Hamazaki, S.; Takashige, M. *J Appl Phys* 2006, *99,* 074108.

[39] Louër, D.; Audebrand, N. *Adv X-ray Anal* 1999, *41,* 556-565.

[40] Garbarczyk, J. E.; Jozwiak, P.; Wasiucionek, M.; Nowinski, J. L. *Solid State Ionics* 2004, *175,* 691-694.

[41] El-Desoky, M. M.; Al- Hajry, A.; Tokunaga, M.; Nishida, T.; Hassaan, M. Y. *Hyperfine Interactions* 2004, *156,* 547-553.

[42] Mandal, S.; Ghosh, A. *Phys Rev B* 1993, *48,* 9388-9393.

[43] Holstein, T. *Annals of physics* 1959, *8,* 325-342.

[44] Bogomolov, V. N.; Kudinov, E. K.; Firsov, Y. A. *Soviet Physics Solid State* 1968, *9,* 2502-+.

[45] Mott, N. F.; Davis, E. A. *Electronic Processesin Non-crystalline Materials;* Oxford: Clarendon Press, 1979.

[46] El-Desoky, M. M.; Mostafa, M. M.; Ayoub, M. S.; Ahmed, M. A. *J Mater Sci: Mater Electron* 2015, *26,* 6793-6800.

In: Polarons: Recent Progress and Perspectives
Editor: Amel Laref

ISBN: 978-1-53613-935-8
© 2018 Nova Science Publishers, Inc.

Chapter 11

POLARON IN PEROVSKITE MANGANITES

Abd El-Moez A. Mohamed[1,2,3,] and B. Hernando[2]*
[1]Physics Department, Faculty of Science, University of Sohag, Sohag, Egypt
[2]Physics Department, Faculty of Science, University of Oviedo, Oviedo, Spain
[3]School of Metallurgy and Materials, University of Birmingham, Birmingham, UK

ABSTRACT

The carrier-lattice interaction includes several effects such as structural and magnetic effects, which lead to different polaron kinds (lattice and spin polarons). Polaron plays an important role in the physics of manganites, where the lattice polaron affects carrier transport at the high temperature semiconducting/paramagnetic region. Polaron, also, is one of the suggested proposals that has explained the colossal magnetoresistance phenomenon. Lattice and magnetic polarons are the most common types found in manganites. The lattice polaron is promoted by structural defects as the ionic size mismatch, the Jahn-Teller distortion, and the octahedra tilting. Magnetic polarons affect the electrical resistivity due to the interaction between the conduction carriers and the localized spins. In this chapter, we will discuss the polaron role in the magneto-transport properties of perovskite manganites.

1. CONCEPT

In 1933, Landau introduced the polaron concept for the first time to describe conduction in ionic semiconductors. It is a quasi-particle result from the mutual interaction of a carrier and a lattice. The behavior of a free carrier in a crystal is determined by the surrounding electronic structure. However, it is hard to neglect the mutual interaction between carriers and lattice (the carrier-lattice coupling), which adds an additional effect on the carrier behavior. When Landau introduced the polaron concept, he suggested that the carrier wave function can be localized in its own polarization field [1]. In a similar way, a carrier can also be self-trapped in a potential well induced by a deformed lattice [2]. Later, the word 'polaron' has

---

* Corresponding Author: abdmoez_hussien@science.sohag.edu.eg;   a.a.m.a.hussein@bham.ac.uk.

been used as a general concept to describe any change e.g., structural and magnetic changes arising from the carrier-lattice interaction. Accordingly, two main kinds of polaron have been introduced, the lattice and the magnetic polarons.

The lattice polaron can be visualized as a quasi-particle consisting of a charge and a local distortion as illustrated in Figure 1. The existence of a charge carrier (electron or hole) near an ion for some time deforms the lattice. In *3d* metals, the average lifetime of a carrier near an ion is about 10^{-15} seconds, which is a very short time for the crystal polarization. However, if the carrier lifetime increases by any mechanism to 10^{-12} seconds, then the carrier can be attracted by the positive ions and pushed away by the negative ones (in case of an electron). In this case, the resultant lattice polaron is called the weak coupling lattice polaron because of the weak mutual forces between carriers and ions that cannot induce ion displacement. In contrast, if the mutual forces between carriers and ions are strong enough to displace ions away from their equilibrium positions (see Figure 2), a potential well is produced, affecting the carrier path, and this is the case with the strong coupling lattice polaron. It should be noted that polaron formation depends on the binding energy of the self-trapped carrier, where the potential well is induced when the carrier binding energy exceeds the strain energy spent in ions displacement.

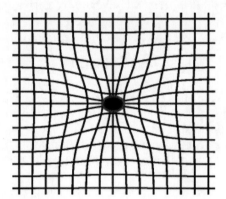

Figure 1. A schematic plot for the lattice polaron concept. The black circle represents a charge and the crossed lines represent a lattice.

Figure 2. A strong coupling lattice polaron due to an extra electron (large white circles are the initial position and the black circles are the new positions, and the small white circle is the electron).

The lattice polaron has been able to describe the physical properties of carriers in polarized solids, and the strong coupling polaron has been extended to small and large systems according to the electronic wave function dimension. In other words, the small polaron wave function is localized in the lattice parameters range, while the large polaron wave function is localized in a radius larger than the lattice parameters. The small polaron model describes the conduction in ionic semiconductors, where the strong interaction of carrier-lattice lowers the electron mobility. In this case, the electron is being "self-trapped" in its own polarization field [5]. It is worth noting that the polarons can interact with each other in various ways like Coulomb interaction, where the same sign self-trapped carriers are repelled and those with opposite signs are attracted. Also, polaron can interfere, where polarons with the same sign are paired, producing a bipolaron.

Regarding polaron motion, a polaron moves via different mechanisms, according to the ambient temperature. For example, a polaron moves via hopping mechanism at high temperatures, leading to the thermally activated mobility [3] that requires a mixed valence state for occurrence [4], while at low temperatures, it moves via quantum tunneling.

Another kind of polaron is the magnetic polaron or the spin polaron. This polaron kind is induced in magnetic materials due to the mutual interaction between the carrier spin and the lattice magnetization. In other words, the magnetic/spin polaron is a cloud of local ferromagnetic moments that is affected by a localized carrier spin. The magnetic/spin polaron may be found as a bounded or a free polaron; however, the concept of the bounded spin polaron is still more acceptable [22]. The bounded magnetic/spin polaron was studied in Mn semi-magnetic materials and its occurrence was attributed to the exchange interaction between the unpaired localized carriers and the Mn surrounding spins. For example in manganites, the resistivity behavior is affected by the magnetic oscillation due to the interaction between the conduction carriers and the localized spins.

To consider the magnetic/spin polaron formation, let us suggest a spin polaron confinement in a sphere with radius r, which is the case of the zero point motion energy. In this case, the confinement energy is $h^2\pi^2/2m^*r^2$, where h is the Planck constant and m^* is the effective mass of polaron, and the total spin polaron energy (E_{sp}) can be expressed as:

$$E_{sp} = \frac{h^2\pi^2}{2mr^2} + \frac{4\pi}{3}\left(\frac{r}{a}\right)^3 J_N - J_H \tag{1}$$

where a is the lattice parameter, J_N is the moment flipping energy and J_H is the Hund coupling energy, and it is worth mentioning that the magnetic/spin polaron is formed only if the E_{sp} is a negative value.

2. POLARON IN PEROVSKITE MANGANITES

The $A_{1-x}B_xMnO_3$ perovskite manganites, where A is a rare earth element and B is a divalent or a monovalent ion, have attracted considerable attention due to their interesting properties such as (I) the huge change in resistivity under the effect of magnetic field application, which is known as the colossal magnetoresistance (CMR), and (II) the metal-semiconductor resistivity transition at a certain temperature (T_{ms}). Manganites exhibit complex transport behavior due to the charge, spin and lattice interactions. Several theoretical and experimental studies were performed trying to illustrate the CMR phenomenon and the transport properties of manganites, and the magnetic/phononic polarons were the most important suggested proposals at this time [6-8]. The polaron hypothesis was considered in manganites to describe the observed anomalous behavior in the transport properties at high temperatures, which was attributed earlier to some extrinsic defects such as grain boundaries and impurities. Later, the high-quality pulsed laser deposited thin films proved that this anomalous behavior arises from intrinsic factors due to carriers localization as small polarons [9]. Several recent experimental studies such as thermoelectric power measurements [10] and isotopic effects [11] have asserted the polaron presence in the paramagnetic phase of manganites materials.

3. Carrier Localization and Lattice Coupling in Manganites

Carrier localization in manganites arises from several factors that determine the polaron transport such as the structural distortions and the exchange interactions. Here we will detail the structural factors responsible for carrier localization and lattice polaron formation.

3.1. Electron-Phonon Coupling

The electron-phonon coupling results from lattice distortion, and lattice distortion is mainly induced by several factors such as the ionic size mismatch, the crystal field, and the cation displacement. These factors play a crucial role in the perovskite structure formation in manganite materials, and in this part, we will discuss these factors.

3.1.1. MnO_6 Octahedra Tilting

Ionic size mismatch between A, B and Mn ions results in a distorted perovskite unit cell and a structural instability. As a measurement of this ionic size mismatch, V. Goldschmidt suggested the tolerance factor (F) in Eq. 2 [12]

$$F = \frac{(\langle r_{A,B} \rangle + r_O)}{\sqrt{2}(\langle r_{Mn} \rangle + r_O)} \tag{2}$$

where r_A is the A ionic radius, r_B is the B ionic radius, r_O is the Oxygen ionic radius and r_{Mn} is the Mn ionic radius. The ideal perovskite structure shows $F = 1$. This is due to the ionic radius matching and the good equilibrium distances among ions. In fact, different ions show a different ionic size, and to maintain the minimum free energy of a system, ions should move away from their essential positions to new equilibrium ones. This process creates free spaces in the lattice and the MnO_6 octahedra have to rotate or tilt to fill this space, leading to a change in the Mn-O and the (A, B)-O distances. Therefore, F changes from unity leading to a deformation in the perovskite unit cell and a change in the Mn-O-Mn angle.

3.1.2. Jahn-Teller (JT) Distortion

In an MnO$_6$ octahedron, the Mn ion is surrounded by six oxygen ions as seen in Figure 3a. Four oxygen ions are located in the Mn ion plane as a tetragonal shape and one oxygen ion is above and the other is below the Mn ion. The crystal field splits the five d levels of the Mn ion into two sets, the e_g set (with two levels) and the t_{2g} set (with three levels), which is an unstable system with an energy excess. The system tends to get rid of this energy excess by spontaneous deformation through an additional splitting in both e_g and t_{2g} sets [13] (see Figure 3a). This effect is known as the JT effect/distortion, which controls the e_g electron localization and the insulating phase stabilization. It is worth stating that the JT distortion is mainly related to the Mn^{3+} ions and not to the Mn^{4+} ions (see Figure 3b). There are three types of the JT distortion Q$_1$, Q$_2$, and Q$_3$ modes, which are classified according to the distortion shape and direction (see Figure 3c).

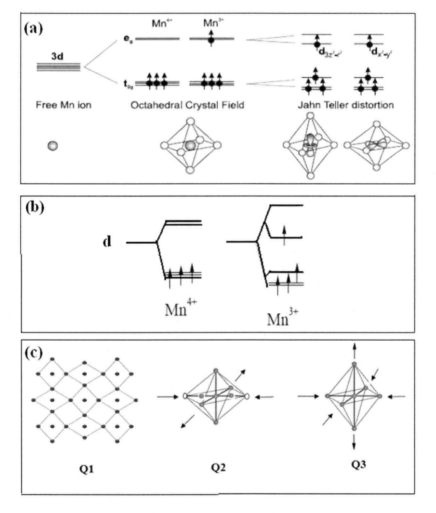

Figure 3. (a) The deformation stages in MnO$_6$ Octahedra, (b) effect of Jahn-Teller distortion on Mn^{3+} and Mn^{4+} ions and (c) Jahn-Teller distortion modes.

The Q_3 distortion mode is a longitudinal distortion occurring along the $d_{3z^2-r^2}$ orbital and results in the oxygen tetragonal contraction in the $d_{x^2-y^2}$ orbital direction. This increases the two long Mn-O bond length and decreases the four short Mn-O bonds length [14]. The Q_2 distortion is an orthorhombic distortion mode with a constant bond length for the two long Mn-O bonds. The variation occurs only in the oxygen tetrahedral shape, where every two opposite oxygen ions move towards each other and simultaneously the other opposite two ions move outwards. The Q_1 distortion mode is called the breathing distortion; this is because it results from the difference in the octahedra sizes due to the ionic size difference of Mn^{3+} and Mn^{4+} ions. Octahedra tilting and JT distortions develop the electron-phonon coupling, and carriers are localized as small polarons, which is the main carrier localization factor in manganites [15].

3.1.3. Coulomb Force

The Coulomb force/potential arises from the difference in valence state between the rare earth element (A^{3+}) and the divalent or monovalent cation (B^{2+} or B^{+1}) leading to Anderson localization [16]. The electron-phonon interaction illustrates how the carrier energy is affected by changing the position of the surrounding ions. In the absence of the electron-phonon coupling, carriers can be localized only as large spin polarons [16].

4. POLARON IN THE PARAMAGNETIC REGION

4.1. Small Polaron Hopping (SPH)

The conduction in the paramagnetic region of manganites occurs through several mechanisms such as the variable range hopping (VRH) and small polaron hopping (SPH) [17]. SPH is applicable in the temperature range of $T > \theta_D/2$, where θ_D is the Debye temperature and $\theta_D/2$ is the linearity deviation point from this model as seen in Figure 4. This model is described mathematically by Eq. 3 [18], where A is a constant equal to k_B/υ_{ph} $Ne^2R^2C(1-C) \exp(2\alpha R)$, k_B is the Boltzmann constant, T is the absolute temperature, N is the number of ion sites per unit volume,

$$\rho/T = A \exp (E_p/k_B T) \tag{3}$$

$$E_\rho = W_H + W_D/2 \text{ (for } T > \theta_D/2) \tag{4}$$

$$= W_D \text{ for } T < \theta_D/4 \tag{5}$$

$R \sim (1/N)^{1/3}$ is the average inter-site spacing, C is the fraction of sites occupied by a Polaron, α is the electron wave function decay constant, υ_{ph} is the optical phonon frequency (estimated from $h\upsilon_{ph} = k_B\theta_D$ relation) and E_ρ is the total activation energy which is the summation disorder energy (W_D) and the polaron hopping energy (W_H) (W_H is the difference between the electric and thermopower activation energies).

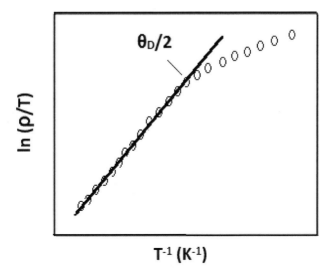

Figure 4. SPH model, the black line is the best-fitted points with this model.

If the carrier hopping frequency is too fast in comparison with the lattice distortion, then, the small polaron hopping is an adiabatic process. In this case, the conductivity can be described by Eq. 5.

$$\sigma = ne\mu = \frac{3}{2}\frac{ne^2a^2v_{ph}}{k_B}\frac{1}{T}\exp\left(-\frac{W_H-2J}{2k_BT}\right) \qquad (5)$$

$$\sigma = \frac{ne^2a^2\pi J^2}{k_B h}\left(\frac{2\pi}{W_H k_B}\right)^{\frac{1}{2}}\frac{1}{T^{\frac{3}{2}}}\exp\left(-\frac{W_H}{2k_BT}\right) \qquad (6)$$

On the other hand, if the carrier hopping frequency is very slow in comparison with the lattice distortion, the hopping is a non-adiabatic process, and in this case, the conductivity can be described by Eq. 6.

Holstein simplified this concept of the adiabatic and the non-adiabatic polaron hopping in two criteria [19], where, $J > H$ is the condition of adiabatic conduction and $J < H$ is the case of the non-adiabatic conduction, where $J(T)$ is the polaron bandwidth $\approx 0.67\, hv_{ph}\left(\frac{T}{\theta_D}\right)^{1/4}$ and $H = \left(\frac{2K_BT W_H}{\pi}\right)^{1/4}\left(\frac{hv_{ph}}{\pi}\right)^{1/2}$.

4.2. Variable Range Hopping Model (VRH)

The VRH theory was proposed to describe the electrical transport in doped semiconductors, where carriers are localized by the dopant potential. VRH occurs when a carrier cannot hop to the nearest site due to the small activation energy, and instead, it has to find another site with a smaller potential difference. The low-temperature transport phenomenon in perovskite and other semiconducting oxides can be described by the VRH model in the temperature range $T_{ms} < T < \theta_D/2$ [20], where θ_D is the Debye temperature and the mathematical expression for VRH was introduced by Mott and Davis in Eq. 7 [18]:

$$\sigma = \sigma_0 \exp(-T_0/T)^{1/4} \tag{7}$$

where T_0 is a constant (Mott characteristic temperature) equals to $18/k_B$ N (E_f) a^3, $N(E_f)$ is the density of states near the Fermi energy and a is the localization length.

4.3. VRH of SPH

In mixed valence manganites, sometimes, carriers are deeply trapped as small polarons. This is due to the low values of their activation energies that disable these carriers to hop from the initial positions. In this case, carriers hop to an intermediate state whose energy is higher than the previous one, after which, the activation energy enables carriers to hop to the nearest neighbor site. This condition of the non-nearest neighbor hopping is known as the variable range hopping of the small polaron due to carrier localization as a small polaron in both sites (initial and intermediate) [21]. The transport of this small polaron occurs in two steps, the thermal activation to an intermediate state and the hopping to a suitable site with a smaller energy. In other words, it can be said that the binding energy of the small polaron decreases by the first step, and these polarons are modified to an extended wavepacket that can be observed as large spin polarons [16]. The carrier can then hop or tunnel between this wavepacket.

As discussed by Y. Sun et al. [21], the resistivity for this model is derived as follows. The carrier in the initial position is localized with a wave function of $\psi = \psi_0 \exp(-\alpha_0 r)$ and with an activation probability to an intermediate state of $P = \upsilon_0 \exp(-\Delta E/kT)$, where $1/\alpha_0$ is the localization length (it is less than the lattice parameter), υ_0 is the phonon characteristic frequency and ΔE is the energy difference between the intermediate and the initial sites. The carrier wave function in the intermediate state is $\psi = \psi_0 \exp(-\alpha r)$, $1/\alpha > 1/\alpha_0$ because of the weak localization. The carrier hopping rate to another site with a distance R is $\gamma = \gamma_0 \exp(-2\alpha R - \Delta E/kT)$, where γ_0 is a constant. Let us now suppose that a pair of sites with an energy difference between 0 and ΔE are in a sphere with volume $(4\pi/3)R^3$. Considering the density of states near the Fermi level, N(E), then the energy difference is expressed as $\Delta E = [(4\pi/3)R^3 N(E)]^{-1}$. Substituting R in γ equation we get:

$$\gamma = \gamma_0 \exp(-T_0/T)^{1/4}, \; T_o = 18\alpha^3/N(E) \tag{8}$$

The total hopping rate from the initial to the final state can be determined by Eq. 8

$$\gamma \, P = \upsilon_0 \exp(-E_1/kT) \times \gamma_0 \exp(-T_0/T)^{1/4} \tag{9}$$

Leading to

$$\rho = CT \exp(E_1/kT + (T_0/T)^{1/4}) \tag{10}$$

5. POLARON ROLE IN THE MAGNETO-TRANSPORT CORRELATION

The electrical resistivity of manganites shows a metal-semiconductor transition near the Curie temperature (T_c), which suggests a correlation between the magneto-transport properties. This correlation has been expressed as $\rho = \rho_0 \exp(-M/\alpha)$ [23], where, M is the magnetization and α is a constant. This equation was modified to be $\rho = \rho_0 \exp(-M^2/\alpha T)$ [24] for small and intermediate magnetization values around T_c.

To understand this correlation, we will highlight the origin of the magnetocaloric effect (MCE) and the CMR phenomena as an example and then try to explain their correlation. Resistivity suppression with the application of magnetic field results in the CMR effect around the T_c. The CMR can be estimated by the relation MR = $\rho(H)$-$\rho(0)/\rho(0)$, where $\rho(H)$ is the resistivity under the application of a magnetic field and $\rho(0)$ is the zero-field resistivity. The CMR peak around the T_c is an intrinsic effect arising from the ferromagnetic spin ordering due to magnetic field application. The ferromagnetic ordering is simultaneously accompanied by a decrease in the magnetic spin entropy of the system that is compensated by an increase in the lattice entropy (ΔS) also with a peak around the T_c [25]. The similar behavior of ΔS, ρ and CMR around T_c supports the magneto-transport correlation concept and reveals the important role of the spin order/disorder feature.

The magnetocaloric effect-resistivity correlation was described previously using Eq. 11 [26], H is the applied magnetic field. The spin-disorder that characterizes ΔS, around T_c leads to the formation of magnetic polarons, which in turn affects carrier transport and resistivity [26]. At low temperatures, magnetic polarons are suppressed due to the ferromagnetic ordering; therefore, we expect the validity of Eq. 10 only around T_c.

$$\Delta S = -\alpha \int_0^H \frac{\delta \ln \rho}{\delta T} dH \qquad (11)$$

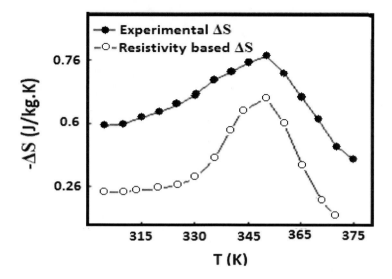

Figure 5. Experimental ΔS and resistivity based ΔS calculations according to Eq. 11.

Figure 5 shows the similar behavior of resistivity and ΔS. This similarity illustrates the effect of magnetic polarons on electrical transport around T_c, where magnetic polaron are driven by the spin disorder feature (ΔS) [26]. To conclude this, the applied magnetic field enhances the ferromagnetic ordering due to the parallel alignment of the spins with the direction of the applied magnetic field. The spin alignment decreases its entropy, which is compensated by the increase in lattice entropy (ΔS) [27] leading to heat release, formation of magnetic polarons and resistivity peak around T_c due to carriers scattering by spin-disorder [28].

CONCLUSION

Polarons play an important role in manganites and so affect the related phenomena. The lattice polaron arises from the electron-phonon coupling, which results from lattice distortions such as the octahedral tilting, the Jahn-Teller distortion, and ionic size mismatch. The lattice polaron is a composition dependent type and exists at the high-temperature paramagnetic state, which helps us to understand the CMR phenomenon. On the other hand, the magnetic polaron arises from the mutual interaction between the carrier spin and the surrounding lattice magnetization. Magnetic polaron plays a role in the magnetocaloric effect-transport properties correlation, where, the similar behavior of resistivity transition and the MCE around T_c refers to the magnetic polaron, which is driven by the magnetic disorder.

ACKNOWLEDGMENTS

The first author would like to acknowledge Basma F for continuous support, and M. A. Ahmed for help.

REFERENCES

[1] Landau L. D. *Physikalische Zeitschrift der Sowjetunion* [Physical Journal of the Soviet Union] 3, 664 (1933) (English translation in Collected Papers (Gordon and Breach, New York, 1965, pp. 67-68).

[2] S. I. Pekar, *J. Phys.* (Moscow) 10, 341 (1946), Untersuchungen über die Electronentheorie der Kristalle [Investigations on the electron theory of crystals] (Akademie-Verlag, Berlin, 1954).

[3] Larson E. G., Arnott R. J., Wickham D. G. *J. Phys. Chem. Solids* 1962, 23, 1771.

[4] Moulson A. J., Herbert J. M. *Electroceramics*, 2nd Edition. John Wiley and Sons Ltd, West Sussex, England, 2003.

[5] Tallan N. M., *Electrical Conductivity in Ceramics and Glass Part A*. Marcel Dekker, Inc., New York, 1974.

[6] Kusters R. M., Singleton J., Keen D. A., McGreevy R., Hayes W. *Physic a B* 1989, *155, 362*.

[7] Von Helmolt R., Wecker J., Holzapfeil B., Schultz L., Samwer K. *Phys. Rev. Lett.* 1993, *71, 2331.*

[8] Peker S. I., Zh. Eksp. *Teor. Fiz.* 1964, 16, 341.

[9] Ohtaki M., Koga H., Tokunaga T., Eguchi K., Arai H. *J. Solid State Chem.* 1995, 120, 105.

[10] Jaime M., Salamon M. B., Pettit K., Rubinstein M., Treece R. E., Horwitz J. S., Chrisey D. B. *Appl. Phys. Lett.* 1996, 68, 1576.

[11] Zhao G. M., Conder K., Keller H., Müller K. A. *Nature* 1996, 381, 676.

[12] Goldschmidt V. *Geochemistry,* Oxford University Press, London, 1958.

[13] Harrison W. A. *Solid State Theory*, Dover Publications Inc., New York, 1979.

[14] Griffith J. S. *The theory of transition metal ions*, Cambridge University Press, Cambridge, 1964.

[15] Millis A. *J Phil. Trans. R. Soc. Lond. A,* 1998, 356, 1473.

[16] Coey J. M. D., Viret M., Ranno L., Ounadjela K. *Phys. Rev. Lett.* 1995, 75, 3910.

[17] Ramirez A. P. *J Phys.: Condens. Matter.* 1997, 9, 8171.

[18] Mott N. F., Davis E. A. *Electronics Process in Non Crystalline Materials*, Clarendon Press, Oxford, 1979.

[19] Holstein T., *Ann. Phys.* (NY) 1959, 8, 343.

[20] Mollah S., Huang H. L., Yang H. D., Pal S., Taran S., Chaudhuri B. K. *J. Magn. Magn. Mater.* 2004, 284, 383.

[21] Sun Y., Xu X., Zhang Y. *J. Phys.: Condens. Matter.* 2000, 12, 10475.

[22] Guillaume B. A, *Phys. Stat. Sol.* B 1993, 175, 369.

[23] Hundley M. F., Hawley M., Heffner R. H., JiaQ. X., Neumeier J. J., Tesmer J, Thompson J. D., Wu X. D., *Appl. Phys. Lett.* 1995, 67, 860.

[24] Chen B., Uher C., Orelli D. T., Mantese J. V., Mance A. M., Micheli A. L. *Phys. Rev. B, Condens. Matter*1995, 53, 5094.

[25] S. Lee, M. S. Anwar, F. Ahmed, B. H. KOO, *Trans. Nonferrous Met. Soc.* China 24 (2014) 141.

[26] Xiong C. M., Sun J. R., Chen Y. F., Shen B. G., Du J., Li Y. X. *IEEE Trans. Magn.* 2005, 41, 122.

[27] Lee S., Anwar M. S., Ahmed F., Koo B. H. *Trans. Nonferrous Met. Soc. China* 2014, 24, 141.

[28] Tripathi T. S., Mahendiran R., Rastogi A. K. *J. Appl. Phys.* 2013, 113, 233907.

In: Polarons: Recent Progress and Perspectives
Editor: Amel Laref

ISBN: 978-1-53613-935-8
© 2018 Nova Science Publishers, Inc.

Chapter 12

CORRELATED POLARONS IN MIXED VALENCE OXIDES

C. M. Srivastava[1] and N. B. Srivastava[2,]*

[1]Department of Physics, Indian Institute of Technology, Powai, Mumbai, India
[2]Department of Physics, R Jhunjhanwala College, Ghatkopar, Mumbai, India

ABSTRACT

The anomalous dc transport properties of magnetite and normal state high temperature copper oxide superconductors are shown to follow from the correlated polaron theory that accounts for the colossal magnetoresistance in manganites [*J.Appl.Phys.* (2009) 105 1]. In magnetite the IR active Γ-centered T1u mode phonon with frequency 343 cm-1 seems to form the correlated polaron state. This then predicts a semimetal to semiconductor Verwey type transition at 117.42 K and a dc conductivity close to200 $(\Omega\text{-cm})$-1 that is nearly temperature -independent in the range 660 to 1600 K. The correlated polaron model also explains the anomalous dc conductivity in the normal state of the high temperature superconductors in which the spin and charge fluctuations over a wide range of excitations are independent of the wave vector q. In particular this gives the reason the ρ versus T curve in the range $Tc \leq T \leq 300$ K shows an upturn for some while for others there is a linear dependence, $\rho = c1\ T + c2$.

Keywords: Verwey transition, CMR, HTSC, transport mechanism, IR spectra, correlated polaron, dilute ferromagnetic oxide film, electron phonon interaction

1. INTRODUCTION

The existence of mixed cation valence, $M^{+(p+\delta)}$ and $M^{+(p-\delta)}$ on energetically equivalent sites permits with the help of longitudinal optic phonons, electronic conduction through valence exchange. When the time τ for charge transfer is large enough for the mobile electrons to get trapped by a local lattice deformation the two metal ions sites $M^{+(p+\delta)}$ and

---

*Corresponding Author Email: neeta6t3@gmail.com.

$M^{+(p-\delta)}$ become energetically inequivalent. In this case the electron carries the local lattice deformation with it and the dressed electron is called a small polaron.

The presence of mixed valence ions on energetically equivalent crystal sites creates a small number of electrons and holes that on coupling to a longitudinal optic phonon of the system enhances the Madelung energy of the solid by a small amount. This was first used by Srinivasan and Srivastava [1] to account for the Verwey transition in magnetite that had remained unexplained despite extensive investigation [2-5]. Here Fe exists in Fe^{2+} and Fe^{3+} states on octahedral B sites of the spinel lattice. This model was subsequently extended to manganites in which Mn atom exist in Mn^{3+} and Mn^{4+} states of the perovskite lattice on identical crystallographic sites. It was shown that in both cases, magnetite and manganite the charge carriers form two n-particle quantum states, ψ_1 and ψ_2, that are degenerate and oscillate from one state to another at the phonon frequency in a phase coherent manner [5]. The anomalous transport properties like the colossal magnetoresistance in manganites and the first order phase transition in magnetite at Verwey transition that could not be explained as Mott transition or Anderson localization have been accounted on the basis of correlated polaron formation. The model was extended to transport in the normal state of high temperature copper oxide superconductors in which it was observed that resistivity $\rho(T)$ was given by an expression linearly dependent on temperature, $\rho(T) = c_1 T + c_2$ [6]. This extends over a wide temperature range and was not understood on the conventional theories of transport in a metal, semiconductor or a semi-metal. A proposal was made by Varma and his collaborators [7] that this arises due to the existence of marginal Fermi-liquid like state in the system. In this the spin and charge excitations behaved differently from the Fermi liquid present in normal metals. The theory is based on the assumption that the spin and charge fluctuations over a wide spectrum of excitations do not depend on the wave number $\mathbf{q}$. This follows from the correlated polaron model as the phonons that couple charge carriers and form the ψ_1 and ψ_2 states are Γ-centered [8].

The presence of magnetism in oxide films and nanoparticles without 3d elements and long range magnetic order at concentrations of magnetic cations below the percolations limit with Curie temperature exceeding 300K is not explained on the basis of the present theories of magnetism in solids [9]. It is shown that these arises when two degenerate spin-polarized macro states, ψ_1 and ψ_2, through coupling to zero-point phonons increases the binding energy of the system by a small amount, ε_p. The coupling through quantum tunneling persists even in the presence of thermal fluctuations until the fraction of recoilless transitions in the ground state of the lattice vanishes. This accounts for the weak intensity universality and existence of small polarons over a wide range of temperature in these systems.

2. CORRELATED POLARON (CP) MODEL

Consider nmolecules M_2N_2 placed on an infinite linear chain with each molecule forming the basis of a one-dimensional unit cell of length l. Assume the two cations M to be in mixed valence state with charges $+(p+\delta)e$ and $+(p-\delta)e$ and the anions N to have the charge $(-p)$ for charge neutrality. There always exists two ways, ψ_1 and ψ_2, in which the charges can be arranged in the unit cell; $M^{(p+\delta)}$ at the boundary and $M^{(p-\delta)}$ at the center of the unit cell (ψ_1)

and $M^{(p-\delta)}$ at the boundary and $M^{(p+\delta)}$ at the center of the cell (ψ_2). The two states are degenerate and can be represented as

$$\psi_1 = \sqrt{n}\,exp(i\phi_1) \text{ and}$$

$$\psi_2 = \sqrt{n}\,exp(i\phi_2) \tag{1}$$

These are related by

$$(\Phi_1 - \Phi_2) = \pi/2 \tag{1a}$$

and are coupled with the phonon energy $h\nu_{ph}$

$$\frac{d\psi_1}{dt} = -\frac{i}{\hbar}(\varepsilon_p\psi_1 + h\nu_{ph}\psi_2)$$

$$\frac{d\psi_2}{dt} = -\frac{i}{\hbar}(\varepsilon_p\psi_2 + h\nu_{ph}\psi_1) \tag{2}$$

In this case the dc conductivity can be expressed as [5]

$$\sigma = (f_+ - f_-)ne^2 l^2 \nu_0 / kT \tag{3}$$

where n is the number of charge carriers per unit cell in the mixed valence state, l is the distance between the two ions in the mixed valence states, $M^{(p+\delta)}$ and $M^{(p-\delta)}$ and ν_0 is the frequency of jump. The probability function is derived from a random walk expression for the one-dimensional problem [5],

$$f_\pm = \frac{(kT/eEl)\,exp[\varepsilon_p \pm eEl/2kT]}{exp\{[\varepsilon_p \pm eEl/2kT\} \pm exp\{[-\varepsilon_p \mp eEl/2kT\}} \tag{4}$$

$$So\,(f_+ - f_-) = \frac{1}{4}sech^2(\varepsilon_p/2kT) \tag{5}$$

Here E is the applied electric field and the expressions are obtained in the limit $eEl \ll kT$. These expressions were obtained by Srinivasan and Srivastava [1] while analyzing the conductivity data of magnetite on the basis of correlated charge transfer from Fe^{2+} to Fe^{3+} on alternate sites of a linear chain of edge-shared octahedra of the spinel structure. From Eq.(3) the conductivity in the temperature range $T_v < T < 660K$ is

$$\sigma = \sigma_0(\frac{h\nu_0}{kT})sech^2(\varepsilon_p/2kT) \tag{6}$$

where,

$$\sigma_0 = \frac{n}{4h} e^2 l^2 \tag{6a}$$

This expression shows Einstein diffusional motion with l as the nearest neighbor jump distance and v_0 as an attempt frequency determined by the optic-mode vibration. The expression in Equation (6) does not apply for $T < T_v$ and also for $T > 660$ K for reasons discussed in section 3. Likewise there are other mixed valence oxide systems like the manganites, $Re_{1-x}A_xMnO_3$ (R = rare earth, A = divalent ion, Ca, Sr, Ba …) with perovskite structure, where the expression for conductivity is considerably modified. In this the CP model is based on the Holstein Hamiltonian for a molecular crystal. For the treatment of these cases we follow Reik [10].This consists of a free field Hamiltonian for the polarons (l_i^+, l_i), the displaced phonons ($b_{q\lambda}^+$, $b_{q\lambda}$) and a small interaction term that describes the Franck-Condon (FC) transitions,

$$H = \sum_i (\varepsilon_i - \varepsilon_p) \, \ell_i^+ \ell_i + \sum_{q\lambda} \hbar\omega_{q\lambda} b_{q\lambda}^+ b_{q\lambda} + \sum_{i,j} t_{ij} (\ell_i^+ \ell_j X_{i,j} + HC) \tag{7}$$

Here ε_i is the on-site energy of the e_g electron, ε_p is the small polaron stabilizing energy, t_{ij} is the transfer integral, X_{ij} describes the FC factor and $\omega_{q\lambda}$ is the phonon frequency of wave vector $\mathbf{q}$ and polarization λ. The polaron hopping conductivity for $T > \theta_D/4$ is then given by [11],

$$\sigma_{hop} = (\frac{\sqrt{\pi}}{2} ne^2 l^2)(\frac{v_0}{kT}) sech^2 (\varepsilon_p / 2kT) \exp(-U/kT) \tag{8}$$

where v_0 is the jump frequency, l is the jump distance and n is the number of ions in the mixed valence state. Further U is the activation energy in the Frank-Condon factor,

$$\exp - (U/kT),$$
$$U = \frac{1}{4} <\eta>< hv_{ph}> \tag{9}$$

Here $<\eta>$ and $< hv_{ph}>$ are the average number and energyof phonons, respectively, that accompany the charge carrier in the hopping process. For $T < \theta_D/4$ the zero-point lattice vibrations take place of thermal phonons. These we next show play a dominant role in the transport and magnetic properties of manganite, magnetite, high T_c copper oxide superconductors and nano-structured materials.

3. TRANSPORT IN MAGNETITE IN THE LOW AND HIGH TEMPERATURE PHASES

The dc conductivity of Fe_3O_4 measured by Miles et al. [12] from 40K to 1600K is shown in Figure 1for the high temperature phase, $T_v < T < 1600$ K and in Figure 2for the low temperature phase,40K< T< 120K. In the temperature range $T_v < T < 660$ K, σ is given by Eq.(6). The first order phase change at T_v when the crystal structure changes from cubic to monoclinic, the conductivity changes from itinerant to hopping type, as below T_v the charge carriers are localized. The transition occurs when the energy of the radiation oscillators that comprises of two parts, thermal and zero-point, become equal. The average energy of the oscillator is given by Boltzmann-Maxwell statistics

$$\bar{\varepsilon}=\hbar\omega \ [\frac{1}{e^{\hbar\omega/kT}-1} + \frac{1}{2}] \tag{10}$$

For

$$\hbar\omega \ /kT_m = \ln 3 \tag{11}$$

the thermal energy of the oscillator equals the zero-point energy. It is generally believed [14] that for $T < \theta_D/4$ where θ_D is the Debye temperature, the zero-point energy dominates the thermal energy. In Eq.(11) $\hbar\omega$ is then $k\theta_D/4$. The Verwey transition temperature is then

$$T_v = T_m = \frac{\hbar\omega}{\ln 3} = \frac{k\theta_D/4}{\ln 3} \tag{12}$$

Recent studies of charge order and phonons in magnetite [8] have shown that conduction electrons are localized on B1 and B4 sites in orbitally ordered t_{2g} states. These are occupied by Fe^{2+} ions while the sites B2 and B3 are occupied by Fe^{3+} ions. There are two formula units of Fe_3O_4 in a primitive unit cell. The 8 oxygen atoms constitute 4 oscillators for the unit cell that forms the states ψ_1 and ψ_2. Displacement of oxygen ions about octahedral coordinate Fe^{2+} and Fe^{3+} ions is given by Rowan et al. [8] for the IR active phonon modes. They have studied the Γ point phonon modes of the cubic ($Fd\bar{3}m$) phase. One of these occurs at 343 cm^{-1}. This appears as a weak shoulder in the stronger mode at 350 cm^{-1}. On using $\omega=343$ cm^{-1} in Eq. (12) we obtain

$$T_m = T_v = 129/\ln3 = 117.42 \text{ K} \tag{13}$$

This is close to the observed Verwey transition.

Using Eq.(6) we plot conductivity against temperature in Figure 1 for the high temperature region $T_v<T<1600K$ and show it by the solid line. We take here $v_0 =343$cm^{-1}. The value of σ_0 is 282.5$(\Omega$-cm$)^{-1}$. The curve is split in two parts (i) $T_v <T <1.28$ hv_0/k and (ii) $1.28T_0 < T< 1600K$. The reason for the split is that when the energy of a harmonic oscillator placed in a lattice whose temperature T exceeds the energy of the radiation oscillator, such that $T> 1.28hv_0/k$ the frequency of the oscillator increases and is given by $v_0'= 0.78kT/h$. The molecule Fe_3O_4 is in thermal equilibrium with the lattice at temperature T when it emits or absorbs a phonon in the exchange between the state functions ψ_1 and ψ_2. The primitive unit cell of magnetite comprises of two molecules. While vibrating in the 343 cm^{-1} T_{1u} mode of the spinel structure it comprises of four identical oscillators, from eight oxygen ions, each with frequency 85.75 cm^{-1}.In the presence of an assembly of bosons in thermal equilibrium with the set of radiating Fe_3O_4 harmonic oscillators, the frequency of jump of the electron from Fe^{2+} to Fe^{3+} is $v_0/4$ for one pair of B site ions. The frequency v_0 does not change as long as the temperature of the lattice is less than 1.28 hv_0/k. When the lattice temperature exceeds this limit the frequency of jump v_0' increases, to $kT/1.28$ h. This arises from the principle of equipartition of energy. At large T, the energy for each degree of freedom of the oscillator is ½ kT including the zero-point energy. This gives the Dulong and Petit law. We then get a discontinuity in the σ vs T curve at T = 1.28 hv_0/k,

$$\sigma \,(\Omega-cm)^{-1} = 282.5(\frac{hv_0}{kT})\operatorname{sech}^2(\frac{\varepsilon_p}{2kT}), \qquad T_v < T < 1.28\, hv_0\,/k = 1.28T_0 \qquad (14a)$$

$$= 282.5(0.78)\operatorname{sech}^2(\frac{\varepsilon_p}{2kT}), \qquad 1.28T_0 \le T \le 1600K \qquad (14b)$$

$$v_0 = 343cm^{-1} = \varepsilon_p\,/h \qquad (14c)$$

From Eq.(6a), $\sigma_0 = 282.5(\Omega\text{-cm})^{-1} = \dfrac{n}{4h}e^2l^2$ $\qquad\qquad$ (14d)

The jump distance l can be obtained from Figure (10b) of Ref. [8]. This is the distance between

$Fe^{2+}(B4)$ and $Fe^{3+}(B3)$ located on neighboring planes,

$$l = \pm\frac{a}{4}(\mathbf{i}+\mathbf{j}+2\,\mathbf{k}) \qquad (15)$$

where i, j, k are unit vectors along x, y, and z directions and a is the lattice constant,

a= 8.3770 Å. The value of l is 5.1298 Å.

The $T_{1u}343cm^{-1}$ phonon mode enables an electron parity-reversed pair $(k\uparrow, -k\uparrow)$, placed on neighboring (110) planes to jump from the site B4 to B3 with the help of a pair of parity-reversed zero-point phonons, $(q_\lambda, -q_\lambda)$. Here q_λ is a zone boundary phonon with a wave vector,

$$q_\lambda = 2\pi\,(2/a)(2i+2j+k) \qquad (16)$$

In Figure 1 we plot σ vs T using Eq.(14a) and (14b) and $\varepsilon_p=hv_0$, $v_0= 343cm^{-1}$. Here $T_0 = hv_0/k$.

We obtain, n = 6.5 charges per unit cubic cell from Eq. (14d). This is close to 8, the number of Fe^{2+} ions on the B-site of the cubic unit cell of Fe_3O_4.

The transition between ψ_1 and ψ_2 then leads to the dc conductivity σ which is proportional to ω/T [Eq.(14a)] for ω less than or equal to T and equal to a constant for T > ω as in Eq.(14b). This is because the term $\operatorname{sech}^2(\varepsilon_p/2kT)$ in the high temperature limit is of order of 1. This is the type of behavior of the marginal Fermi liquid assumed by Verma et al. [7]

while accounting for the anomalous dc transport properties of high temperature oxide superconductors. This we discuss further in section 4. The ion exchange, $Fe^{2+} \leftrightarrow Fe^{3+}$, in the high temperature phase ceases in the low temperature phase and leads to the distortion of the crystal from cubic to orthorhombic symmetry, P2/c, with lattice constant of a =5.9444Å, b= 5.9247Å, c=16.7750 Å and γ=90.236°. The transport in this phase is of an insulating semiconductor type. The conductivity occurs in the small polaron hopping mode. The zero-point energy dominates here over the thermal energy. The conductivity including Franck-Condon transitions from Eq.(8) is then given by

$$\sigma_{hop} (\Omega-cm)^{-1} = 0.511 \left(\frac{T}{T_V}\right)^2 \exp\{-f(T) U\}/\frac{1}{2}\hbar\omega\} \tag{17}$$

$$U = \frac{1}{4} <\eta> <\hbar\omega> \tag{18a}$$

$$f(T) = \{1-\left(\frac{T}{T_v}\right)^{0.5}\} \tag{18b}$$

The constant factor $\sigma_0 = 0.511$ $(\Omega\text{-cm})^{-1}$ is obtained from the experimental data of Miles et al. [12]. The exponential term arises from the Franck-Cordon transition in the localized state. In Figure 2 we show by open circles the measured conductivity by Miles et al. [12] and compared it with Eq. (17) which is shown as a solid line. The value of $(U/\frac{1}{2}\hbar\omega)$ is taken as 30 and T_v is 117.4K. The agreement is satisfactory.

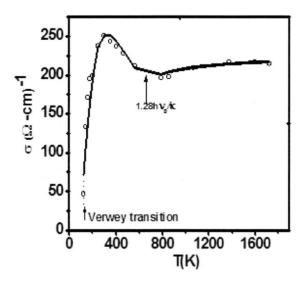

Figure 1. The observed σ(T) curve for Fe_3O_4 is shown in hollow circles for the high temperature phase $T_v < T < 1600K$. The theoretical curve in black solid line is from Eq.(14a) for $T_v < T < 1.28T_0$ and from Eq.(14b) for $1.28T_0 \leq T \leq 1600K$.

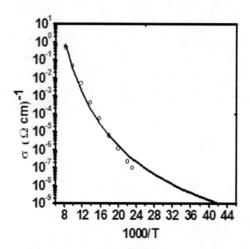

Figure 2. The observed σ(T) curve for Fe₃O₄ is shown in hollow circles for the low temperature phase 45K< T< T_v. The theoretical curve in black solid line is from Eq.(17).

4. TRANSPORT IN MANGANITE

The transport of charge in manganites on the basis of the correlated polaron model is discussed in detail in ref. [11]. In magnetite we observe three different regions of transport (a) $0<T<T_v$, (b) $T_v<T<1.28T_0$ and (c) $1.28T_0<T$. In (a) the small polaron is localized and the transport is by polaron hopping with an activation energy, U, that depends on the number of phonons surrounding the electron and thereby distorting the lattice. When the electron jumps from X_i to a neighboring site X_{i+j} the phonon cloud gets transferred but on a time scale, τ_{ph}, much larger than τ_e, the time for electron transfer, leading to Franck-Condon transition. In (b) the transport is by Einstein type diffusional motion so the conductivity varies inversely with temperature. In (c) the conductivity is independent of T as the frequency of the radiation oscillator v_0' becomes proportional to the temperature of the lattice.

The transport in manganite though primarily governed by Eq. (6), the hopping conductivity in Eq. (8) is different from magnetite in the following respects. In Fe₃O₄ the mixed cations, Fe^{2+} and Fe^{3+}, are on spinel lattice and the two ions are on crystallographic equivalent B-sites with parallel spins due to the strong Weiss field produced by Fe^{3+} ions on the A-sites. On the other hand, in $La_{1-x}A_xMnO_3$ there is no mixed state of manganese for x=0 and the compound is an antiferromagnetic insulator. A Fe₃O₄ type ferromagnetic metal state is obtained only for x > 0.15 for Sr-doped manganite [13].The unusual colossal magnetoresistance of manganite is obtained only within certain values of x that extend between 0.15 and 0.45 for this compound.We have shown earlier that the transportand magnetoresistance in $Re_{1-x}A_xMnO_3$ (R= rare earth, A=divalent ion, Ca, Sr, Ba...) is not fully understood despite numerous attempts [11, 13]. Usually manganites show a ferromagnetic to paramagnetic transition at T_c that lies near the room temperature. The resistivity has a peak at T_c and shows a large drop on the application of a magnetic field giving rise to the colossal magnetoresistance (CMR) effect. The crystallographic properties of the manganite changes with hole doping x. The parent compound LaMnO₃ is a Jahn–Teller distorted perovskite. The three strong local interactions that determine the dynamics of the e_g electrons are (i) the Jahn-

Teller (JT) phonon modes that split the two-fold orbital degeneracy, (ii) the ferromagnetic spin coupling between the e_g and t_{2g} orbitals originating from the Hund's rule (J_H), and (iii) the Coulomb repulsion, U. The magnitude of these energies are $E_{JT} \sim 0.5$ eV, $E_H \sim 2$eV, and U~ 5eV.In the background of these interactions, the itinerancy of the charge carrier is achieved through the e_g electron intersite hopping, t' ~ 0.2 eV, that determines the energy scale of the transport phenomena. The observed energy scale is smaller by an order of magnitude. Only when these are examined on the correlated polaron model that we obtain the correct magnitude for the hopping energy integral t' that accounts for the CMR effect in these compounds.

On doping with strontium the room temperature structure of $La_{1-x}A_xMnO_3$ that is orthorhombic for x < 0.16 changes to rhombohedral for large x. The transport properties depend on the number of charge carriers / Mn atom. In $La_{1-x}Sr_xMnO_3$, there is a large number of parameters that determine the transport properties [13]. For x < 0.08 the phase is an anti-ferromagnetic insulator (AFI). Between 0.08 < x < 0.16, it is a FMI insulator and for 0.16 < x < 0.48, it is a Ferromagnetic metal (FMM). The number of charge carriers per Mn atom (n/Mn) in AFI is ~ 0.1, in FMM phase, for x= 0.175, it is ~ 0.872. It is largest in single crystal where it reaches a maximum of 2.86, for x= 0.40. In Figure 3 we give the resistivity curves for x= 0.05, 0.10, 0.20 and 0.40 in the temperature range 0 < T < 500K. The expression for resistivity is given in Eq.(19) while the physical parameterused for the fit are shown in Table 1,

$$\rho(T) = \rho(0)[(1-f)\frac{\theta_D}{4T_0}\cosh^2(2\varepsilon_p/k_B\theta_D) + f\frac{T}{T_0}\cosh^2(\varepsilon_p/2k_BT)][1+c(1-m^2)\sigma_a^2]\exp[U/k_BT]$$

$$\tag{19}$$

$$\rho(0) = \frac{A}{n}T_0 \; , \; A = \frac{k}{l^2e^2v_0} \tag{19a}$$

$$kT_0 = hv_0 \; , \tag{19b}$$

$$f = \frac{1}{e^{\varepsilon_p/kT}+1} \tag{19c}$$

$$U = U_0(1-m^k)\sigma_a^2 \tag{19d}$$

$$m = \frac{M(T)}{M(0)} \tag{19e}$$

v_0 is the frequency of the active phonon mode that couple the states ψ_1 and ψ_2,
ε_p= gain in the crystal binding energy due to exchange i.e., polaron binding energy,
θ_D= Debye temperature,
σ_a is the short range atomic order parameter that varies with T as

$$\sigma_a = (1-0.75\,t_{ca}^3)^{1/2} \; , \; t_{ca} = T/T_{ca} \tag{19f}$$

From the study of the Sr-doped manganite the following conclusions can be drawn: (i) for AFI and FMI phases at low temperature the behavior of ρ(T) is similar to a semiconducting compound with a band gap $E_g \sim U_0 = 0.18$ eV in the AFI phase while in the FMI phase it is 0.08 eV, (ii) in both the FMI and FMM phases a metal ($\partial\rho/\partial T > 0$) to insulator ($\partial\rho/\partial T < 0$) transition occurs at T_c which is well represented by Eq. (19), (iii) in the metallic phase the low temperature transport is dominated by the zero-point lattice vibration so the residual resistivity is small in the 0.10-1mΩ cm range and is almost independent of hole concentration and (iv) in the metallic phase the number of charge carrier per Mn/atom n is large of the order of 2 compared to the nominal hole concentration of 0.30-0.40 (Table 1). It is well known that large magnetoresistance is obtained in the FMM phase. It is shown that in the metallic phase the transport is due to the valence exchange (VE) through the CP transport that is similar to the conventional double exchange but differs from it in two ways, it has an order of magnitude smaller energy scale (t~.02 eV) and the dynamics is described by the collective behavior of the charge carrier. Further at low temperature the polaron transport is dominated by the zero-point lattice vibrations.

Table 1. The parameters for $La_{1-x}Sr_xMnO_3$ obtained from fit to observed resistivity vs temperature, ρ(T), curves to Eq. (19). The samples are single crystal. The crystal structure is assumed to be cubic with a= 3.858Å and the energy of the active phonon mode is $h\nu_0$=22meV. This gives A/n = 4.35×10^{-5} Ω cm K^{-1} for n=0.10/Mn. The Debye temperature, θ_D, is taken as 425K

x	Phase	T_C/T_N (K)	A/n (Ω-cm K^{-1})	n/Mn	T_{ca} (K)	ε_p (K)	U_0 (K)	k	c	Ref
0.05	AFI	139	2.4×10^{-5}	0.18	---	---	2300	---	---	13
0.10	FMI	150	5.2×10^{-4}	0.0082	425	300	950	---	---	13
0.20	FMM	310	8.46×10^{-6}	0.504	865	150	625	0.75	2	13
0.40	FMM	371	1.49×10^{-6}	2.86	1400	100	400	0.75	2	13

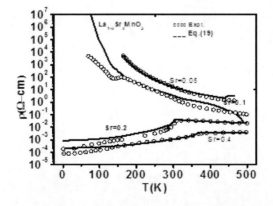

Figure 3. Variation of the resistivity with temperature ρ(T) in $La_{1-x}Sr_xMnO_3$ x= 0.05 (antiferromagnetic insulator, AFI), 0.10 (ferromagnetic insulator, FMI), and 0.20 and 0.40 (both ferromagnetic metal, FMM) are shown by hollow circles. These are for single crystal and are obtained from ref [13]. The theoretical curves in black solid line is from Eq.(19) with parameters given in Table (1). In all cases the Debye temperature θ_D is 525K and the frequency of the active phonon mode, ν_0=176cm^{-1}. Note that T_C denotes the Curie temperature for both the FMI and FMM phases. Also n/x is the ratio of the carrier concentration data from Eq.(19) and the nominal hole concentration from composition. The reason it exceeds 1 in FMM is discussed in [13].

We have also studied the correlated polaron transport in La$_{2/3}$Ca$_{1/3}$MnO$_3$ in polycrystalline, single crystal and thin film forms. For calcium doped manganite, n/Mn is close to 2.5 for all single crystal and thin films [11]. In this case ε_p is small and varies between 40 K and 75 K. For polycrystalline solids n/Mn is 3 to 4 orders of magnitude smaller and ε_p increases by a factor of seven to eight. Since resistivity varies as cosh^2 ($\varepsilon_p/2k_BT$) it shows a change from FMI to FMM as the number of charge carriers increases on large doping x. In Figure (4a) & (4b) we have shown ρ(T) vs T curves for a Ca doped thin film with and without the external field, respectively, from ref. [11]. The theoretical curve is from Eq. (19) with parameters given in Table 2. The contribution to magnetoresistance comes mainly from the exp term in Eq.(19) and is given by

$$MR = (3U_0/2kT)(1/T_C)\, \sigma_a^2\, k\, m^{k-2} \Delta T(H) \tag{20}$$

where $\Delta T(H) = T_{MI}(H) - T_{MI}(0)$ is the temperature separation between the resistivity peaks with and without H. The close fit to experiment shows that Eq.(19) describes well the mobility of the charge carriers over the range 0 to 400K. In the low temperature range (0 <T <150K) it is like in a polaron band that is unaffected by a field of the order of 7T. Here the transport is due to quantum tunneling [11]. The charge carriers begin to get localized when temperature exceeds 150K [15]. The transport is then with an activation energy U that is dependent on both the spin and atomic orders. This gives rise to the large magnetoresistance in the system. We conclude that the transport in manganites can be reasonably well explained on the basis of the correlated polaron model.

Table 2. The parameter for La$_{1-x}$Ca$_{1/3}$MnO$_3$ thin film obtained from fit to observed resistivity vs temperature, ρ(T) curve to Eq. (19) for (a) H= 0 and (b) H=7T from Ref [11]

Compound	H (T)	T$_C$ (K)	A/n (Ω-cm-K^{-1})	n/Mn	T$_{ca}$ (K)	ε_p (K)	U$_o$ (K)	k	c	Ref
La$_{2/3}$Ca$_{1/3}$MnO$_3$(thin film)	0	260	1.76×10^{-6}	2.42	825	40	545	1.5	2.0	11
La$_{2/3}$Ca$_{1/3}$MnO$_3$(thin film)	7	330	1.59×10^{-6}	2.68	825	70	545	1.5	2.0	11

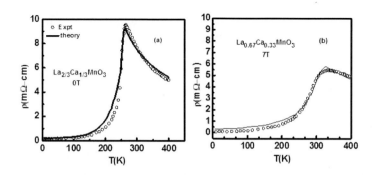

Figure 4. (a & b). Variation of the resistivity with temperature, ρ(T), for thin film of La$_{2/3}$Ca$_{1/3}$MnO$_3$ from Ref [11] are shown by hollow circles for H= 0 in (a) and H= 7T in (b) The theoretical curves in black solid line are from Eq.(19) with parameters given in Table(2). The analysis covers the range 0≤T≤400K. Note that on application of field only three parameters in Table (2) change, T$_c$, n and ε_p. The dominant change ρ(T) is from change in T$_c$ and n [11].

5. NORMAL STATE ELECTRON TRANSPORT IN CU-O HIGH TEMPERATURE SUPERCONDUCTORS

Varma and his collaborators have proposed a phenomenological model called the marginal Fermi liquid that successfully describes the unusual transport properties of high T_c oxide superconductors in the normal state [7]. Srivastava has shown that their treatment follows from the CP model [6].The theory is based on the assumption that the spin and charge fluctuations that occur in the system do not depend on the wave number q since these are related to the fluctuations created by the longitudinal optical phonons that are in Brillouin zone-boundary modes. In the CP model the fluctuations are created by Γ-point phonons as discussed in section 2 for magnetite.In this case the conductivityσ (0,0) for ω=0 and q=0 is given by Eq.(6). In the correlatedpolaron model the electron oscillates around a mean position in the self-trapped field as discussed by Landau [14]. This happens also in the correlated polaron model in Eq.(3) when a pair of electrons oscillates within a unit cell comprising of four ions in dynamic ordering correlation leading to a small enhancement of the Madelung energy, ε_p. From this we obtain the self-energy operator [6]

$$\Sigma(0,\omega) = g^2 N^2(0) [\omega\ln\{\frac{\{\omega^2 + T^2 \cosh^2(\varepsilon_p\beta/2)\}^{1.2})}{\omega_{0p}}\} - i\frac{\pi}{2}\{\omega^2 + T^2 \cosh^2(\varepsilon_p\beta/2)\}^{1/2}] = \Sigma_1 + i\Sigma_2$$

$$(21)$$

Here g is the coupling constant and N(0) is the density of states at the bottom of the band and is equal to $1/\hbar\omega_0$.We have

$$gN(0) = \alpha = \varepsilon_p / \hbar\omega_0 \tag{21a}$$

The plasma frequency v_{0p} is given by

$$v_{0p}^2 = \frac{ne^2}{4\pi m} \tag{21b}$$

where m is the effective mass of the free electron. The Drude conductivity is given by

$$\sigma_D(0,\omega) = (4\pi v_{0p}^2)\frac{\Gamma_D}{\omega^2 + \Gamma_D^2}, \tag{21c}$$

$$\Gamma_D = 2\Sigma_2(0,0) \tag{21d}$$

This gives the dc conductivity in Eq.(6) as $\sigma_D(0,0)$.The electrons in the correlated phase coherent state are in localized state below the bottom of the polaron band [15].The expression for σ in Eq.(6) does not apply in the limit $T\to 0$ when zero-point phonon begins to dominate. If the charge carriers get localized as in magnetite there is a first order phase transition and conductivity has the form obtained in Eq.(17). For HTSC we need an expression that is valid in the low temperature region between T_c and room temperature when the charge carriers are itinerant. For this we examine the transport properties of a material whose T_c is so low that superconductivity can be suppressed by the application of a strong enough steady magnetic field such that the transport property of the charge carriers in the normal state can then be examined. As these compounds show strong anisotropy in their transport properties only results on high quality single crystal sample can be analyzed. A peculiar feature of the resistivity is that the in-plane resistivity shows linear dependence over a wide temperature range and is expressed as [16]

$$\rho_{ab}(T) = c_1 T + c_2 \tag{22}$$

where c_1 and c_2 are constants. The out- of- plane resistivity ρ_c is often non-linear, it increases when the temperature is lowered from room temperature towards the critical temperature T_c and disappears abruptly at T_c. Any transport theory should account for both the type of observations. Daou et al. [17] have measured the transport properties of $La_{1.6-x}Nd_{0.4}Sr_xCuO_4$ (x = 0.2, 0.24) in which it is possible to suppress superconductivity by applying a steady magnetic field. For example, for x= 0.2, T_c is 20 K. With a field of 35 T the resistivity measured at 1 K is 245 $\mu\Omega$-cm. This is shown in Figure 5. When resistivity is measured between 1 and 150 K this sample shows a deviation from linearity at 80 K and a minimum at 37 K. In sharp contrast to this the sample with x= 0.4 shows a linear drop from 150 K to T→ 0 K and can be expressed as in Eq.(22). Similar linear dependence of ρ (T) is observed in $YBa_2Cu_3O_{7-\delta}$, in all the three directions a, b, c of the orthorhombic structure with a = 3.828 Å, b=3.888 Å and c = 11.65 Å [18]. Here $\rho_b < \rho_a$ and ρ_c is three orders of magnitude larger than $\rho_b \sim \rho_a$. These are shown in Figure 6 for resistivity along a and b directions and in Figure 7 for the c direction.

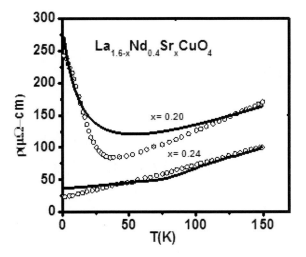

Figure 5. The observed variation of resistivity $\rho(T)$ of $La_{1.6-x}Nd_{0.4}Sr_xCuO_4$ single crystal is shown by open circles (i) for x= 0.20 and (ii) for x=0.24 from ref [17] between temperature 0 and 150K. The theoretical curves are plotted using Eq. (23) are shown by solid black line with parameters given in Table 3.

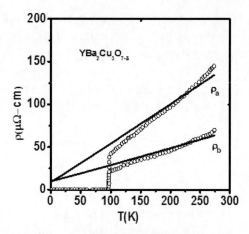

Figure 6. The observed variation of resistivity ρ(T) of YBa$_2$Cu$_3$O$_{7-\delta}$ single crystal between T$_c$ and 300K along the a and b axes is shown by open circles from ref [18]. The theoretical curves in black solid lines are plotted using Eq. (23) with parameters in Table 3.

We have shown that zero point vibrations play a significant role in the formation of the phase coherent state in Eq.(2). For HTCS we consider the region T$_c$≤ T ≤ 300 K for which most of the resistivity data is reported. On physical grounds extending the arguments used in deriving Eq.(6) to T→ 0 we can write based on Eq.(10) that includes the zero-point energy,

$$\rho(T) = \rho(0) \left[\frac{1}{2} + \frac{T}{T_0}\right] \cosh^2\left[\frac{\alpha}{2(0.5 + T/T_0)}\right] \tag{23}$$

where, $\rho(0) = \dfrac{\hbar}{ne^2 l^2}$, $\alpha = \dfrac{\varepsilon_p}{kT_0}$, $kT_0 = \hbar\omega$

The resistivity curves for La$_{1.6-}$Nd$_{0.4}$Sr$_x$CuO$_4$ (x = 0.2, 0.24) are plotted in Figure 5 with the parameters ρ(0), T$_0$ and α given in Table (3). For x = 0.2 Daou et al. [17] find that the curve ρ(T) deviates from linearity at T* = 80K when the sample is cooled from 150K. This is the temperature T$_0$ in Eq.(23). The minimum in this equation for ρ(T) occurs at T$_{min}$ such that,

$$\coth\theta = 2\theta, \tag{24}$$

$$\theta = \frac{\alpha}{1 + \dfrac{2T_{min}}{T_0}} = 0.77$$

In Figure 6 we give the variation of ρ(T) with T for YBa$_2$Cu$_3$O$_{7-\delta}$ for a and b directions using Eq.(23) and the parameters ρ(0), α, and T$_0$ from Table (3). Figure 7 gives the ρ(T) plot for c direction. Here α is nearly the same for all the three axes while ρ(0) is about the same for a and b but increases by two orders of magnitude for the c direction. Further T$_0$ is smallest for the a axis and highest for the c axis. Both ρ(0) and T$_0$ increases with the lattice constant. In both cases the experimental data from [18] has been used.

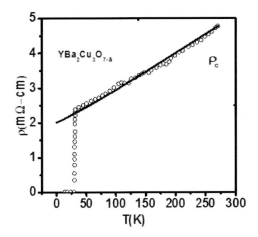

Figure 7. The observed variation of resistivity ρ(T) of YBa$_2$Cu$_3$O$_{7-\delta}$ single crystal between T$_c$ and 300K along the c axis is shown by open circles from ref [18]. The theoretical curves in black solid line is plotted using Eq. (23) with parameters in Table 3.

Table 3. The constants ρ$_0$, T$_0$ and α to get the dc conductivity using Eq.(23) for the high temperature oxide superconductors. For comparison these constants are given for magnetite if Eq.(23) is used instead of Eq.(14a) in the temperature range T$_v$< T <1.28hv$_0$/k

HTSC	ρ(0) (μΩ−cm)	T$_0$(K)	α = ε$_p$/T$_0$
Y123(a)	15.3	33	0.5
(b)	16.88	84	0.4
(c)	3160	282	0.5
214 (0.24)	51.25	92	0.6
214 (0.20)	60	80	1.8
magnetite	T$_v$< T < 1.25T$_0$		
Fe$_3$O$_4$	0.0316	516	1

It follows from Eq.(24) that T$_{min}$ is non-vanishing and positive only when α exceeds 0.77. For α < 0.77 there is no upturn and Eq.(23) gives a monotonous change in ρ(T) with T if we measure resistivity between room temperature and T$_c$ or if superconductivity is suppressed on the application of a steady magnetic field between 300 and 0 K. In [17] the resistivity ofLa$_{1.6-}$Nd$_{0.4}$Sr$_x$CuO$_4$ (x = 0.2, 0.24) single crystal is measured. This is shows in Figure 5 by open circles. For x = 0.20 at 1K it is 245(μΩ-cm). From Table 3 using Eq.(23) we obtain, ρ(1K) =273.30 (μΩ-cm) which is in reasonable agreement with experiment. For the sample x = 0.24, α=0.6< 0.77, so there is no upturn. It may be noted that the magnitude of ρ(0) and T$_0$ for (214) are not very different for x=0.20 and 0.24 but α isdifferent and determines the nature of the curve.

We have included also the parameters for Fe$_3$O$_4$ in Table 3 for comparison when ρ(T) is expressed as in Eq.(23) instead of Eq.(6) in the temperature range T$_v$<T<1.28hv$_0$/k. Despite the fact that the transition in Fe$_3$O$_4$ is of first order while in superconductors it is of second order, expressions in Eq.(6) and Eq.(23) represent similartype of transition when α exceeds 0.77 and ρ(T) has an upturn at T$_{min}$. In both cases the zero-point phonons play an important

424   *C. M. Srivastava and N. B. Srivastava*

role since a group of n particles that form phase coherent states, ψ_1 and ψ_2, between which they oscillate about their mean position to enhance the ionic Coulomb binding energy by a small amount ε_p begin to get localized.

6. MAGNETIC POLARONS IN DILUTE FERROMAGNETIC OXIDE FILMS AND NANOPARTICLES

A large group of dilute magnetic oxide and nitrides like doped ZnO and GaN exist today that show magnetism with T_c in some cases exceeding 900K. In the pure form these are wide gap semiconductors and are diamagnetic in the bulk form [19, 20].The electrical transport and magnetic studies in some undoped- and doped- oxides at nanoscale have presently invoked immense interest due to their ability to switch magnetization under small magnetic fields which is central to the technology of spintronic applications. The earlier suggestions that Mn-doped ZnO and GaN could produce room temperature ferromagnetism (RTF) based on Zener model was found to be inadequate as it is based on unrealistic assumptions. We have suggested that CP model through coupling of electrons to phonons in mixed valence ionic compounds can explain the transport and ferromagnetism in nano-structured oxides [9]. For nano structured solids we express Eq.(23) using the approximation that only zero-point phonons produce the one-dimensional phase coherent states. In this case the resistivity varies as [9]

$$\rho(T) = \frac{A}{n}\left(\frac{T_0}{2}\right)\cosh^2\left[\frac{\varepsilon_p}{2\left(T+\frac{T_0}{2}\right)}\right][1-f]^{1/3} \qquad (25)$$

$$= \rho(0)\cosh^2\left[\frac{\varepsilon_p}{2\left(T+\frac{T_0}{2}\right)}\right][1-f]^{1/3}$$

where

$$\frac{A}{n} = \frac{k}{e^2 a^2 v_{ph}}, \; f = \frac{1}{e^{\varepsilon_p/kT}+1}, \; \rho(0) = \frac{A}{n}\left(\frac{T_0}{2}\right) \qquad (25a)$$

HerekT_0 is the energy of the radiation oscillator that produces the exchange $\psi_1 \leftrightarrow \psi_2$. In nano-structured solids due to the one dimensional nature of the crystal structure and the mixed valence state of oxygen that produces the mobility of charge carrier, the function $(1-f)^{1/3}$is used in Eq.(25).Consider a doped material $[A_{1-x}M_x][O_{1.6}]$. The magnetization depends on (i) dopant, M, (ii) its concentration, x, (iii) the matrix A and (iv) the oxygen deficiency, δ. Al and Ni co-deposited ZnO films are not magnetic but develop room temperature ferromagnetism on magnetic annealing [21]. In Figure 8 we give the plot of $\rho(T)$ curve for the film of $Zn_{0.91}Al_{0.07}Ni_{0.02}O_2$. Here oxygen is in mixed valence state and the valence exchange takes place from site i to j through

$$O_i^{1-}\uparrow + O_j^{2-} \leftrightarrow O_i^{2-} + O_j^{1-}\uparrow$$

The spin on O^{1-} remains unchanged on phonon induced exchange. The density functional calculations of magnetism in ZnO thin film and nano wires show that the unpaired 2p electrons at oxygen sites exist at the surface of the film [22, 23]. In Figure 8 ρ(T) for $Zn_{0.91}Al_{0.07}Ni_{0.02}O$ film with the parameters given in Table 4 is plotted along with the experimental point from [21]. The agreement with theory is satisfactory.

Table 4. The parameters for $Zn_{0.91}Al_{0.07}Ni_{0.02}O$ thin film from ref [9] obtained from fit to observed resistivity vs temperature, ρ(T), curve to Eq.(25)

Composition Film(F)	A/n Ω-cm K^{-1}	T_0 (K)	ε_p(K)	ρ(0) (Ω-cm)	d	Ref.
$Zn_{0.91}Al_{0.07}Ni_{0.02}O_2$(F)	2.818x10^{-5}	500	300	0.01409	1	9

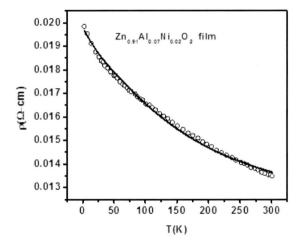

Figure 8. The observed variation in the resistivity ρ(T) of $Zn_{0.91}Al_{0.07}Ni_{0.02}O_2$ thin film shown by open circles from ref [9] along with theoretical curve in black solid line from Eq. (25) with parameters given in Table 4.

CONCLUSION

The transport properties of some mixed valence oxides compounds exhibit anomalous behavior that is not understood on the present conventional theories. It is shown that these arise from the electron-phonon interaction between two mixed valence cations/anions that on coupling to a particular phonon mode exchange their charges in a phase coherent manner. This increases the binding energy of the ionic solid by a small amount. In magnetite, it leads to an order-disorder transition at the Verwey temperature, T_V, at which there is a steep fall in dc conductivity arising from a first order phase transition. The unusual behavior of resistivity between 660K and 1600K that is very nearly independent of temperature and is close to 200μΩ-cm, arises as the charge carriers begin to jump from one site to another at a frequency ω_0 that is proportional to T, so ω_0/T that appears in Einstein relation for diffusional motion becomes independent of temperature. Likewise the first order phase transition at T_v is shown to arise from the Franck-Condon transition in the small polaron model.

The colossal magnetoresistance in hole doped manganite is shown to arise from a mechanism of exchange between two degenerate states when both magnetic and atomic order affects the activation energy for transport. In the Ca-doped manganite discussed here at low temperature (<150K) the exchange is without the Franck-Condon transition while these are present at high temperature (T>150K). The activation energy is then magnetization dependent in the ferromagnetic metallic phase resulting in the colossal magnetoresistance in the material near room temperature. The observed variation of resistivity by two orders of magnitude in thin films of manganites when temperature is varied from near zero to 400 K is in close agreement with the theory based on correlated polarons. Likewise on application of small external field the change in the transport properties is well explained. The normal state oxide superconductors whose conductivity behavior is equally anomalous are well explained on the same principle of enhancement of ionic binding energy on charge exchange induced by electron-phonon coupling. The widely different behavior of the variation of resistivity with temperature between some of the high T_c oxide compounds arises from the dimensionless parameter α that measures the electron-hole binding energy with respect to the phonon energy. In doped oxide films without any 3d elements the unusual behavior of magnetic order is suggested to arise from the oxygen ion being in the mixed valence state.

REFERENCES

[1]   Srinivasan, G. and Srivastava, C.M., (1981) *Phys. Stat. Sol. (b)*103 665.

[2]   Verwey, E.J.W., *Nature* (1939)144, 327.

[3]   Verwey, E.J.W. and Haayman, P.W., (1941) *Physica* 8979.

[4]   Verwey, E.J.W., Haayman, P.W. and Romeijn, F.C., (1947) *J.Chem. Phys.* 15 174 and 181.

[5]   Goodenough, J.B., in *Recent Advances in Materials Research* ed. C.M. Srivastava, (1982)Oxford & IBH Publishing Co. New Delhi, p. 1-20.

[6]   Srivastava, C.M., (1991) *Physica C* 176, 481.

[7]   Varma, C.M., Schmitt-Rink, S and Abrahams E. (1987) *Solid State Commun.* 62 681.

[8]   Rowan, A.D., Patterson, C.H., Gasparov, L.V., (2009) *Phys. Rev. B* 79, 205103.

[9]   Srivastava, N.B., Bahadur, D., and Srivastava, C.M., (2014) *J Jpn. Soc. Powder Powder Metallurgy* Vol 61 Supplement No. S1.

[10]   Reik, H. G. in *Polaron in ionic crystals and polar semiconductors* edited by J T Derve (North-Holland Amsterdam, 1972) p.679-714.

[11]   Srivastava, C.M., Srivastava, N.B., Singh, L.N., Bahadur, D., (2009), *Jour. Appl. Phys.*105, 1.

[12]   Miles, P.A., Westphal, W.B. and Von Hippel, A., (1957) *Rev. Mod. Phys.* 29 279.

[13]   Srivastava, N.B., Singh, L.N. and Srivastava, C.M., (2009) *Journal of Applied Physics*10507D704.

[14]   Landau, L.D. *Physik Z. Sowjetunion* [*Physics Z. Soviet Union*] (1965) 3 A 338.

[15]   Appel, J., (1968) *Solid State Physics,* vol. 21, 365.

[16] Plakida, N.M., (1995) *High-Temperature Superconductivity, Experiment and Theory*, Springer-Verlag, Berlin. pp- 127-132

[17] Daou, R., Leyraud, N. D., LeBoeuf, D., Li, S. Y., Laliberté, F., Cyr-Choiniére, O., Jo, Y. J., Balicas, L., Yan, J. Q., Zhou, J. S., Goodenough, J. B., Taillefer, L., (2009) *Nature Physics* 5 January p.31.

[18] Friedmann, T.A., Rabin, M.W., Giapintzakis, J., Rice, J.P., Ginsberg, D.M., (1990) *Phys. Rev. B* 42 6217.

[19] Yamada, Y., et al. (2011), *Science*, 332, 1065.

[20] Hann, Q., et al. (2011) *J. Phys. Chem. C*, 115 (8) 3447.

[21] Schoenhaiz, A.L., et al. (2009) *Appl. Phys. Lett.* 94 162503.

[22] Yu, M., et al. (2011), *Materials Chemistry and Physics* 126, 797.

[23] Wang, Q., et al. (2008) *Phys. Rev. B*77 205411.

In: Polarons: Recent Progress and Perspectives
Editor: Amel Laref

ISBN: 978-1-53613-935-8
© 2018 Nova Science Publishers, Inc.

Chapter 13

POLARONS AND BIPOLARONS IN COLOSSAL MAGNETORESISTIVE MANGANITES

Guo-Meng Zhao[1,2,*], *J. Labry*[1] *and Bo Truong*[1]
[1]Department of Physics and Astronomy,
California State University, Los Angeles, CA, US
[2]Research Institute of Zhejiang University—Taizhou,
Taizhou, China

Abstract

The microscopic mechanism for colossal magnetoresistance (CMR) and ferromagnetism in doped manganites has not been fully understood although it is generally accepted that strong Jahn-Teller electron-phonon interaction plays an important role in the basic physics of manganites. It appears to reach a consensus that double-exchange is the primary cause of the ferromagnetism in doped manganites. Here we show that the d-p exchange rather than the double-exchange is the primary origin of the ferromagnetism due to the fact that the undoped parent manganites are not Mott but charge-transfer insulators. We further show that the theoretical model based on the d-p exchange and strong electron-phonon interactions can quantitatively explain the CMR and the unconventional oxygen-isotope effects in doped ferromagnetic manganites. The unconventional oxygen-isotope effects on the Curie temperature and on electrical transport provide indisputable evidence for the existence of the small polarons in the ferromagnetic state and the coexistence of small polarons and bipolarons in the paramagnetic state in doped ferromagnetic manganites.

Keywords: polarons, bipolarons, colossal magnetoresistance, manganites

1. Introduction

1.1. Concept of Polarons

For the last three decades, high-temperature superconductivity and colossal magnetoresistance have been discovered in several doped perovskite oxides (e.g., cuprates [1], bis-

---

*E-mail address: gzhao2@calstatela.edu.

muthates [2] and manganites [3]). These doped oxides are characterized by strong electron-phonon interactions, significant carrier densities ($\geq 10^{21}$ cm^{-3}), and low mobility of the order or even less than the Mott-Ioffe-Regel limit ($ea^2/\hbar \sim 1$ cm^2/Vs) (where a is the lattice constant). The very nature of the low-temperature 'metallic' state of these materials cannot be understood within the framework of the canonical theory of metals. Since the electron-phonon interactions in these perovskite oxides are much stronger than in normal metals, new theoretical approaches based on the formation of polarons/bipolarons are required to understand the basic physics of these materials.

The concept of polarons was first introduced by Landau in 1933 [4]. If an electron is placed into the conduction band of an ionic crystal, the electron is 'trapped by digging its own hole' due to a strong Coulombic interaction of the electron with its surrounding positive ions. The electron together with the lattice distortions induced by itself is called polaron (lattice polaron). Lattice polarons are not 'bare' charge carriers, but are the carriers which are 'dressed' by lattice distortions. Later on, the polaron problem was treated in great detail. One of the examples is the Holstein's treatment where an electron is trapped by self induced deformation of two-atomic molecules (Holstein polaron) [5, 6]. In this case, the polaron moves by thermally activated hopping at high temperatures with a diffusion coefficient $\omega a^2 \exp[-(E_p/2 - t)/k_B T)]$, where ω is the characteristic vibration frequency, E_p is the polaron binding energy, and t is the bare hopping integral. Further extensive theoretical studies (for review see [7, 8]) have shown that polarons behave like heavy particles, and can be mobile with metallic conduction at sufficiently low temperatures. Under certain conditions [9] they form a polaronic Fermi liquid with some properties being different from ordinary metals.

It is well known that electrons can change their mass in solids due to the interactions with ions, spins, and themselves. The renormalized (effective) mass of electrons is independent of the ion mass M in ordinary metals where the Migdal adiabatic approximation is believed to be valid. However, the effective mass of polarons m^* will depend on M. This is because in the strong-coupling limit the polaron mass $m^* = m \exp(\gamma E_p/\hbar\omega)$ [7, 10], where m is the band mass in the absence of the electron-phonon interaction, γ is a constant, and ω is a characteristic phonon frequency which depends on the masses of ions. Hence, there is a large isotope effect on the polaronic carrier mass, in contrast to the zero isotope effect on the effective carrier mass in ordinary metals.

The total exponent of the isotope effect on the effective carrier mass is defined as $\beta = \sum -d\ln m^*/d\ln M_i$ (M_i is the mass of the ith atom in a unit cell). For polaronic carriers, this definition leads to

$$\beta = -\frac{1}{2}\ln(m^*/m). \tag{1}$$

It is interesting that the simple relation $m^* = m \exp(\gamma E_p/\hbar\omega)$ is even valid in the weak coupling region in the case of the long-range Fröhlich electron-phonon interaction [10]. Then the polaron mass enhancement factor m^*/m in this case is simply equal to $\exp(-2\beta)$.

Therefore, if electron-phonon coupling in a solid is strong enough to form polarons and/or bipolarons, one will expect a substantial isotope effect on the effective mass of carriers. In this article, we will review some unconventional oxygen-isotope effects in doped manganites. These include oxygen-isotope effects on the Curie temperature and electrical

Polarons and Bipolarons in Colossal Magnetoresistive Manganites

transport in doped ferromagnetic manganites. The observed large unconventional isotope effects clearly demonstrate that the formation of polarons/bipolarons due to strong electron-phonon coupling is relevant to the basic physics of manganites and important for the occurrence of colossal magnetoresistance.

1.2. Background of Doped Maganites

Doped manganites $Ln_{1-x}A_xMnO_3$ (where Ln is a trivalent rare earth ion and A is a divalent ion) have been found to exhibit some remarkable features. The undoped parent compound $LaMnO_3$ (with Mn^{3+}) is an insulating antiferromagnet [11]. When Mn^{4+} ions are introduced by substituting a divalent ion (e.g., Ca) for La^{3+}, the materials undergo a transition from a high-temperature paramagnetic and insulating state to a ferromagnetic and metallic ground state for $0.2 \leq x \leq 0.5$ [12]. For $0.5 \leq x \leq 0.8$, the materials exhibit an insulating, charge-ordered and antiferromagetic ground state. The temperature at which the insulator-metal transition occurs can be increased by applying a magnetic field. As a result, the electrical resistance of the material can be decreased by a factor of 1000 or more [3], if the temperature is held in the region of the transition. This phenomenon is now known as colossal magnetoresistance (CMR).

The coexistence of ferromagnetism and metallic conduction in doped manganites has long been explained by the double-exchange (DE) model [13, 14]. Crystal fields split the Mn 3d orbitals into three localized t_{2g} orbitals, and two higher energy e_g orbitals which are hybridized with the oxygen p orbitals. Each manganese ion has a core spin of $S = 3/2$, and a fraction $(1 - x)$ have extra electrons in the e_g orbitals with spin parallel to the core spin due to strong Hund's exchange. The electron can hop to an adjacent Mn site with unoccupied e_g orbitals without loss of spin polarization, but with an energy penalty that varies with the angle between the core spins. This double-exchange model accounts qualitatively for ferromagnetic ordering and carrier mobility that depends on the relative orientation of Mn moments which near T_C will therefore be strongly dependent on the applied field. However, Millis, Littlewood and Shraiman [15] have pointed out that double-exchange alone cannot fully explain the resistivity data of $La_{1-x}Sr_xMnO_3$. They suggest that lattice-polaronic effects due to strong electron-phonon coupling (arising from a strong Jahn-Teller effect) should be involved. The basic argument [16] is that in the high-temperature paramagnetic state the electron-phonon coupling constant is large and the carriers are small polarons, while the growing ferromagnetic order increases the bandwidth and thus decreases the coupling constant sufficiently for metallic behavior to occur below the Curie temperature T_C. The nature of charge carriers is switched from small polarons in the paramagnetic state to large polarons in the ferromagnetic state. On the other hand, Alexandrov and Bratkovsky (AB) [17, 18] show that, in order to explain CMR quantitatively, one needs to consider the formation of small bipolarons (pairs of small polarons) in the paramagnetic state. They pointed out that the calculated magnetoresistance based on the model of Millis, Littlewood, and Shraiman (MLS) is about one order of magnitude smaller than the observed MR values.

2. The Origin of Ferromagnetism in Doped Manganites

2.1. Double-Exchange

In the DE model, it is implicitly assumed that doped carriers are Mn e_g electrons, that is, the undoped parent compounds are Mott insulators. This assumption is not justified by both electron-energy-loss [19] and photoemission spectroscopies [20], which have shown that the ferromagnetic manganites ($x<0.4$) are doped charge-transfer insulators with doped carriers mainly residing on the oxygen orbitals. Now a question arises: Does the ferromagnetism of doped manganites really originate from the DE interaction? If not, what causes the ferromagnetism in these compounds? One way to address this fundamental issue is to make a quantitative comparison between the predicted properties of the DE model and experiment [21].

There are two important parameters in the DE model, namely, the bare bandwidth W of the e_g bands, and the Hund's rule coupling J_H between e_g and t_{2g} electrons. These parameters are related to an optical transition between the exchange splitted e_g bands [22, 23], and thus can be determined from optical data. With these unbiased parameters, one can calculate the zero-temperature spin stiffness $D(0)$ and the Curie temperature T_C. Here $D(0)$ is defined as $\omega_q = D(0)q^2$ with ω_q being the magnon frequency. When one introduces interactions such as electron-phonon and electron-electron interactions, the magnitudes of both $D(0)$ and T_C are generally reduced.

Now we start with a Kondo-lattice type Hamiltonian [24], which leads to Zener's DE model when $J_H \rightarrow \infty$,

$$H = -\frac{1}{2} \sum_{<ij>ab\alpha} t_{ij}^{ab}(d_{ia\alpha}^{\dagger}d_{jb\alpha} + h.c.)$$

$$-J_H \sum_{ia\alpha\beta} \vec{S}_i \cdot d_{ia\alpha}^{\dagger}\vec{\sigma}_{\alpha\beta}d_{ia\beta} + H_{INT}. \tag{2}$$

Here $d_{ia\alpha}^{\dagger}$ creates an electron in e_g orbital a with spin α, t_{ij}^{ab} is the direction-dependent amplitude for an electron to hop from orbital a to orbital b on a neigboring site, and H_{INT} represents the other interactions. The calculated band structure is well fit by a t_{ij}^{ab}, which involves only nearest-neighbor hopping that is only nonzero for one particular linear combination of orbitals, i.e., $t_{ij}^{ab} \propto t$, where t is a characteristic hopping amplitude that is related to the bare bandwidth W by $W = 4t$. Here we still call Eq. 2 as 2-orbital DE model rather than 2-orbital Kondo-lattice model for convience. The quantum and thermal average of the hopping term in Eq. 2, defines a quantity K:

$$K = (1/6N_{site}) \sum_{<ij>ab\alpha} t_{ij}^{ab}\langle d_{ia\alpha}^{\dagger}d_{jb\alpha} + h.c.\rangle, \tag{3}$$

The quantity K is related to the optical spectral weight by a famillar sum rule,

$$K = \frac{2a}{\pi e^2} \int_0^{\infty} d\omega\sigma_1(\omega), \tag{4}$$

where σ_1 is the real-part optical conductivity contributed only from the e_g electrons. The quantity K generally consists of the Drude part K_D, and incoherent part K_I which, in

Polarons and Bipolarons in Colossal Magnetoresistive Manganites 433

general, involves interband and intraband optical transitions. The Drude part K_D can be related to the plasma frequency Ω_p as

$$K_D = \frac{a}{4\pi e^2}(\hbar\Omega_p)^2. \tag{5}$$

On the basis of Eq. 2, Quijada et al. [24] showed that, to the order of $1/J_H$, the spin stiffness $D(0)$ is,

$$D(0) = \frac{Ka^2}{4S^*}[1 - \frac{\eta t^2}{J_H SK}], \tag{6}$$

where $S = 3/2$, $S^* = S + (1 - x)/2$, and $\eta = 1.04$ when $H_{INT} = 0$. The presence of interactions may change the value of η. A similar result was obtained by Furukawa [22] for an 1-orbital DE model using the dynamical mean field method, but the value of η is doping dependent and less than 1.

From the above equations, one can calculate K and $D(0)$ using realistic values of the bare bandwidth $W = 4t$ and the Hund's rule coupling J_H. Both the local density approximation (LDA) [25] and 'constrained' LDA [26] calculations show that $J_H \simeq 1.5$ eV. The calculated J_H value is very close to the atomic values for $3d$ atoms. The values of both W and J_H can be determined from an optical transition between the exchange splitted e_g bands [22, 23]. The peak position of this optical transition is about $2J_H$, and the width of the peak contains information about the bare bandwidth [22, 23]. From the optical data of Ref. [24, 27], one finds $J_H \simeq 1.6$ eV and $W = 1.6$-1.8 eV by comparing the data with the theoretical predictions [22, 23]. The value of J_H obtained from the optical data is in excellent agreement with the calculated one. This implies that the feature appeared at about 3 eV in the optical data indeed arises from the optical transition between the exchange splitted e_g bands.

The quantity K° for noninteracting 2-orbital model can be evaluated when the bare bandwidth W is known. Takahashi and Shiba [28] have calculated the optical conductivity using a tight binding (TB) approximation of the band structure. From their calculated result for the interband optical conductivity, we evaluate that $K_I^\circ = 0.088t$ for $x \sim 0.3$. One should also note that the magnitude of t defined in Ref. [28] is 1.5 times smaller than the t defined here. Since $K_D^\circ = 1.2K_I^\circ$ [28], then $K_D^\circ = 0.106t$ and $K^\circ = 0.194t$. The LDA calculation for a cubic and undistorted structure shows that [29] $W = 3$ eV and $\hbar\Omega_p^\circ = 1.9$ eV. Using Eq. 5 and $\hbar\Omega_p^\circ = 1.9$ eV, one yields $K_D^\circ = 78.6$ meV. Since $t = W/4 = 0.75$ eV, one readily finds that $K_D^\circ = 0.105t$, in remarkably good agreement with that $(K_D^\circ = 0.106t)$ estimated from the TB approximation. This justifies the relation $K^\circ = 0.194t$ obtained from the TB approximation. It is interesting to compare the present result with those reported in Ref. [23] and [24]. In Ref. [23], it is found that $K^\circ = 0.34t$ for $x = 0.3$ using the dynamic mean field method (DMF). In Ref. [24], Quijada et al., claimed $K^\circ = 0.46t$ for $x = 0.3$, which might be true if the two e_g bands have the same dispersion. Therefore the quantity K° is significantly overestimated in Refs. [23] and [24].

When the Hund's coupling J_H is turned on, the quantity K is reduced compared with K°. For $J_H = \infty$ and $x = 0.3$, $K = 0.77K^\circ$ [23, 30]. It was also shown that [23] the reduction factor (0.77) is basically the same for $J_H \geq t$. Therefore, we have $K = 0.77K^\circ = 0.147t$ for $x = 0.3$. Using $t = 0.4$ eV, $J_H = 1.6$ eV, and $K = 0.147t$, we obtain $D(0) = -25$ mV $Å^2$ from Eq. 6. The negative value of $D(0)$ suggests that the ferromagnetism should

434 Guo-Meng Zhao, J. Labry and Bo Truong

not be sustainable. It is possible that Eq. 6 is not valid for small K values. Even if we would ignore the second term of Eq. 6 (that is, $J_H = \infty$), the calculated $D(0)$ would be 125 mV Å^2. This upper-limit value is even much smaller than the measured ones (160-190 mV Å^2) [31, 32, 33].

Now we turn to the calculation of T_C for $x = 0.3$. For an 1-orbital DE model, the DMF calculation shows that $T_C^{DMF} = 0.02W/k_B$ for $J_H = \infty$ and $T_C^{DMF} = 0.01W/k_B$ for $J_H = 1.6$ eV (Ref. [22]). The finite J_H value reduces T_C by a factor of 2. Monte Carlo simulations yield $T_C = 0.01W/k_B$ [34] or $T_C = 0.0075W/k_B$ [35] in the case of $J_H = \infty$. This implies that the DMF calculation overestimates T_C by a factor of about 2 due to the neglecting of spatial fluctuations. For realistic parameters $W = 1.6$ eV, $J_H = 1.6$ eV, the DMF calculation obtains $T_C^{DMF} = 0.01W/k_B = 180$ K [22]. Considering the fact that the DMF method can overestimate T_C by a factor of 2, one has $T_C \sim 100$ K.

It is essential to consider the two-orbital DE model with inclusion of a strong electron-electron correlation. This two-orbital model has been generally accepted to be more relevant to the physics of manganites. From Monte Carlo simulation on this model with realistic onsite Coulomb interactions and J_H, Sheng and Ting [36] find that $T_C = 0.005W/k_B$ in the case of an anisotropic spin or orbital configuration. This leads to $T_C = 90$ K with $W = 1.6$ eV. Therefore, with the unbiased parameters, both the one- and two-orbital DE models predict T_C values, which are about a factor of 4 smaller than the maximum T_C (~ 400 K) observed in manganites.

The above calculations have not taken into account any electron-phonon interaction. The electron-phonon interaction can substantially reduce the K value and thus $D(0)$ if the coupling constant $E_p/2t >> 1$ (Refs. [37, 38]). In reality, the polaron binding energy E_p in manganites is estimated to be about 1 eV [39], which suggests that the electron-phonon coupling does not lead to a large decrease in $D(0)$.

The question is why the DE model cannot explain the ferromagnetism in doped manganites. As mentioned above, the DE model implicitly assumes that doped carriers are Mn e_g electrons, which is not the case according to the electron-energy-loss and photoemission spectra [19, 20]. Furthermore, the 'constrained' LDA calculation [26] shows a large onsite Coulomb repulsion of about 8-10 eV, in agreement with the photoemission data [20]. The simple LDA calculation which ignores the strong correlation effect shows a bare plasma energy of about 1.3 eV for $x = 0.33$ with a distorted structure determined by neutron scattering [40]. The bare plasma energy calculated is much smaller than the one observed in $Nd_{0.7}Sr_{0.3}MnO_3$ (3.3 eV) [41]. The large bare plasma energy observed in this material is consistent with the fact that doped holes reside mainly on the oxygen orbitals with a large bandwidth. The bare plasma energy of about 3.3 eV for single conduction band (oxygen band) implies a bare K° of about 0.24 eV, which gives an upper limit for the K value in the presence of interactions. The electron-phonon interaction with a coupling constant $E_p/2t \sim 1$ will reduce the K value slightly [37, 38], but can significantly decrease the Drude weight. Optical data indeed show that the K value for $Nd_{0.7}Sr_{0.3}MnO_3$ is about 0.2 eV [24], which is about 20% smaller than the expected 'bare' $K^\circ = 0.24$ eV. The effective plasma energy at low temperatures (in the ferromagnetic state) is found to be about 0.57 eV [42], which is reduced by a factor of 5.6 compared with the bare one. This implies that the polaronic effect is strong even in the low-temperature ferromagnetic state, in agreement with the observed polaronic electrical transport [41].

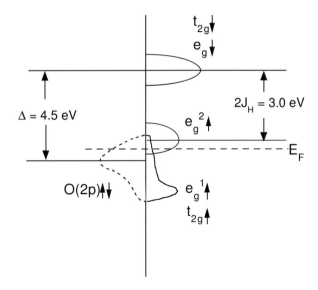

Figure 1. The schematic band structure for doped manganites ($x<0.4$) constructed from the LDA + U calculation [26]. The energy scales are consistent with the optical data [24, 42, 46]. The density of the oxygen holes is equal to x per cell while the densities of the Mn^{2+} and Mn^{4+} ions are the same due to the charge disproportion ($2Mn^{3+} \rightarrow Mn^{2+} + Mn^{4+}$) [45].

In Fig. 1, we plot a schematic band structure for doped manganites ($x< 0.4$), which is extracted from the LDA + U calculation. Here we have assumed that the local Jahn-Teller distortions still survive upon doping, but the average magnitude of the distortions decreases, in agreement with the experiments [43, 44]. Doping with a divalent element shifts down the e_g^2 band due to the decrease of the Jahn-Teller distortions. The density of the oxygen holes is equal to x per cell, while the electron carrier density in the majority e_g^2 band (corresponding to the density of the Mn^{2+} ions) is the same as the hole carrier density in the majority e_g^1 band (corresponding to the density of the Mn^{4+} ions). In other words, the doping does not change the average valence of the Mn ions, but lead to the charge disproportion ($2Mn^{3+} \rightarrow Mn^{2+} + Mn^{4+}$). This is because the quench of the static Jahn-Teller distortions by doping makes the Mn^{2+}-Mn^{4+} pairs more stable than the Mn^{3+}-Mn^{3+} pairs [45]. The current band structure is consistent with the optical transitions at the photon energies of about 1.5 eV, 3.0 eV and 4.5 eV [24, 42, 46]. The optical spectral weight for the 4.5 eV transition should be much larger than for the 1.5 eV transition, as observed [46]. This is because the unoccupied state density for the former optical transition (i.e., 3 minority t_{2g} and 2 minority e_g states per cell) is at least 5 times larger than that for the latter one (i.e., less than 1 majority e_g states per cell). The optical transition at about 3 eV is related to the transition between the exchange splitted e_g bands.

For $x > 0.4$, a significant fraction of doped holes should get into the d-orbitals and eventually the electronic band structure should cross over to the one predicted from the LDA calculation for $x > 0.5$. In this case, the double-exchange should be the primary cause of the ferromagnetism. One should consider the realistic two-orbital DE model with realistic onsite Coulomb interactions and J_H to calculate the Curie temperature apart from the DMF

436 Guo-Meng Zhao, J. Labry and Bo Truong

method which tends to significantly overestimate T_C. In the absence of any electron-phonon interaction, Sheng and Ting [36] find that $T_C = 0.005W/k_B$ from Monte Carlo simulation. With W = 3 eV (Ref. [29]), we have T_C = 174 K. Strong polaronic effect will reduce T_C so that $T_C < T_N \simeq 150$ K, where T_N is the Néel temperature of $CaCuO_3$. This naturally explains why the ground state is not ferromagnetic for $x > 0.5$.

2.2. d-p Exchange

What is an alternative model for the ferromagnetism in doped ferromagnetic manganites? Since doped holes mainly reside on the oxygen orbitals for $x < 0.4$ according to EELS and photoemission spectra [19, 20], there should be d-p exchange between the oxygen-holes and Mn spins. If we consider an oxygen-hole (spin 1/2) sitting in between two Mn ions, an exchange interaction between the oxygen-hole and Mn spins (d-p exchange) will lead to a ferromagetic interaction between Mn spins [47]. In this case, the ferromagnetic exchange energy between two Mn spins is [47] $J \propto t_{pd}^4/E_g^3$, where t_{pd} is a hybridization matrix element between the d and p orbitals, and E_g is a charge transfer gap. In addition, the exchange interaction between Mn and oxygen-hole spins is given by $J_{pd} \propto t_{pd}^2/E_g$ (Ref. [47]). Using a scaling relation: $t_{pd} \propto d^{-4}$ (where d is the bonding length) [48] and the parameters: d =1.97 Å and $E_g \simeq 4$ eV for doped manganites [46], d = 1.89 Å and $E_g \simeq 2$ eV for La_2CuO_4 (Ref. [49]), we find that J_{pd} for the former should be smaller than for the latter by a factor of about 2.8. Since $J_{pd} = 0.17$ eV for La_2CuO_4 (Ref. [50]), J_{pd} for doped manganites should be about 0.06 eV.

Alexandrov and Bratkovsky [17] have proposed that, in addition to the d-p exchange interaction, there is a strong electron-phonon interaction that may lead to the formation of small polarons or bipolarons. In the paramagnetic state, the singlet bipolarons (spin zero) are stable and the ferromagnetic interaction is produced by the thermally excited polarons (spin 1/2). Thus the Curie temperature T_C is self-consistently determined by the polaron density at T_C. In this model, there are three coupled mean-field equations which can quantitatively explain the observed colossal magnetoresistance [17] and the isotope shift of T_C (Ref. [18]). Here we want to generalize these equations in Ref. [17] to a more realistic case where $T_C << W_p/k_B$, where W_p is the polaron bandwidth. For $T_C << W_p/k_B$ and in zero magnetic field, one easily finds that

$$n = 2(\frac{k_B T}{1.05W_p})^{3/2} \exp(-\Delta/2k_B T) \cosh(J_{pd}S\sigma/2k_B T), \tag{7}$$

$$m = n \tanh(J_{pd}S\sigma/2k_B T), \tag{8}$$

$$\sigma = B_S(J_{pd}Sm/4k_B T). \tag{9}$$

Here n is the polaron density per cell, Δ is the bipolaron binding energy, σ and m are the magnetizations of Mn^{3+}ions and oxygen holes, respectively, and $B_S(y) = (1 + 1/2S) \coth[(S+1/2)y] - (1/2S) \coth(y/2)$ is the Brillouin function. The above equations are the same as those in Ref. [17] except that the prefactor of Eq. 7 is different from that of the corresponding Eq. 5 in Ref. [17]. Linearizing these equations with respect to m and σ

near T_C and taking $S = 2$, we find that the Curie temperature T_C in zero magnetic field is given by

$$k_B T_C = \frac{J_{pd}^4}{(1.05 W_p)^3} \exp(-\Delta/k_B T_C). \qquad (10)$$

One should note that Eq. 10 is valid only for the second-order phase transition where the polaron density has no discontinuity at T_C. Furthermore, in order for Eq. 10 to have a physically meaningful solution, it is required that $dT_C/d\Delta < 0$ (Ref. [17]). In the end of the chapter, we will show that Eq. 10 can quantitatively explain the observed Curie temperature and its isotope effect.

3. Metallic Electrical Conduction of Small Polarons

A number of experiments have provided strong evidence for the existence of small polaronic charge carriers [51, 52, 53] and their hopping conduction in the paramagnetic state of manganites [54, 55]. However, the nature of the charge carriers and the electrical transport mechanism in the low-temperature metallic state remains elusive. At low temperatures, a dominant T^2 contribution in resistivity is generally observed, and has been ascribed to electron-electron scattering [56]. Jaime et al. [57] have shown that the resistivity is essentially temperature independent below 20 K and exhibits a strong T^2 dependence above 50 K. In addition, the coefficient of the T^2 term is about 60 times larger than that expected for electron-electron scattering. They thus ruled out the electron-electron scattering as the conduction mechanism and proposed single magnon scattering with a cutoff at long wavelengths. In their scenario [57], they considered a case where the manganese e_g minority (spin-up) band lies slightly above the Fermi level (in the majority spin-down band) with a small energy gap of about 1 meV. This is in contradiction with optical data [27] which show that the manganese e_g minority band is well above the Fermi level. This suggests that the conduction mechanism proposed in Ref. [57] is not relevant.

Alternatively, one should consider a contribution from electron-phonon scattering. At low temperatures, the acoustic phonon scattering would give a T^5 dependence, which is not consistent with the data [57]. Within the CMR theory of Alexandrov and Bratkovsky [17], small polaronic transport is the prevalent conduction mechanism below the ferromagnetic ordering temperature T_C, in contrast to large polaronic carriers predicted from the CMR theories based on double-exchange and strong Jahn-Teller electron-phonon coupling [15, 16]. If the resistivity data at low temperatures are consistent with small polaron conduction mechanism, the data strongly support the CMR theory of Alexandrov and Bratkovsky [17].

Although a theory of small polaron conduction at low temperatures was developed more than 50 years ago [58, 59], no experimental data had been used to compare with the theoretical prediction until 2000 [41, 61]. The theory shows that [58, 59], for $k_B T < 2t_p$, the resistivity is given by

$$\rho(T) = (\hbar^2/ne^2 a^2 t_p)(1/\tau), \qquad (11)$$

where t_p is the effective hopping integral of polarons, n is the carrier density and $1/\tau$ is the relaxation rate:

$$1/\tau = \sum_\alpha A_\alpha / \sinh^2(\hbar\omega_\alpha/2k_B T), \qquad (12)$$

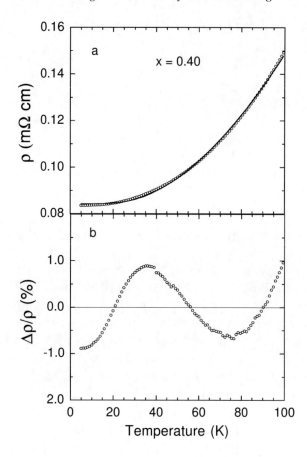

Figure 2. a) Low temperature resistivity $\rho(T)$ for a high-quality film of $La_{0.60}Ca_{0.40}MnO_3$. The data can be fitted by Eq. 13 with $\hbar\omega_s/k_B = 100(2)$ K. The solid line is the fitted curve. b) The relative deviation $\Delta\rho/\rho$ between the data and the fitted curve in a. The maximum deviation is about 1%. After Ref. [61].

where ω_α is the average frequency of one optical phonon mode, A_α is a constant, depending on the electron-phonon coupling strength. It is worth noting that the above expression for $1/\tau$ has been generalized from one optical phonon mode to multiple modes since complex compounds such as manganites contain several optical phonon modes. From the above equations, one can see that only the low-lying optical modes with a strong electron-phonon coupling contribute to the resistivity at low temperatures due to the factor of $1/\sinh^2(\hbar\omega_\alpha/2k_BT)$. As discussed below, among the low-lying optical modes, only the softest optical phonon branch that is related to the tilt/rotation of the oxygen octahedra is strongly coupled to the carriers. The high-frequency phonon modes such as the Jahn-Teller modes also have a strong coupling with carriers, but these modes have negligible contributions to the resistivity below 100 K because of an exponentially small factor in Eq. 12. By inclusion of impurity scattering, the resistivity at low temperatures is

$$\rho(T) = \rho_0 + C/\sinh^2(\hbar\omega_s/2k_BT), \tag{13}$$

where ω_s is the average frequency of the softest optical mode, and C is a constant, being proportional to m^*/n (where m^* is the effective mass of carriers and n is the mobile carrier concentration).

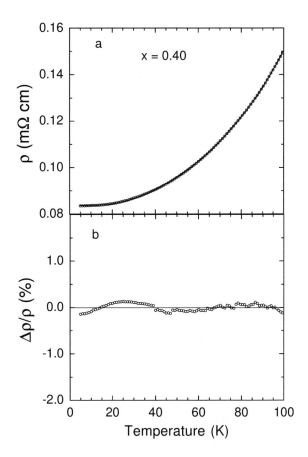

Figure 3. a) Low temperature resistivity $\rho(T)$ for a high-quality film of La$_{0.60}$Ca$_{0.40}$MnO$_3$. The data are fitted by Eq. 14 with $\hbar\omega_s/k_B = 80.0(8)$ K. The solid line is the fitted curve. b) The relative deviation $\Delta\rho/\rho$ between the data and the fitted curve in a. The maximum deviation is about 0.1%. After Ref. [61].

Now we consider the contribution from magnon scattering. Kubo and Ohata [30] have studied the magnon scattering for half metals where the spin-up (minority) and spin-down (majority) bands are well separated. In this case, one-magnon scattering is forbidden, and thus two-magnon process is responsible for the low temperature resistivity, which gives a contribution that is proportional to $T^{4.5}$ [30]. If we include this contribution, we have

$$\rho(T) = \rho_0 + BT^{4.5} + C/\sinh^2(\hbar\omega_s/2k_BT), \tag{14}$$

In Fig. 2a, we show the low temperature resistivity $\rho(T)$ for a high-quality epitaxial thin film of La$_{1-x}$Ca$_x$MnO$_3$ with $x = 0.40$. The films of La$_{1-x}$Ca$_x$MnO$_3$ were prepared by pulsed laser deposition using a KrF excimer laser [60]. The films were finally annealed for

10 h at about 940 °C and oxygen pressure of about 1 bar. The resistivity was measured using the van der Pauw technique, and the contacts were made by silver paste [41]. The residual resistivity is 84 $\mu\Omega$cm which is even smaller than that for single crystalline samples [62]. This indicates that the quality of the film is high, which allows one to study the intrinsic electrical transport properties of this system.

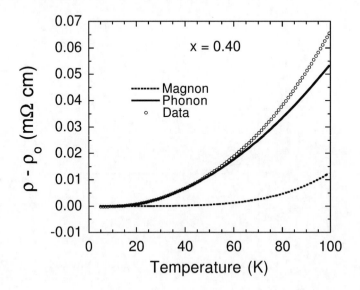

Figure 4. The respective contribution to the resistivity from the phonon or magnon scattering in $La_{0.60}Ca_{0.40}MnO_3$. It is clear that the phonon scattering makes a dominant contribution to the resistivity at low temperatures. After Ref. [61].

The data can be fitted by Eq. 13 with $\hbar\omega_s/k_B = 100(2)$ K. It is evident that the fitting is quite good, but still shows a systematic deviation as seen more clearly from Fig. 2b where the relative difference $\Delta\rho/\rho$ between the data and the fitted curve is plotted. This suggests that an additional contribution should be included in the fitting. In Fig. 3a, we fit the data with Eq. 14 which comprises the contribution from two-magnon scattering. It is striking that the fit is nearly perfect with a negligible systematic deviation (see Fig. 3b). The fitting parameter $\hbar\omega_s/k_B = 80.0(8)$ K. In Fig. 4, we plot the respective contribution to the resistivity from the phonon or magnon scattering. It is apparent that the phonon scattering makes a dominant contribution to the resistivity at low temperatures. This is consistent with the fact that there is a negligible magnetoresistance effect below 100 K [62]. We would like to mention that we have fitted the data only below 100 K. This is because n/m^* is temperature independent below 100 K [63], so that Eq. 14 is valid only in this temperature range.

Fig. 5a shows $\rho(T)$ for another epitaxial thin-film of $La_{1-x}Ca_xMnO_3$ with $x = 0.25$. The data can be perfectly fitted by Eq. 14 with $\hbar\omega_s/k_B = 74.5(4)$ K. The systematic deviation is very small, as seen from Fig. 5b. If we allow the power in the second term of Eq. 14 to be a fitting parameter, the best fit gives a power of 4.6(4), very close to 4.5 expected from 2-magnon scattering [30].

The excellent agreement between the data and Eq. 14 implies the presence of small polarons and their metallic conduction in the low temperature ferromagnetic state. The present

Polarons and Bipolarons in Colossal Magnetoresistive Manganites

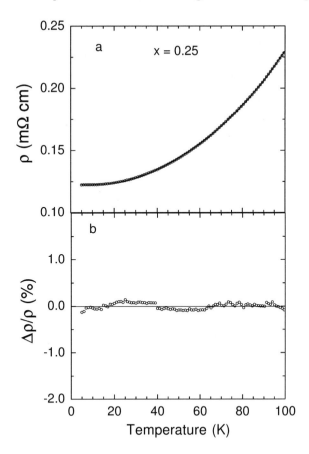

Figure 5. a) Low temperature resistivity $\rho(T)$ for a high-quality film of $La_{0.75}Ca_{0.25}MnO_3$. The data are fitted by Eq. 14 with $\hbar\omega_s/k_B = 74.5(4)$ K. The solid line is the fitted curve. b) The relative deviation $\Delta\rho/\rho$ between the data and the fitted curve in a. The maximum deviation is less than 0.1%. After Ref. [61].

result thus gives a strong support to the theory of colossal magnetoresistance proposed by Alexandrov and Bratkovsky [17]. The result also indicates that the low-lying mode with $\hbar\omega_s/k_B \simeq 80$ K has a strong coupling with the conduction electrons.

Now a question arises: What is the origin of such a soft mode that is strongly coupled to the carriers? We are not aware of any measurements for the phonon dispersions of the manganites by inelastic neutron scattering. However, it is known that the soft mode with $\hbar\omega_s/k_B \simeq 100$ K commonly exists in perovskite oxides, and is associated with the tilt/rotation of the oxygen octahedra [64, 65]. Tunneling experiment on the superconducting $Ba(Pb_{0.75}Bi_{0.25})O_3$ [66] directly reveals a strong electron-phonon coupling at a phonon energy of about 6 meV. Inelastic neutron scattering experiment on the same compound shows a soft mode at 5.9 meV, which is attributed to rotational vibrations of the oxygen octahedra at M point of the cubic Brillouin zone [65]. These results clearly demonstrate that the rotational mode in this perovskite oxide has a strong coupling with the carriers.

One should also note that the phonon energy ($\sim$6 meV) of the rotational mode observed in $Ba(Pb_{0.75}Bi_{0.25})O_3$ is very close to that (6.4-6.9 meV) deduced for the manganite films from the resistivity data. Moreover, the theoretical investigations [67, 68] show that a static distortion of the tilting/rotational mode in both cuprates and manganites can open pseudo-gaps in the conduction bands, implying a strong electron-phonon coupling [67, 68].

4. Oxygen-Isotope Effects in Manganites

In most materials, magnetic phenomena at room temperature and below are essentially un-affected by lattice-vibrations because the electronic and lattice subsystems are decoupled according to the Born-Oppenheimer adiabatic approximation. The atoms can usually be considered as infinitely heavy and static in theoretical descriptions of the magnetic phe-nomena. However, this approximation will break down and (bi)-polarons can be formed in compounds where the electron-phonon interactions are strong and bare conduction band-widths are narrow. In this case, magnetic properties may be influenced by lattice vibrations and thus by the masses of nuclear ions (isotope effect).

To be more specific, the polaronic nature of the conduction carriers can be demonstrated by the isotope effect on the effective bandwidth W_{eff} of polarons, which in turn depends on the isotope mass M as discussed in the introduction. If certain electrical or magnetic quantity depends on the effective conduction bandwidth in a material with strong lattice polaronic effects, this quantity may have a large isotope effect.

4.1. Giant Oxygen-Isotope Shift of the Curie Temperature

The first observation of the oxygen isotope effect on the Curie temperature was made in $La_{1-x}Sr_xMnO_{3+y}$ system by Zhao and Morris in 1995 [69]. In Fig. 6, the normalized magnetizations for the ^{16}O and ^{18}O samples of $La_{0.9}Sr_{0.1}MnO_{3+y}$ are plotted as a function of temperature. The oxygen isotope shifts of T_C were determined from the differences between the midpoint temperatures on the transition curves of the ^{16}O and ^{18}O samples. It is clear that the ^{18}O sample has a lower T_C than the ^{16}O sample by $\sim$6.7 K. It should be noted that since the value of y is substantial (>0.05) when samples are prepared below 1100 °C, the curie temperatures in these samples are much higher than that for the corresponding single crystal samples where y is close to zero. Actually the extra oxygen in the above chemical formula is caused by the existence of cation vacancies.

On the other hand, the oxygen-isotope shift of T_C in $La_{0.8}Ca_{0.2}MnO_{3+y}$ is very large [51], as seen from Fig. 7. The samples with a heavier oxygen isotope mass (about 95% of ^{18}O) have a much lower T_C. The relative isotope shift of T_C is as large as 10%. Such a large oxygen isotope shift of the ferromagnetic transition is very unusual since lattice vibrations were believed to play no role in the magnetic interactions of most magnetic materials. It is a clear-cut experiment to establish what many have suspected that atomic motion must be included in any viable description of the manganites. It was also the first experiment in condensed matter physics to demonstrate that there can be a giant isotope shift of a magnetic transition temperature. The oxygen isotope exponent is defined as usual, $\alpha_O = - (\Delta T_C/T_C)/(\Delta M_O/M_O)$, where both T_C and the oxygen isotope mass M_O are referred

Polarons and Bipolarons in Colossal Magnetoresistive Manganites 443

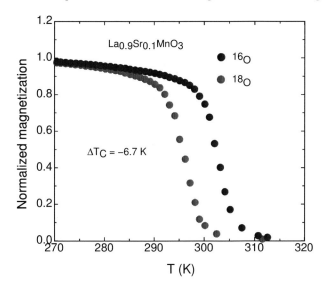

Figure 6. Oxygen isotope effect on the Curie temperature of $La_{0.9}Sr_{0.1}MnO_{3+y}$. The figure is reproduced from Ref. [69].

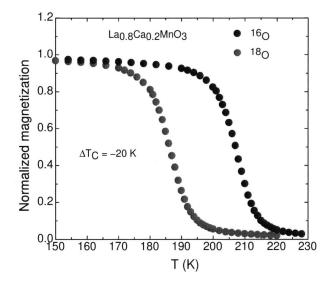

Figure 7. Oxygen isotope effect on the Curie temperature of $La_{0.8}Ca_{0.2}MnO_{3+y}$. The figure is reproduced from Ref. [51].

to a ^{16}O sample. With this definition, we obtain $\alpha_O = 0.85$ for $La_{0.8}Ca_{0.2}MnO_{3+y}$ and 0.142 for $La_{0.9}Sr_{0.1}MnO_{3+y}$.

The oxygen isotope effect on the Curie temperature has also been studied in other manganite systems [70, 71, 72]. It is found that [72] the α_O value increases rapidly as the Curie temperature decreases. In Fig. 8, we plot α_O vs the Curie temperature of ^{16}O samples. The data can be well fit by an equation

$$\alpha_O = 21.9 \exp(-0.016 T_C). \tag{15}$$

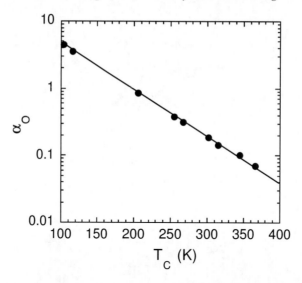

Figure 8. α_O vs the Curie temperature of ^{16}O samples. The data are from Ref. [72].

The above simple empirical relation between α_O and T_C is quite unexpected. From a simple argument based on the double-exchange model and the small polaron theory discussed above, one would expect that [51] $T_C \propto \exp(-2\alpha_O)$. At least this scenario cannot quantitatively explain Eq. 15 possibly because it is too simple.

It is also interesting that Eq. 15 is very similar to the relation between the pressure effect ($d\ln T_C/dP$) and T_C. Plotted in Fig. 9 is $d\ln T_C/dP$ (for fixed $x = 0.3$) as a function of T_C. The data can also be well fit by an equation

$$d\ln T_C/dP = 4.4\exp(-0.016T_C). \qquad (16)$$

Combining Eq. 15 and Eq. 16, one has $\alpha_O = 5.0\,d\ln T_C/dP$. Such a simple relation between the oxygen isotope exponent and the pressure effect implies that the oxygen isotope and pressure effects have the same origin, and that the observed oxygen isotope effect must be intrinsic.

4.2. Oxygen-Isotope Effects on Electrical Transport below the Curie Temperature

We have shown that the temperature dependent part of the resistivity at low temperatures are in excellent agreement with small polaron conduction mechanism. The nature of small polarons in the ferromagnetic state of doped manganites can be further proved by the oxygen-isotope effect on the residual resistivity.

The oxygen-isotope effect on the low-temperature resistivity has been studied [73] in high-quality epitaxial thin films of $La_{0.75}Ca_{0.25}MnO_3$ and $Nd_{0.7}Sr_{0.3}MnO_3$. The residual resistivity of these compounds shows a strong dependence on the oxygen-isotope mass. The quantitative data analyses suggest that the nature of the charge carriers in the ferromagnetic state of doped manganites are intermediate-size polarons.

Polarons and Bipolarons in Colossal Magnetoresistive Manganites 445

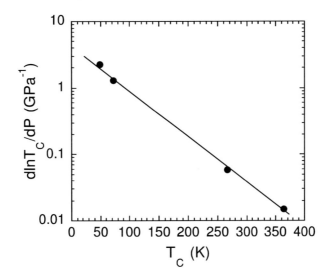

Figure 9. $d\ln T_C/dP$ vs Curie temperature. The quantity $d\ln T_C/dP$ is referred to the pressure effect in the low pressure region. The figure is reproduced from Ref. [70].

Fig. 10 shows the resistivity of the oxygen-isotope exchanged $La_{0.75}Ca_{0.25}MnO_3$ (LCMO) and $Nd_{0.7}Sr_{0.3}MnO_3$ (NSMO) films over 100-300 K. It is apparent that the ^{18}O samples have lower metal-insulator crossover temperatures and much sharper resistivity drop. The Curie temperature T_C normally coincides with a temperature where $d\ln\rho/dT$ exhibits a maximum. We find that the oxygen-isotope shift of T_C is 14.0(6) K for LCMO, and 17.5(6) K for NSMO, in excellent agreement with the results for the bulk samples [72].

In Fig. 11 we plot the low-temperature resistivity of the oxygen-isotope exchanged films of (a) LCMO; (b) NSMO. In both cases, the residual resistivity ρ_o for the ^{18}O samples is larger than for the ^{16}O samples by about 15%. Repeating the van der Pauw measurements at 5 K several times with different contact configurations indicates that the uncertainty of the difference in ρ_o of the two isotope samples is less than 3%. We should mention that the intrinsic resistivity cannot be obtained from ceramic samples where the boundary resistivity is dominant. Thus one cannot use ceramic samples to study the isotope effect on the intrinsic resistivity. Moreover, the van der Pauw technique is particularly good to precisely measure the resistivity difference between the oxygen-isotope exchanged films which have the same thickness. Thus the data shown in Figs. 10 and 11 represent the first precise measurements on the intrinsic resistivity of the isotope substituted samples.

It is known that [29, 74] the residual resistivity $\rho_o \propto m^*/n\tau_o$. Here $\hbar/\tau_o$ is the scattering rate which is associated with the random potential produced by randomly distributed trivalent and divalent cations [40], and/or with impurities; m^* is the effective mass of carriers at low temperatures, and n is the mobile carrier concentration. If the chemical potential is far above the mobility edge, $\rho_o \propto (m^*)^2$, that is, $\hbar/\tau_o \propto m^*$. This is what one expects from the simple Born approximation. On the other hand, one can show [75] that $\rho_o \propto m^*$ if the chemical potential is slightly above the mobility edge. Therefore, the large oxygen-isotope effect on ρ_o implies that the effective mass of the carriers strongly depends on the oxygen mass, as expected for polaronic charge carriers.

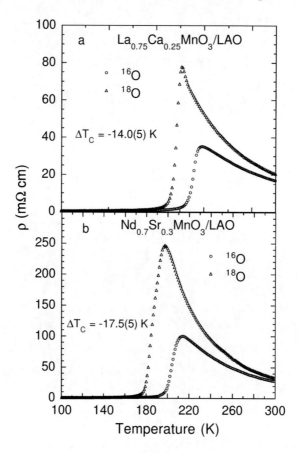

Figure 10. The resistivity of the oxygen-isotope exchanged films of (a) $La_{0.75}Ca_{0.25}MnO_3$; (b) $Nd_{0.7}Sr_{0.3}MnO_3$. After Ref. [73].

From the oxygen-isotope effects on the residual resistivity and the thermoelectric power at low temperatures, one can deduce the exponent of the oxygen-isotope effect on m^* based on a polaronic Fermi liquid model [75]. The exponent β_O of the oxygen-isotope effect on m^* was estimated [75] to be about -0.7 for $La_{0.75}Ca_{0.25}MnO_3$ and about -1.1 for $Nd_{0.7}Sr_{0.3}MnO_3$. From Eq. 1, one finds that the polaronic mass enhancement factor $m^*/m = \exp(-2\beta) \simeq \exp(-2\beta_O) = 4\text{-}9$.

Angle-resolved photoemission spectroscopic data [76] of a layered manganite $La_{1.2}Sr_{1.8}Mn_2O_7$ ($T_C = 126$ K) suggest the coexistence of bipolaronic carriers (pseudo-gapped state) along the Mn-O binding direction and polaronic charge carriers along the diagonal direction. The polaron mass is enhanced by a factor of 5.6 (Ref. [76]), which is in good agreement that (4-9) obtained from our isotope experiments.

The coexistence of polaronic and bipolaronic carriers in the ferromagnetic state of the layered manganite [76] is also consistent with our earlier results [77, 78] for a 3-dimensional manganite $(La_{0.5}Nd_{0.5})_{0.67}Ca_{0.33}MnO_3$ with a similar T_C (~100 K). The ground state of these low T_C manganites is phase-separated into ferromagnetic metallic regions where mo-

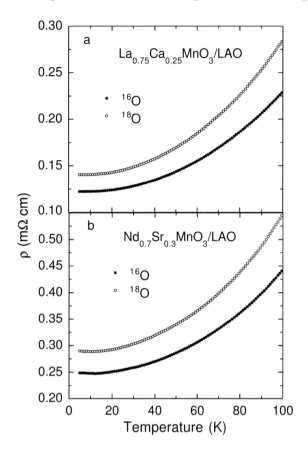

Figure 11. The low-temperature resistivity of the oxygen-isotope exchanged films of (a) La$_{0.75}$Ca$_{0.25}$MnO$_3$; (b) Nd$_{0.7}$Sr$_{0.3}$MnO$_3$. After Ref. [73].

bile polarons reside and charge-ordered insulating regions—where localized bipolarons sit.

4.3. Oxygen-Isotope Effects on Electrical Transport above the Curie Temperature

The first experimental evidence for small polaronic charge carriers in the paramagnetic state was provided by transport measurements [54]. It was found that the activation energy E_ρ deduced from the conductivity data is one order of magnitude larger than the activation energy E_s obtained from the thermoelectric power data. Such a large difference in the activation energies is the hallmark of the small-polaron hopping conduction. However, the data cannot make a distinction between small polarons and small bipolarons. Small bipolarons are normally much heavier than small polarons, and should be localized in the presence of small random potentials. In order to discriminate between polarons and bipolarons and to place constraints on the CMR theories, it is essential to study the oxygen-isotope effects on the intrinsic electrical properties. This is because the activation energies in resistivity and thermoelectric power will depend on the oxygen isotope mass if there coexist localized bipolaronic charge carriers and mobile polaronic carriers.

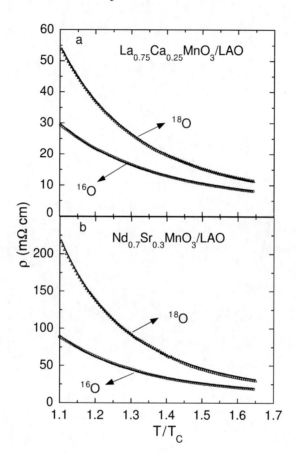

Figure 12. The resistivity of the oxygen-isotope exchanged films of $La_{0.75}Ca_{0.25}MnO_3$ and $Nd_{0.7}Sr_{0.3}MnO_3$. The maximum temperature of the data points for the ^{16}O film of $La_{0.75}Ca_{0.25}MnO_3$ is 380 K. The solid lines are the fitted curves by Eq. 18. As in Ref. [54], we excluded the data points below $1.1T_C$ for the fitting. After Ref. [79].

Zhao et al. [79] have precisely measured resistivity for the oxygen-isotope exchanged high-quality epitaxial thin films of $La_{0.75}Ca_{0.25}MnO_3$ and $Nd_{0.7}Sr_{0.3}MnO_3$, and the thermoelectric power for the oxygen-isotope exchanged ceramic samples of $La_{0.75}Ca_{0.25}MnO_3$. The data cannot be explained by a simple small-polaron model, but are in quantitative agreement with a model where the formation of localized small bipolarons is essential.

Fig. 12 shows the resistivity of the oxygen-isotope exchanged films of LCMO and NSMO above $1.1T_C$. From the figure, one can see that there is a large difference in the intrinsic resistivity between the two isotope samples. Such a large isotope effect is reversible upon the oxygen isotope back-exchange.

It is known that the resistivity can be generally expressed as $\rho = 1/\sigma = 1/ne\mu$, where n is the mobile carrier concentration and μ is the mobility of the carriers. For adiabatic small-polaron hopping, the mobility is given by [6]

$$\mu = \frac{ed^2}{h}\frac{\hbar\omega_o}{k_BT}\exp(-E_a/k_BT). \qquad (17)$$

Here d is the site to site hopping distance, which is equal to $a/\sqrt{2}$ in manganites since the doped holes in this system mainly reside on the oxygen sites [19]; ω_o is the characteristic optical phonon frequency; $E_a = (\eta E_p/2)f(T) - t$; $f(T) = [\tanh(\hbar\omega_o/4k_BT)]/(\hbar\omega_o/4k_BT)$ for $T > \hbar\omega_o/4k_B \simeq 200$ K [80]; E_p is the polaron binding energy; t is the "bare" hopping integral; $\eta \leq 1$ [6, 39]. In the harmonic approximation, E_p is independent of the isotope mass M.

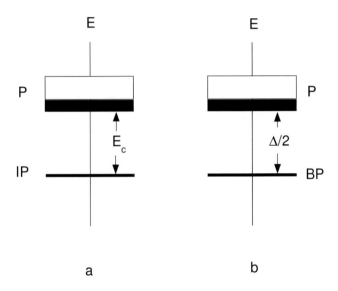

Figure 13. A schematic diagram of the polaron band, and polaron trapping into impurity (IP) states (a) (Ref. [80]), or into localized bipolaron (BP) states (b) (Ref. [17]). The bipolaron binding energy Δ is isotope-mass dependent [39], while E_c is independent of the isotope-mass M [80]. After Ref. [79].

The mobile polaron density n can be calculated for the two possible cases shown in Fig. 13. For a simple parabolic band, $n = 2(k_BT/1.05a^2W_p)^{3/2}\exp(-E_s/k_BT)$ at $T \ll W_p/k_B$ [17, 80]. Here $E_s = E_c$ if polarons are trapped into impurity (IP) states [80], while $E_s = \Delta/2$ if polarons are bound into localized bipolaron (BP) states [17]. The bipolaron binding energy $\Delta = 2(1-\gamma)E_p - V_c - W_p$, where V_c is the Coulombic repulsion between bound polarons [39]. In fact, the above $n(T)$ expression is the same as that for semiconductors when the chemical potential is pinned to the impurity levels. Using the above $n(T)$ expression and Eq. 17, we finally have

$$\rho = \frac{C}{\sqrt{T}}\exp(E_\rho/k_BT), \qquad (18)$$

where $E_\rho = E_a + E_s$, and $C = (ah/e^2\sqrt{k_B})(1.05W_p)^{1.5}/\hbar\omega_o$. The quantity C should strongly depend on the isotope mass M and decrease with increasing M. This is because

W_p decreases strongly with increasing M according to $W_p \propto \exp(-\gamma E_p/\hbar\omega_o) = \exp(-g^2)$ [39, 10]. It is worthy of noting that Eq. 18 is valid ony if $T<<W_p/k_B$. For $T>>W_p/k_B$, the prefactor in Eq. 18 should be proportional to T/ω_o [54].

The thermoelectric power is given by [80]

$$S = E_s/eT + S_o, \qquad (19)$$

where S_o is a constant depending on the kinetic energy of the polarons and on the polaron density [80]. One should note that Eq. 19 is valid only if there is one type of carriers (e.g., holes).

One can make a distinction between the two cases shown in Fig. 13. If small polarons are bound to impurity centers, there will be no isotope effect on E_s since E_c is independent of M [80]. On the other hand, if small polarons are bound to localized bipolaron states, both E_ρ and E_s in Eq. 18 and Eq. 19 will depend on M due to the fact that Δ is M dependent. In general, the isotope shift of E_ρ will be larger than the isotope shift of E_s. This is because $E_a = (\eta E_p/2)f(T) - t$, and $f(T) = [\tanh(\hbar\omega_o/4k_BT)]/(\hbar\omega_o/4k_BT)$, which may depend on M if the temperature is not so high compared with $\hbar\omega_o/k_B$.

The data are now fitted by Eq. 18 (see solid lines in Fig. 12). It is apparent that the fits are quite good for both isotope samples. The fitting parameters are summarized in Table 1. From Table 1, one can see that, upon replacing ^{16}O with ^{18}O, the parameter C for LCMO/NSMO decreases by 35(5)/40(7)%, while E_ρ increases by 13.2(3)/14.2(8) meV. The huge oxygen-isotope effect on the parameter C is consistent with Eq. 18.

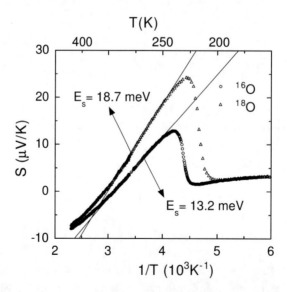

Figure 14. The thermoelectric power $S(T)$ of the oxygen-isotope exchanged ceramic samples of $La_{0.75}Ca_{0.25}MnO_3$. After Ref. [79].

One can obtain the isotope shift of E_s by measuring the thermoelectric power for two isotope samples according to Eq. 19. In Fig. 14, the thermoelectric power S as a function of $1/T$ is plotted for the ^{16}O and ^{18}O samples of $La_{0.75}Ca_{0.25}MnO_3$. Both T_C's and the isotope shift of the ceramic samples [72] are the same as those in the corresponding thin

Polarons and Bipolarons in Colossal Magnetoresistive Manganites 451

films. Since the grain-boundary effect on S is negligible, the thermoelectric power obtained in ceramic samples should be intrinsic. From the slops of the straight lines in Fig. 14, one finds E_s = 13.2±0.3 meV for the ^{16}O sample and 18.7±0.3 meV for the ^{18}O. The isotope shift is δE_s = 5.5±0.6 meV, which is about half the isotope shift of E_ρ. The observed oxygen-isotope effects on both E_ρ and E_s do not support a simple small-polaron model [80], but provide evidence that small polarons are bound into localized bipolaron states.

One can use the values of the parameter C to calculate the polaron bandwidth W_p according to the relation: $C = (ah/e^2\sqrt{k_B})(1.05W_p)^{1.5}/\hbar\omega_o$. The calculated W_p values are listed in Table 1. In the calculation, $\hbar\omega_o$ = 74 meV is used for the ^{16}O samples [39], and $\hbar\omega_o$ for the ^{18}O samples is 5.7% lower than for the ^{16}O samples. From the W_p values (see Table 1), one can see that the data satisfy $T < W_p/k_B$, which justifies the use of Eq. 18.

Furthermore, one can quantitatively explain the isotope dependence of E_s if small polarons form localized bipolarons. In this scenario [39], $\delta\Delta = -\delta W_p$. From Table 1, δW_p= −11.2±1.6 meV for $La_{0.75}Ca_{0.25}MnO_3$. So $\delta E_s = \delta\Delta/2 = 5.6\pm0.8$ meV for $La_{0.75}Ca_{0.25}MnO_3$, in quantitative agreement with the value (5.5±0.6 meV) deduced from the thermoelectric power data.

Table 1. Summary of the fitting and measured parameters for the ^{16}O and ^{18}O films of $La_{0.75}Ca_{0.25}MnO_3$ (LCMO) and $Nd_{0.7}Sr_{0.3}MnO_3$ (NSMO). The errors of the parameters come from the fitting and from the van der Pauw measurement. The absolute uncertainty of the thickness of the films was not included in the error calculations since it only influences the absolute values of the resistivity

Compounds	LCMO(^{16}O)	LCMO(^{18}O)	NSMO(^{16}O)	NSMO(^{18}O)
T_C (K)	231.5(3)	216.5(3)	204(1)	186(1)
C (mΩcmK$^{0.5}$)	17.3(5)	12.9(3)	23.2(8)	16.2(7)
E_ρ (meV)	72.8(2)	86.0(1)	78.8(4)	92.9(4)
E_s (meV)	13.2(3)	18.7(3)		
W_p (meV)	49.0(9)	38.8(7)	60(2)	45(2)
J_{pd} (meV)	56.56(60)	55.35(60)		

Therefore, the oxygen isotope effects on the intrinsic electrical transport in the paramagnetic state of doped manganites can be quantitatively explained by a scenario [17] where the small polarons are bound into localized bound pairs (bipolarons) in the paramagnetic state. The coexistence of small polarons and bipolarons in the paramagnetic state may lead to a dynamic phase separation into the insulating antiferromagnetically coupled region where the bipolarons reside, and into the ferromagnetically coupled region where the polarons sit. This simple picture can naturally explain the observation of the ferromagnetic clusters (intrinsic electronic inhomogeneity) in the paramagnetic state [81].

Using the deduced parameters (Table 1) of the ^{16}O and ^{18}O films of LCMO, we can determine the values of J_{pd} from Eq. 10. The calculated J_{pd} values are listed in the last row of Table 1. It is remarkable that the deduced J_{pd} values for the ^{16}O and ^{18}O samples are the same within the experimental uncertainty. The absolute J_{pd} value of about 57 meV is also in quantitative agreement with that (60 meV) estimated above from the measured bond-

452 Guo-Meng Zhao, J. Labry and Bo Truong

ing lengths and charge-transfer gaps of both La_2CuO_4 and $LaMnO_3$. The parameter-free agreement between theory and experiment provides indisputable proof for the microscopic theory developed for doped manganites.

Acknowledgments

The authors would like to thank the National Science Foundation of USA for financial support under the PREM grant (DMR 1523588).

Conclusion

We have shown that the d-p exchange is the primary origin of the ferromagnetism in doped manganites due to the fact that the undoped parent manganites are charge-transfer insulators so that doped holes mainly reside on the oxygen p-orbitals. The theoretical model based on the d-p exchange plus the coexistence of oxygen-hole polarons and bipolarons can quantitatively explain the CMR and the unconventional oxygen-isotope effects in doped manganites ($x < 0.4$). As x further increases, more and more holes get into the Mn d-orbitals, leading to the reduction in the Curie temperature. This can naturally explain why in the overdoped region ($x \geq 0.5$), the antiferromagnetic correlation is dominant over the ferromagnetic one. The observed unconventional oxygen-isotope effects on the Curie temperature and on electrical transport provide indisputable evidence for the existence of the small polarons in the ferromagnetic state and the coexistence of small polarons and bipolarons in the paramagnetic state of doped ferromagnetic manganites.

References

[1] J. G. Bednorz and K. A. Müller, Z. *Phys. B* **64**, 189 (1986).

[2] L.F. Mattheiss, E.M. Gyorgy, D.W. Johnson et al., *Phys. Rev. B***37**, 3715 (1988).

[3] R. von Helmolt, J. Wecker, B. Holzapfel, L. Schultz, and K. Samwer, *Phys. Rev. Lett.* **71**, 2331 (1993); S. Jin, T. H. Tiefel, M. McCormack, R. A. Fastnacht, R. Ramesh, and L. H. Chen, Science **264**, 413 (1994).

[4] L. D. Landau, *Phys. Z. Sowjetunion* **3**, 664 (1933).

[5] T. Holstein, *Ann. Phys.* (N. Y.), **8**, 325 (1959).

[6] D. Emin and T. Holstein, Ann. *Phys.* (N. Y.), **53**, 439 (1969).

[7] A. S. Alexandrov and N. F. Mott, *Polarons and Bipolarons*, 67-106 (World Scientific, Singapore, 1995).

[8] J. T. Devreese, in Encyclopedia of Applied Physics, vol. 14, p. 383, VCH Publishers (1996); J. Tempere and J. T. Devreese, *Phys. Rev. B* **64**, 104504 (2001); J. T. Devreese (in the present volume).

[9] A. S. Alexandrov, *Phys. Rev. B* **61**, 12315 (2000).

Polarons and Bipolarons in Colossal Magnetoresistive Manganites 453

[10] A. S. Alexandrov and P. E. Kornilovitch, *Phys. Rev. Lett.* **82**, 807 (1999).

[11] E. O. Wollan, and W. C. Koeler, *Phys. Rev.* **100**, 545 (1955).

[12] G. H. Jonker, and J. H. van Santen, *Physica* **16**, 337 (1950).

[13] C. Zener, *Phys. Rev.* **82**, 403 (1951).

[14] P. W. Anderson, and H. Hasegawa, *Phys. Rev.* **100**, 675 (1955).

[15] A. J. Millis, P. B. Littlewood, and B. I. Shraiman, *Phys. Rev. Lett.* **74**, 5144 (1995).

[16] A. J. Millis, B. I. Shraiman, and R. Mueller, *Phys. Rev. Lett.***77**, 175 (1996); H. Röder, J. Zang, and A. R. Bishop, *Phys. Rev. Lett.***76**, 1356 (1996).

[17] A. S. Alexandrov and A. M. Bratkovsky, *Phys. Rev. Lett.***82**,141 (1999).

[18] A. S. Alexandrov and A. M. Bratkovsky, *J. Phys. Condens. Matter*, **11**, 1989 (1999).

[19] H. L. Ju et al., *Phys. Rev. Lett.* **79**, 3230 (1997).

[20] T. Saitoh et al., *Phys. Rev. B* **51**, 13942 (1995).

[21] G. M. Zhao, *Phys. Rev. B* **62**, 11 639 (2001).

[22] N. Furukawa, *Physics of manganites*, edited by T. A. Kaplan and S. D. Mahanti (Kluwer Academic/Plenum Publisher, 1999) page 1.

[23] A. Chattopadhyay, A. J. Millis, and S. Das Sarma, *Phys. Rev. B* 61, 10738 (2000).

[24] M. Quijada, J. Cerne, J. R. Simpson, H. D. Drew, K. H. Ahn, A. J. Millis, R. Shreekala, R. Ramesh, M. Rajeswari, and T. Venkatesan, *Phys. Rev. B* **58**, 16 093 (1998).

[25] W. E. Pickett and D. J. Singh, *Phys. Rev. B* **53**, 1146 (1996).

[26] S. Satpathy, Z. S. Popvic, and F. R. Vukajlovic,*Phys. Rev. Lett.***76**, 960 (1996).

[27] A. Machida, Y. Moritomo, and A. Nakamura, *Phys. Rev. B* **58**, R4281 (1998).

[28] A. Takahashi and H. Shiba, *Eur. Phys. J. B* **5**, 413 (1998).

[29] W. E. Pickett and D. J. Singh, *Phys. Rev. B***55**, R8642 (1997).

[30] K. Kubo and N. Ohata, *J. Phys. Soc. Jpn.* **33**, 21 (1972).

[31] M. C. Martin, G. Shirane, Y. Endoh, K. Hirota, Y. Moritomo, and Y. Tokura, *Phys. Rev. B* **53**, R14285 (1996).

[32] J. A. Fernandez-Baca, P. Dai, H. Y. Hwang, C. Kloc, and S. W. Cheong, *Phys. Rev. Lett.***80**, 4012 (1998).

[33] J. W. Lynn, R. W. Erwin, J. A. Borchers, Q. Huang, A. Santoro, J. L. Peng, and Z. Y. Li, *Phys. Rev. Lett.***76**, 4046 (1996).

454   *Guo-Meng Zhao, J. Labry and Bo Truong*

[34] M. J. Calderon and L. Brey, *Phys. Rev. B* **58**, 3286 (1998).

[35] S. Yunoki, J. Hu, A. L. Malvezzi, A. Moreo, N. Furukawa, and E. Dagotto,*Phys. Rev. Lett.* **80**, 845 (1998).

[36] L. Sheng and C. S. Ting, *Phys. Rev. B* **60**, 14809 (1999).

[37] Yu. A. Firsov, V. V. Kabanov, E. K. Kudinov, and A. S. Alexandrov, *Phys. Rev. B* **59**, 12 132 (1999).

[38] M. Capone, M. Grilli, and W. Stephan, Eur. *Phys. J.* B **11**, 551 (1999).

[39] A. S. Alexandrov and A. M. Bratkovsky, *Phys. J. Condens. Matter*, **11**, L531 (1999).

[40] W. E. Pickett, D. J. Singh, and D. A. Papaconstantopoulos, *Physics of manganites*, edited by T. A. Kaplan and S. D. Mahanti (Kluwer Academic/Plenum Publisher, 1999) page 87.

[41] G. M. Zhao, V. Smolyaninova, W. Prellier, and H. Keller, *Phys. Rev. Lett.* **84**. 6086 (2000).

[42] H. J. Lee, J. H. Jung, Y. S. Lee, J. S. Ahn, T. W. Noh, K. H. Kim , and S. W. Cheong, *Phys. Rev. B* **60**, 5251 (1999).

[43] D. Louca, T. Egami, E. L. Brosha, H. Röder, and A. R. Bishop, *Phys. Rev. B* **56**, R8475 (1997).

[44] A. Lanzara, N. L. Saini, M. Brunelli, F. Natali, A. Bianconi, P. G. Radelli, and S. W. Cheong, *Phys. Rev. Lett.* **81**, 878 (1998).

[45] M. F. Hundley and J. J. Neumeier, *Phys. Rev. B***55**, 11 511 (1997).

[46] J. H. Jung, K. H. Kim, T. W. Noh, E. J. Choi, and J. J. Yu, *Phys. Rev. B***57**, R11 043 (1998).

[47] D. I. Khomskii and G. A. Sawatzky, *Solid State Commun.* **102**, 87 (1997).

[48] A. K. McMahan, R. M. Martin, and S. Satpathy, *Phys. Rev. B* **38**, 6650 (1988).

[49] J. P. Falck, A. Levy, M. A. Kaster, and R. J. Birgeneau, *Phys. Rev. Lett.* **69**, 1109 (1992).

[50] E. B. Stechel and D. R. Jennison, *Phys. Rev. B* **38**, 4632 (1988).

[51] G.-M. Zhao, K. Conder, H. Keller, and K. A. Müller, *Nature* **381**, 676 (1996).

[52] S. J. L. Billinge, R. G. DiFrancesco, G. H. Kwei, J. J. Neumeier, and J. D. Thompson, *Phys. Rev. Lett.* **77**, 715 (1996).

[53] C. H. Booth, F. Bridges, G. H. Kwei, J. M. Lawrence, A. L. Cornelius, and J. J. Neumeier, *Phys. Rev. Lett.* **80**, 853 (1998).

Polarons and Bipolarons in Colossal Magnetoresistive Manganites 455

[54] M. Jaime, M. B. Salamon, M. Rubinstein, R. E. Treece, J. S. Horwitz, and D. B. Chrisey, *Phys. Rev. B* **54**, 11914 (1996).

[55] D. C. Worledge, L. Mieville, and T. H. Geballe, *Phys. Rev. B* **57** 15267 (1998).

[56] A. Urushibara, Y. Moritomo, T. Arima, A. Asamitsu, G. Kido, and Y. Tokura, *Phys. Rev. B* **51** 14103 (1995).

[57] M. Jaime, P. Lin, M. B. Salamon, and P. D. Han, *Phys. Rev. B* **58** R5901 (1998).

[58] I. G. Lang and Yu. A. Firsov, Sov. *Phys. -JETP* **16**, 1301 (1963).

[59] V. N. Bogomolov, E. K. Kudinov, and Yu. A. Firsov, Sov. *Phys.* - Solid State **9**, 2502 (1968).

[60] W. Prellier et al., Appl. *Phys. Lett.* **75**, 1446 (1999).

[61] Guo-meng Zhao, H. Keller, and W Prellier, J. *Phys.: Condens. Matter* **12**, L361 (2000).

[62] G. J. Snyder, R. Hiskes, S. DiCarolis, M. R. Beasley, and T. H. Geballe, *Phys. Rev. B***53** 14434 (1996).

[63] J. R. Simpson, H. D. Drew, V. N. Smolyaninova, R. L. Greene, M. C. Robson, A. Biswas, and M. Rajeswari, *Phys. Rev. B* **60**, R16 263 (1999).

[64] P. Böni, J. D. Axe, G. Shirane, R. J. Birgeneau, D. R. Gabbe, H. P. Jessen, M. A. Kastner, C. J. Peters, P. J. Picone, and T. R. Thurston, *Phys. Rev. B***38**, 185 (1988).

[65] W. Reichardt, B. Batlogg, and J. P. Remeika, *Physica B* **135**, 501 (1985).

[66] B. Batlogg, *Physica B* **126**, 275 (1984).

[67] W. E. Pickett, R. E. Cohen, and H. Krakauer, *Phys. Rev. Lett.* **67**, 228 (1991).

[68] D. J. Singh and W. E. Pickett, *Phys. Rev. B***57**, 88 (1998).

[69] Zhao G.-M. and Morris, D. E. 1995 (unpublished).

[70] G. M. Zhao, M. B. Hunt and H. Keller, *Phys. Rev. Lett.* **78**, 955 (1997).

[71] J.-S. Zhou and J. B. Goodenough, *Phys. Rev. Lett.* **80**, 2665 (1998).

[72] G. M. Zhao et al., *Phys. Rev. B* **60**, 11 914 (1999).

[73] G. M. Zhao et al., *Phys. Rev. B* **63**, 060402R (2000).

[74] D. A. Papaconstantopoulos and W. E. Pickett, *Phys. Rev. B* **57**, 12751 (1998).

[75] A. S. Alexandrov, G. M. Zhao, H. Keller, B. Lorenz, Y. S. Wang, and C. W. Chu, *Phys. Rev. B* **64**, R140404 (2001).

[76] N. Mannella et al., *Nature* **438**, 474 (2005).

[77] G. M. Zhao et al., Solid State Commun. **104**, 57 (1997)

[78] M. R. Ibarra, G. M. Zhao, J. M. De Teresa, B. Garcia-Landa, Z. Arnold, C. Marquina, P. A. Algarabel, and H. Keller, *Phys. Rev. B* **57**, 7446 (1998).

[79] G. M. Zhao et al., *Phys. Rev. B* **63**, 11949R (2000).

[80] I. G. Austin and N. F. Mott, Adv. *Phys.* **18**, 41 (1969).

[81] J. M. De Teresa, M. R. Ibarra, P. A. Algarabel, C. Ritter, C. Marquina, J. Blasco, J. Garcia, A. del Moral, and Z. Arnold, *Nature* (London), **386**, 256 (1997).

In: Polarons: Recent Progress and Perspectives
Editor: Amel Laref

ISBN: 978-1-53613-935-8
© 2018 Nova Science Publishers, Inc.

Chapter 14

POLARONIC HIGH-TEMPERATURE SUPERCONDUCTIVITY IN BISMUTHATES AND CUPRATES

Guo-Meng Zhao[1,2,*], *N. Derimow*[1,3], *J. Labry*[1]
and A. Khodagulyan[1]
[1]Department of Physics and Astronomy,
California State University, Los Angeles, CA, US
[2]Research Institute of Zhejiang University–Taizhou, Taizhou, China
[3]Department of Physics and Astronomy,
University of California, Riverside, CA, US

Abstract

The pairing mechanism for high-temperature superconductivity (HTS) in copper and iron-based superconductors remains elusive despite tremendous efforts. Bacause the electron-phonon coupling constant predicted from the density functional theory (DFT) is too small to explain HTS and because their insulating phases are antiferromagnetic, it is natural to believe that antiferromagnetic fluctuations may play an essential role in HTS. On the other hand, HTS in the non-magnetic $Ba_{1-x}K_xBiO_3$ cannot arise from any magnetic interaction. Surprisingly, the electron-phonon coupling constant predicted from DFT is also too small (about 0.3-0.4) to explain HTS. Therefore, the conventional phonon-mediated pairing mechanism fails to explain HTS in all these materials. Here we will present experimental evidences for the polaronic Cooper pairs in these superconductors. We show that the significant enhancement of the effective density of states due to the lattice polaronic effect in bismuthates increases the effective electron-phonon coupling constant to 1.41, which is sufficient to explain the observed superconducting transition temperature. We also show that, in addition to the polaronic enhancement of superconductivity in cuprate superconductors, antiferromagnetic correlation enhances significantly the electron-phonon coupling constant. The combination of the substantial polaronic effect and the strong antiferromagnetic correlation in cuprates can naturally explain an order-of-magnitude discrepancy between the DFT calculated electron-phonon constant and the measured one. This combination is the key for achieving high-temperature superconductivity in cuprates.

---

*E-mail address: gzhao2@calstatela.edu.

Keywords: polaronic superconductivity, bismuthates, cuprates

1. Introduction

Developing the microscopic theory for high-T_c superconductivity (HTS) is one of the most challenging problems in condensed matter physics. Over thirty years after the discovery of HTS by Bednörz and Müller [1], no consensus on the microscopic pairing mechanism has been reached despite tremendous experimental and theoretical efforts. Because copper- and iron-based superconductors are in the proximity of antiferromagnetic instability, it is natural to believe that antiferromagnetic fluctuations play an essential role in bringing about high-temperature superconductivity in these two systems. In contrast, high-temperature superconductivity in $Ba_{1-x}K_xBiO_3$ (BKBO) and MgB_2 cannot arise from antiferromagnetic fluctuations because they are not magnetic.

The first-principle calculations of the electron-phonon coupling constant and the superconducting properties were made for MgB_2 within the density-functional theory (DFT) and the multi-band anisotropic Eliashberg formalism [2]. The calculated results are in quantitative agreement with the experimental ones. This implies that the first-principle calculation within the DFT is able to accurately predict electron-phonon coupling constant at least in nonmagnetic materials. On the other hand, the electron-phonon coupling constant of optimally doped BKBO is predicted to be about 0.3-0.4 from the first-principle calculations [3, 4, 5]. This calculated electron-phonon coupling constant can only lead to superconductivity of about 1 K within the single-band Eliashberg formalism. The conventional phonon-mediated theory is thus difficult to explain 30 K high-temperature superconductivity in the nonmagnetic BKBO. Similar calculations for optimally doped cuprates also show that the electron-phonon coupling constant is significantly smaller than 1 [e.g., <0.5 (Ref. [6])], which may give rise to superconductivity of less than 2 K, in disagreement with the observed T_c of 100 K.

The theoretical predictions of the weak retarded electron-phonon interaction for both cuprates and bismuthates are not supported by the experimental results. Extensive studies of unconventional oxygen-isotope effects in hole-doped cuprates have clearly shown strong electron-phonon interaction and the existence of polarons/bipolarons in both normal and superconducting states [7, 8, 9, 10, 11, 12, 13, 14, 15, 16]. Neutron scattering [17, 18, 19], angle-resolved photoemission (ARPES)[20, 21], pump probe [22], and optical specroscopies [23, 24, 25, 26] have also demonstrated strong electron-phonon interaction. Furthermore, ARPES data [27] and tunneling spectra [28, 29, 30, 31, 32, 33] have consistently provided direct evidence for strong coupling to multiple-phonon modes in hole-doped cuprates. The effective electron-phonon coupling constants extracted from the tunneling spectra in optimally doped BKBO and cuprates are about 3 to 10 times as large as the values calculated from DFT. Now a question arises as to what mechanism causes such a large enhancement in the effective electron-phonon coupling constants of both cuprate and bismuthate superconductors.

The polaron-bipolaron theory of superconductivity [34], which is based on strong electron-electron correlation and significant electron-phonon interaction, has gained strong support from various experimental results. It has been shown that the superconductivity of underdoped cuprates is in quantitative agreement with the bipolaron theory of superconductivity [16]. For optimally doped cuprates, high-temperature superconductivity arises from the Cooper pairing of polarons mediated by both retarded and non-retarded electron-phonon interactions [12, 35]. The same mechanism can quantitatively explain superconductivity in BKBO (see below).

Within this polaronic strong-coupling phonon-mediated mechanism, the strong electron-phonon coupling with high-energy optical phonon modes leads to the formation of lattice polarons, and the polarons are bound into the Cooper pairs through the retarded and non-retarded electron-phonon interactions. The mass of a polaron and the effective density of polaronic states are both enhanced by a factor $f_p = A \exp(\gamma E_p/\Omega)$, where E_p is the polaronic binding energy, Ω is the characteristic optical phonon energy, γ is a constant of less than 1, depending on E_p, Ω, and the bare bandwidth W_0, and A is a constant of order of 1, depending on E_p, Ω and W_0. The high-energy optical phonon modes are responsible for the formation of lattice polarons and for the retarded electron-phonon interaction. The effective coupling constant for the retarded electron-phonon interaction is increased by a factor of f_p while the direct Coulomb potential μ is reduced by the non-retarded electron-phonon interaction. The μ value may be significantly lower than the bare Coulomb potential μ_0 and the effective retarded Coulomb potential μ^* is further reduced by a factor of $1 + \ln(E_F/\hbar\omega_{\ln})$, where E_F is the effective Fermi energy of polarons and $\hbar\omega_{\ln}$ is the logarithmically-averaged energy of the phonons participating in the retarded electron-phonon interaction.

Within this approximation, the superconductivity can be effectively described by the conventional strong-coupling Eliashberg formalism except for renormalization of some parameters by the polaronic enhancement factor f_p. Since the effective density of states is enhanced by a factor of f_p, the effective retarded electron-phonon coupling constant λ, which is proportional to the effective density of states, should also increase by the the same factor, that is, $\lambda = f_p\lambda_b$, where λ_b is the bare electron-phonon coupling constant determined from the first-principle calculation within the DFT. The total mass enhancement factor f_t within this model is given by $f_t = f_p(1 + \lambda) = f_p(1 + f_p\lambda_b)$. The enhancement factor $1 + \lambda$ arises from the retarded electron-phonon interaction, which is treated within the Migdal approximation.

This mechanism may be oversimplified and needs to be justified by both theory and experiment. Theoretical justification for the approximation may be difficult because electrons are strongly coupled to multiple phonon modes with different characteristics. Here we would like to justify it by the experimental results of the simplest BKBO system where the conduction band is single isotropic $s - p$ hybridized band. We will show that the experimental results for optimally doped BKBO are in quantitative agreement with this simplified model. We also show that, in addition to the polaronic enhancement of superconductivity in cuprate superconductors, antiferromagnetic correlation enhances significantly the electron-phonon coupling constant. The combination of the substantial polaronic effect and the strong antiferromagnetic correlation in cuprates can naturally explain an order-of-magnitude discrepancy between the DFT calculated electron-phonon constant and the measured one. This combination is the key for achieving high-temperature superconductivity in cuprates. Due to the antiferromagnetic correlation, the pairing gap function becomes highly anisotropic and even changes signs [36, 37, 38] to minimize the pair-breaking effect of the antiferromagnetic fluctuation.

2. Superconductivity in Optimally-Doped Bismuthates

2.1. Upper Critical-Field at Zero Temperature

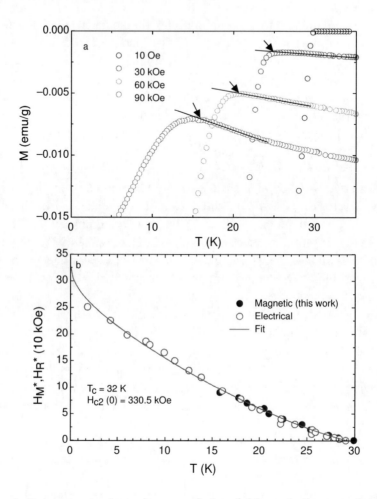

Figure 1. a) Temperature dependences of the field cooled magnetizations of the Ba$_{0.63}$K$_{0.37}$BiO$_3$ sample in different magnetic fields (up to 90 kOe). b) Temperature dependences of critical fields measured magnetically H_M^* and electrically H_R^*, respectively. The data for H_R^* were taken from Ref. [39]. The solid line is the curve with T_c = 32 K and $H_{c2}(0)$ = 330.5±4.6 kOe, which was predicted from the model based on a large superconducting fluctuation [40].

Figure 1a shows the temperature dependences of the field cooled magnetizations of the Ba$_{0.63}$K$_{0.37}$BiO$_3$ in different magnetic fields (up to 90 kOe). The superconducting transition in 10 Oe magnetic field is rather sharp, indicating high-quality of the sample. We define the critical temperature as the point of the onset of drop in magnetization (see arrows in Fig. 1a). With this definition, we obtain phase diagram of the magnetically determined critical field $H_M^*(T)$ for this bismuthate superconductor, as shown in Fig. 1b. It is interesting that the magnetically determined critical field $H_M^*(T)$ coincides with the electrically determined critical field $H_R^*(T)$ for a high-quality single-crystalline sample with a compo-

Polaronic High-Temperature Superconductivity in Bismuthates ...

sition of $x = 0.36$ and the lowest residual resistivity of about 70 $\mu\Omega$cm (Ref. [39]). This suggests that the critical fields obtained from both magnetic and electrical measurements are associated with the same physical phenomenon. The great variations in the resistivity and Hall effect for several single-crystalline samples with the same average composition of $x = 0.36$ (Ref. [39]) suggest that the insulating phase should correspond to $x \leq 0.35$ and the optimally-doped pure superconducting phase may correspond to $0.36 \leq x \leq 0.38$.

Intriguingly, the critical-field curve of the optimally-doped bismuthate superconductor shows an upward curvature, which does not follow the conventional Ginzburg-Landau model. This unusual behavior was well explained in terms of a large superconducting fluctuation proposed by Cooper *et al.* [40]. This model was also confirmed by the results reported for the sample irradiated by heavy ions [41]. Because of the superconducting fluctuation, the critical fields determined from electrical [39], magnetic (this work), and thermal [42] measurements are not true thermodynamic upper critical fields at any finite temperature. Only the critical field at zero temperature is the true thermodynamic upper critical field, which can be used to determine the intrinsic thermodynamic quantities. Because of the large superconducting fluctuation, the superconducting phase transition is not of second order, but of third or even fourth order [43, 44]. The higher-order superconducting transition could make the specific-heat anomaly negligibly small, in agreement with experiments [45, 46] (also see discussion below).

The solid line in Figure 1b is the best fitted curve by the equation derived from the theory of thermodynamic fluctuations [40]. The zero-field T_c is fixed to be 32 K for fitting and the best fit leads to the intrinsic zero-temperature upper critical field $H_{c2}(0) = 330.5\pm4.6$ kOe. At any finite temperature, the measured critical field is significantly suppressed compared with the intrinsic upper critical field (e.g., at 6 K, $H_{c2} = 300$ kOe and $H_R^* = 200$ kOe).

2.2. Data Interpretation

Having reliably determined $H_{c2}(0) = 330.5$ kOe and $\lambda(0) = 198.5$ nm (Ref. [47]), we should be able to determine the thermodynamic quantities of the superconductor if we can accurately determine the reduced energy gap $2\Delta(0)/k_BT_c$. Fortunately, a high-quality tunneling spectrum [48] was measured earlier for BKBO with $T_c = 32$ K. The point-contact tunneling spectrum is reproduced in Fig. 2. The solid line is the best fitted curve using the Blonder-Tinkham-Klapwijk (BTK) theory [49]. The fitting parameters are displayed in the figure. From the inferred gap $\Delta(0) = 5.95$ meV and the measured $T_c = 32$ K, we obtain $2\Delta(0)/k_BT_c = 4.31$, indicating a quite large electron-boson coupling constant within the conventional strong-coupling theory. The result also rules out the unconventional pairing mechanism based on the interaction with high-energy charge excitations [50], which would predict $2\Delta(0)/k_BT_c \simeq 3.53$.

From the reduced energy gap, we determine $k_BT_c/\hbar\omega_{\ln}$ to be 0.107 using a standard expression for conventional superconductors [51]:

$$\frac{2\Delta(0)}{k_BT_c} = 3.53[1 + 12.5(\frac{k_BT_c}{\hbar\omega_{ln}})^2 \ln(\frac{\hbar\omega_{ln}}{2k_BT_c})]. \tag{1}$$

From $k_BT_c/\hbar\omega_{\ln} = 0.107$, $T_c = 32$ K, and assuming a typical Coulomb pseudo-potential $\mu^* = 0.1$, we obtain $\hbar\omega_{\ln} = 25.78$ meV and $\lambda = 1.41$. The experimentally inferred numbers

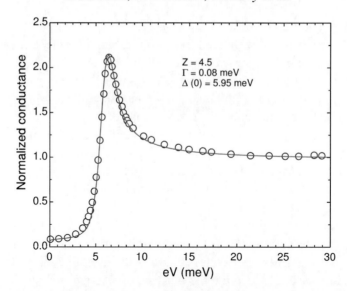

Figure 2. Point-contact tunneling spectrum of optimally doped $Ba_{1-x}K_xBiO_3$ with T_c = 32 K. The data were digitized from Ref. [48]. The solid line is fitted by the BTK theory with the following fitting parameters: barrier strength Z = 4.5, life-time broadening parameter Γ = 0.08 meV, and zero-temperature superconducting gap $\Delta(0)$ = 5.95 meV.

based on the conventional strong-coupling theory would imply the conventional phonon-mediated pairing mechanism in $Ba_{0.63}K_{0.37}BiO_3$.

We can determine the electronic Sommerfeld coefficient γ, Ginzburg-Landau parameter κ, and the specific-heat jump ΔC at T_c using the standard expressions for the conventional strong-coupling superconductors. Assuming that $Ba_{0.63}K_{0.37}BiO_3$ is a clean superconductor, as justified below, and using $k_B T_c/\hbar\omega_{ln}$ = 0.107 and empirical formulas for conventional superconductors [51], we determine the slope dH_{c2}/dT and $d(\Phi_0/\lambda^2)/dT$ near T_c to be $1.344 H_{c2}(0)/T_c$ and $2.882\Phi_0/[\lambda^2(0)T_c]$, respectively. The specific heat jump $\Delta C/\gamma T_c$ is calculated to be 2.416 with $k_B T_c/\hbar\omega_{ln}$ = 0.107. Using $\lambda(0)$ = 198.5 nm and $H_{c2}(0)$ = 330.5 kOe and the above relations, we obtain $d(\Phi_0/\lambda^2)/dT$ = −47.3 Oe/K and dH_{c2}/dT = −13880 Oe/K. We then determine κ = 42.92 using the standard expressions [52]: $2\kappa^2/\ln\kappa = (dH_{c2}/dT)/(dH_{c1}/dT)$ and $dH_{c1}/dT = (\ln\kappa/4\pi)[d(\Phi_0/\lambda^2)/dT]$. Finally from $\Delta C/T_c = (1/8\pi\kappa^2)(dH_{c2}/dT)^2$ (Ref. [53]) and $\Delta C/\gamma T_c$ = 2.416, we find $\Delta C/T_c$ = 19.77 mJ/moleK^2 and γ = 8.18 mJ/moleK^2.

The total mass enhancement factor f_t is calculated to be 7.54 using the relation $f_t = \gamma/\gamma_b$, where γ_b is the bare electronic Sommerfeld coefficient, which was found to be 1.08 mJ/molK^2 for $Ba_{0.6}K_{0.4}BiO_3$ (Ref. [54]). The f_t value inferred from the superconducting properties is in quantitative agreement with the theoretically predicted enhancement factor based on the polaronic model [55]. Within this model, $f_t = W_{opt}(g=0)/W_D(g)$, where $W_{opt}(g=0)$ is the total optical weight without electron-phonon coupling and $W_D(g)$ is the Drude weight in the presence of electron-phonon coupling. In Fig. 3, we plot the theoretically calculated [55] Drude weight, W_D, normalized by total weight W_{opt}, as a function of the doping level x together with $1/f_t$ for x = 0.37, which is inferred above from the thermodynamic quantities. From Figure 3, it is clear that our inferred mass enhancement factor f_t is in quantitative agreement with the theoretical prediction [55].

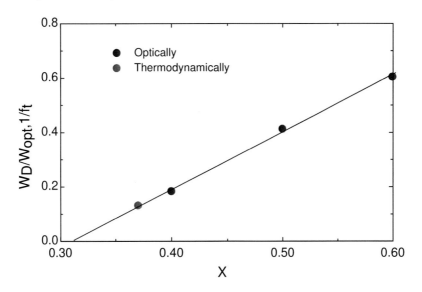

Figure 3. Theoretically calculated [55] Drude weight, W_D, normalized by total optical weight W_{opt}, as a function of the doping level x together with $1/f_t$ for $x = 0.37$, which is inferred from the thermodynamical quantities.

In contrast, this enhancement factor f_t is too large to be consistent with the conventional phonon-mediated mechanism. According to the conventional model, $\gamma = (1+\lambda)\gamma_b$. This would imply $\lambda = 6.54$, in serous contradiction with the inferred $\lambda = 1.41$ from the tunneling spectrum. Therefore, the conventional phonon mediated pairing mechanism cannot consistently explain the experimental results.

As discussed above, the conventional phonon-mediated mechanism cannot explain the giant difference in the total mass enhancement factors inferred from the tunneling spectrum (2.41) and from the thermodynamic quantities (7.54). Even the effective electron-phonon coupling constant (1.41) inferred from the tunneling spectrum is too large compared with the calculated value of 0.3-0.4 based on the DFT. Therefore, the conventional phonon-mediated pairing mechanism is unable to consistently explain all the data and also contradicts the result predicted from the DFT.

Alternatively, we can quantitatively explain the data in terms of a modified strong-coupling phonon-mediated mechanism [12, 35] where the strong electron-phonon coupling with high-energy optical phonon modes leads to the formation of lattice polarons, and the polarons are bound into the Cooper pairs through the retarded electron-phonon interaction with other phonon modes. Within this simplified model, the polaronic effect simply enhances the effective density of states by a polaronic mass enhancement factor f_p so that the effective retarded electron-phonon coupling constant λ increases by the same factor, that is, $N(0) = f_p N_b(0)$, $\lambda = f_p \lambda_b$. The total mass enhancement factor f_t within this model is given by

$$f_t = f_p(1+\lambda) = f_p(1+f_p\lambda_b). \qquad (2)$$

464   G.-M. Zhao, N. Derimow, J. Labry et al.

Substituting the value of $f_t = 7.54$ and $\lambda = 1.41$ into Eq. 2 yields $f_p = 3.26$ and $\lambda_b = 0.40$. The inferred $\lambda_b = 0.40$ is in quantitative agreement with the first-principle calculation [4].

The coherence length near T_c within the BCS model is given by [56]: $\xi(T) = 0.74\xi_{BCS}\sqrt{G(0.882\xi_{BCS}/l)}/\sqrt{1 - T/T_c}$ where $G(x)$ is the Gor'kov function defined by $G(x) = \sum_{n=0}^{\infty} \frac{0.95}{(1+2n)^2(1+2n+x)}$, $\xi_{BCS} = \frac{\hbar v_F}{\pi\Delta(0)}$, and v_F is the effective Fermi velocity. Due to strong electron-boson coupling, v_F is reduced by a factor of f_t compared with the bare v_F^b, which is calculated to be 1.08×10^6 m/s for $Ba_{0.60}K_{0.40}BiO_3$ [57]. Because of strong electron-boson coupling, the slope of the upper critical field near T_c is modified as [58]

$$\frac{dH_{c2}}{dT} = -\frac{\phi_0}{2\pi T_c} \frac{\eta_{H_{cs}}(T_c)}{0.74^2\xi_{BCS}^2 G(0.882\xi_{BCS}/l)}, \tag{3}$$

where $\eta_{H_{c2}}(T_c) = 1.18$ for a clean superconductor with $k_B T_c/\hbar\omega_{ln} = 0.107$ [51]. Substituting $\Delta(0) = 5.95$ meV, $v_F^b = 1.08 \times 10^6$ m/s, and $f_t = 7.54$ into $\xi_{BCS} = \frac{\hbar v_F^b}{f_t\pi\Delta(0)}$ yields $\xi_{BCS} = 50.6$ Å. Substituting $\xi_{BCS} = 50.6$ Å, $dH_{c2}/dT = -13880$ Oe/K, and $\eta_{H_{c2}}(T_c) = 1.18$ into (3), we get $l = 70.77$ Å. The fact that $l > \xi_{BCS}$ implies that $Ba_{0.60}K_{0.40}BiO_3$ is a clean superconductor, which justifies the above analyses.

Finally, the bare plasma energy $\hbar\Omega_p$ is calculated to be 3.75 eV from the relation $\hbar\Omega_p = [\hbar c/\lambda(0)]\sqrt{f_t(1 + (\pi^2/8)\xi_{BCS}/l)}$ (Eq. 145 of Ref. [59]) and $\lambda(0) = 198.5$ nm. The inferred bare plasma energy of 3.75 eV is slightly smaller than the value (4.03 eV) predicted from the first-principle calculation [57]. Since charge carriers in the band edge can be localized by random potential arising form dopants and/or defects, the free carrier density should be lower than the total carrier density.

With the bare plasma energy of 3.75 eV, $l = 70.77$ Å, and $v_F^b = 1.08 \times 10^6$ m/s, we can readily calculate the residual resistivity to be 53.6 $\mu\Omega cm$ which is slightly lower than the measured values (72, 79, and 92 $\mu\Omega cm$) of the three best crystals (labeled by F, E, D) in Ref. [39]. This may be consistent with the fact that the crystals contain mixed superconducting and insulating phases due to possible composition inhomogeneity. The effective conductivity σ_{eff} of the mixed system is reduced compared with the intrinsic conductivity of the metallic component σ_M. For a three-dimensional system, σ_{eff} is given by [60]

$$\sigma_{eff} = \frac{3}{2}\sigma_M(p_M - \frac{1}{3}), \tag{4}$$

where p_M is the fraction of the metallic (superconducting) phase. Using the calculated $1/\sigma_M = 54.6$ $\mu\Omega cm$, we can evaluate p_M to be 0.829, 0.785, 0.721 for crystals F, E, D with residual values of 72, 79, and 92 $\mu\Omega cm$, respectively. For crystal D, which was used for the magnetization and specific-heat measurements in Ref. [42], the superconducting fraction is 72%.

The inferred $\Delta C/T_c = 19.77$ mJ/mol K^2 is over one order of magnitude larger than the specific-heat anomaly observed earlier [45, 46]. This inconsistency can be explained in terms of a higher-order superconducting phase transition [43, 44]. Considering a weak superconducting fluctuation for a second-order phase transition, the free energy can be written as $F(T) = -f_0(1 - T/T_c)^{2-\alpha}$, where α is a small number which can be calculated by

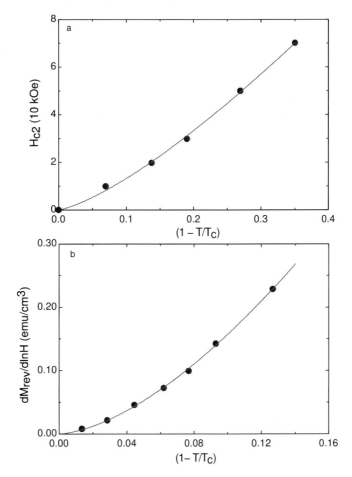

Figure 4. a) Upper critical field as a function of $(1 - T/T_c)$ for a BKBO crystal. The data were taken from Ref. [42]. The solid line is the curve proportional to $(1 - T/T_c)^p$ with $p = 1.33 \pm 0.04$. b) $dM_{rev}/d\ln H$ versus $(1 - T/T_c)$ for the crystal (where M_{rev} is the reversible magnetization). The data were taken from Ref. [42]. The solid line is the curve proportional to $(1 - T/T_c)^q$ with $q = 1.59 \pm 0.03$.

pseudo-pertubative schemes (such as Gaussian approximation) [43]. However, when the superconducting fluctuation is large, the magnitude of α should be large and beyond the realm of a perturbative approach. Then the unperturbed ground state could be a transition of order corresponding to the nearest integer, that is, a third order or even a fourth order depending on how large is the superconducting fluctuation. A very small $\Delta C/T_c$ value observed earlier [45, 46], is consistent with a fourth order phase transition, as clearly demonstrated by experiments [43, 44].

On the other hand, the specific-heat measurement on a high-quality single-crystalline sample [42] showed a much larger specific-heat anomaly, that is, $\Delta C/T_c \simeq 6$ mJ/moleK^2, which is about 30% of the value inferred above from the upper critical-field and magnetic penetration depth at zero temperature. This implies that the superconducting fluctuation in this crystal should be much smaller than in the other samples, but still substantial. We can check the critical behaviors to see how large is the superconducting fluctuation in this crystal. In Fig. 4a, we display the upper critical fields near T_c, which were determined by specific heat measurements [42]. It is apparent that $H_{c2} \propto (1 - T/T_c)^p$ with $p = 1.33 \pm 0.04$. In Fig. 4b, we plot $dM_{rev}/d\ln H$ ($\propto H_{c1}$) as a function of $(1 - T/T_c)$ near T_c. The solid line is the curve proportional to $(1 - T/T_c)^q$ with $q = 1.59 \pm 0.03$. The values of p and q imply that $F(T) \propto H_{c2}H_{c1} \propto (1 - T/T_c)^{2-\alpha}$ with $\alpha = -0.92 \pm 0.07$. The value of $\alpha \simeq -1.0$ suggests a large superconducting fluctuation, which would lead to a third-order phase transition. The fourth order superconducting transition found in other crystals [43, 44] may be due to the fact that those crystals have more disorders, leading to shorter mean-free paths, to shorter coherence lengths, and thus to larger superconducting fluctuations.

2.3. Tunneling Evidence for Polaronic Superconductivity

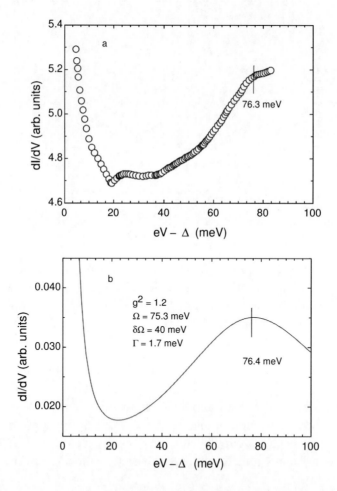

Figure 5. a) Point-contact tunneling spectrum of optimally doped $Ba_{1-x}K_xBiO_3$ with $T_c = 32$ K. The data were digitized from Ref. [48]. b) Point-contact tunneling spectrum predicted from the theory of polaronic superconductivity with the parameters displayed in the figure.

Polaronic High-Temperature Superconductivity in Bismuthates ... 467

An important signature of the polaronic superconductivity is the coexistence of the coherent (Drude) and incoherent spectral weights in both optical conductivity [55] and superconducting tunneling conductance dI/dV. Optical data [61] have clearly demonstrated the polaronic nature of superconductivity. In the superconducting tunneling conductance dI/dV spectra, there exist peak-dip-hump features, which provide information about the energies of the bosonic modes involved in the formation of polaronic clouds and the polaronic coupling strength g^2 (Ref. [62]). According to the polaron-bipolaron theory of superconductivity, the hump is the incoherent broad feature that reflects the local hopping of electrons in various frozen lattice configurations [62]. The theory predicts that the peak-hump energy separation $E_{hp} - E_{pk}$ in superconductor-insulator-superconductor (SIS) tunneling spectra is close to the characteristic energy Ω of the phonon modes in the polaronic cloud when the electron-phonon coupling strength g^2 is not too large. It is also true for the tunneling spectrum of a superconductor-insulator-normal (SIN) junction. The tunneling conductance dI/dV of an SIS junction was derived based on the polaron-bipolaron theory of superconductivity [62]. By replacing the parameters of a superconductor in an SIS junction by those of a normal-metal, we can readily deduce dI/dV for an SIN junction as

$$dI/dV \propto \Sigma_0^\infty \frac{g^{2n}(\Gamma + n\delta\Omega)}{n![(\Gamma + n\delta\Omega)^2 + (e|V| - \Delta - n\Omega)^2]}, \tag{5}$$

where Ω is the characteristic energy of phonon modes in the polaron cloud, $\delta\Omega$ is the phonon dispersion, g^2 is proportional to the polaronic binding energy E_p, Γ is the life-time broadening parameter, $\Delta = \sqrt{\Delta_p^2 + \Delta_c^2}$, Δ_p is the single-particle gap (normal-state gap), and Δ_c is the superconducting gap that closes at T_c. For the polaronic superconductivity where polarons are bound into the Cooper pairs, the norma-state gap $\Delta_p = 0$ so that $\Delta = \Delta_c$.

In Fig. 5a, we show SIN tunneling spectrum of an optimally doped BKBO crystal. It is apparent that there exists a peak-dip-hump feature, which is consistent with the polaronic superconductivity. The energy of the hump feature is about 76 meV, which is close to the energy (75.3 meV) of the bosonic mode participating in the formation of polarons because the coupling strength g^2 is not too large (1.2). It is known that the oxygen breathing mode (75.3 meV) is strongly coupled to electrons, leading to the formation of polarons. Fig. 5b displays the theoretical prediction of the tunneling spectrum with $g^2 = 1.2$ and $\Omega = 75.3$ meV (identical to the energy of the breathing mode). The theoretical curve predicts a hump feature at about 76 meV, in quantitative agreement with the experimental tunneling spectrum.

3. Superconductivity in Cuprates

3.1. Tunneling Spectra of Single-Layered Cuprates

Fine structures in tunneling spectra of a superconductor have been very useful in identification of boson modes responsible for superconductivity. Because of some highly publicized experimental papers [63, 64, 65], the d-wave magnetic pairing mechanism of high-temperature superconductivity in cuprates appears indisputable. In one of the papers [64], the authors used an unrealistic parameter, which overestimated the magnetic coupling constant by two orders of magnitude (for detailed arguments, see Ref. [66] and references therein). Other two papers [63, 65] reported the combined neutron and tunneling data for two electron-doped $Pr_{0.88}LaCe_{0.12}CuO_{4-y}$ (PLCCO) crystals with different superconducting transition temperatures (21 and 24 K). These data seemingly suggest that the energies of the magnetic resonance modes are the same as those of the bosonic modes revealed in the second derivative (d^2I/dV^2) of electron tunneling current (I) with respective to bias voltage (V). They thus concluded that the magnetic resonance modes rather than phonons mediated electron pairing in electron-doped cuprates [63, 65]. However, Zhao [32] has pointed out a basic mistake in the data analyses of the tunneling spectra in Ref. [63] and shown that the combined neutron and tunneling data actually disprove this magnetic pairing mechanism. Despite citing Ref. [32], the same basic mistake was repeated in Ref. [65]. Based on the repeated incorrect analyses of the tunneling spectra, these authors conclude that their data support d-wave magnetic pairing mechanism.

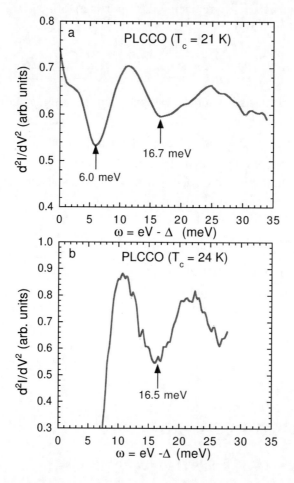

Figure 6. a) d^2I/dV^2 spectrum for an electron-doped PLCCO sample with T_c = 21 K, which is reproduced from Ref. [65]. b) d^2I/dV^2 spectrum for an electron-doped PLCCO sample with T_c = 24 K, which is reproduced from Ref. [63]. The energy positions of the d^2I/dV^2 spectra are measured from their superconducting gaps.

Polaronic High-Temperature Superconductivity in Bismuthates ... 469

It is well known [67, 68] that the energies of the *dip* positions in d^2I/dV^2 correspond to the energies of the bosonic modes strongly coupled to electrons while the authors in Refs. [63, 65] have assigned the energies of the *peak* positions in d^2I/dV^2 to the energies of the bosonic modes. Using this incorrect assignment, these authors find that the energy of a bosonic mode happens to be close to the energy of the magnetic resonance mode. They thus conclude that the magnetic resonance mode in electron-doped cuprates is strongly coupled to electrons and responsible for high-temperature superconductivity. However, this extremely important conclusion is based on their incorrect analyses of the tunneling spectra. Their incorrect data analyses have led to a wrong conclusion. The wrong conclusion of these works has seriously hindered the development of the correct microscopic theory of HTS because of the high publicity of these papers.

In fact, when we adopt the well-established correct protocol to identify the mode energies from their d^2I/dV^2 spectra, the conclusion is just opposite. In Fig. 6, we show our correct analyses of the d^2I/dV^2 spectra for two electron-doped samples. For the electron-doped sample with $T_c = 21$ K, the mode energies identified from the d^2I/dV^2 spectrum are 6.0 meV and 16.7 meV (indicated by the arrows marking the *dip* positions). For the electron-doped sample with $T_c = 24$ K, the mode energy identified from the d^2I/dV^2 spectrum is 16.5 meV. It is apparent that the mode energies revealed in the d^2I/dV^2 spectra of the two electron-doped cuprates are nearly the same (16.7 and 16.5 meV) and significantly different from the magnetic resonance energies (9.0 and 10.5 meV) revealed by inelastic neutron scattering experiments [63, 65]. The mode energy of about 16.5 meV, which is independent of doping and T_c, agrees with the energies of the two transverse optical (TO) phonon modes (15.6 meV for the E_u mode and 17.0 meV for the A_{2u}) [69]. The simple average of these two phonon energies is 16.3 meV, which is very close to the mode energy consistently seen in the tunneling spectra of the two electron-doped cuprates. This implies that the phonon modes rather than the magnetic-resonance modes are the origin of the fine structures in the tunneling spectra of the electron-doped cuprates. Therefore, their combined neutron and tunneling data [63, 65] actually disprove rather than support their claim that magnetic resonance modes mediate d-wave pairing in electron-doped cuprates.

In the structurally similar single-layer hole-doped $La_{1.84}Sr_{0.16}CuO_4$, there is a similar mode with an energy of 17.5 meV shown in the tunneling spectrum (see Fig. 7a) and 16.8 meV in the electron-boson spectral function of $La_{1.97}Sr_{0.03}CuO_4$ determined from angle-resolved photoemission spectrum (see Fig. 7b). The energies (6.0, 16.8, and 26.9 meV) of the three boson modes in the spectral function are nearly identical to those (5.9, 17.5, and 27.3 meV) of the dip positions in the d^2I/dV^2 spectrum. This consistency further proves that the dip positions (rather than peak positions) in d^2I/dV^2 spectra are associated with the bosonic modes strongly coupled to electrons. The mode energy of 6 meV seen in the spectral function determined from ARPES is also seen in the dip energy position of the d^2I/dV^2 spectra of the 21 K electron-doped PLCCO and hole doped LSCO, as well as hole-doped BSCCO (see below). This lowest phonon mode at 6 meV is also seen in the high-resolution neutron data (see below).

Since the magnetic-resonance mode energy was found to be proportional to T_c (Ref. [70]), insensitivity of the mode energies to T_c further argues against the magnetic origin of the modes. The similar mode energies across different doping levels (Fig. 6) and across different single-layer structures (PLCCO and LSCO) are consistent with the phonon

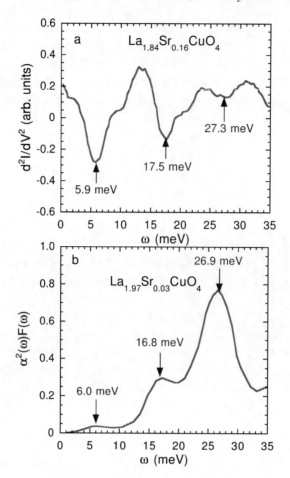

Figure 7. a) d^2I/dV^2 spectrum at 7.2 K for a hole-doped La$_{1.84}$Sr$_{0.16}$CuO$_4$ with T_c = 24 K, which is reproduced from Fig. 4 of Ref. [29]. The energy position of the d^2I/dV^2 spectrum is measured from the superconducting gap. b) Electron-boson spectral function of an underdoped La$_{1.97}$Sr$_{0.03}$CuO$_4$, which is determined from angle-resolved photoemission spectrum (ARPES). The spectrum is reproduced from Ref. [27].

mode assignment and contradict the magnetic resonance mode assignment. Therefore, the tunneling spectra and neutron data of the single-layered cuprates have unambiguously disproved d-wave magnetic pairing mechanism, in contrast to their own claims [63, 65], which are based on incorrect analyses of their tunneling spectra.

3.2. Tunneling Spectra of Double-Layered Cuprates

For double-layer hole-doped Bi$_2$Sr$_2$CaCu$_2$O$_{8+y}$ (BSCCO or Bi-2212), strong coupling features are also seen in the d^2I/dV^2 spectrum (see Fig. 8a). It is striking that the energies of all the dip features in the d^2I/dV^2 spectrum precisely match the energies of the peak features in the phonon density of states obtained from high-resolution inelastic neutron scattering [72]. This suggests strong coupling to multiple phonon modes in this double-layer hole-doped cuprate. Therefore, strong coupling to multiple phonon modes is universally seen in

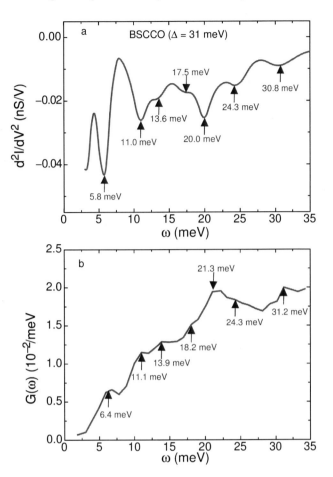

Figure 8. a) d^2I/dV^2 spectrum of double-layer hole-doped $Bi_2Sr_2CaCu_2O_{8+y}$. The d^2I/dV^2 spectrum is obtained by numerically taking the derivative of the dI/dV spectrum of Ref. [71] after the spectrum is smoothened by cubic spline. The energy position of the d^2I/dV^2 spectrum is measured from the superconducting gap. b) Phonon density of states for slightly overdoped $Bi_2Sr_2CaCu_2O_{8+y}$. The curve is reproduced from Ref. [72].

single-layered electron- and hole-doped cuprates as well as in double-layered hole-doped cuprates. It is the multiple phonon coupling that leads to a large electron-phonon coupling constant and to HTS. In contrast, the Cu-O buckling mode and the oxygen half-breathing mode, which have been considered to the most important phonon modes in contributing to a d-wave pairing, cannot explain high-temperature superconductivity.

Within the d-wave channel, the phonon contribution to the pairing is too small, so magnetic contribution must be dominant in order to explain HTS. But within an s-wave channel, all the phonon modes contribute to the pairing, so the total electron-phonon coupling constant could be large although the coupling constant of single phonon mode could be small. The high-quality tunneling spectrum [30] indicates that the total electron-phonon coupling constant in Bi-2212 could reach to a value of about 3 in certain portion of the Burillounin zone. In particular, low energy phonon modes below 40 meV make dominant contributions

[30] and the HTS appears to be well explained in terms of the strong coupling phonon-mediated mechanism [30].

One of the major problems within the phonon-mediated mechanism is the d-wave pairing symmetry, which has been accepted by the most scientists in the field. In fact, based on the quantitative analyses of many bulk-sensitive experiments in optimally doped and overdoped cuprates, Zhao has concluded that the intrinsic pairing gap symmetry in the bulk of cuprates is not d-wave, but extended s-wave (having eight line nodes) in hole-doped cuprates [36, 37] and nodeless s-wave in electron-doped cuprates [36, 37, 38]. Some phase-sensitive experiments and ARPES data that appear to support d-wave order-parameter symmetry can be naturally explained in terms of the d-wave symmetry of the Bose-Einstein superconducting condensation in underdoped cuprates [36, 37, 38].

The second major problem is that the calculated electron-phonon coupling constant from local density approximation (LDA) is about one order of magnitude smaller than that inferred from the tunneling data. It is important to notice that LDA could significantly underestimate electron-phonon coupling due to a large overestimate of electron screening in quasi-two-dimensional electronic systems. In fact, theoretical studies [73, 74] show that the optical phonons are strongly coupled to electrons due to the unscreened long-range interaction along the c-axis. Strong electron-electron (or antiferromagnetic) correlation [75] and nonadiabatic effects [74] further enhance electron-phonon coupling. Moreover, strong long-range coupling to high-energy phonon modes can lead to Fröhlich polarons, which can significantly enhance the density of states and the effective coupling constant. If one considers all these enhancement effects together, the effective electron-phonon coupling constant is likely to be enhanced by an order of magnitude compared with that calculated from LDA.

3.3. Peak-Dip-Hump Feature in Tunneling Spectra of Cuprates

The above analyses indicate that the d^2I/dV^2 tunneling spectra do not reveal strong coupling features of magnetic resonance modes in both electron- and hole-doped cuprates. Now a question arises as to whether strong coupling to magnetic resonance modes can be seen in the dI/dV tunneling spectra and angle-resolved photoemission spectra (energy distribution curves). The observed peak-dip-hump (PDH) structures in the dI/dV tunneling spectra and ARPES spectra (along the antinodal direction) have been used to support the d-wave magnetic pairing mechanism. Fig. 9 shows tunneling spectra of optimally doped double-layer $Bi_2Sr_2CaCu_2O_{8+y}$ and triple-layer $Bi_2Sr_2Ca_2Cu_3O_{10+y}$ (Bi-2223). There are clear peak-dip-hump features in both compounds and the triple-layer compound has a more pronounced hump feature than the double-layer one. The energy separation $E_{dip} - E_{pk}$ between the dip and peak features is about 30 meV in the double-layer compound and about 45 meV in the triple-layer compound. It is clear that $E_{dip} - E_{pk}$ is not linearly proportional to T_c.

The phenomenological spin-fermion model of Abanov and Chubukov [78] showed that strong coupling to the magnetic resonance mode yields a peak-dip-hump structure in ARPES spectra along the antinodal direction. This theory is based on the one-loop correction to the $t - t' - J$ mean-field theory. This PDH structure was also predicted to be present in dI/dV tunneling spectra based on the conventional strong coupling theory [79]. Within

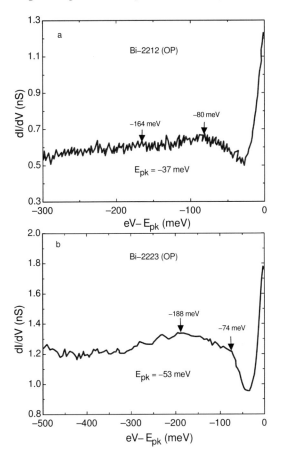

Figure 9. a) Tunneling spectra of optimally doped double-layer $Bi_2Sr_2CaCu_2O_{8+y}$. The spectrum is reproduced from Ref. [76]. b) Tunneling spectra of optimally doped triple-layer $Bi_2Sr_2Ca_2Cu_3O_{10+y}$. The spectrum is reproduced from Ref. [77].

both approaches, the energy separation between the dip and peak features is exactly equal to the energy E_r of the magnetic resonance mode, that is, $E_{dip} - E_{pk} = E_r$. Since E_r was found [70] to be proportional to T_c, $E_{dip} - E_{pk}$ should also be proportional to T_c. This is in sharp contrast to the experimental results shown in Fig. 10. The $E_{dip} - E_{pk}$ values obtained from ARPES (Fig. 10a) and tunneling spectra (Fig. 10b) are nearly independent of doping or T_c, in contradiction with the d-wave magnetic pairing mechanisms based on the spin-fermion model [78] and on the conventional approach [79].

Furthermore, the mode energies inferred from the tunneling spectra are about 27 meV, which are significantly lower than the measured E_r (43 meV) for the optimally doped Bi-2212. Only for the optimally doped Bi-2212, $E_{dip} - E_{pk}$ in the APPES spectrum is close to E_r. This single accidental agreement has been used as a smoking gun for the magnetic pairing mechanism. The systematical ARPES and tunneling data disprove the magnetic pairing mechanism.

Alternatively, the tunneling spectra shown in Fig. 9 are in quantitative agreement with the calculated SIN tunneling spectra (Fig. 11), predicted from the polaronic superconduc-

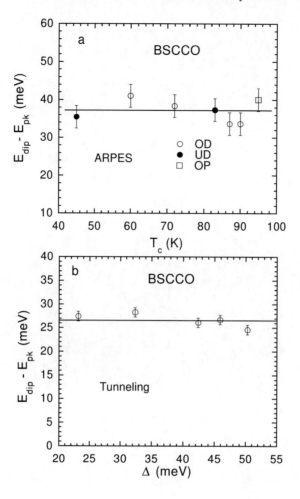

Figure 10. a) Dependence of $E_{dip} - E_{pk}$ on T_c in $Bi_2Sr_2CaCu_2O_{8+y}$, which is reproduced from Ref. [81]. The $E_{dip} - E_{pk}$ values were determined from ARPES data. b) Dependence of $E_{dip} - E_{pk}$ on the superconducting gap. The $E_{dip} - E_{pk}$ values are obtained from Supplementary Figure 2 of Ref. [80].

tivity (Eq. 5). It is remarkable that the energy positions of the double hump features in the measured tunneling spectra (marked by arrows) are in quantitative agreement with those in the theoretical curves (marked by vertical lines). The characteristic energy of the bosonic modes participating in the formation of polarons in Bi-2212 is about 72 meV, which is very close to the highest energy (72 meV) of the peak in the generalized phonon density of states measured by inelastic neutron scattering [72]. A much higher T_c in Bi-2223 compared to that in Bi-2212 may be associated with a much stronger polaronic coupling strength g^2 in Bi-2223. We do not believe that any other theoretical model can quantitatively explain the double hump features clearly observed in the tunneling spectrum of Bi-2223.

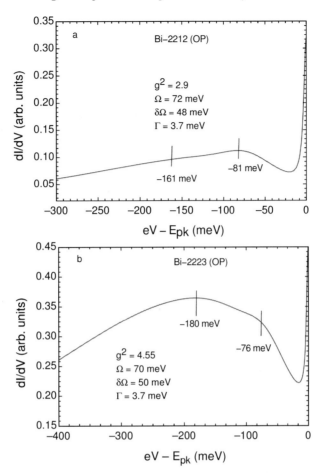

Figure 11. a) Point-contact tunneling spectrum of optimally doped Bi-2212, predicted from the theory of polaronic superconductivity with the parameters displayed in the figure. b) Point-contact tunneling spectrum of optimally doped Bi-2223, predicted from the theory of polaronic superconductivity with the parameters displayed in the figure.

3.4. Polaronic Enhancement Factor of an Optimally Doped Bi-2212

The polaronic mass enhancement factor f_p for optimally doped Bi-2212 can be extracted from the ARPES and tunneling data. The band dispersion along the diagonal direction (Γ-X) is flattened by a factor of 2 (Ref. [82]), suggesting that the polaronic mass enhancement factor should be 2. The band dispersion along the antinodal direction is more strongly renormalized, suggesting a much larger polaronic enhancement factor. The effective electron-boson coupling constant λ along the diagonal direction is found to be about 0.8 for optimally doped Bi-2212, which is determined from the ARPES data [83]. The tunneling spectrum of optimally doped Bi-2212 shows fine coupling structures in addition to a peak-dip-hump feature [30]. The observation of the peak-dip-hump feature in the tunneling spectrum indicates that the spectrum mainly probes the states near the antinodal direction. From the tunneling spectrum, the authors extract an electron-boson spectral function which matches well with the generalized phonon density of states measured by inelastic neutron scattering. The effective electron-boson coupling constant λ along the antinodal direction is deduced to be about 3, leading to a T_c of about 90 K with $\mu^* = 0.1$.

The bare electron-phonon coupling constant λ_b along the diagonal direction is equal to $\lambda/f_p = 0.8/2 = 0.4$, which is a factor of 1.74 larger than that (0.23) calculated from LDA [6]. As discussed above, antiferromagnetic correlation can significantly enhance the electron-phonon coupling constant. By taking into account the antiferromagnetic correlation in the double-layer Fe-based superconductor $Ba_{1-x}K_xFe_2As_2$, Boeri $et\ al.$, show that the electron-phonon coupling constant can be enhanced by a factor of about 1.6 compared with the case of the nonmagnetic correlation [84]. Therefore, the enhancement factor of 1.74 in the diagonal direction should mainly arise from the strong antiferromagnetic correlation. If we use the same enhancement factor of 1.74 for the antinodal direction (which should be underestimated due to stronger antiferromagnetic correlation along this direction), we calculate $\lambda_b = 0.31 \times 1.74 = 0.54$. Here we have used the LDA calculated electron-phonon coupling constant of 0.31 for the antinodal direction in the absence of the antiferromagnetic correlation [6]. Then the polaronic enhancement factor $f_p = \lambda/\lambda_b = 6.18$, which is much larger than that (2.0) for the diagonal direction. The large polaronic enhancement factor for the antinodal direction is consistent with the very flat band seen from ARPES [82] and with the pronounced hump feature observed in the tunneling spectrum and ARPES.

Optical data [85] for a slightly overdoped Bi-2212 indicate an average mass enhancement factor of 4.5. The result can be understood in terms of two-components of the carriers. The very flat band in the proximity of the antinodal direction occupies about 20% of the Burillounin zone [82], suggesting that the bare plasma energy contributed from this region is less than 10% of the total bare plasma energy due to the fact that the effective bare mass in this region is larger than that in the other region. In the antinodal region, the total mass enhancement factor is: $f_p(1 + \lambda) = 6.18 \times (1 + 3) = 25$. In the diagonal direction, the total mass enhancement factor is $2 \times (1 + 0.8) = 3.6$. Then the renormalized plasma energy contributed from the antinodal region is less than 3.7%. Thus the measured plasma energy mainly contributes from the non-antinodal region. The total mass enhancement factor along the diagonal direction (3.6) should be smallest and thus smaller than the average enhancement factor of 4.5 measured from the optical data [85]. If the average electron-phonon coupling constant $\lambda = 1.0$ for the non-antinodal region, then the average polaronic enhancement factor in the same region is 2.15.

Because in-plane magnetic penetration depth $\lambda_{ab}(0)$ is proportional to $\sqrt{f_t} = \sqrt{f_p(1 + \lambda)} = \sqrt{f_p(1 + f_p\lambda_b)}$ and because $f_p = \exp(\gamma E_p/\Omega) = \exp(g^2)$, there exist an isotope effect on the magnetic penetration depth. Using $f_p = 2.15$, we obtain $g^2 = 0.765$. With $\lambda = f_p\lambda_b = 1.0$, the exponent of the isotope effect on f_p is $-d\ln f_p/d\ln M = -\frac{1}{2}g^2 = -0.383$. The exponent of the isotope effect on f_t is $-\frac{1}{2}g^2[1 + \lambda/(1+\lambda)] = -0.574$. Finally, the exponent of the isotope effect on $1/\lambda_{ab}^2(0)$ is calculated to be 0.574. This is in quantitative agreement with the measured exponents of the oxygen-isotope effect on $1/\lambda_{ab}^2(0)$ in optimally doped $YBa_2Cu_3O_{6.93}$ [7, 15] and Bi-2223 [12]. The earlier magnetic data of the oxygen-isotope exchanged samples of $YBa_2Cu_3O_{6.93}$ [7, 12] suggest the oxygen-isotope exponent to be 0.54(4) and the later μSR data [15] indicate a similar value of 0.47(16).

Acknowledgments

The authors would like to thank the National Science Foundation of USA for financial support under the CREST grant (HRD 1547723).

Conclusion

We have presented experimental evidences for the polaronic Cooper pairs in both bismuthate and cuprate superconductors. The significant enhancement of the effective density of states due to the lattice polaronic effect increases the effective electron-phonon coupling constant to a value necessarily for the observed superconducting transition temperature in bismuthates. We also show that, in addition to the polaronic enhancement of superconductivity in cuprate superconductors, antiferromagnetic correlation enhances significantly the electron-phonon coupling constant. The combination of the substantial polaronic effect and the strong antiferromagnetic correlation in cuprates can naturally explain an order-of-magnitude discrepancy between the non-magnetic LDA calculated electron-phonon constant and the measured one. This combination is the key for achieving high-temperature superconductivity in cuprates.

References

[1] J. G. Bednorz and K. A. Müller, Z. *Phys. B* **64**, 189 (1986).

[2] J. M. An and W. E. Pickett, *Phys. Rev. Lett.* **86**, 4366 (2001); A. Y. Liu, I. I. Mazin, J. Kortus, *Phys. Rev. Lett.* **87**, 087005 (2001); H. J. Choi, D. Roundy, H. Sun, M. L. Cohen, and S. G. Louie, *Phys. Rev. B* **66**, 020513(R) (2002); I. I. Mazin and V.P. Antropov, *Physica C* **385**, 49 (2003).

[3] A. I. Liechtenstein, I. I. Mazin, C. O. Rodriguez, O. Jepsen, O. K. Andersen, and M. Methfessel, *Phys. Rev. B* **44**, 5388 (1991).

[4] N. Hamada, S. Massidda, and A. J. Freeman, *Phys. Rev. B* **40**, 4442 (1989). The electron-phonon coupling constant of $Ba_{0.7}K_{0.3}BiO_3$ is calculated to be 0.41 when one uses the Debye temperature $\theta_D = 513$ K, which corresponds to $\hbar\sqrt{<\omega^2>} = 31.29$ meV or $\hbar\omega_{ln} = 25.78$ meV.

[5] V. Meregalli and S.Y. Savrasov, *Phys. Rev. B* **57**, 14453 (1998).

[6] Rolf Heid, Klaus-Peter Bohnen, Roland Zeyher, and Dirk Manske, *Phys. Rev. Lett.* **100**, 137001 (2008).

[7] G. M. Zhao and D. E. Morris, *Phys. Rev. B* **51**, 16487(R)(1995).

[8] G. M. Zhao, K. K. Singh, A. P. B. Sinha, and D. E. Morris, *Phys. Rev. B* **52**, 6840 (1995).

[9] G. M. Zhao, M. B. Hunt, H. Keller, and K. A. Müller, *Nature* **385**, 236 (1997).

[10] G. M. Zhao, K. Conder, H. Keller, and K. A. Müller, *J. Phys. Condens. Matter* **10**, 9055 (1998).

[11] G. M. Zhao, H. Keller, and K. Conder, *J. Phys. Condens. Matter* **13**, R569 (2001).

[12] G. M. Zhao, V. Kirtikar, and D. E. Morris, *Phys. Rev. B* **63**, 220506(R) (2001).

[13] G. M. Zhao, *Phil. Mag. B* **81**, 1335 (2001).

[14] G. M. Zhao, V. Kirtikar, and D. E. Morris, *Phys. Rev. B* **63**, 220506(R) (2001).

[15] R. Khasanov, D. G. Eshchenko, H. Luetkens, E. Morenzoni, T. Prokscha, A. Suter, N. Garifianov, M. Mali, J. Roos, K. Conder, and H. Keller, *Phys. Rev. Lett.* **92**, 057602 (2004).

[16] A. S. Alexandrov and G. M. Zhao, *New Journal of Physics* **14**, 013046 (2012).

[17] R. J. McQueeney, Y. Petrov, T. Egami, M. Yethiraj, G. Shirane, and Y. Endoh, *Phys. Rev. Lett.* **82**, 628 (1999).

[18] T. R. Sendyka, W. Dmowski, T. Egami, N. Seiji, H. Yamauchi, and S. Tanaka, *Phys. Rev. B* **51**, 6747 (1995).

[19] D. Reznik, L. Pintschovius, M. Ito, S. Iikubo, M. Sato, H. Goka, M. Fujita, K. Yamada, G. D. Gu, and J. M. Tranquada, *Nature* **440**, 1170 (2006).

[20] A. Lanzara et al., *Nature* **412**, 510 (2001).

[21] O. Rösch, O. Gunnarsson, X. J. Zhou, T. Yoshida, T. Sasagawa, A. Fujimori, Z. Hussain, Z.-X. Shen, and S. Uchida, *Phys. Rev. Lett.* **95** 227002 (2005).

[22] C. Gadermaier, A. S. Alexandrov, V. V. Kabanov, P. Kusar, T. Mertelj, X. Yao, C. Manzoni, D. Brida, G. Cerullo, and D. Mihailovic, *Phys. Rev. Lett.* **105**, 257001 (2010).

[23] D. Mihailovic, C. M. Foster, K. Voss, and A. J. Heeger, *Phys. Rev. B* **42**, 7989 (1990).

[24] O. V. Misochko, E. Ya. Sherman, N. Umesaki, K. Sakai, S. Nakashima, *Phys. Rev. B* **59**, 11495 (1999).

[25] Y. S. Lee, K. Segawa, Z. Q. Li, W. J. Padilla, M. Dumm, S. V. Dordevic, C. C. Homes, Yoichi Ando, and D. N. Basov, *Phys. Rev. B* **72**, 054529 (2005).

[26] A. S. Mishchenko, N. Nagaosa, Z.-X. Shen, G. De Filippis, V. Cataudella, T. P. Devereaux, C. Bernhard, K. W. Kim, and J. Zaanen, *Phys. Rev. Lett.* **100**, 166401 (2008).

[27] X. J. Zhou et al., *Phys. Rev. Lett.* **95**, 117001 (2005).

[28] S. I. Vedeneev, P. Samuely, S. V. Meshkov, G. M. Eliashberg, A. G. M. Jansen, and P. Wyder, *Physica* C **198**, 47 (1992).

[29] D. Shimada, Y. Shiina, A. Mottate, Y. Ohyagi, and N. Tsuda, *Phys. Rev. B* **51**, 16495 (1995).

Polaronic High-Temperature Superconductivity in Bismuthates ... 479

[30] R. S. Gonnelli, G. A. Ummarino, and V. A. Stepanov, *Physica C* 275, 162 (1997).

[31] G. M. Zhao, *Phys. Rev. B* 75, 214507 (2007).

[32] G. M. Zhao, *Phys. Rev. Lett* **103**, 236403 (2009).

[33] H. Shim, P. Chaudhari, G. Logvenov and I. Bozovic, *Phys. Rev. Lett.* 101, 247004 (2008).

[34] For a recent review see A. S. Alexandrov and J. T. Devreese, *Advances in Polaron Physics* (Springer, Berlin 2009); P. E. Kornilovitch, *Phys. Rev. Lett.* **81**, 5382 (1998); A. S. Alexandrov and P. E. Kornilovitch, *Phys. Rev. Lett.* **82**, 807 (1999).

[35] A. S. Alexandrov, V. V. Kabanov, *Phys. Rev. B* **54**, 3655 (1996).

[36] G. M. Zhao, *Phys. Rev. B* **64**, 024503 (2001).

[37] G. M. Zhao, *Phys. Rev. B* **82**, 012506 (2010).

[38] G. M. Zhao and J. Wang, *J. Phys. Condens. Matter* **22** 352202 (2010).

[39] M. Affronte *et al.*, *Phys. Rev. B* **49**, 3502 (1994).

[40] J. R. Cooper, J. W. Loram, and J. M. Wade, *Phys. Rev. B* **51**, 6179 (1995).

[41] T. Klein, C. Marcenat, S. Blanchard, J. Marcus, C. Bourbonnais, R. Brusetti, C. J. van der Beek, and M. Konczykowski, *Phys. Rev. Lett.* **92**, 037005 (2004).

[42] S. Blanchard, T. Klein, J. Marcus, I. Joumard, A. Sulpice, P. Szabo, P. Samuely, A. G. M. Jansen, and C. Marcenat, *Phys. Rev. Lett.* **88**, 177201 (2002).

[43] P. Kumar, D. Hall, and R. G. Goodrich, *Phys. Rev. Lett.*, **82**, 4532 (1999).

[44] D. Hall, R. G. Goodrich, C. G. Grenier, P. Kumar, M. Chaparala, and M. L. Norton, *Philos. Mag. B* **80**, 61-79 (2000).

[45] J. E. Graebner, L. F. Schneemeyer, and J. K. Thomas, *Phys. Rev. B* **39**, 9682 (1989).

[46] B. F. Woodfield, D. A. Wright, R. A. Fisher, N. E. Philips, and H. Y. Tang, *Phys. Rev. Lett.* **83**, 4622 (1999).

[47] G. M. Zhao, *Phys. Rev. B* **76**, 020501(R) (2007).

[48] P. Samuely, P. Szabo , A. G. M. Jansen, P. Wyder. J. Marcus, C. Escribe-Filippini, and M.Affronte, *Physica B* **194-196**, 1747 (1994).

[49] G. E. Blonder, M. Tinkham, and T. M. Klapwijk, *Phys. Rev. B***25**, 4515 (1982).

[50] B. Batlogg *et al.*, *Phys. Rev. Lett.* **61**, 1670 (1988).

[51] J. P. Carbotte, *Rev. Mod. Phys.* **62**, 1027 (1990).

[52] M. Tinkham, *Introduction to Superconductivity*, (McGraw-Hill, New York, 1996).

[53] W. K. Kwok *et al.*, *Phys. Rev. B* **40**, 9400 (1989).

[54] L. F. Mattheiss and D. R. Hamann, *Phys. Rev. Lett.* **60**, 2681 (1988).

[55] R. Nourafkan, F. Marsiglio, and G. Kotliar, *Phys. Rev. Lett.* **109** , 017001 (2012).

[56] J. R. Tucker and B. I. Halperin, *Phys. Rev. B* **3**, 3768 (1971).

[57] L. F. Mattheiss, *Phys. Rev. B* **28**, 6629 (1983).

[58] T. P. Orlando, E. J. McNiff, Jr., S. Foner, and M. R. Beasley, *Phys. Rev. B***19**, 4545 (1979).

[59] F. Marsiglio and J. P. Carbotte, *arXiv:cond-mat/0106143v1*.

[60] Bertrand I. Halperin and David J. Bergman, *Physica B* **405**, 2908 (2010).

[61] M. A. Karlow, S. L. Cooper, A. L. Kotz, M. V. Klein, P. D. Han, and D. A. Payne, *Phys. Rev. B* **43**, 6499 (1993).

[62] A. S. Alexandrov and C. Sricheewin, *Europhys. Lett.* **58**, 576 (2002).

[63] F. C. Niestemski, S. Kunwar, S. Zhou, S. L. Li, H. Ding, Z. Q. Wang, P. C. Dai, and V. Madhavan, *Nature* (London) **450**, 1058 (2007).

[64] T. Dahm, V. Hinkov, S. V. Borisenko, A. A. Kordyuk, V. B. Zabolotnyy, J. Fink, B. Buchner, D. J. Scalapino, W. Hanke, and B. Keimer, *Nature Phys.* **5**, 780 (2009).

[65] J. Zhao, F. C. Niestemski, S. Kunwar, S.-L. Li, P. Steffens, A. Hiess, H. J. Kang, S. D. Wilson, Z.-Q. Wang, P. C. Dai, and V. Madhavan, *Nature Phys.* **7**, 719 (2011).

[66] G. M. Zhao, *Phys. Scr.* **83**, 038304 (2011).

[67] W. L. McMillan and J. M. Rowell, *Phys. Rev. Lett.* 14, 108 (1965).

[68] J. P. Carbotte, *Rev. Mod. Phys.* **62**, 1027 (1990); W. L. McMillan and J. M. Rowell, in *Superconductivity*, edited by R. D. Parks, (Marcel Dekker, New York, 1969), Vol. 1, p. 561; E. L. Wolf, *Principles of electron tunneling spectroscopy*, (Oxford University Press, New York, 1985).

[69] M. K. Crawford, G. Burns, G. V. Chandrashekhar, F. H. Dacol, W. E. Farneth, E. M. McCarron, III and R. J. Smalley, *Solid State Commun.* **73**, 507 (1990).

[70] J. Zhao, P. C. Dai, S. L. Li, P. G. Freeman, Y. Onose, and Y. Tokura, *Phys. Rev. Lett.* **99**, 017001 (2007).

[71] E. W. Hudson, S. H. Pan, A. K. Gupta, K.-W. Ng, and J. C. Davis, *Science* **285**, 88 (1999).

[72] B. Renker, F. Gompf, D. Ewert, P. Adelmann, H. Schmidt, E. Gering, and H. Mutka, *Z. Phys. B: Condens. Matter* **77**, 65 (1989).

Polaronic High-Temperature Superconductivity in Bismuthates ... 481

[73] W. Meevasana, T. P. Devereaux, N. Nagaosa, Z.-X. Shen, and J. Zaanen, *Phys. Rev. B* **74**, 174524 (2006).

[74] T. Bauer and C. Falter, *Phys. Rev. B* **80**, 094525 (2009).

[75] P. Zhang, S. G. Louie, and M. L. Cohen, *Phys. Rev. Lett.* **98**, 067005 (2007).

[76] Y. DeWilde, N. Miyakawa, P. Guptasarma, M. Iavarone, L. Ozyuzer, J. F. Zasadzinski, P. Romano, D. G. Hinks, C. Kendziora, G. W. Crabtree, and K. E. Gray, *Phys. Rev. Lett.* **80**, 153 (1998).

[77] M. Kugler, G. Levy de Castro, E. Giannini, A. Piriou, A. A. Manuel, C. Hess, and O. Fischer, *J. of Phys. Chem. Solids* **67**, 353 (2006).

[78] A. Abanov and A. V. Chubukov, *Phys. Rev. Lett.* bf 83, 1652 (1999).

[79] J. F. Zasadzinski, L. Coffey, P. Romano, and Z. Yusof, *Phys. Rev. B* **68**, 180504(R) (2003).

[80] J.-H. Lee *et al.*, *Nature* **442**, 546 (2006).

[81] G. M. Zhao, *Phil. Mag. B* **81**, 1335 (2001).

[82] D. S. Dessau, Z.-X. Shen, D. M. King, D. S. Marshall, L. W. Lombardo, P. H. Dickinson, A. G. Loeser, J. DiCarlo, C.-H Park, A. Kapitulnik, and W. E. Spicer, *Phys. Rev. Lett.* **71**, 2781 (1993).

[83] P. D. Johnson, T. Valla, A.V. Fedorov, Z. Yusof, B. O. Wells, Q. Li, A. R. Moodenbaugh, G. D. Gu, N. Koshizuka, C. Kendziora, Sha Jian, and D. G. Hinks, *Phys. Rev. Lett.* **87**, 177007 (2001).

[84] L. Boeri, M. Calandra, I. I. Mazin, O. V. Dolgov, and F. Mauri, *Phys. Rev. B* **82**, 020506R (2010).

[85] D. B. Tanner, H. L. Liu, M. A. Quijada, A. M. Zibold, H. Berger, R. J. Kelley, M. Onellion, F. C. Chou, D. C. Johnston, J. P. Rice, D. M. Ginsberg, J. T. Markert, *Physica B*, **244**, 1 (1998).

In: Polarons: Recent Progress and Perspectives
Editor: Amel Laref

ISBN: 978-1-53613-935-8
© 2018 Nova Science Publishers, Inc.

Chapter 15

POLARONS IN FERRITES

Madhuri Wuppulluri[*]
Centre for Crystal Growth,
VIT University, Vellore, Tamil Nadu, India

ABSTRACT

This chapter deals with Polarons in ferrites, as the title suggest. Various sections of the chapter will elaborate on different conduction mechanisms found in ferrites with emphasis on polaron models. Appropriate latest results on MgCuZn, NiMg and NiMgZn ferrite systems will be presented and discussed.

Keywords: ferrites, small polaron, overlapping large polaron tunnelling, conduction

INTRODUCTION

Dielectric and magnetic characteristics of ceramic materials are of increasing importance in the fields of radio electronics, optoelectronics, microwave electronics and modern communication devices. Ferrites find vast applications in these fields. An important property of these materials is their high electrical resistance compared to other elemental magnetic materials, and generally offer low eddy current losses at high frequencies. The order of magnitude of the conductivity greatly influences the electric and magnetic behaviour of ferrites [1-3]. Ferrites attracted the attention of physicists and technologists because of magnetic semiconductoring behavior. Both semiconductors and magnetic materials exhibit interesting properties that could be used in electronic devices. Ferrites which are ferromagnetic semiconductors opened a new vista in the physics of material and the need for high resistivity ferrites led to the synthesis of various ferrite compositions. The increasing demand for low loss ferrites resulted in detailed investigations on the conductivity, thermoelectric power, dielectric behaviour etc.

---

[*] Corresponding Author Email: madhuriw12@gmail.com.

The conduction mechanism in ferrites is quite different from that of normal semiconductors. In ferrites, the temperature dependence of mobility affects the conductivity and the carrier concentration is almost unaffected by temperature variation. Unlike in normal semiconductors where the charge carriers occupy states in wide energy band, the charge carriers in ferrites are localized at the magnetic ions. In ferrites, the cations are surrounded by close-packed oxygen anions and to a first approximation; they can well be treated as isolated from each other. There will be little overlap of the anion charge clouds or orbital. Alternatively, the electrons associated with particular ion will largely remain isolated and hence a localized electron model is more appropriate in the case of ferrites rather than the band (collective electron band) model. This accounts for the basically insulating nature of these materials.

Localized charge carriers may contribute to dielectric relaxation in two fundamentally different ways which result in a delayed response of current to the applied field. On the other hand, these carriers may suffer a delayed release from the localized levels into the free band where they take part in the ordinary conduction process. The localized charges displaced from their original positions by the action of an external field by means of hopping transitions between localized levels, not involving excitations into the respective free bands. This hopping conductivity corresponds to an extension of two potential well model.

The fact remains that most commonly used dielectric materials, including virtually all polymers and the majority of glasses and ceramics, do not show any evidence of the presence of free charge carriers, which would give rise to much higher levels of dc conductivity than are actually observed. The inference is that the only carriers that may be present are necessarily very low mobility carriers that may be hopping electronic charges inevitably are, ionic charges or by their very nature hopping carriers, since the concept of a free band for ions has no physical sense, and their corresponding mobility are extremely low, even in materials with very high ionic conductivities such as the so-called "fast ion conductors." In the forthcoming sections various electrical conduction models are discussed with a special reference to models including polarons.

VARIOUS ELECTRICAL CONDUCTION MODELS

1. Two-Potential Well Model

The low-frequency dielectric responses are to be expected from any localized charge carriers that may be represented in a semiconductor, especially in conditions where the effects of the free charge carriers are not dominant, for example at sufficiently low temperatures or in the space charge regions of p-n junctions.

These hopping transitions have very different probabilities according to the relative distances between the localized sites and to their separation in energy so that certain easy transitions will be executed many times in both directions, as in a two – well system, while the more difficult ones are only traversed relatively less frequently. This shows clearly that a hopping charge carrier shows both dielectric characteristics; insofar as the site behaves like a jumping dipole in its reciprocating motions, and simultaneously conducting characteristics resulting from its extended hopping over many sites. The dielectric properties are determined by the easiest transitions, while the conducting properties are determined by the most difficult transitions which limit the free percolation of charges from one electrode to the other.

In crystalline semiconductors, any such point defects give rise to localized states which, if their energy separation from the edges of the conduction and valence bands is sufficiently large, act as effective traps for electrons or holes. If the density of these traps is sufficiently large for tunnelling between them to be possible, transport may occur in these states by thermally assisted transport i.e., hopping. The density of these traps increases with increasing disorder and a limiting situation is attained in the case of completely disordered or amorphous materials.

Localization may also take place under conditions of apparent structural order if the medium is sufficiently strongly polarizable. The polarization is accompanied by a significant local distortion which leads to lowering of potential energy of the carrier with respect to the undistorted medium. This is referred to as the "carrier digging its own potential well" and the resulting quasi-particle consisting of the charge carrier and its surrounding distortion is known as a *polaron*. Polarons are necessarily localized and may move by hopping when sufficient energy is available for their excitation out of the local potential well.

2. Electron Hopping Model

In cobalt ferrites, Jonker [4] has observed that the transport properties differ considerably from those of normal semiconductors, as the charge carriers are not free to move through the crystal lattice but jump from ion to ion. It was also noted that in this type of materials, the possibility exists for changing the valence of a considerable fraction of metal ions and specially that of iron ions. Assuming the number of electron contribution to be equal to the number of Fe^{2+} ions and the number of holes to be equal to the number of Co^{3+} ions, Jonker has calculated the mobilities from the resistivity data and obtained extremely low values of mobilities; $\mu_e = 10^{-4}$ cm^2/Vsec for electrons and $\mu_h = 10^{-8}$ cm^2/Vsec for holes. Further, even for samples with a large concentration of Fe^{2+} or Co^{3+} ions, a fairly strong exponential dependence of resistivity on temperature was found. From the ordinary band theory of conduction, one would expect metallic behaviour for such high concentration of charge carriers, i.e., a high mobility with only slight temperature dependence. Such behaviour is indeed found in some cases such as ferromagnetic mixed crystal series of $LaMnO_3$-$SrMnO_3$ and $LaCoO_3$-$SrCoO_3$ [5, 6].

In cobalt ferrites, the behaviour is similar to NiO and Fe_2O_3 and the ordinary theory based on simple band picture does not apply. For the equilibrium lattice, there is a little overlap between the wave functions of ions on adjacent octahedral site, with the result that the electrons/holes are not free to move through the crystal lattice.

In the presence of lattice vibrations, however, the ions occasionally come close together for transfer to occur with a high degree of probability. Thus the conduction is induced only by the lattice vibrations and as a consequence, the carrier mobility shows a temperature dependence characterized by an activation energy. For such a process of jumping of electrons and holes the mobilities μ are given by

$$\mu_e = \frac{ed^2\gamma_e \exp\left(\frac{-E_e}{kT}\right)}{kT} \qquad (1)$$

$$\mu_h = \frac{ed^2\gamma_h \exp\left(\frac{-E_h}{kT}\right)}{kT} \qquad (2)$$

where the subscripts 'e' and 'h' represent the parameters for electrons and holes, d represents the jumping length, γ_e and γ_h are the jump frequencies, E_e and E_h are the activation energies involved in the required lattice deformation.

Following the usual convention, the energy levels are denoted by symbols representing the valence states of the metal ions. In the case where these levels are occupied by Fe^{2+} and Co^{2+}, the energies needed to remove an electron from these levels are determined by the ionization potential. Jonker [4] has given an energy level diagram for this case.

The activation energy does not represent the energy required for the electrons but to the crystal lattice around the site of electrons. The Co^{3+} and Fe^{2+} are distributed randomly and there will be a certain spread of individual ionic levels. The general expression for the total conductivity in this case where we have two types of charge carriers can be given as

$$\sigma = n_e e\mu_e + n_h e\mu_h \qquad (3)$$

The temperature dependence of conductivity arises only due to mobility and not due to the number of charge carriers in ferrites.

Based on the discussion given the evidence for hopping conduction may be summarized as follows:

1. The low value of mobility: The mobility has a value much lower than the limiting value ($0.1 cm^2/Vsec$) taken as the minimum for band conduction [7].
2. Independence of Seebeck coefficient on temperature: This property is due to the fact that in hopping model the number of charge carriers are fixed.
3. Thermally activated process: with an activation energy, E_a called the hopping activation energy.
4. Conduction transition: Occurrence of n-p transition with changes in the Fe^{2+} or oxygen concentration in the system.

Conduction in Mn-doped MgCuZn ferrites synthesized by conventional and microwave sintering is proposed due to electron hopping between different valence states of the same ion by Bhaskar et al. [8-10]. According Bhaskar et al. electron hopping between different valance states is possible due to the co-existence of Fe^{2+} and Fe^{3+} ions within the hopping radius of electrons. Such proximity coexistence of Fe^{2+} and Fe^{3+} ions is possible only at B-sites which are closely located in the spinel lattice. Bharati [11] also suggested same conduction mechanism in MgCuZn nano crystalline ferrite powders synthesized by sol-gel auto combustion technique. Temperature variation of Seebeck coefficient of MgCuZn ferrites exhibited electron hopping up to a certain temperature. Thermally activated charge carriers being the dominant factor in the increase in Seebeck coefficient [12]. Sustenance of electron hopping between octahedral ferrous and ferric ions is the dominant conduction mechanism in Ni ferrite nano particles [13].

3. Small Polaron Hopping Model

A small polaron is a defect created when an electron carrier becomes trapped at a given site as a consequence of the displacement of adjacent atoms or ions. The entire defect (carrier plus distortion) then migrates by an activated hopping mechanism. Small polaron formation can take place in materials whose conduction electron belong to incomplete inner d or f shells due to electron orbital overlap, this tends to form extremely narrow bands. The possibility for the occurrence of hopping conductivity in certain low mobility semiconductors specially oxides has been widely recognized for some time, and extensive theoretical literature has been developed which considers the small polaron model and its consequences [14 - 19].

The migration of small polaron requires the hopping of both the electron and the polarized atomic configuration from one site to an adjacent one [20]. For an fcc lattice, the drift mobility takes the form

$$\mu = (1-c)ea\frac{2\Gamma}{kT} \tag{4}$$

where e is the electronic charge, a is the lattice parameter, c is the fraction of the site which contains an electron ($c = n/N$), n is the number of electrons per unit volume and N is the number of available sites per unit volume. The quantity Γ which is the jump rate of the polaron from one site to a specific neighbouring site is given by

$$\Gamma = P\mu_o \exp\left(\frac{-E_H}{kT}\right) \tag{5}$$

Here μ_o is the appropriate optical mode phonon frequency, E_H is the activation energy, and P is a factor which gives the probability that the electrons will transfer after the polarized configuration has moved to the adjacent site. In evaluating P there are two cases to consider depending on the relative value of the electron transfer time t_{el} and the t_{at}, the atomic transfer time which characterizes the transfer of atomic polarization between adjacent sites. Specifically in the adiabatic case for which $t_{el} < t_{at}$ and $P = 1$. On the other hand for the non-adiabatic case, for which $t_{el} > t_{at}$ and $P < 1$ and takes the form $P \propto J^2/(kT)^2$ where J is the current density.

For the hopping of polarons in this model, we get the expression for conductivity as

$$\sigma = \frac{A}{T}\exp\left(\frac{-E_H}{kT}\right) \tag{6}$$

488   *Madhuri Wuppulluri*

where the factor A is

$$A = NPc(1-c^2)e^2a^2\gamma_o / k \tag{7}$$

Under the assumption that only one electron is permitted on a given site that the other interaction effects are negligible we can get an expression for the thermoelectric coefficient α as

$$\alpha = -\frac{k}{e}\left\{\ln\beta\left[\frac{(1-c)}{c}\right] + \frac{S_T^*}{k}\right\} \tag{8}$$

Here S_T^* is the vibrational entropy associated with the ions surrounding a polaron on a given site and β is the degeneracy factor, including both spin and orbital degeneracy of the electronic carrier. For the case β = 1 the expressions reduce to the Heiks formula for hopping of electrons. It can be noted that α is temperature independent and a number of charge carriers could be determined if β is known.

The small polaron model also explains the low value of mobility, temperature independent Seebeck coefficient and thermally activated hopping. In addition to these properties if the hopping electron becomes localized by virtue of its interaction with phonons, then a small polaron is formed and the electrical conduction is due to hopping motion of small polarons. The treatment of conduction by polarons is discussed by several workers [21-23]. The small polarons conduct in the band–like manner up to a certain temperature, the conductivity showing an increase with frequency. At higher temperatures, the conduction is by thermally-activated hopping mechanism. Small polaron hopping mechanism at high temperatures was proposed by Petrov et al. [24].

As an evidence for small polaron conduction, the variation of conductivity with frequency on the basis of relaxation process as applied to bound state Adler and Feinleib [21, 25] have shown that for conduction by small polarons the conduction increases with frequency and the following relation holds good

$$\sigma_\omega - \sigma_{dc} = \frac{\omega^2\tau^2}{1+\omega^2\tau^2} \tag{9}$$

where ω is the angular frequency and τ the staying time (~10^{-10} sec) for frequencies $\omega^2\tau^2 < 1$. log ($\sigma_\omega - \sigma_{dc}$) should be a straight line. If the plots are straight line it indicates that the conduction is due to small polarons. For small polarons the polaron radius r_p should be less than the inter-ionic distance and the coupling constant should be greater than 6 given by Bosman and van Daal [26].

Penchal et al. [27] have investigated frequency dependence of ac conductivity of MgCuZn ferrites and on the basis of the influence of log ω^2 on log σ_{ac} has proposed small polaron conduction mechanism in MgCuZn ferrites. According to Penchal et al. log σ_{ac} linearly increases with frequency suggesting inactive conducting grain and active grain boundaries at low frequency keeps the conductivity low at lower frequencies and conducting grains become active at high frequencies increasing ac conductivity. Investigations on the transport properties of microwave processed MgCuZn and glass doped MgCuZn ferrites is found to be in accordance with small polaron hopping probably with different hopping frequencies [28]. DC, ac and thermoelectric investigations on the similar system by Varalakshmi et al. [29] revealed small polaron hopping mechanism is applicable only at lower frequencies and mixed polaron hopping at high frequency. This conclusion is supported by the deviation of log ($\sigma_{AC}-\sigma_{DC}$) versus log ω^2 plot from linearity. Polaron hopping conduction mechanism is suggested in the iron-deficient MgCuZn system from the thermal spectra of dc conductivity and Seebeck coefficient [30]. NiCuZn-MgCuZn composite ferrites exhibited four regions of conduction with an increase in temperature indicating mixed valence and small polaron conduction mechanism [31].

The temperature variation of bulk conduction in microwave processed NiMg ferrites exhibited small polaron hopping mechanism in the low temperature or extrinsic region. Above the critical temperature, the activation energies are on par with electron exchange in octahedral ferrous and ferric ions resulting in a band like conduction [32]. Zheng et al. [33] could estimate the hopping range using the relation

$$
R = \frac{\sqrt[4]{3}}{\sqrt[4]{2\pi\alpha N(E_F)kT}} \tag{10}
$$

where $1/\alpha$ is Bohr radius of localized centre and $N(E_F)$ can be deduced from the fitting value of T_o. The hopping range is found to decrease with increase in temperature and the conduction is due to localized polaron hopping. Investigations on NiMgZn ferrites also revealed small polaron conduction [34]. Frequency independent (temperature dependent) and frequency dependent conduction regions have different activation energies indicating two different conduction mechanisms in the two regions.

4. Phonon Induced Tunneling Model

Electrical properties of ferrites have been explained on the basis of tunneling of electrons amongst Fe^{2+} and Fe^{3+} atoms on B sites by Srinivasan and Srivastava [20]. It has been shown by Bates and Seggels [35] that the paramagnetic Fe^{2+} ions in Al_2O_3 are strongly coupled to the lattice and due to this coupling, some unusual properties are observed. Along the same line, it has been assumed that the electrons which participate in the $Fe^{2+} \leftrightarrow Fe^{3+} + e$, exchange process are strongly coupled to the lattice and tunnel from one site to the other due to a phonon– induced transfer mechanism developed by Srivastava [20].

Under lattice displacements, the state Fe^{3+} on the B site is more stable than the Fe^{2+} state due to the Coulomb interaction in the system. This can be illustrated by the simple model of a linear chain. Consider a state Ψ_1 in which the Fe^{2+} ions are located at sites Γ_{N+4n} and Fe^{3+} ions at Γ_{N+2+4n} where n can take values from 0 to infinity. The oxygen ions are located at Γ_{N+1+2n}. In another state Ψ_2 the sites Γ_{N+4n} are occupied by Fe^{3+} ions and Γ_{N+2+4n} sites are occupied by Fe^{2+} ions. Ψ_1 and Ψ_2 are degenerate for a state lattice. In the presence of phonons, however, the degeneracy between the Ψ_1 and Ψ_2 states is removed.

Considering now a longitudinal optical mode of lattice vibration and assume the oxygen ions to be moving and the metal ions to be fixed. Let each of the two oxygen ions Γ_{N+1} move towards the Fe^{2+} ion at Γ_N at $t = 0$ and let this be repeated for each of the Fe^{2+} ions at sites Γ_{N+4n}, such an approximation is valid for lattice waves of wave number $q \rightarrow 0$. We now calculate the change in energy as a function of the ionic displacement and compare it with the change in energy at $t = \tau/2$ when the oxygen ions are moving towards the site occupied by the Fe^{3+} ions.

A unit cell for the linear lattice of this model would consist of four ions, one Fe^{2+} and one Fe^{3+} and two oxygen ions. As the unit cell should be neutral, each oxygen ion is assigned -2.5e charge. The total lattice energy U_o of the crystal composed of N unit cells is given by:

$$U_o = N(U_{Fe^{2+}} + U_{Fe^{3+}}) \tag{11}$$

where

$$U_{Fe^{2+}} = -2\frac{\ln 2}{R}e^2 + \frac{\lambda_1}{R^n} \tag{12}$$

and $$U_{Fe^{3+}} = -4.5\frac{\ln 2}{R}e^2 + \frac{\lambda_2}{R^m} \tag{13}$$

here λ_1, n and λ_2, m are the constants of the repulsive interaction terms for Fe^{3+} and Fe^{2+} ions respectively and R is the nearest neighbour distance. This theory yields an expression for the conductivity as

$$\sigma = \sigma_o \frac{(v/kT)}{\left\{\exp\left(\frac{v}{2kT}\right) + \exp\left(\frac{-v}{2kT}\right)\right\}^2} \tag{14}$$

where

$$\sigma_o = \frac{4N_o \varepsilon a^2 e^2}{h} \tag{15}$$

where N_o is the density of Fe^{2+} ions, ε is the computing coefficient, υ is the energy of the single particle, a is the nearest Fe^{2+}-Fe^{3+} distance.

The values of σ calculated using this model for $ZnFe_2O_4$ and $CuFe_2O_4$ agree well with the experimental data [35].

5. Overlapping Large Polaron Tunneling (OLPT) Model

Large polaron tunneling was first proposed by Long in amorphous semiconductors [36]. According to Long the polaron energy when derived from polarization changes in the

Polarons in Ferrites

deformed lattice, as observed in ionic crystals; the resultant excitation is called a large polaron or dielectric polaron. As its name suggests, its well will extend over many interatomic distances, because of the long range of the Coulomb interaction. Here it should be noted that the large polaron radius is larger than the lattice constants. This has important consequences for the a.c. loss because when the wells of two sites overlap, the activation energy W_H associated with particle transfer between them will be reduced according to

$$W_H = W_{Ho}\left(1 - \frac{r_o}{R}\right) \tag{16}$$

where r_0 is the polaron radius, R is intersite separation which varies randomly and W_{HO} is assumed to be constant for all the sites and is given by

$$W_{HO} = \frac{e^2}{4\varepsilon\varepsilon_o r_o} \tag{17}$$

The total conductivity as the sum of dc and ac conductivity is given by [37]

$$\sigma_{tot} = \sigma_{dc} + \left[\pi^4 e^2 k_B T\left(\frac{(N(E_F))^2}{12}\right)\right]\left[\frac{\omega R_\omega^4}{2\alpha k_B T + \frac{W_{Ho} r_o}{R_\omega^2}}\right] \tag{18}$$

where $N(E_F)$ is the density of states at the Fermi level and R_ω the tunneling distance can be evaluated from the quadratic equation at the frequency ω

$$R_\omega'^2 + [\beta W_{Ho} + \ln(\omega\tau_o)]R_\omega' - \beta W_{Ho} r_o' = 0 \tag{19}$$

where $R_\omega' = 2\alpha R_\omega, r_o' = 2\alpha r_o, \beta = \frac{1}{k_B T}$ and α is the spatial extent of the large polaron.

The frequency exponent n for the model can be written as

$$n = 1 - \left[\frac{8\alpha R_\omega + \left(\frac{6\beta W_{Ho} r_o}{R_\omega}\right)}{\left(2\alpha R_\omega + \left(\frac{\beta W_{Ho} r_o}{R_\omega}\right)\right)^2}\right] \tag{20}$$

From the above expression for n, it is understood that n decrease with increase in temperature and attains a minimum at a certain temperature for small r_o and then increases with temperature and r_o.

Several workers could evidence large polaron tunneling in Mn, NiZn, Co, CoMn and AlCd ferrite nanoparticles [37–43].

CONCLUSION

In summary MgCuZn, NiMg and NiMgZn ferrites predominantly exhibit small polaron hopping in alternating fields. The frequency of hopping is found to be different in different temperature regions. Thermal variation of conductivity in static and dynamic fields is mostly explained on the basis of electron hopping between different valance states of an element. Jahn-Teller shift of Cu/Mn ions in the spinel crystal structure must result in the formation of the potential well or a charge trap centre depending on concentration and oxidation state at thermal equilibrium. Formation of small polarons must mainly be attributed to the presence of Jahn Teller ion copper in MgCuZn and MgCuZn-NiCuZn composites. However Jahn Telller ions effect is not reported in the literature. Overlapping large polaron tunneling is mostly observed in manganese ferrites.

ACKNOWLEDGMENTS

The author would like to express thanks to our research supervisor and collaborators Prof. K. V. Siva Kumar Professor (Rtd), Department of Physics, Sri Krishnadevaraya University, Anantapur, Prof. N. Ramamanohar Reddy Department of Materials and Nano technology, Yogi Vemana University, Kadapa, Prof. N. Varalaxmi, Department of Physics, Kakathiya University, Warangal for their constant support in developing this manuscript.

REFERENCES

[1] R. C. Morris, J. E. Christopher and R. V. Coleman, "Conduction phenomenon in thin layers of iron oxide" *Phys. Rev.*, 184 (1969) 565-573.

[2] J. Peters and K. J. Stanley, "The dielectric behavior of magnesium manganese ferrite" *Proc. Phys. Soc.*, 71 (1958) 131-133.

[3] L. G. van Uitert, "Dielectric properties of and conductivity in ferrites" *Proc. IRE*, 44 (1956) 1294-1303.

[4] G. H. Jonker, "Analysis of semiconducting properties of cobalt ferrite" *J. Phys. Chem. Solids*, 9 (1959) 165-175.

[5] Tsuda, Nobuo, and Nasu, Keiichiro, and Yanase, Akira and Siratori, Kiiti, Chapter - "Representative Conducting Oxides" Book *Electronic Conduction in Oxides* 1991, Springer Berlin Heidelberg, (1991) 105-286.

[6] G. H. Jonker and J. H. van Santen, "Ferromagnetic compounds with manganese with perovskite structure" *Physica*, 16 (1950) 337-347.

[7] J. H. van Santen and G. H. Jonker, "Magnetic compounds with perovskite structure III. Ferromagnetic compounds of cobalt" *Physica*, 19 (1953) 120-130.

[8] A. Bhaskar, B. Rajini Kanth, S. R. Murthy "Electrical properties of Mn added MgCuZn ferrites prepared by microwave sintering method" *Journal of Magnetism and Magnetic Materials* 283 (2004) 109–116.

[9] A. Bhaskar and S. R. Murthy "Influence of sintering temperature on electrical and dielectric properties of Mn (1%) added MgCuZn ferrites" *Indian J Phys* 88(2) (February 2014)151–156.

[10] Ankam Bhaskar S. R. Murthy "Effect of sintering temperature on the electrical properties of Mn (1%) added MgCuZn ferrites by microwave sintering method" *J Mater Sci: Mater Electron* 24 (2013)3292–3298.

[11] M. R. Barati "Influence of zinc substitution on magnetic and electrical properties of MgCuZn ferrite nanocrystalline powders prepared by sol–gel, auto-combustion method" *Journal of Alloys and Compounds* 478 (2009) 375–380.

[12] W. Madhuri, M. Penchal Reddy, N. Rammanohar Reddy, K. V. Siva Kumar, "Thermoelectric studies of MgCuZn ferrites," *Int. J. Chem Tech Research*, Vol. 6, No. 3, (2014) pp 1771-1774.

[13] K. Chandra Babu Naidu and W. Madhuri "Hydrothermal synthesis of $NiFe_2O_4$ nanoparticles: structural, morphological, optical, electrical and magnetic properties," *Bull. Mater. Sci.*, Vol. 40, No. 2, (2017) pp. 417-425.

[14] J. Yamashita and T. Kurosawa, "On electronic current in NiO" *J. Phys. Chem. Solids*, 5 (1958) 34-43.

[15] T. Holstein, "Studies in polaron motion: Part II. The "small" polaron" *Ann. Phys., (New York)* 8 (1959) 343-389.

[16] J. Apple, *Solid State Physics*, (Eds. F. Seitz, D. Turnbull and H. Ehrenreich), New York, Academic Press 18 (1969) 41.

[17] I. G. Austin and N. F. Mott, "Polarons in crystalline and non-crystalline materials" *Adv. Phys.*, 18 (1969) 41-102.

[18] H. L. Tuller and A. S. Nowick, "Small polaron electron transport in reduced CeO_2 single crystal" *J. Phys. Chem. Solids*, 38 (1977) 859-867.

[19] B. Gillot, J. F. Ferriot, G. Dupre, Abel R, "Study of the oxidation kinetics of finely-divided magnetites. II-Influence of chromium substitution" *Mater. Res. Bull.*, 11 (1976) 843-849.

[20] G. Srinivasan and C. M. Stivastava, *Phys. Stat. Solidi.*, (b) 108 (1981) 665.

[21] J. Apple, *Solid State Physics*, (Eds. F. Seitz, D. Turnbull and H. Ehrenreich), New York, Academic Press, 21 (1968) 193.

[22] I. G. Austin and N. F. Mott, "Polarons in crystalline and non-crystalline materials" *Adv. Phys. Suppl.*, 30 (1969) 299S.

[23] N. F. Mott and E. A. Davis, *Phonons and Polarons in Electronics -Processing of Non–Crystalline Materials*, Oxford Clarendon Press (1971).

[24] A. N. Petrov, G. V. Densiov and U. M. Zhukovskii, *Inorg. Mater.*, (US) 22 (1986) 579.

[25] D. Adler and Feinleib, "Electrical and optical properties of narrow-band materials" *Phys. Rev., B* 2 (1970) 3112.

[26] A. J. Bosmann and H. J. van Daal, "Small polaron versus band conduction in some transition-metl oxides" *Adv. Phys.*, 19 (1970) 1-117.

[27] M. Penchal Reddy, W. Madhuri, N. Ramamanohar Reddy, K. V. Siva Kumar, V. R. K. Murthy, R. Ramakrishna Reddy, "Influence of copper substitution on magnetic and electrical properties of MgCuZn ferrite prepared by microwave sintering method," 2010 *Material Science Engineering:C* 30 pp1094-1099.

[28] W. Madhuri, M. Penchal Reddy, Il Gon Kim, N. Rama Manohar Reddy, K. V. Siva Kumar, V. R. K. Murthy, "Transport properties of microwave sintered pure and glass added MgCuZn ferrites" *Materials Science and Engineering B*, 178 (2013) 843 – 850.

[29] N. Varalaxmi, K. V. Sivakumar "Structural, magnetic, DC–AC electrical conductivities and thermo electric studies of MgCuZn Ferrites for microinductor applications" *Materials Science and Engineering C* 33 (2013) 145–152.

[30] Madhuri. W., Roopas Kiran. S., Penchal Reddy. M., Ramamanohar Reddy. N., Siva Kumar K. V. "DC conductivity and Seebeck coefficient of nonstoichiometric MgCuZn ferrites" *Materials Science-Poland*, Vol. 13 (6) (2017) P. No. 40 – 44.

[31] N. Varalaxmi and K. V. Sivakumar "Studies on Structural and Electrical Properties of Ball-Milled NiCuZn-MgCuZn Nanocomposites Ferrites" *Metallurgical And Materials Transactions A Volume* 45A (2013) March 2014—1579.

[32] K. Chandra Babu Naidu, S. Roopas Kiran, W. Madhuri "Investigations on transport, impedance and electromagnetic interference shielding properties of microwave processed NiMg ferrites" *Mater. Res. Bull.* 89 (2017) 125–138.

[33] Hui Zheng, Wenjian Weng, Gaorong Han, and Piyi Du "Colossal Permittivity and Variable-Range-Hopping Conduction of Polarons in Ni0.5Zn0.5Fe2O4 Ceramic" *J. Phys. Chem.* C 117 (2013) 12966−12972.

[34] Chandra Babu Naidu K, Madhuri W "Microwave assisted solid state reaction method: Investigations on electrical and magnetic properties NiMgZn ferrites," *Materials Chemistry and Physics* 181 (2016) 432-443.

[35] C. A. Bates and P. Steggels, "The Jahn Teller theory and calculations of the relaxation time and resonance lineshape for Fe^{2+} ions in Al2O3" *J. Phys. C.*, 8 (1975) 2283.

[36] Long A. R. "Frequency-dependent loss in amorphous semiconductors" *Adv. Phys.* 31 553 – 637.

[37] E. Veena Gopalan, K. A. Malini, S. Saravanan, D. Sakthi Kumar, Yasuhiko Yoshida and M. R. Anantharaman "Evidence for polaron conduction in nanostructured manganese ferrite" *J. Phys. D: Appl. Phys.* 41 (2008) 185005 (9pp).

[38] Muhammed Tana, Yuksel Koseoglua, Furkan Alana, Erdogan Senturk "Overlapping large polaron tunneling conductivity and giant dielectric constant in Ni0.5Zn0.5Fe1.5Cr0.5O4 nanoparticles (NPs)" *Journal of Alloys and Compounds* 509 (2011) 9399– 9405.

[39] Atta ur Rahman, M. A. Rafiq, S. Karim, K. Maaz, M. Siddique and M. M. Hasan "Semiconductor to metallic transition and polaron conduction in nanostructured cobalt ferrite" *J. Phys. D: Appl. Phys.* 44 (2011) 165404 (6pp).

[40] R. K. Panda, R. Muduli, S. K. Kar, D. Behera "Dielectric relaxation and conduction mechanism of cobalt ferrite Nanoparticles" *Journal of Alloys and Compounds* 615 (2014) 899–905.

[41] Arifa Jamila, M. F. Afsara, F. Sherb, M. A. Rafiq "Temperature and composition dependent density of states extracted using overlapping large polaron tunnelling model in MnxCo1−xFe2O4 (x = 0.25, 0.5, 0.75) nanoparticles" *Physica B* 509 (2017) 76–83.

[42] Dev K. Mahatoa, Sumit Majumderb, S. Banerjee "Large polaron tunneling, magnetic and impedance analysis of magnesium ferrite nanocrystallite" *Applied Surface Science* 413 (2017) 149–159.

[43] T. Sasmaz Kuru, E. Senturk, V. Eyupoglu "Overlapping Large Polaron Conductivity Mechanism and Dielectric Properties of Al0.2Cd0.8Fe2O4 Ferrite Nanocomposite" *J Supercond Nov Magn* 30 (2017)647–655.

In: Polarons: Recent Progress and Perspectives
Editor: Amel Laref

ISBN: 978-1-53613-935-8
© 2018 Nova Science Publishers, Inc.

Chapter 16

MAGNETIC POLARONS IN EUTE: EARLY AND MODERN DISCOVERIES IN MAGNETIC ORDER CONTROL IN SEMICONDUCTORS

Flavio C. D. de Moraes
Physics Institute, University of Sao Paulo, Brazil

ABSTRACT

Only two years after the discovery of magnetic semiconductors, all Eu-chalcogenides (EuO, EuS, EuSe and EuTe) had been successfully synthesized, each exhibiting different types of magnetic order. Eu-chalcogenides form a class of face-centered materials and are nearly ideal Heisenberg magnetic systems with localized magnetic moments carried by the Eu atoms. Due to their simple structure, high intrinsic magnetic moment ($7\,\mu_b$ for each Eu) and the possibilities of doping, which may be used to tune both electric and magnetic properties of Eu-chalcogenides, these materials have gained attention in the emerging fields of spintronics and magneto-optics.

EuTe has an intrinsic type-II antiferromagnetic (AFM) order below Néel temperature ($TN = 9.6\,°K$), while above TN, the AFM order is suppressed by thermal fluctuation of the magnetic sublattices. Nevertheless, first experiments observed formation of ferromagnetic (FM) domains around dopants (Bound Magnetic Polarons) slightly above TN, changing EuTe magneto-optical and transport properties. More recently, the formation of FM domains has been observed due to localized photoexcited electrons (Photoexcited Magnetic Polarons), which can survive even well above TN, drawing renewed attention to this material.

This chapter provides a short review on magneto-optical studies and theoretical modeling for magnetic polarons in EuTe, from its synthesis in 1962 to recent measurements in highly crystalline samples grown by molecular beam epitaxy (MBE).

1. INTRODUCTION

The first discovered magnetic semiconductor was most likely $CrBr_3$, in 1960 [1]. In 1961, Matthias, Bozorth and van Vleck discovered the first Europium chalcogenide [2], EuO. Only a year later, all the other Eu-chalcogenides (EuS, EuSe and EuTe) had been synthesized and studied [3, 4]. However, it would be almost ten years before the proposal of the magnetic polaron model explaining the uncommon shift in luminescence spectrum of these chalcogenides [5, 6].

The concept of magnetic polarons was first proposed by De Gennes in 1960 [7]. It is a quasi-particle with magnetic moment, formed due to the strong ferromagnetic exchange interaction between a charge carrier and its surrounding localized spins. By polarizing the localized spins, the carrier reduces its energy and forms a cloud of ferromagnetic aligned spins. However, due to the formation of magnetic polarons, it holds that this energy reduction must be higher than the kinetic Fermi energy. Magnetic semiconductors are materials conducive for magnetic polaron formation owing to the combination of strong exchange coupling with small characteristic kinetic energy of free carriers, associated with small carrier concentration [8].

It was P. Streit and P. Wachter who first came up with the notion of magnetic polaron formation in EuSe semiconductors [5]. According to them, the photoabsorption process can generate an electron in a narrow conduction band, while the holes left in the valence band are strictly localized. These phenomena affect the crystal in two ways: (i) the exchange interaction aligns the localized spins where the electron wave function is non-zero, yielding a magnetic polaron; (ii) the localized hole distorts the crystal lattice around itself. Both effects generate a minimum potential, where the electron becomes "self-trapped" in such a way that the net polarization, lattice deformation and electron wave function are correlated. The consequence is a complex photoluminescent behavior in which the emission line shifts with magnetic field and temperature, experimentally observed in the near infra-red region.

In EuS semiconductors, magnetic polaron formation had also been expected, but the corresponding near infra-red was not simultaneously observed because, between ~4.2 K and ~16 K, the EuS localized spins are ferromagnetically oriented and the magnetic polaron disappear, although it still affects charge transport in the trapped electron vicinity. The magnetic polaron model for EuS was proposed, together with the model for EuTe, by G. Busch in 1970 [6].

Improvements in MBE growth techniques and spectroscopy measurement resolution led to further experimental evidence of magnetic polarons, from which much relevant information was gleaned, allowing the development of an ever more complete mathematical model. In this chapter, we present the evolution of this mathematical modeling for magnetic polarons and the background experimental measurements, with an emphasis on the EuTe magnetic polaron.

2. Types of Magnetic Polarons

2.1. Lattice Magnetic Order

Because magnetic polarons are regions of a material, where the localized spins are ferromagnetically aligned around a charge carrier, they cannot exist in ferromagnetically saturated materials. Nevertheless, they can suppress thermal fluctuation of spins in ferromagnets and even exist in the paramagnetic regime, far above the critical temperature. At low temperature, however, magnetic polarons can only exist in antiferromagnetic systems. Therefore, this system configuration is more fitting for magnetic polarons studies.

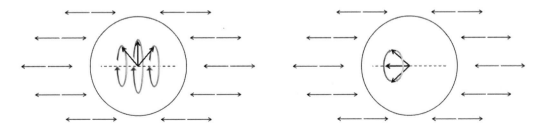

Figure 1. Possible initial orientation of polarons. In the left image the central electron is oriented perpendicular to the magnetic sublattices, while in the right image it is oriented in the direction of one of the sublattices.

There are two possibilities for the alignment of the magnetic polaron in antiferromagnetic systems: the central charge spin can point in a direction perpendicular to both antiferromagnetic sublattices and tilt the spins around it, or it can point in the direction of one of the sublattices and flip the spins within the magnetic polaron (Figure 1). In the first case, the central charge is free to point in any direction on a plane perpendicular to the antiferromagnetic sublattices, while in the second case the localized spins are free to form new sublattices in any direction on this perpendicular plane. Although similar, there is a significant difference between the two alignments. In the first, the polaron magnetic moment can point in any direction within the perpendicular plane, while in the second case it can only point in the direction of the initial sublattices (two possible directions). This difference can affect not only the material magnetic moment behavior under the effect of a weak external magnetic field, but also the electronic transport, since the direction of the localized spins is relevant for hopping.

It is also important to consider that, for the second possible alignment, in order to form the magnetic polaron, the effective field of the exchange interaction acting on each localized spin, must be higher than the spin-flop field, i.e., the field necessary to flop the spins' sublattices in antiferromagnetic systems.

2.2. Spin Configuration

In antiferromagnetic systems, there is a competition between the ferromagnetic exchange interaction and the lattice's antiferromagnetic exchange interaction. The relationship between these interactions determines the extent to which the localized spins are aligned. Since the strength of the exchange interaction depends on the overlap of the wave function of the particle involved, it is natural to expect the localized spins to be more aligned, the closer they are to the center of the magnetic polaron. If the central charge wave function is spherically symmetric, the magnetic polaron should also be spherical. This means that the angle distribution of the localized spins may have spherical symmetry, but not the spins themselves. All localized spins inside the magnetic polaron will be tilted in the same direction.

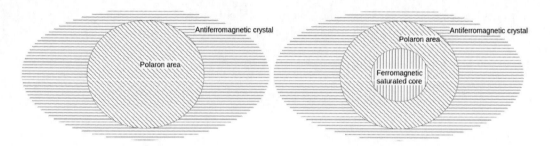

Figure 2. Figure to the left depicts a type-I magnetic polaron, while the figure to the right represents a type-II magnetic polaron with a ferromagnetic core.

It is also important to take into account that, if the exchange interaction close to the center of the polaron is sufficiently strong to fully align the localized spins, the magnetic polaron will have a core of ferromagnetically saturated spins and an external shell of partially-aligned spins. A. Mauger and D. L. Mills [8] referred to this kind of magnetic state as a type-II localized state, while the type-I localized state denotes a polaron without a saturated core (Figure 2).

2.3. Magnetic Polaron Stability

After proposing the magnetic polaron model for the Eu-chalcogenides, the Solid State Group from the EHT, Zurich, began studying optical processes in doped EuTE [9]. They found that the presence of doped donor impurities strongly increased the magnetic moment of EuTe crystals, when external field is applied. This is explained by formation of the magnetic polaron around the impurity electron introduced in EuTe. A similar effect was also observed in EuO semiconductors with oxygen vacancy [10]. A lightly bound electron in the crystal defect/donor, forms a bound magnetic polaron (BMP). This kind of magnetic polaron differs from the self-trapped magnetic polaron (STMP) for which no impurity is necessary. Umehara also distinguished the magnetic polaron bound to a photoexcited hole, which he referred to as a PILMP (photo-induced localized magnetic polaron), from the STMP [11].

The discrimination between different kinds of magnetic polarons gave rise to a discussion about which polarons can exist in EuTe crystals. This discussion gained momentum after the observation of two luminescence bands above 1.2 eV [12].

3. EuTe Magnetic Polaron

In 1984, Mauger and Mills [8] published a mathematical model for the magnetic polaron in EuTe at low temperature. They described the total-energy for the excited state as a sum of a Coulomb potential, the conduction band electron kinetic *energies*, and the exchange energy of the crystal due to interaction between the conduction band and the localized spins.

In 1995, Umehara [11] resorted to a more complex model that also included electron and hole-optical-phonon interactions, to analyze the Stokes shift of a recently discovered photoluminescence band in EuTe above 1.2 eV in photon energy [12]. By constructing the EuTe magnetic structure using a computer program and making some assumptions to solve his equations, he computed the total energy for the Photo-Induced Localized Magnetic Polaron and compared this with the energy of the Self-Trapped Magnetic Polaron, for which the Coulomb potential and the hole-optical-phonon interaction were set to zero. This led to the conclusion that the exchange energy required to form the STMP in EuTe is unrealistically high compared to the expected values obtained from Faraday Rotation measurements. However, uncertainty remained regarding this value and theoretical model predictions were far from a perfect match with experimental results.

The subject gained renewed attention some years later when, following improvements in the structural quality of crystals grown by MBE, one of the EuTe bands above 1.2 eV was resolved into two excitonic peaks. Both peaks demonstrated an expected magnetic polaron behavior in function of external magnetic field and temperature, called MX1 and MX2 [13]. This seminal study of Heiss et al. led to the discovery of another two new photoluminescence lines: one visible only with very low excitation power ($\sim$0.1 mW/cm^2), denoted MX0 [14]; and another visible only when high excitation power ($\sim$0.2 MW/cm^2) was applied, referred to as HE [15], both consistent with the polaron model.

The importance of the MX0 line lies in the fact that it is not a single line but a structure of lines, consistent with the coupling of one single electronic transition to vibrational modes. This means that the most energetic line of the structure is actually associated with a pure electronic transition, which was accurately measured as a function of the external magnetic field and temperature. The results can be compared with the magnetic polaron model of Mauger and Mills [8], from which the strength of the conduction band electron exchange interaction can be derived.

4. MATHEMATICAL MODEL

In order to calculate the exchange energy of the EuTe crystal lattice, first and second neighbor interaction must be taken into account. In all of the Eu-chalcogenide crystals, the Eu-ions have a total spin $S = 7/2$ and are organized in a face-centered cubic (FCC) unity cell. Considering a spin in the cube vertex, its first neighbors are located in the center of the cube face and the interaction between them is ferromagnetic (see Figure 3). The second neighbors are the spins in the closest vertex and the interaction is ferromagnetic. Each Eu-spin has 12 first neighbors and 6 second neighbors.

In EuTe crystals, the second neighbor exchange interaction is dominant and forces the spins to align antiparallel. However, there is no configuration where each spin can be aligned antiparallel with its second neighbors while also aligned in parallel with all first neighbors. Thus, half of the first neighbors of a certain localized spin are aligned parallel with it, while the other half is aligned parallel with its second neighbor (Figure 4), forming parallel planes containing the spins from each magnetic sublattice.

If we consider there are no spin fluctuations and no external forces, all crystal spins will be aligned over the same axis and the first neighbor exchange interaction will vanish due to the alignment symmetry. The exchange energy of the crystal in this case can be calculated from the Heisenberg Hamiltonian:

$$\langle H_c \rangle = -\sum_{i\,j} J_{ij} \langle \mathbf{S_i} \cdot \mathbf{S_i} \rangle = -\sum_i^Z 6 J_2 S^2 \,, \tag{1}$$

where J_2 is the second neighbor exchange factor, Z is the number of spins in the crystal and $S = 7/2$ is the Eu spin angular moment.

If an external magnetic field is applied perpendicular to the initial magnetic sublattices, spins will be canted in the direction of the magnetic field (Figure 5) and first neighbor interaction will not vanish. In this case, the energy of each crystal spin will be a combination of the exchange energy

$$\langle H_c \rangle_i = -6J_2 S^2 + 12(J_1 + J_2)S^2 \sin^2 \phi \qquad (2)$$

and the Zeeman energy

$$\langle H_Z \rangle_i = -g\mu_b SB \sin \phi, \qquad (3)$$

where ϕ is the canted angle, g is the electron *g-factor*, μ_b is the Bohr magneton and B is the external magnetic field intensity.

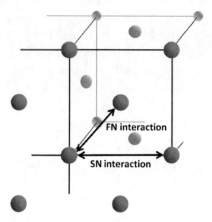

Figure 3. Schematic diagram of the EuTe crystal structure depicting first and second neighbors (FN and SN).

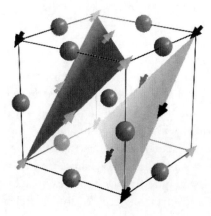

Figure 4. Schematic diagram of EuTe unity cell, where arrows represent Eu spins and balls represent Te atoms. The blue and orange planes contain spins in opposite magnetic sublattices.

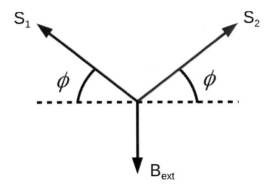

Figure 5. Representation of EuTe magnetic sublattice alignment in the presence of an external magnetic field.

The angle that minimizes the system energy is then

$$\sin\phi = \frac{g\mu_b SB}{24(J_1+J_2)S^2}, \qquad (4)$$

which saturates when B reaches the critical field given by $\sin\phi = 1$:

$$B_c = \frac{24(J_1+J_2)}{g\mu_b S}. \qquad (5)$$

In the presence of a photoexcited electron, each localized spins feels the force of an effective position-dependent exchange magnetic field, and the canted angle will be different for each spin depending on the distance from the photoexcited electron. The energy obtained from Heisenberg Hamiltonian for the exchange interaction between a photoexcited electron in position $\mathbf{r}$ and a localized spin in position $\mathbf{R_i}$ is [8]:

$$E_{exc}(\mathbf{R_i},\mathbf{r}) = \langle\Psi|H_{exc}|\Psi\rangle = -2\langle\Phi|\mathbf{s}\cdot\mathbf{S_i}|\Phi\rangle \int d^3r \psi^*(\mathbf{r}) I(\mathbf{r}-\mathbf{R_i})\psi(\mathbf{r})$$

$$= -I\frac{1}{\rho}|\psi(\mathbf{r}-\mathbf{R_i})|^2 S \sin\phi(R_i), \qquad (6)$$

where I is the exchange factor between the photoexcited electron and the localized spin, $\mathbf{s}$ is the electron spin, $\phi(R_i)$ is the canted angle of the localized spin in position $\mathbf{R_i}$, and $\rho = 4/a^3$ is the localized spin density, where $a = 6.6$ Å is the EuTe lattice parameter.

The energy for each localized spin is then:

$$E_{s_i} = -6J_2 S^2 + 12(J_1+J_2)S^2 \sin^2\phi(R_i) - \left(g\mu_b SB + I\frac{1}{\rho}|\psi(R_i)|^2 S\right)\sin\phi(R_i), \qquad (7)$$

which gives us the position-dependent condition for the angle saturation:

$$\sin\phi(R_i) = \frac{g\mu_b SB + I\frac{1}{\rho}|\psi(R_i)|^2 S}{24(J_1+J_2)S^2}. \qquad (8)$$

Including the kinetic $\langle T \rangle = -\int d^3 r \psi^*(\mathbf{r}) \frac{\nabla^2}{2m^*} \psi(\mathbf{r})$ and the potential $\langle V \rangle = -\int d^3 r \psi^*(\mathbf{r}) \frac{-ke^2}{\epsilon r} \psi(\mathbf{r})$ electron energies, and assuming a radial distribution with the photoexcited electron in the axis origin, the total energy for the magnetic polaron in EuTe is:

$$E_{pol} = \langle T \rangle + \langle V \rangle - 6ZJ_2 S^2 + \sum_i^Z \Big[12(J_1 + J_2)S^2 \sin^2 \phi(R_i) - \Big(g\mu_b SB + I\frac{1}{\rho}|\psi(R_i)|^2 S\Big) \sin \phi(R_i)\Big]. \tag{9}$$

with the photoexcited electron effective mass m^* and dielectric constant ϵ and k being the Boltzmann constant. Since the localized spin alignment is dependent on the photoexcited electron wave function, the problem is self-consistent and the electron can adjust its radius to minimize energy [16].

Considering a continuous distribution of localized spins with the same density, the sum of the localized ions can be substituted by an integral:

$$E_{pol} = \langle T \rangle + \langle V \rangle - 6ZJ_2 S^2 + \frac{16\pi}{a^3} \int_0^{R_{pol}} dr \, [12(J_1 + J_2)S^2 \sin^2 \phi(R_i) - (g\mu_b SB + I|\psi(R_i)|^2 S) \sin \phi(R_i)]. \tag{10}$$

It is also important to consider the possibility of forming a type-II localized state, which has a ferromagnetically saturated core, where $\sin \phi(R_{core}) = 1$. This splits the integral into two regions

$$E_{pol} = \langle T \rangle + \langle V \rangle - 6ZJ_2 S^2 + \frac{16\pi}{a^3} \int_0^{R_{core}} dr \, [12(J_1 + J_2)S^2 - (g\mu_b SB + I|\psi(R_i)|^2 S)] + \frac{16\pi}{a^3} \int_{R_{core}}^{R_{pol}} dr \, [12(J_1 + J_2)S^2 \sin^2 \phi(R_i) - (g\mu_b SB + I|\psi(R_i)|^2 S) \sin \phi(R_i)], \tag{11}$$

where the core radius R_{core} is determined by the electron wave functions that exactly match the conditions:

$$|\psi(R_{core})|^2 = \rho \left(\frac{24(J_1 + j2)S^2 - g\mu_b SB}{IS}\right). \tag{12}$$

The magnetic polaron total energy, however, cannot be directly related to the photoluminescence line. Since the crystal lattice carries part of the magnetic polaron energy, the energy in the absence of the photoexcited electron must also be considered. The photoluminescence energy is hence the difference:

$$E_{PL} = \langle T \rangle + \langle V \rangle + \frac{16\pi}{a^3} \int_0^{R_{core}} dr \, [12(J_1 + J_2)S^2 - (g\mu_b SB + I|\psi(R_i)|^2 S)] + \frac{16\pi}{a^3} \int_{R_{core}}^{R_{pol}} dr [12(J_1 + J_2)S^2 \sin^2 \phi(R_i) - (g\mu_b SB + I|\psi(R_i)|^2 S) \sin \phi(R_i)] - \frac{16\pi}{a^3} R_{pol}^3 [12(J_1 + J_2)S^2 \sin^2 \phi_0 - g\mu_b SB \sin \phi_0, \tag{13}$$

where ϕ_0 is the canted angle of magnetic sublattices due to the external magnetic field alone. Here we assume that the external magnetic field is smaller in intensity than the critical field, else there would be no contribution of the magnetic polaron formation to the photoluminescence energy.

A self-consistent solution was compared with a variational solution, which assumes that the electron has a Bohr wave function $\psi(r) = (\pi a_b^3)^{-1/2}(-2r/a_b)$ with the Bohr radius a_b as the variational parameter [16]. Both solutions produce similar results and can be used to extract the Polaron radius and magnetic moment in function of the external magnetic field, an average effective field for the photoexcited electron and the exchange factor to match the Stokes shift.

A more up-to-date approach to the problem uses a Monte Carlo simulation to introduce thermal fluctuation into the magnetic polaron system [17]. The J_1 and J_2 constants are chosen to reproduce the EuTe Néel temperature and critical field. These constants differ from theoretical calculation, which assumes the first neighbor interaction always vanishes, which is not necessary the case above $T = 0\,°K$ [18]. The Monte Carlo method reproduces the 0.3 eV Stoke-shift of the MX0 zero-phonon line with $I = 0.103$ eV. It also gives us a magnetic moment of approximately $\mu_{pol} \approx 600\mu_b$ for the polaron at $T = 0\,°K$ and $\mu_{pol} \approx 100\mu_b$ at $T = 100\,°K$. These results are in good agreement with theoretical predictions of Henriques et al. [14].

CONCLUSION

The aim of this chapter was to briefly introduce readers to the subject of magnetic polarons by focusing on the most important research studies in EuTe. The subject of magnetic polarons is, however, far more complex than would be possible to present here. Despite the large number of studies in the literature, there remains much to be elucidated. While most investigations about the subject seek to explore the possibility of optical control of magnetism in semiconductors and the development of spintronic devices, magnetic polarons are also of great interest from a more fundamental standpoint, conferring information about the physics of the exchange interactions and their dynamics, remaining very much an open field in material science and solid state physics. Furthermore, many intriguing studies about magnetic polarons in materials other than EuTe, including different Eu-chalcogenides, have been conducted, exploring interaction between polarons, polaron formation and extinction dynamics, polarons in 2D system, among others.

ACKNOWLEDGMENT

The author would like to extend thanks to Alysson F. Morais for suggestions made and for stimulating discussion on the topic.

REFERENCES

[1]   Tsubokawa, I. *J. Phys. Soc. Jap.* 1960, 15 1664-1668.
[2]   Matthias, B. T.; Bozorth R. M. and Van Vleck. *Phys. Rev. Lett.* 1961, 7 160-161.

504   *Flavio C. D. de Moraes*

[3]   McGuirre, T. R.; Argyle B. E.; Shafer, M. W. and Smart, J. S. *App. Phys. Lett.* 1962, 1 17-18.

[4]   Van Houten, S. *Phys. Lett.* 1962, 2 215-216.

[5]   Streit, P. and Wachter, P. *Phys. Kond. Mat.* 1970, 11 231-242.

[6]   Busch, G.; Streit, P. and Wachter, P. *S. S. Comm.* 1970, 8 1759-1763.

[7]   De Gennes, P. G. *Phys. Rev.* 1960, 118 141-154.

[8]   Mauger, A. and Mills, D. L. *Phys Rev. B.* 1985, 31 8024-8033.

[9]   Vitins, J. and Wachter, P. *S. S. Comm.* 1973, 13 1273-1277.

[10] Torrance, J. B.; Shafer, M. W. and McGuire, T. R. 1972.

[11] Umehara, M. *Phys Rev. B.* 1995, 52 8140-8149.

[12] Akimoto, R.; Kobayashi, M. and Suzuki, T. *J. Phys. Soc. Jap.* 1994, 63 4616-4628.

[13] Heiss, W.; Prechtl, G. and Springholz, G. *Phys. Rev. B.* 2001, 63 165323.

[14] Henriques, A. B.; Galgano, G. D.; Abramof, E.; Diaz, B. and Rappl, P. H. O. *App. Phys. Lett.* 2011, 99 091906.

[15] Heredia, E.; Motisuke, P.; Rappl, P. H.; Brasil, M. J. S. P. and Iikawa, F. *App. Phys. Lett.* 2012, 101 092108.

[16] Henriques, A. B.; Moraes, F. C. D.; Galgano, G. D.; Meaney, A. J.; Christianen, P. C. M.; Maan, J. C.; Abramof, E. and Rappl, P. H. *Phys. Rev. B.* 2014, 90 165202.

[17] Moraes, F. C. D. *Optically induced magnetization in magnetic semiconductors.* University of Sao Paulo, 2017.

[18] Söllinger, W.; Heiss, W.; Lechner, R. T; Rumpf. K.; Granitzer, P.; Krenn, H. and Springholz, G. *Phys. Rev. B.* 2010, 81 155213.

ABOUT THE EDITOR

Amel Laref, PhD
Physics Department, King Saud University
Riyadh, Saudi Arabia
Email: amel_la06@yahoo.fr

Amel Laref is lecturer and researcher working in the field of material science at King Saud University, Riyadh, Saudi Arabia. She received her PhD in Physics from the University of Algeria in collaboration with France. More than 70 scholarly articles, and six book chapters are published by Dr. Laref. She is a member of the physical, chemical, and materials society and she assisted as a referee for various prestigious journals of material science.

INDEX

A

absorption, 10, 13, 16, 19, 23, 26, 32, 39, 42, 48, 51, 55, 59, 61, 71, 76, 82, 83, 84, 90, 104, 132, 146, 171, 177, 179, 182, 186, 187, 188, 189, 190, 191, 192, 199, 202, 267, 268, 300, 301, 303, 304, 307, 308, 310, 331, 332, 335

absorption bands, 55, 171, 190, 310

absorption term, 61

ac conduction, 310, 330

ac conductivity, 12, 21, 28, 39, 83, 305, 310, 312, 316, 319, 321, 325, 327, 328, 329, 331, 333, 334, 335, 336, 337, 341, 344, 345, 346, 347, 348, 349, 488, 491

acceptor interface, 5

activation energy(ies), 20, 36, 37, 38, 40, 47, 57, 81, 143, 144, 146, 152, 158, 159, 160, 161, 162, 163, 164, 165, 167, 199, 309, 310, 328, 329, 332, 333, 334, 335, 339, 340, 344, 347, 367, 369, 370, 373, 386, 387, 389, 390, 391, 402, 403, 404, 412, 416, 419, 426, 447, 485, 486, 487, 489, 491

amorphous phase, 26, 82, 383, 389, 390, 393

amplimodes, 320, 321, 322, 326, 359

anharmonicity, v, viii, 197, 199, 200

anion radical(s), 8, 10, 43, 45, 46, 47, 48, 50, 65, 66, 75, 79

anisotropic, 3, 9, 11, 19, 23, 31, 43, 45, 52, 53, 55, 66, 70, 72, 74, 89, 99, 181, 201, 360, 434, 458, 459

anisotropy, 15, 25, 35, 36, 43, 56, 57, 68, 72, 88, 182, 201, 314, 421

anisotropy of spin dynamics, 35, 56, 68

annihilation of charges, 75

anti-ferromagnetic interaction, 30

Arrhenius, 149, 158, 159, 288, 328, 333, 335, 339, 340, 348, 349, 367, 370

Arrhenius plot(s), 158, 159, 288, 328, 333, 335, 339, 340, 348, 370

asymmetry factor, 12, 39, 52, 82, 83

B

background illumination, 46, 48, 58, 63

ballistic flight, 197, 214, 217

ballistic particle, 200

band structure, 5, 6, 67, 69, 109, 177, 184, 192, 201, 225, 253, 254, 256, 432, 433, 435

bandgap, 5, 41, 51, 67, 69, 95, 171, 175, 179, 180, 183, 184, 186, 188, 189, 309, 331, 352

bands theory, 4

bimolecular recombination, 7, 48, 73

binding energy, 5, 132, 149, 159, 176, 187, 202, 268, 270, 275, 276, 277, 281, 282, 284, 285, 287, 289, 293, 294, 295, 296, 297, 298, 366, 398, 404, 410, 417, 424, 425, 426, 436, 449, 467

bipolaron bands, 5, 186, 187, 188, 190, 191, 192, 195

bipolaron decay, 40

bipolaron dissociation, 40

bipolaron(s), vii, viii, ix, 1, 2, 4, 5, 6, 11, 13, 15, 18, 21, 22, 23, 24, 30, 31, 33, 38, 40, 42, 88, 89, 90, 101, 104, 131, 132, 133, 135, 140, 141, 171, 172, 175, 176, 178, 179, 182, 183, 184, 185, 186, 187, 188, 189, 190, 191, 192, 193, 195, 196, 198, 218, 334, 353, 367, 399, 429, 430, 431, 433, 435, 436, 437, 439, 441, 443, 445, 447, 448, 449, 450, 451, 452, 453, 455, 458, 467

bis-methanofullerene, 43

bisolectron, 200

block-scheme, 85

Bohr magnetons, 52

bound magnetic polaron (BMP), 495, 498

bulk heterojunctions, 5, 95, 99

C

carrier concentration, 74, 135, 156, 162, 163, 164, 169, 181, 183, 310, 335, 344, 348, 418, 445, 448, 484, 496

508   *Index*

cell, 104, 120, 177, 194, 201, 202, 203, 205, 273, 307, 308, 309, 312, 313, 316, 317, 319, 320, 321, 322, 324, 326, 332, 336, 344, 345, 346, 348, 349, 400, 410, 411, 413, 414, 420, 430, 435, 436, 490, 499, 500

cell parameter(s), 312, 313, 319, 321, 324, 326, 345, 346, 349

chain librations, 21, 37

charge carrier mobility, 56, 73, 82, 148, 159, 160, 161, 163, 164, 165, 169, 366

charge carriers, ix, 1, 2, 3, 4, 7, 8, 9, 10, 11, 13, 15, 18, 21, 24, 28, 29, 34, 37, 40, 42, 43, 44, 45, 46, 47, 48, 49, 50, 55, 58, 59, 60, 63, 64, 65, 67, 68, 69, 73, 74, 75, 78, 80, 94, 95, 101, 102, 105, 125, 130, 131, 132, 133, 134, 142, 145, 146, 155, 160, 165, 168, 169, 175, 176, 178, 182, 183, 187, 189, 190, 223, 224, 228, 230, 231, 265, 266, 272, 273, 276, 288, 310, 311, 312, 327, 330, 334, 335, 336, 342, 343, 344, 347, 348, 366, 389, 410, 411, 413, 417, 419, 420, 425, 430, 431, 437, 444, 445, 446, 447, 448, 464, 484, 485, 486, 487, 488

charge dynamics, 37, 69, 80, 95

charge recombination, 48, 73

charge separation, 5, 10, 60, 65, 67, 93, 98, 192

charge transfer, 1, 3, 5, 9, 10, 11, 13, 21, 26, 28, 29, 30, 33, 34, 36, 37, 40, 54, 57, 59, 66, 69, 70, 72, 73, 80, 81, 95, 96, 189, 335, 354, 381, 409, 411, 436

charge transfer mechanisms, 36

charge transport, v, vi, viii, 2, 11, 26, 36, 42, 57, 66, 68, 70, 73, 80, 93, 95, 97, 100, 103, 107, 108, 130, 139, 140, 143, 144, 145, 150, 157, 158, 160, 161, 164, 166, 167, 169, 171, 176, 181, 182, 184, 198, 214, 312, 313, 350, 352, 376, 429, 457, 496

charge-separated states, 5, 59

colossal magnetoresistance (CMR), vii, ix, 179, 397, 399, 405, 406, 409, 410, 416, 426, 429, 431, 436, 437, 441, 447, 452

concentration of carrier(s), 335, 342, 349

concentration of charge carriers, 18, 22, 74, 135, 330, 343, 344, 485

conduction, vi, ix, 5, 36, 42, 62, 93, 123, 130, 131, 146, 168, 174, 175, 176, 181, 182, 185, 188, 189, 198, 213, 214, 215, 253, 254, 255, 256, 261, 269, 270, 271, 272, 289, 309, 310, 312, 327, 328, 331, 333, 334, 335, 336, 337, 339, 340, 341, 342, 344, 348, 349, 352, 353, 354, 355, 358, 359, 360, 362, 366, 369, 370, 373, 375, 379, 380, 381, 386, 389, 390, 391, 393, 397, 399, 402, 403, 413, 430, 431, 434, 437, 440, 441, 442, 444, 447, 459, 483, 484, 485, 486, 487, 488, 489, 492, 493, 494, 496, 498, 499

conduction band, 5, 36, 62, 123, 131, 174, 175, 183, 185, 188, 189, 214, 215, 253, 254, 255, 256, 269, 309, 335, 430, 434, 442, 459, 496, 498, 499

conduction mechanism(s), vi, ix, 130, 310, 312, 328, 331, 334, 335, 336, 337, 339, 340, 341, 342, 344, 349, 352, 354, 358, 359, 360, 362, 366, 369, 379, 390, 391, 437, 444, 483, 484, 486, 488, 489, 494

conductivity mechanisms, 13, 22, 327, 335, 337, 348, 349, 494

conjugated polymers, v, viii, 1, 2, 3, 4, 9, 10, 11, 12, 14, 18, 19, 22, 24, 32, 37, 38, 40, 41, 49, 53, 54, 55, 56, 57, 69, 70, 71, 72, 73, 74, 78, 80, 86, 87, 88, 90, 91, 93, 94, 96, 99, 101, 103, 105, 108, 109, 115, 125, 130, 131, 132, 133, 139, 140, 141, 142, 157, 166, 169, 171, 173, 174, 175, 178, 179, 181, 182, 184, 185, 187, 189, 192, 193

contribution, ix, 9, 12, 18, 21, 22, 27, 29, 43, 45, 52, 55, 59, 72, 82, 92, 129, 144, 146, 149, 158, 167, 176, 198, 210, 228, 230, 249, 266, 268, 277, 281, 282, 284, 285, 287, 288, 293, 294, 295, 296, 332, 336, 350, 367, 419, 437, 439, 440, 471, 485, 503

correlated barrier hopping, 329, 330, 346

correlated barrier hopping (CBH) model, 329, 330, 331, 333, 334, 346, 348

correlated polaron, vi, ix, 409, 410, 416, 417, 419, 420, 426

Coulomb field, 73

Coulomb interactions, 47, 111, 124, 171, 267, 269, 334, 367, 434, 435

Coulomb radius, 73

Coulombic interaction, 8, 24, 267, 289, 430

counter ions, 62, 64, 180

critical field, 460, 461, 464, 465, 466, 501, 502, 503

crossover frequency, 327, 328, 330, 339, 348, 350

cross-relaxation, 10

crystal structure(s), ix, 145, 305, 306, 308, 309, 311, 312, 313, 314, 316, 317, 319, 320, 321, 324, 326, 329, 336, 341, 344, 345, 347, 348, 349, 350, 356, 357, 359, 361, 362, 412, 418, 424, 492, 500

crystalline domains, 20, 21, 26, 29, 30, 49, 57, 80, 83

crystalline phase, 25, 26, 380

crystalline structure, 4, 311, 312, 313, 344, 349

cyclic voltammetry (CV), 172, 190, 196

D

dc and ac conductivities, 333

dc conductivity, 3, 15, 18, 28, 29, 30, 41, 51, 57, 58, 311, 321, 327, 328, 332, 334, 335, 336, 344, 346, 347, 348, 350, 351, 370, 371, 372, 374, 380, 386, 390, 391, 393, 409, 411, 412, 414, 420, 423, 425, 484, 489, 494

deconvolution, 46, 102

Index 509

deep traps, 43, 59, 60, 66, 67

detrapping, 48, 75

diamagnetic bipolarons, 6, 15, 18, 38, 42, 69

dielectric spectroscopy, 309, 310, 336

differential scanning calorimeter (DSC), 379, 381, 382, 384, 393

diffusion coefficients, 35, 50, 68, 78, 81

dihedral angle, 35

dilute ferromagnetic oxide film, 409

dipole moment(s), 23, 80, 150, 151, 225, 227, 228, 229, 230, 239, 240, 248, 259, 366, 376

dipole-dipole broadening, 23

dipole-dipole interaction, 22, 23, 24, 45, 49, 57, 76, 79, 80

discrete breather (DB), 207, 208, 209

disordered semiconductor, 48, 50, 75

dispersion, 13, 19, 21, 32, 33, 34, 51, 55, 83, 103, 185, 186, 269, 273, 331, 335, 352, 433, 467, 475

dispersion signal, 33

dispersion spectra, 19, 32, 33, 34, 51, 55

dispersion spectrum, 13, 55

domestic, 58, 61, 63, 65, 80

donor, 5, 6, 7, 8, 9, 10, 50, 93, 157, 158, 173, 268, 276, 289, 290, 291, 292, 293, 294, 295, 296, 297, 301, 304, 375, 498

doping level, 5, 6, 11, 22, 25, 28, 30, 38, 42, 69, 102, 171, 182, 184, 186, 187, 188, 189, 190, 195, 462, 463, 469

double exchange, 193, 418

Drude conductivity, 420

Dulong-Petit plateau, 198, 199, 207

dynamic structure factor, 207, 208

dynamics, x, 1, 2, 4, 7, 8, 10, 11, 15, 19, 21, 30, 31, 34, 38, 47, 49, 50, 53, 55, 57, 58, 66, 68, 69, 70, 73, 80, 81, 82, 92, 94, 95, 101, 102, 104, 105, 119, 124, 126, 127, 129, 135, 140, 141, 142, 172, 184, 192, 193, 194, 195, 199, 203, 208, 212, 264, 268, 311, 330, 334, 344, 351, 352, 367, 416, 418, 503

dynamics parameters, 2, 11, 21, 58, 66, 69

Dysonian contribution, 25

Dysonian line, 82, 83

Dysonian shape, 25

Dysonian spectra, 28

Dysonian term, 83

E

effective paramagnetic susceptibility, 27, 28, 46, 52, 63

electric field, 9, 24, 73, 80, 119, 124, 125, 131, 135, 149, 150, 152, 153, 154, 157, 158, 163, 164, 166, 169, 184, 185, 197, 199, 203, 204, 212, 213, 227, 229, 236, 239, 240, 246, 247, 248, 253, 254, 255, 256, 257, 258, 268, 290, 301, 326, 327, 330, 411

electric properties, 344, 346

electric transport, 306, 311, 327, 332

electrical, v, ix, 6, 90, 91, 97, 98, 108, 131, 138, 144, 147, 168, 171, 172, 173, 176, 181, 182, 186, 193, 194, 226, 227, 259, 305, 306, 310, 311, 312, 313, 315, 319, 325, 327, 328, 333, 335, 336, 344, 345, 347, 348, 349, 352, 353, 354, 357, 359, 360, 361, 362, 366, 367, 368, 373, 375, 379, 380, 381, 383, 389, 393, 397, 403, 405, 406, 424, 430, 431, 434, 437, 440, 442, 444, 447, 451, 460, 461, 483, 484, 488, 489, 492, 493, 494

electrical properties, ix, 193, 194, 325, 333, 344, 348, 352, 359, 366, 379, 380, 389, 393, 447, 492, 493, 494

electrical properties of glasses, ix, 366

electrical transport, 97, 306, 335, 336, 353, 354, 362, 403, 406, 424, 434, 437, 440, 444, 447, 451

electrochemical properties, 308

electron (hole) polarons, 310

electron acceptor, 5, 6, 7, 73, 179

electron donor, 5, 6, 7, 11, 14, 30, 175

electron energy loss spectroscopy (EELS), 172, 192, 436

electron hopping model, 485

electron paramagnetic resonance (EPR), v, viii, 1, 2, 9, 10, 11, 12, 13, 15, 16, 18, 19, 20, 21, 22, 23, 25, 26, 27, 28, 30, 31, 32, 33, 34, 37, 38, 39, 40, 41, 42, 43, 45, 48, 51, 52, 53, 54, 55, 56, 57, 58, 59, 60, 61, 64, 65, 66, 67, 70, 71, 75, 77, 78, 80, 82, 83, 84, 85, 86, 88, 90, 91, 92, 93, 94, 95, 96, 98, 99, 103, 104, 105

electron paramagnetic resonance (EPR) spectroscopy, viii, 1, 10, 42, 58, 71, 84, 85, 86, 91, 94, 99

electron phonon interaction, 409

electron relaxation, 13, 31, 49, 53, 79

electron scattering, 40, 202, 437

electron spin resonance (ESR), 87, 89, 90, 92, 93, 94, 95, 96, 97, 99, 100, 101, 102, 103, 105, 172, 192

electron surfing, v, viii, 197, 199, 200, 206, 209, 212, 217

electron transfer, 59, 80, 104, 175, 186, 199, 335, 370, 416, 487

electron-hole pairs, 5, 269

electronic and ionic conductivity, 326

electronic conduction, 310, 346, 348, 354, 366, 379, 381, 389, 409, 492

electronics properties, 4, 30

electron-phonon coupling, ix, 121, 141, 149, 171, 272, 298, 303, 392, 400, 402, 406, 426

510

Index

electron-phonon interaction, 2, 81, 131, 149, 198, 297, 298, 302, 391, 392, 402, 425
electro-soliton, 198
electrostatic field, 24
Elliott mechanism, 38, 54
emeraldine base, 17, 25, 26
emeraldine salt, 17, 25, 26
energy barrier, 47, 61, 64, 65, 74, 310, 332, 341, 373
$Eu_2(MoO_4)_3$, 312, 318, 319, 321, 322, 323, 325, 326, 335, 336, 338, 339, 341, 342, 343, 344, 345, 347, 354, 357, 358
exchange coupling, 52, 73, 496
exchange interaction, x, 8, 18, 40, 58, 61, 63, 64, 65, 66, 69, 70, 76, 77, 178, 224, 268, 303, 399, 400, 436, 496, 497, 498, 499, 501, 503
exchange interactions, x, 8, 58, 400, 503
exciton(s), v, viii, 5, 7, 8, 9, 10, 48, 66, 73, 87, 91, 96, 103, 132, 133, 134, 135, 140, 141, 142, 192, 193, 196, 202, 265, 266, 267, 268, 269, 270, 271, 273, 274, 275, 276, 277, 278, 279, 281, 282, 284, 285, 287, 288, 289, 290, 292, 294, 296, 297, 298, 299, 300, 301, 302, 303
exciton-phonons, v, viii, 265, 266, 273

F

fergusonite, 306, 308, 314, 315, 318, 324, 357
Fermi energy, 28, 30, 58, 62, 404, 459, 496
Fermi level, 26, 28, 52, 58, 72, 146, 153, 159, 172, 176, 184, 185, 335, 392, 404, 437, 491
Fermi velocity, 37, 464
ferrites, vi, vii, ix, 483, 484, 485, 486, 488, 489, 491, 492, 493, 494
field, vii, 2, 9, 10, 14, 22, 26, 33, 42, 49, 54, 55, 58, 70, 71, 73, 77, 79, 80, 82, 84, 85, 90, 94, 98, 104, 108, 117, 120, 125, 135, 140, 141, 144, 147, 149, 151, 152, 154, 158, 163, 164, 165, 166, 167, 168, 172, 176, 177, 178, 184, 197, 198, 199, 200, 203, 204, 205, 209, 210, 212, 213, 214, 216, 217, 218, 227, 228, 229, 230, 231, 235, 249, 251, 252, 253, 267, 268, 272, 273, 283, 288, 289, 298, 299, 300, 306, 311, 327, 345, 348, 358, 368, 369, 397, 399, 400, 401, 405, 406, 412, 416, 419, 420, 421, 423, 426, 431, 433, 436, 437, 460, 461, 465, 472, 484, 496, 497, 498, 499, 500, 501, 502, 503, 505
flip-flop probability, 47, 75, 77
Franck-Condon transition, 416, 426
free electron, 10, 16, 28, 42, 54, 71, 174, 181, 201, 202, 217, 269, 270, 282, 420
frequency exponent, 327, 328, 329, 330, 331, 334, 335, 341, 344, 349, 491
frequency exponent s, 328, 329, 330, 331, 334, 335, 341, 344

frequency prefactor, 349
Friedman-Holstein model, 81
Fullerene, 5, 58

G

Gaussian, 25, 42, 43, 45, 51, 53, 55, 59, 77, 78, 150, 153, 155, 156, 157, 166, 212, 290, 465
Gaussian distribution, 45, 77, 157, 212
geminate, 5, 7, 74
geminate recombination, 7, 74
g-factor, 10, 11, 12, 15, 16, 17, 25, 31, 39, 42, 43, 45, 52, 53, 54, 57, 58, 67, 71, 72, 85, 500
glass-ceramic nanocomposites, vi, ix, 379, 380, 381, 385, 386, 387, 388, 389, 390, 391, 393
graphene nanoribbons, 136, 137, 142
g-tensor, 12, 14, 31, 54
guest charge carrier, 80
g-value, 43, 72

H

harmonic vibration, 54, 206, 208
Heisenberg Hamiltonian, 499, 501
heteroatoms, 25, 43, 53, 57, 71, 147
heterojunction, 7, 90, 98, 99
highest occupied molecular orbital, 7, 145, 175
high-frequency/field CW EPR spectroscopy, 85
high-frequency/field EPR spectroscopy, 10
Hoesterey-Letson formalism, 50, 81
hole, 5, 42, 74, 84, 102, 126, 131, 132, 133, 145, 149, 176, 177, 178, 179, 196, 197, 199, 201, 202, 217, 224, 225, 230, 236, 238, 249, 253, 254, 255, 256, 259, 266, 269, 270, 271, 272, 273, 274, 275, 276, 277, 280, 283, 285, 288, 299, 303, 308, 309, 315, 328, 331, 343, 356, 366, 367, 398, 416, 418, 426, 430, 435, 436, 452, 458, 469, 470, 471, 472, 496, 498
hole conduction, 328
homogeneous, 9, 43, 47, 236, 248, 251, 253, 255, 256, 259, 260, 268, 298, 300
hopping, ix, 5, 24, 30, 36, 37, 38, 40, 51, 57, 61, 63, 65, 73, 75, 76, 78, 80, 81, 93, 102, 121, 122, 125, 144, 145, 146, 149, 152, 153, 154, 155, 156, 157, 160, 161, 162, 164, 167, 168, 169, 177, 184, 185, 210, 255, 310, 321, 328, 329, 330, 332, 333, 334, 335, 348, 352, 357, 360, 361, 362, 366, 367, 368, 369, 373, 379, 386, 389, 390, 391, 392, 393, 399, 402, 403, 404, 412, 413, 416, 430, 432, 437, 447, 449, 467, 484, 485, 486, 487, 488, 489, 491, 494, 497
hopping integrals, 210

HR-TEM, 384
HTSC, 409, 420, 423
hyperconduction, 197, 199
hyperfine broadening, 59, 77
hyperfine constant, 24

I

immobile radicals, 34
impedance spectroscopy, 327, 332, 345, 349, 351, 355, 357
impurities, viii, 42, 47, 109, 119, 124, 145, 265, 266, 289, 301, 304, 310, 322, 336, 380, 400, 445, 498
inhomogeneous, 9, 43, 72, 236, 307, 332
insulators, 3, 147, 172, 173, 179, 181, 335, 429, 432, 452
interacting spins, 73
interchain diffusion, 36, 57, 68
interchain transfer integral, 37
intersite interactions, 208, 210
intrachain mobility, 64
intrachain transfer integral, 45, 78
intrinsic conductivity, 21, 26, 39, 40, 41, 57
intrinsic localized mode (ILM), 207, 208, 209
inverter, 214, 217, 218
ionic, 120, 148, 175, 177, 193, 197, 218, 227, 272, 302, 307, 308, 309, 310, 312, 315, 316, 317, 319, 326, 327, 328, 330, 333, 343, 345, 346, 348, 349, 350, 353, 357, 365, 366, 367, 368, 374, 397, 399, 400, 402, 406, 424, 425, 426, 430, 484, 486, 488, 490, 491
ionic conduction, 328, 330, 333, 343, 349, 350
ionic conductivities, 315, 319, 484
ionic conductivity, 310, 346, 349, 365, 366, 367, 374
ion-radical pairs, 76
IR spectra, 409
isotropic g-factor, 12, 16, 43
isotropic spectra, 52

J

JT effect/distortion, 401

K

Kivelson-Heeger theory, 36

L

$La_{1.6-x}Nd_{0.4}Sr_xCuO_4$, 421
Landé factor, 71

large polaron(s), 159, 164, 177, 273, 309, 311, 329, 331, 336, 346, 348, 349, 399, 431, 437, 476, 491, 492, 494
laser irradiation, 16, 17, 18, 19
lattice parameters, 120, 177, 201, 273, 320, 321, 323, 334, 344, 345, 347, 399, 404, 487, 501
lattice phonons, 20, 21, 29, 36, 40, 49, 55, 57, 70, 77, 78, 80
lattice polaron, ix, 37, 397, 398, 399, 400, 406, 430, 457, 459, 463, 477
Le Bail, 320, 321, 357
Le Bail refinement, 320, 321
lead tungstate, 307, 332, 333
libron-exciton interactions, 37
line asymmetry parameter, 83
line broadening, 12, 17, 43, 45, 76
line shape, 10, 12, 31, 38, 51, 53, 59, 71, 76, 77, 78, 82
linewidth, 11, 14, 15, 18, 19, 22, 25, 26, 27, 32, 33, 38, 40, 41, 42, 43, 45, 48, 49, 52, 53, 54, 55, 59, 61, 62, 64, 66, 76, 77, 78, 79, 83, 85, 92, 100, 297, 307
local magnetic fields, 76
localized polarons, 32, 43, 44, 46, 48, 60, 72, 75
localized states, 57, 69, 145, 146, 152, 155, 156, 185, 236, 307, 311, 332, 390, 393, 485
longitudinal elongations, 207
Lorentzian, 15, 25, 42, 43, 44, 45, 51, 52, 53, 59, 76, 77, 78
Lorentzian line, 43, 53
Lorentzian spectrum, 15
lowest unoccupied molecular orbital, 7, 145, 175, 189
luminescence properties, 302, 306, 355, 356, 362
luminescent properties, 317, 353, 360

M

macromolecular dynamics, 21, 37, 55, 57, 64, 94
Madelung energy, 410, 420
magnetic field fluctuating, 76
magnetic polarons, vi, x, 178, 397, 398, 399, 405, 406, 424, 495, 496, 497, 498, 499, 502, 503
magnetic semiconductor, vii, 261, 483, 495, 504
magnetic susceptibility, 18, 22, 28, 40, 41, 92, 98, 357
magnetite, 409, 410, 411, 412, 413, 416, 420, 423, 425
magnetoresistance, 2, 416, 418, 419, 426, 431, 440
manganite, ix, 400, 410, 412, 416, 418, 419, 426, 442, 446
marginal Fermi liquid, 414, 420
master curve, 330, 342, 344, 349

mechanical control of electros, 212

mechanism, ix, 11, 13, 20, 30, 38, 47, 57, 69, 79, 80, 105, 130, 133, 138, 144, 145, 146, 160, 167, 186, 189, 195, 198, 199, 248, 259, 308, 310, 326, 329, 332, 333, 334, 337, 340, 342, 357, 366, 367, 369, 381, 391, 398, 399, 409, 426, 429, 437, 457, 458, 459, 461, 462, 463, 467, 468, 470, 472, 473, 487, 488, 489

mechanism of charge transport, 38, 105, 326

mechanism of conduction, 334

mechanisms, vii, ix, 11, 54, 69, 102, 108, 119, 305, 310, 312, 319, 320, 328, 329, 333, 336, 341, 342, 344, 346, 348, 349, 350, 355, 399, 402, 473

metal-insulator transition, 30, 87, 358

metal-like domains, 21, 29, 30

method of Rietveld, 320

method of spin probe, 22

microwave, viii, 1, 9, 61, 70, 79, 82, 85, 88, 309, 326, 347, 352, 356, 359, 483, 486, 489, 492, 493, 494

mixed conductivity glasses, vi, ix, 365

mixed ionic-electronic conduction, 305

mobile polarons, 13, 17, 18, 19, 20, 21, 30, 32, 34, 41, 43, 49, 59, 61, 308

mobilities, 37, 164, 165, 194, 309, 310, 336, 343, 485

mobility, 1, 11, 12, 15, 18, 21, 22, 25, 30, 31, 32, 37, 42, 43, 45, 49, 62, 68, 69, 72, 73, 90, 93, 109, 139, 146, 149, 154, 157, 158, 159, 160, 161, 162, 163, 164, 166, 168, 169, 176, 177, 178, 181, 185, 189, 199, 202, 257, 310, 330, 335, 336, 342, 343, 344, 348, 349, 353, 366, 367, 389, 399, 419, 424, 430, 431, 445, 448, 449, 484, 485, 486, 487, 488

modulated scheelites, v, ix, 305, 306, 311, 313, 316, 317, 318, 319, 335, 336, 342, 345, 347, 348, 358, 360

modulated structures, 305, 308, 311, 312, 316, 322, 334, 346, 350, 351, 358

molecular axis, 75, 79

molecular dynamics, 53, 69, 78, 126, 141

molecular electronics, 1, 2, 3, 11, 30, 51, 66, 70, 87, 88, 96, 144, 219

molecular-lattice polaron, 37, 70

molybdates, 305, 307, 308, 312, 313, 317, 318, 319, 320, 321, 322, 324, 325, 333, 334, 335, 347, 348, 350, 351, 353, 354, 356, 357, 358, 359, 361, 362, 363

morphology, 3, 4, 13, 24, 31, 41, 49, 65, 66, 69, 91, 103, 138, 139, 333

Morse anharmonic interactions, 199

Morse potential, 206, 207, 208, 214, 217

Mott formula, 386

multifrequency EPR, 11, 21, 31, 69, 72, 86, 105

multifrequency EPR spectroscopy, 11, 21, 69, 105

multispin composite, 66, 67, 68

multistep tunneling, 74

N

nanocrystallization, 379, 380, 389, 390

nano-structured oxides, 424

nanostructures, v, viii, 179, 261, 265, 266, 267, 288, 289, 297, 298, 302

$Nd_2(MoO_4)_3$, 312, 318, 319, 320, 321, 322, 323, 325, 336, 338, 339, 341, 342, 343, 345, 348, 349, 355

neutron thermodiffraction, 347

nitroxide radical, 23, 24

non-conventional oxide glasses, 365

non-geminate, 7, 74

non-geminate recombination, 74

non-planar morphology, 5

nuclear magnetic resonance, 15, 89

O

OLPT model, 342

onsite potential, 207, 208

onsite vibrations, 207

optic phonon, 409, 410

optical and electrical properties, 306

optical phonon frequency, 278, 380, 391, 393, 402, 449

optical phonon modes, 265, 266, 272, 291, 294, 438, 459, 463

optical phonons, ix, 57, 176, 177, 265, 271, 272, 277, 297, 392, 420, 472

optical properties, 41, 108, 172, 193, 224, 263, 267, 268, 269, 270, 271, 276, 288, 297, 299, 300, 303, 307, 311, 312, 319, 327, 334, 348, 352, 381, 493

orbit interaction, 10

organic conjugated polymers, 1, 2, 5, 16, 71, 79, 86

organic metal, 32, 103

organic molecular electronics, 3, 105

organic semiconductor, v, vi, vii, viii, 2, 8, 9, 10, 58, 66, 90, 102, 107, 108, 119, 125, 130, 131, 132, 135, 143, 144, 145, 146, 148, 152, 158, 160, 162, 163, 167, 168, 169, 173, 429, 457

overlapping integral, 16, 24, 35, 214

overlapping large polaron (OLP), 305, 329, 330, 331, 342, 344, 346, 347, 483, 490, 494

overlapping large polaron model (OLP model), 331, 341, 342, 344, 346, 347

overlapping large polaron tunneling (OLPT) model, 490

Index

P

paramagnetic centers, 11, 21, 23, 25, 58, 59, 60, 64, 65, 67, 71, 72, 77, 79, 82

paramagnetic susceptibility, 21, 25, 27, 34, 40, 45, 46, 52, 72, 73, 76

parameter, 3, 12, 15, 19, 22, 32, 33, 34, 36, 41, 49, 60, 61, 62, 67, 72, 73, 77, 78, 116, 135, 153, 157, 158, 161, 177, 202, 203, 207, 210, 213, 217, 230, 232, 233, 234, 235, 236, 237, 238, 241, 243, 245, 246, 248, 271, 276, 277, 288, 290, 313, 321, 322, 323, 325,326, 327, 329, 330, 333, 342, 344, 345, 348, 349, 417, 419, 426, 440, 450, 451, 452, 462, 467, 468, 472, 503

PEDOT:PSS, 171, 185, 186, 187, 188, 189, 190, 191, 195

phase transitions, 52, 53, 64, 98, 240, 257, 259, 260, 308, 311, 312, 313, 317, 318, 334, 346, 347, 348, 350, 356, 357, 358, 359, 410, 420, 425, 437, 461, 464, 465

phonon-assisted hopping, 37

phonon-assisted hopping model of Mott, 393

photo-induced localized magnetic polaron, 498, 499

photo-induced localized magnetic polaron (PILMP), 498, 499

photon energy, 15, 44, 46, 47, 48, 49, 50, 67, 68, 69, 75, 498

photonics devices, 41

photons, 5, 7, 20, 46, 47, 48, 49, 50, 66, 68, 69, 70, 182

photovoltaic systems, 9

planar morphology, 5, 13, 17, 25, 54

polarizability, 224, 327, 342, 345, 346, 347, 349

polarization, 2, 8, 37, 86, 101, 144, 150, 151, 177, 197, 198, 199, 209, 210, 217, 227, 230, 236, 238, 272, 273, 311, 319, 320, 335, 352, 366, 368, 369, 381, 397, 398, 399, 412, 485, 487, 490, 496

polaron band, 22, 187, 190, 368, 403, 419, 420, 436, 449, 451

polaron binding energy, 5, 149, 150, 158, 417, 430, 434

polaron conduction, vi, ix, 309, 361, 365, 489, 494

polaron diffusion, 20, 32, 37, 47, 49, 63, 76, 79

polaron dynamics, 49, 50, 58, 68, 77, 140, 142, 184

polaron effect, v, vi, viii, ix, 143, 144, 152, 153, 158, 160, 161, 162, 165, 166, 167, 168, 248, 367, 429, 457

polaron excitations, 48, 55

polaron motion, 46, 62, 90, 99, 144, 168, 353, 355, 399, 493

polaron pairs, 8, 21, 367

polaron radius, 329, 341, 342, 368, 369, 380, 392, 488, 491

polaron sites, 81, 369

polaron transport, v, viii, 45, 148, 169, 177, 197, 360, 366, 367, 400, 418

polaron-fullerene pairs, 7

polaronic, viii, ix, 2, 6, 9, 11, 48, 61, 67, 73, 75, 95, 169, 171, 176, 177, 184, 185, 186, 199, 212, 236, 237, 258, 265, 266, 276, 282, 284, 285, 286, 287, 289, 291, 293, 294, 295, 296, 297, 298, 302, 303, 305, 307, 309, 310, 312, 315, 321, 327, 328, 331, 332, 333, 334, 337, 340, 341, 344, 347, 349, 366, 367, 368, 369, 374, 430, 431, 434, 436, 442, 445, 446, 448, 457, 458, 459, 461, 462, 463, 465, 466, 467, 469, 471, 473, 474, 475, 476, 477, 479, 481

polaronic behaviour, 344

polaronic center, 307, 332

polaronic conduction, 309, 328, 331, 337, 349, 366

polaronic conductivity, 312, 347, 368

polaronic electron transport, 321

polaronic mechanisms, ix, 327, 334

polaronic reservoirs, 61

polaronic transport mechanisms, 331

polaronic type, 312, 333, 340, 341

polaron-phonon interaction, 57

poly(3-alkylthiophene), 7, 87, 89, 95, 97, 102, 104

poly(3-dodecylthiophene), 18, 41, 95, 101, 104

poly(3-hexylthiophene), 41, 42, 86, 90, 96, 98, 104

poly(3-methylthiophene), 42, 97, 101

poly(3-octylthiophene), 41, 51, 88, 93, 95, 96, 97, 100, 102

poly(*bis*-alkylthioacetylene), 4, 14, 15, 95

poly(*p*-phenylene), 4, 6, 11, 14, 91

poly(tetrathiafulvalenes), 16, 30, 54, 91, 100

polyaniline, 4, 14, 17, 18, 19, 20, 21, 24, 25, 26, 35, 45, 52, 53, 62, 64, 78, 87, 90, 91, 92, 93, 94, 95, 96, 97, 99, 100, 102, 103, 104, 175, 185, 194

poly-diacetylene (PDA), 199, 201, 202, 213, 214, 217

polymer:fullerene BHJ, 8, 42, 46

polymer:fullerene composite, 5, 43, 45, 47, 48, 49, 50, 65, 70, 73, 79

polymer backbone, 17, 21, 25, 38, 43, 46, 48, 50, 55, 59, 65, 66, 69, 71, 75, 174, 180, 181, 186, 187, 189

polymer chain, 1, 3, 5, 11, 13, 15, 17, 19, 20, 21, 22, 24, 25, 30, 31, 32, 34, 36, 37, 38, 40, 45, 46, 48, 50, 54, 58, 61, 62, 63, 64, 65, 69, 70, 72, 73, 74, 75, 79, 109, 121, 129, 132, 140, 141, 174, 175, 176, 177, 179, 180, 181, 182, 184, 185, 189, 190, 200

polymer chain librations, 40

polymer lattice, 57, 60, 69, 125, 189

polymer matrix, 5, 12, 13, 20, 24, 29, 31, 34, 45, 47, 48, 49, 50, 59, 65, 69, 70, 72, 75, 81, 83, 191

514 *Index*

polypyrrole, 4, 14, 21, 23, 24, 35, 87, 89, 93, 99, 100, 101, 102, 104, 179, 185, 193, 194

polythiophene, 4, 14, 37, 38, 39, 42, 89, 90, 91, 93, 95, 97, 98, 101, 102, 104, 179, 180, 183, 185, 189

positively charged polaron, 6, 42, 59, 65, 75

pre-exponential factor, 328, 339, 341, 342, 346, 347, 348, 349, 350, 367, 375, 387, 390, 391, 392

prefactor, 327, 329, 341, 436, 450

pressure, 93, 97, 201, 265, 266, 268, 282, 283, 284, 285, 286, 287, 288, 292, 293, 294, 295, 296, 297, 301, 304, 313, 316, 317, 318, 353, 354, 355, 356, 357, 360, 440, 444, 445

Q

quadrature spectrum term, 55

quantum dots, vii, 2, 255, 265, 266, 273, 289, 297, 298, 299, 300, 301, 302, 303, 304

quantum tunneling, 399, 410, 419

quantum-mechanical tunnelling (QMT), 328, 329, 330

quasi-one-dimensional, 3, 103, 132, 282

quasi-particles, viii, 2, 3, 5, 10, 11, 70, 73, 108, 119, 131, 135

R

radiative lifetime, 74, 267, 307

radical microenvironment, 54

radical quasi-pairs, 44, 60, 66, 67

rare earth (RE) molybdates, 307, 308, 311, 316, 317, 318, 319, 321, 322, 323, 324, 325, 326, 332, 336, 351, 354, 361

recombination, viii, 6, 7, 8, 9, 10, 46, 48, 50, 63, 64, 73, 75, 88, 98, 101, 102, 103, 108, 119, 130, 133, 142, 180, 270, 288, 309, 316

recombination rate, 48, 63, 288

relaxation processes, 53, 77, 258

relaxation times, 12, 19, 34, 35, 50, 55, 56, 65, 79, 329, 331

resonant absorption, 9

retrapping, 48, 75

rhombic symmetry, 43, 71

Rietveld refinements, 320, 322, 323, 326

rotational motion, 79

S

samarium and europium molybdates, 319, 322, 344, 348

saturated calomel electrode (SCE), 172, 188, 190

saturated spins, 34, 498

scattering of polarons, 42, 49, 80

scheelite related compounds (SRCs), 306

scheelite related structure, 348

scheelite structure, 307, 308, 309, 311, 312, 313, 314, 317, 318, 321, 326, 331, 333, 336, 348, 355, 358

scheelites, ix, 305, 306, 307, 308, 309, 311, 312, 313, 314, 315, 316, 317, 318, 320, 321, 324, 326, 327, 329, 331, 333, 336, 342, 347, 348, 350, 351, 353, 355, 356, 357, 358, 361

self-trapped magnetic polaron, 498, 499

semiconductor-metal transition, 19

separation, 5, 7, 47, 50, 65, 74, 85, 109, 191, 207, 228, 266, 369, 380, 419, 451, 467, 472, 473, 484, 485, 491

SiFET, 214, 215, 216, 217, 218

skin depth, 82

skin-layer, 27, 82, 83

$Sm_2(MoO_4)_3$, 312, 319, 321, 322, 323, 325, 326, 327, 328, 336, 338, 339, 341, 342, 343, 344, 345, 348, 354

small electron (hole) polarons, 310

small polaron (SP), vi, ix, 31, 69, 81, 90, 99, 149, 158, 168, 177, 273, 309, 315, 326, 329, 330, 331, 333, 334, 335, 341, 342, 346, 348, 350, 353, 355, 356, 360, 367, 379, 380, 381, 386, 390, 392, 393, 399, 400, 402, 403, 404, 410, 412, 415, 416, 426, 429, 431, 436, 437, 444, 447, 450, 451, 452, 483, 487, 488, 489, 491

small polaron hopping, vi, ix, 168, 333, 379, 380, 381, 386, 402, 403, 415, 487, 489, 491

small polaron hopping model, 487

small polarons, 31, 69, 99, 177, 309, 315, 329, 331, 335, 342, 348, 353, 356, 360, 367, 390, 400, 402, 404, 410, 429, 431, 436, 437, 447, 450, 451, 452, 488, 492

solectron, viii, 197, 199, 200, 202, 203, 204, 205, 211, 212, 213, 214, 215, 216, 217

solectron field effect transistor (SFET), viii, 197, 200, 214, 216, 217, 218

solectrons, 197, 200, 205, 213, 214, 215

solid-state synthesis, 321

solitary wave acoustic polaron, 199, 200

soliton, v, viii, 4, 89, 124, 140, 141, 172, 175, 182, 183, 184, 189, 194, 197, 198, 199, 200, 206, 208, 209, 211, 212, 213, 214, 217, 219, 224, 234, 235, 236, 237, 238, 251, 259

soliton acoustic wave polaron (SWAP), 199, 200, 206, 212, 214

solitons, 88, 91, 96, 97, 105, 124, 131, 139, 140, 141, 175, 182, 189, 192, 193, 195, 197, 198, 199, 200, 206, 207, 209, 212, 213, 217, 219, 256, 260

Index

SP model, 334

specific morphology, 67, 69, 70

spectral components, 14, 20, 23, 32, 39, 53, 54, 55, 72

spin carriers, 7, 10, 30, 70

spin charge carriers, 4, 10, 12, 13, 32, 46, 48, 57, 58, 59, 65, 70, 75, 81, 94

spin concentration, 15, 20, 26, 28, 47, 52, 63, 68, 75

spin diffusion, 12, 13, 17, 20, 38, 56, 68, 78, 80

spin dynamics, 18, 29, 37, 39, 56, 57, 66, 70, 77, 79, 80, 92, 93, 94, 95, 98

spin exchange, 13, 47, 64, 70, 72, 77, 95, 98

spin hopping, 10, 20, 50, 70, 81

spin interaction, 61, 66, 76, 104

spin packets, 10, 11, 18, 33, 45, 55, 59, 68, 77

spin pairs, 8, 47, 62, 63, 65, 73, 75

spin polarization, 8, 10, 58, 431

spin precession, 9, 10, 14, 18, 25, 27, 32, 56, 70, 81, 83, 85

spin relaxation, 9, 11, 12, 13, 18, 21, 24, 26, 55, 64, 82, 92, 99

spin reservoir, 58, 71, 79

spin susceptibility, 17, 18, 33, 46, 47, 52, 62, 63, 64, 71, 72, 73, 76

spin traps, 21, 48, 67, 69, 75, 81

spin-assisted charge transfer, 10, 58, 66, 70

spin-controlled charge transport, 68

spin-flop field, 497

spin-lattice relaxation, 8, 30, 42, 56, 64

spin-orbit coupling, 8, 9, 10, 30, 54, 71, 90

spin-orbit coupling constant, 30, 71

spin-orbit interaction, 16, 38, 54

spin-packets, 12, 69

spin-spin exchange, 2, 11, 30, 39, 61, 64

spin-spin interaction, 18, 21, 40, 52, 59, 68, 76, 77

spin-spin interactions, 59

spin-spin relaxation times, 19, 42

spintronic devices, x, 7, 10, 66, 503

spintronics, 1, 2, 3, 70, 97, 101, 103, 105, 495

SRCs, 305, 307, 309, 312, 336, 337

SSH Hamiltonian, 120, 121, 126, 198, 202, 210

steady-state saturation, 49, 79

steady-state saturation effect, 79

STMP, 498, 499

stoichiometric vacancies, 312, 315, 316, 348

structural distortion, 4, 189, 306, 320, 400

subdomains, 65

supercell, 317, 324, 345, 348, 349

supercells, 323

symmetry, 25, 43, 109, 114, 119, 121, 125, 126, 131, 213, 227, 229, 238, 242, 246, 256, 257, 258, 259, 260, 302, 308, 311, 312, 313, 314, 315, 316, 317, 318, 319, 320, 321, 359, 415, 472, 497, 499

symmetry modes, 320

symmetry-mode, 320

system crystallinity, 53, 54

T

theoretical model, vii, viii, x, 49, 80, 143, 150, 152, 158, 167, 272, 277, 328, 331, 341, 429, 452, 474, 495, 499

thermochromism, 41, 97

thermodiffractometry, 321

thermoelectric coefficient, 488

tight binding approximation, 198, 269

Toda potential, 207

topological excitations, 1, 5

torsion (dihedral) angle, 4, 16

torsion angle, 44, 51

transfer integral, 21, 36, 37, 81, 102, 150, 152, 157, 202, 412

transition energy, 171, 172, 183, 188, 195

transition metal oxides (TMO), vii, 308, 309, 365, 366, 380, 390

transport mechanism, 57, 145, 146, 167, 183, 313, 328, 332

trap-assisted spin diffusion, 81

trapped charge carriers, 75

trapped polaron, 45, 46, 179

trapping, 46, 47, 48, 75, 146, 176, 183, 197, 199, 219, 300, 307, 309, 332, 449

trimolybdates, 312, 335

triplet excitations, 58

tungstate(s), 307, 308, 313, 314, 316, 317, 318, 319, 332, 333, 334, 335, 347, 348, 350, 351, 354, 355, 356, 357, 358, 359, 361, 362, 363

tunneling model, 74, 489

tunnelling, 145, 267, 310, 329, 331, 342, 483, 485, 494

tunnelling (OLPT) model, 342

two-potential well model, 484

type-I localized state, 498

type-II localized state, 498, 502

U

ultrahigh mobility, 213

universal dielectric response (UDR), 327

universal dielectric response (UDR) behaviour, 327

universal power law, 305, 310, 337, 348, 349

unpaired electron, 11, 12, 16, 25, 30, 31, 43, 54, 63, 70, 71, 109, 183, 189

516 *Index*

V

valence band, 5, 6, 42, 44, 123, 131, 174, 175, 182, 185, 188, 189, 190, 253, 255, 256, 269, 270, 309, 315, 335, 485, 496
variable range hopping (VHR), 11, 80, 402, 403, 404
variable range hopping (VRH) model, 80
Verwey transition, 409, 410, 413

W

Wannier states, 201
wave functions, 8, 64, 65, 73, 130, 145, 233, 238, 241, 268, 269, 271, 276, 283, 485, 502
wolframites, 306

X

X-ray and neutron diffraction, 311, 316
X-ray diffraction (XRD), 318, 379, 380, 383

Y

$YBa_2Cu_3O_{7-\delta}$, 421

Z

zero-point energy, 413, 414, 415, 422
zero-point lattice vibration, 412, 418
zero-point phonon, 410, 414, 420, 423, 424